P D Morgan

Advanced Diagnostics for Magnetic and Inertial Fusion

Advanced Diagnostics for Magnetic and Inertial Fusion

Edited by

Peter E. Stott
International School of Plasma Physics "P. Caldirola"
Milan, Italy

Alan Wootton
Lawrence Livermore National Laboratory
Livermore, USA

and

Giuseppe Gorini
Elio Sindoni
Dimitri Batani
University of Milan-Bicocca
Milan, Italy

Kluwer Academic / Plenum Publishers
New York, Boston, Dordrecht, London, Moscow

Proceedings of the International Conference on Advanced Diagnostics for Magnetic and Inertial Fusion, held September 3–7, 2001 at Villa Monastero, Varenna, Italy

ISBN 0-306-47297-X

233 Spring Street, New York, New York 10013

http://www.wkap.nl/

10 9 8 7 6 5 4 3 2 1

A C.I.P. record for this book is available from the Library of Congress

Printed in the United States of America

PREFACE

This book of proceedings collects the papers presented at the International Conference on Advanced Diagnostics for Magnetic and Inertial Fusion held at Villa Monastero Varenna Italy from 3 to 7 September 2001. The conference was jointly organised by the "Piero Caldirola" International Centre for the Promotion of Science and International School of Plasma Physics and the Istituto di Fisica del Plasma "Piero Caldirola" of CNR.

This conference was the 8th in a series of meetings on the theme of fusion diagnostics that have been held in Varenna at intervals between two and four years since 1975. Whereas previous meetings have concentrated on diagnostics for magnetic confinement fusion, this meeting had the specific purpose of bringing together experts from Magnetic and Inertial Confinement. The meeting concentrated on advanced diagnostics in these fields.

The number of papers and participants was balanced between the two communities with approximately 35 invited and 60 contributed papers, with nearly 100 participants coming mainly from Europe, USA, Russia, and Japan, with a smaller but significant participation from China, Korea, and Israel.

Throughout the workshop, topics of simultaneous interest to both the magnetic and inertial fields were catalogued. Several topics were identified for further collaboration: solutions to environmental issues, detector technology and calibration techniques. Co-ordinators were appointed to promote future exchange and interaction between the magnetic and inertial confinement communities.

This is probably the first event to explicitly bring together the two communities of Magnetic and Inertial Fusion, highlighting the common problems in plasma diagnostics (and in plasma physics), showing the tremendous advances which have been realised in recent years in fusion, as well as the many tasks which still need to be addressed and solved on the devices of the next generation which promise, with both approaches, the achievement of fusion conditions.

Thanks are due to the institutions that have contributed financially to the organisation of this conference. In particular the Istituto di Fisica del Plasma "Piero Caldirola" of CNR, the DG RTD of the European Commission which provided financial support though the Euroconference scheme.

Peter E Stott, Alan Wootton, Giuseppe Gorini, Dimitri Batani, Elio Sindoni

CONTENTS

FUSION EXPERIMENTS AND RELATED DIAGNOSTIC NEEDS

NEUTRON AND PARTICLE DIAGNOSTICS

OPTICAL AND X-RAY DIAGNOSTICS

RADIATION EFFECTS AND OTHER EMERGING PROBLEMS

INTERFEROMETRY, POLARIMETRY, THOMSON SCATTERING

OTHER DIAGNOSTIC SYSTEMS

REQUIREMENTS AND ISSUES IN DIAGNOSTICS FOR NEXT STEP BURNING PLASMA EXPERIMENTS

Alan E Costley *

1. INTRODUCTION

Next step burning plasma experiments (BPXs) will require an extensive set of plasma and first wall measurements. The number and type of parameters that will have to be measured will be very similar to those measured on today's large devices, but the specification of some of the measurements, such as relative spatial resolution and accuracy, and the required reliability of the measurements, will be more demanding. Further, because of the harsh radiation environment, diagnostic system selection and design will have to cope with a range of phenomena not previously important in diagnostic design: for example, Radiation Induced Conductivity in magnetic sensors mounted in the vacuum vessel, and erosion of diagnostic first mirrors due to energetic particle bombardment. The diagnostic designs also have to satisfy stringent engineering requirements on tritium confinement, vacuum integrity, neutron shielding and remote handling maintainability. This combination of demanding measurement requirements and implementation difficulties constitutes a major challenge, arguably the most difficult challenge ever undertaken in magnetic fusion energy (MFE) diagnostics.

Since 1994, and under the frame of the ITER programme, the various aspects of the problem have been tackled in the preparation of diagnostics for ITER[1]. An extensive database of radiation effects on candidate diagnostic components has been established, and specific diagnostic components and some new diagnostic techniques have been developed[2]. Diagnostic designs have been advanced to the point where implementation of some systems could now begin[3,4]. The work has shown that in most cases it should be possible to meet the desired measurement requirements but not in every case. Where there are shortfalls there will be limitations in the information that can be obtained and possibly on plasma operation.

In this paper, the requirements for measurements are briefly reviewed and the difficulties arising from the harsh environment are outlined. Some examples of successful developments undertaken specifically for ITER are given. Key issues which are presently unresolved are identified and some potential solutions highlighted. Finally, comments are

* Alan Costley, ITER NAKA JWS, 801-1 Mukouyama, Naka, Ibaraki 311-0193, Japan.

Advanced Diagnostics for Magnetic and Inertial Fusion, Edited by Stott *et al.*
Kluwer Academic/Plenum Publishers, New York, 2002

made on the relative feasibility of implementing diagnostics on three current BPX contenders (ITER, FIRE and Ignitor).

2. REQUIREMENTS FOR PLASMA AND FIRST WALL MEASUREMENTS ON A BPX

As on existing machines, first wall and plasma measurements will be required on a BPX for machine protection, plasma control and physics evaluation. Key parameters which must be measured at a very high level of reliability for machine protection are the gap between the plasma edge and the first wall, first wall temperature, line-average density, halo currents and disruption precursors (particularly the detection of locked modes). For machines with divertors, it is also necessary to measure the divertor plate surface temperature, fusion power, and edge localised mode (ELM) type.

For plasma control, the principal measurements made on existing machines will also be needed; plasma shape and position, plasma current, and the spatial profile of the electron density (n_e). A burning plasma operating close to operational limits will also require additional control measurements: for example, plasma beta (β), total radiated power, fuelling ratio n_T/n_D, plasma rotation and helium ash concentration. Key measurements in the divertor will also be needed – in particular the position of the ionisation front must be determined. The kinetic control must also keep the plasma away from the β and density limits and provide sufficient power flow through the separatrix to ensure H-mode plasma operation. This leads to requirements for a range of additional plasma measurements for control. These include the radiative power loss from the plasma core, scrape off layer (SOL), X-point and divertor regions, and the plasma density profile, β, n_T/n_D ratio, rotating MHD modes and n_e and the electron temperature (T_e) at the divertor plates. It is clear that a sophisticated multi-input, multi-actuator feedback control scheme is required for successful operation.

For operating scenarios exploiting advanced high confinement modes, for example reverse shear and optimised shear, control and hence measurement of the spatial profile of key parameters such as the current density, pressure and rotation are likely to be required. For sustained operation near the β limit, suppression of neoclassical tearing modes (NTMs) will probably be needed. Similarly, for steady-state operation at β levels above the ideal limit (without a conducting wall) stabilization of the resistive wall modes (RWMs) will be required. Measurements of the location, phase and amplitude of these modes are therefore required.

The BPX plasma will be the first in which there is significant alpha particle heating, so the experimental program will have an extensive explorative physics component. Key topics to be investigated include energetic alpha particle effects, confinement and transport, MHD, and the physics of the divertor and the boundary. Important required measurements are the energy and number density of confined alphas with space and time resolution, the energy and number density of escaping alphas, MHD modes, the T_e and n_e profiles with high spatial resolution in the edge and in the region of the internal transport barriers, and measurements of the helium ash and other light impurities. Measurements of the amplitude and correlation length of fluctuations are needed in tests of theories and models of transport.

Table 1. Plasma and First Wall Measurements required for a BPX (ITER)

GROUP 1a Measurements For Machine Protection and Basic Control	GROUP 1b Measurements for Advanced Control	GROUP 2 Additional Measurements for Performance Eval. and Physics
Plasma shape and position, separatrix-wall gaps, gap between separatrixes	Neutron and α-source profile	Confined α-particles
Plasma current, q(a), q(95%)	Helium density profile (core)	TAE Modes, fishbones
Loop voltage	Plasma rotation (toroidal and poloidal)	T_e profile (edge)
Fusion power	Current density profile (q-profile)	n_e, T_e profiles (X-point)
$\beta_N = \beta_{tor}(aB/I)$	Electron temperature profile (core)	T_i in divertor
Line-averaged electron density	Electron density profile (core and edge)	Plasma flow (divertor)
Impurity and D,T influx (divertor, & main plasma)	Ion temperature profile (core)	$n_T/n_D/n_H$ (edge)
Surface temp. (div. & upper plates)	Radiation power profile (core, X-point & divertor)	$n_T/n_D/n_H$ (divertor)
Surface temperature (first wall)	Z_{eff} profile	T_e fluctuations
Runaway electrons	Helium density (divertor)	n_e fluctuations
'Halo' currents	Heat deposition profile (divertor)	Radial electric field and field fluctuations
Radiated power (main pla, X-pt & div).	Ionization front position in divertor	Edge turbulence
Divertor detachment indicator (J_{sat}, n_e, T_e at divertor plate)	Impurity density profiles	MHD activity in plasma core
Disruption precursors (locked modes, m=2)	Neutral density between plasma and first wall	Pellet ablation
H/L mode indicator	n_e, T_e of divertor plasma	
Z_{eff} (line-averaged)	Alpha-particle loss	
n_T/n_D in plasma core	Low m/n MHD activity	
ELMs	Sawteeth	
Gas pressure (divertor & duct)	Net erosion (divertor plate)	
Gas composition (divertor & duct)	Neutron fluence	

Studies of the macroscopic stability of a burning plasma involve the investigation of the complex internal coupled processes which will determine the profiles of the pressure, current density and rotation. The stabilization of NTMs and RWMs will be required for the highest performance. Modes that lock can lead to a disruption and so should be measured preferably when they are small.

The global issue to be addressed in studies of the boundary is that of the burning core – plasma boundary interaction. Specific topics will include transport in the SOL, ELMs and edge stability, and the L – H threshold. In the divertor, key topics will be the ELM power handling capability, impurity screening, particle pumping, dissipation of core power efflux, erosion/redeposition and tritium retention. Many measurements will be required to support the study of these topics including measurements of n_e and T_e with high time and spatial resolution, radiated power, divertor helium density and impurities. Measurements of the temperature of the divertor plates and, for long pulse operation, erosion of the plates will also be required. The composition, quantity and location of dust may also have to be determined.

The parameters that will have to be measured are listed in table 1 and categorised according to their role in machine protection and basic plasma control (group 1a), advanced control (group 1b) and physics studies (group 2).

3 ENVIRONMENT OF A BPX: KEY RADIATION EFFECTS

Clearly, from the point of view of implementing diagnostics, the principal difference between a BPX and present-day MFE devices is the environment in which some of the key diagnostic components will have to be installed and operate. The flux levels of all the potentially damaging and influencing radiations - neutrons, gammas, photons and particles - will be higher and in some cases by one or two orders of magnitude. Moreover the pulse lengths will be tens to hundreds of seconds and possibly steady state, and so these conditions will have to be endured for much longer than on present machines. The precise conditions will depend on the BPX and on the diagnostic design.

These high levels of radiation mean that diagnostic selection and design has to cope with a wide range of phenomena not previously important in diagnostic design (figure 1). For example, under gamma irradiation the conductivity of ceramic insulators is enhanced due the generation of charge carriers within the bulk material (Radiation Induced Conductivity (RIC)). This has to be taken into account in the design of all diagnostics which employ cables inside the vacuum vessel, for example magnetics. Similarly, it has been observed that when insulated cable is irradiated a small current can flow between the core and the sheath due the phenomenon of Radiation Induced EMF (RIEMF). This is especially important in the design of coils used to determine the plasma shape and position by measuring the magnetic fields. If there are asymmetries in the coil, this effect can lead to a differential voltage which could lead to an error in the measurement. Both effects have to be dealt with by careful design[5].

For many diagnostics, the plasma facing optical element is a mirror, and the lifetime of these first mirrors is a key parameter. The high levels of energetic neutral particle flux can give rise to erosion. Extensive experimental work has shown that in situations where erosion will be the dominant potentially damaging mechanism it will be possible to design

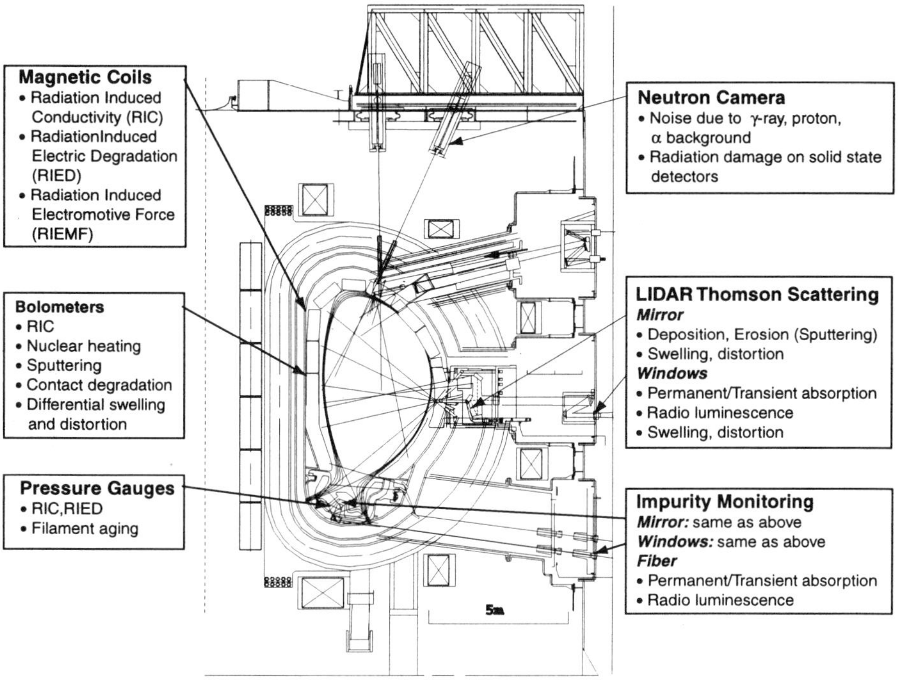

Figure 1. ITER: Location of some representative diagnostic components and the principal, radiation induced, physical effects of interest.

mirrors with an adequate lifetime but the material has to be carefully chosen[6]. On the other hand, erosion of the first wall or viewing ducts can lead to deposition of materials on the mirrors. This can also lead to a significant reduction in the reflecting properties and effectively a shortening of the mirror lifetime. Possible mitigating methods include installation of shutters and baffles and cleaning techniques such as laser ablation and discharge cleaning.

Many diagnostics employ refractive optical materials, for example fused silica in optical windows, and under irradiation by neutrons and gammas two effects can occur: absorption and radioluminescence. The absorption has two components: a permanent component and a prompt component. Radioluminsecence can generate a signal which can compete with the signal being measured especially when the latter is relatively low, for example in Thomson scattering diagnostics. Optical fibres are a very convenient means of transporting the radiation and can simplify the diagnostic design. However, since the length of the fibre is long these radiation induced optical effects can be more severe and limit the application of the fibres.

Other effects can also be important. For example, surface heating due to the plasma radiation and bulk heating due to neutron absorption. The nuclear absorption can also cause dimensional changes and material transmutation. All these effects have to be carefully taken into account in the diagnostic design.

4. THE ITER DIAGNOSTIC SYSTEM: THE MOST DEVELOPED EXAMPLE OF BPX DIAGNOSTICS

The most developed BPX diagnostic system design is that which has been prepared for ITER[3,4]. It is a comprehensive system comprising about 40 individual measurement systems largely based on existing tokamak diagnostics but with appropriate selection of materials and designs to cope with the harsh environment and the special requirements for diagnostic implementation. While the design of individual systems are at different stages of development, representative systems of all the main types of diagnostics - magnetics, neutron, optical, bolometric, spectroscopic and microwave - have been developed to the point where the key problems have been found and design solutions developed . In the designs, steps are taken to minimise the adverse effects of the environment and to produce the necessary diagnostic performance and component lifetime. For example, the magnetics diagnostic utilises mineral insulated cable, which has been the subject of extensive R&D and the radiation effects are well known, and the coils and loops are installed behind the blanket shield modules thereby giving some protection from the radiation fluxes (figure 2). With this design the level of gamma and neutron fluxes experienced by the key components are typically no higher than those already experienced on JET and TFTR, although of course the pulse length will be much longer.

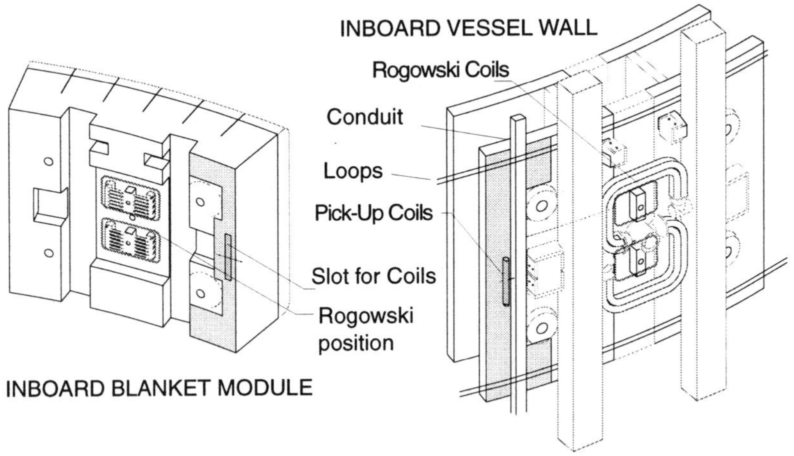

Figure 2. Some magnetic diagnostic components installed behind the blanket shield modules.

Optical designs incorporate shielding labyrinths. In these, mirrors are used to fold the optical paths and the latter is embedded in material which absorbs the radiation (figure 3). The most critical diagnostic component is usually the first mirror and, as explained above, by carefully selecting the materials an adequate lifetime can be obtained. At the diagnostic window, the radiation levels are relatively low and generally not higher than

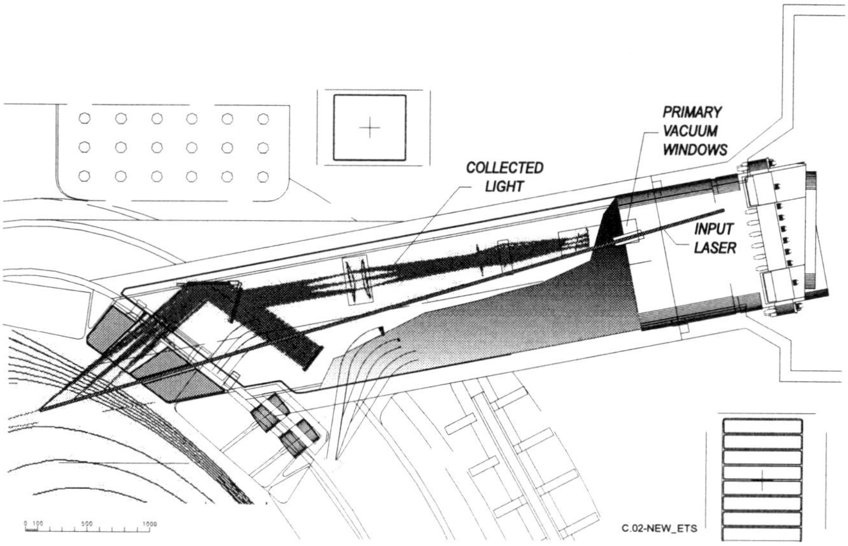

Figure 3. The front end of the edge Thomson scattering system showing the laser line to the last element supported from the diagnostic block, and collection optics / raytrace to the first vacuum boundary (courtesy P Nielsen, Consorzio RFX, Padova).

those already experienced on JET and TFTR. Microwave diagnostics also employ shielding labyrinths with waveguide transmission lines incorporating several bends.

Some spectroscopic systems, however, require a direct vacuum coupling and for these shielding labyrinths and windows cannot be employed. Examples are VUV and X-ray crystal spectroscopy. The neutral particle analyser (NPA) also requires a direct coupling. In these cases, special provisions have to be made to deal with the neutron and gamma radiation. For example, in the X-ray crystal spectrometer a graphite reflector is employed to deviate the radiation by a few degrees so that the detector is not exposed to the direct neutron and gamma fluxes. The NPA device employs a high voltage accelerator to increase the signal relative to that due to the neutron background[7]. Bolometers also have to view the plasma directly and for this application special radiation hard bolometers are being developed[8].

5. KEY OUTSTANDING ISSUES

While good progress has been made, further developments are needed in order to be able to bring the measurements to the point where all the requirements can be met. There are several key areas in the radiation effects work that need further development. An unexpectedly high asymmetric voltage was measured in tests with a prototype magnetic coil apparently due to the phenomenon of RIEMF. However, a recent re-examination of the results suggests that there were problems with the long pulse integrator used in the tests, and/or other sources of voltage, for example thermoelectric effects, and further tests are required. While first mirror designs resistant to erosion exist, for mirrors in the divertor region and possibly in ducts, deposition is likely to be the limiting mechanism and potential cleaning and mitigating methods (baffles and shutters) need to be developed. It is believed that there are window solutions for most diagnostics. However, for systems employing high power lasers, even a small amount of absorption can be serious and more accurate measurements of radiation induced prompt absorption are needed. Significant inconsistencies have been found in tests on optical fibres and need to be resolved.

Key sensors requiring development include compact steady state magnetic sensors, compact efficient neutron detector/spectrometers, and radiation hard bolometers and soft X-ray detectors. In each case candidate sensors exist but development work is required before devices suitable for installation on a BPX can be selected.

To solve some of the measurement problems, new techniques are required. For example, techniques are needed to measure the confined and escaping alpha particles[9], divertor plate erosion, dust and re-deposited layers. In other cases existing techniques appear to be very difficult to implement under BPX conditions and new techniques are highly desirable. Examples are measurements of the plasma position and shape under long pulse (steady state conditions), measurements of the light impurities and especially the Helium ash, and measurements of the internal electric field. In yet other cases techniques have been identified which could make measurements of an important parameter simpler. A good example is Fast Wave Reflectometry which could make measurements of the fuel mix ratio, n_T/n_D, with spatial and temporal resolution but requires development. The hardware is very simple and application on a BPX can be easily envisaged.

Active charge exchange recombination spectroscopy (CXRS) for the measurement of light impurities, ion temperature and plasma rotation requires a conventional neutral beam but this is a high cost item and requires significant space and power. A short pulse intense beam would require less power and space and would give an improved diagnostic performance but requires development.

As the diagnostic designs develop the need for R and D in certain areas becomes apparent. It is important that there is a close interaction between design and R&D.

6. IMPLEMENTATION ON DIFFERENT BPX CONTENDERS

The feasibility of implementing diagnostic systems on a BPX depends on several aspects but especially on the access and the environment. Three current BPX contenders, ITER, FIRE and Ignitor, differ substantially in these two aspects.

All three have large ports at the mid-plane and so good diagnostic access can be expected at this level. However, ITER also has large ports at the upper level. This gives access for many diagnostics including views of the diagnostic neutral beam for CXRS, views of the divertor for spectroscopy, views of the first wall which combined with views from the midplane give a high surface coverage. Further, they give the possibility of installing diagnostics for edge measurements which can take account of the natural flux expansion at this level to achieve high spatial resolution. According to current designs both FIRE and Ignitor have small diameter, long, vertical ports. On a BPX, optical systems have to use reflective optics close to the plasma (because there are no sufficiently radiation hard optical materials for refractive optics), and in narrow tubes it is not possible to achieve high optical throughput. If there is the additional requirement to include shielding in the tubes, the possible optical throughput will be very low. Hence, the vertical ports on FIRE and Ignitor will be of limited use for viewing or collection systems but may be of use for input laser beams.

At the divertor level access is restricted and the lifetime of diagnostic components is potentially limited. Both FIRE and ITER have to provide diagnostics at this level. In the case of FIRE a duplicate set will be needed at both the upper and lower divertor levels. This requirement, combined with the absence of large upper ports, makes FIRE a relatively difficult machine for diagnostic implementation.

The first wall neutron flux on both FIRE and Ignitor is approximately five times higher than on ITER. Moreover neither of these machines has a blanket and so in-vessel diagnostic components and sensors will be subjected to the unattenuated first wall flux. On ITER, diagnostic components will be installed in the gaps between blanket shield modules or behind the modules giving an attenuation of a factor of 10 – 100. The radiation levels on FIRE and Ignitor will therefore be 50 – 500 times higher and so the prompt radiation effects will be more severe. Nuclear heating will also be much higher leading to high operating temperatures of the diagnostic components and/or to a requirement for active cooling.

It is clear that the problems and solutions of implementating diagnostics on a BPX will be machine specific. The diagnostic designs must therefore be developed in close association with the machine design. A fully integrated approach is required.

7. SUMMARY

The combination of demanding measurement requirements and the harsh environment means that the preparation of diagnostics for a BPX will be a major challenge. There is a good basis of experience and knowledge that has been gained on existing machines and this is being supplemented within the ITER framework with dedicated design and R&D, but the step to be made remains substantial. Solutions to many of the measurement requirements have been developed but some key requirements, for example measurements of confined and escaping alpha particles, remain unresolved and must be the focus of future work. The multiple diagnostic/machine interfaces mean that the development of the design of the diagnostic systems must be closely integrated with the machine design. Similarly, there must be a tight coupling between the design and the supporting R&D. The environment and access for diagnostics is different for three different BPX contenders - ITER, FIRE and Ignitor - and this will have a substantial impact on the diagnostic system selection and design.

8. ACKNOWLEDGEMENTS

It is a pleasure to acknowledge useful discussions and input from the members of the ITER Expert Group on Diagnostics, A J Donne, Y Kusama, A Krasilinkov, A Kislyakov, F Orsitto, T Nishitani, T Sugie, and S Strelkov, and from colleagues in the ITER Joint Central Team, particularly K Ebisawa, G Vayakis, C Walker and S Yamamoto.

Contributions from S Kasai, A Malquias, V Zaveriaev of the Japanese, European and Russian Home Teams respectively are also gratefully acknowledged.

9. REFERENCES

1. Chapter 7: Measurement of Plasma Parameters' ITER Physics Expert Group on Diagnostics and ITER Physics Basis Editors, Sect. 3.3.4; Nucl. Fusion Vol. 39, 2559 (1999).
2. A E Costley, D J Campbell, S Kasai, K E Young, and V Zaveriaev, 'ITER R&D Plasma Diagnostics', Fusion Engineering and Design, Vol 55, 331 (2001).
3. K Ebisawa et al, Rev of Sci. Instrum., Vol. 72, No 1, 545 (2001).
4. 'Technical Basis for ITER Final Design Report', to be published by the IAEA in the ITER EDA Documentation Series, IAEA, Vienna (2001)
5. G Vayakis, this conference
6. V Voitsenya, this conference
7. A.I.Kislyakov, V.I.Afanassiev, A.V.Khudoleev, S.S.Kozlovskij and M.P.Petrov, in 'Diagnostics for Experimental Thermonuclear Fusion Reactors 2', ed. P.E. Stott, G. Gorini, and E. Sindoni, Plenum Press, New York, p.353 (1998).
8. R Reichle, this conference
9. K E Young, this conference

INTRODUCTION TO DIAGNOSTICS REQUIREMENTS FOR ICF

Denis Juraszek*

1. INTRODUCTION

During the last ten years considerable progress have been accomplished in all topics related to Inertial Confinement Fusion (ICF): better understanding and modeling of the physical processes, lower cost and reliable drivers, target technology. Of course the development of new or improved diagnostics is of primary importance. Many experiments have been performed on the existing facilities and compared to detailed numerical simulations. Using these sophisticated radiation hydrodynamics codes, it has been demonstrated that the ignition of a small DT filled capsule could be achieved with a 1 to 2 MJ UV laser. So the main objective of the next generation of large ICF facilities, i.e. the National Ignition Facility (NIF)[1] in USA and the Laser Megajoule (LMJ)[2] in France, is the achievement of moderate, i.e. of the order of 10, thermonuclear gain implosions. The gain is defined as the ratio of thermonuclear energy to driver energy.

The diagnostics needed for such ICF experiments is an essential part of these large facilities design. ICF is a several steps process and at each of these steps some critical requirements have to be met to achieve ignition and burn. In spite of the large base of experimental data, uncertainties still remain and parameters such as driver power, beam pointing, etc will have to be adjusted on the way.

After a brief recall of ICF principles and various schemes, we will consider the successive critical points, determine what parameter has to be measured and finally the diagnostics requirements.

2. ICF CRITICAL MILESTONES[3]

Nominal ICF capsule is a millimeter scale shell of solid DT contained in a shell of low Z material, called "ablator" and filled with low-density DT gas. In Direct-Drive scheme (DD)[4], the driver power is used to heat the ablator outer surface. In the Indirect-Drive scenario (ID), the capsule is enclosed in a high Z container, also called cavity or

* CEA/DAM-Ile de France, B.P.12, 91680 Bruyères-le-Châtel, France

Advanced Diagnostics for Magnetic and Inertial Fusion, Edited by Stott *et al.*
Kluwer Academic/Plenum Publishers, New York, 2002

hohlraum; the driver beams are used to produce an intense X-ray radiation bath inside the cavity. In both cases, the ablator surface is strongly heated, it expands rapidly outward and remaining part of the shell is imploded at high velocity (3 to 4×10^7 cm/s). As the convergence increases the DT fuel is compressed and heated. Ignition occurs when the central part of the fuel called hot spot has a temperature $T \geq 6$ keV and an areal density $(\rho R) \geq 0.3$ g/cm^2.

A small fraction, about one to a few percent, of drive energy is transferred to the fuel but as only a fraction of the DT is heated to ignition conditions, the main part of the fuel being compressed along a low isentrope, it is still possible to achieve high gains.

In another scheme, named "fast ignitor"[5], all the fuel is compressed at low temperature. The hot spot is not generated by the implosion process, but is triggered at the end of the compression by the suprathermal particles generated with an ultra-high intensity short laser pulse[6].

An important characteristic of ICF laser driven experiments is the large range of plasma conditions encountered (figure 1) from the corona conditions where laser beams propagate to the burning fuel.

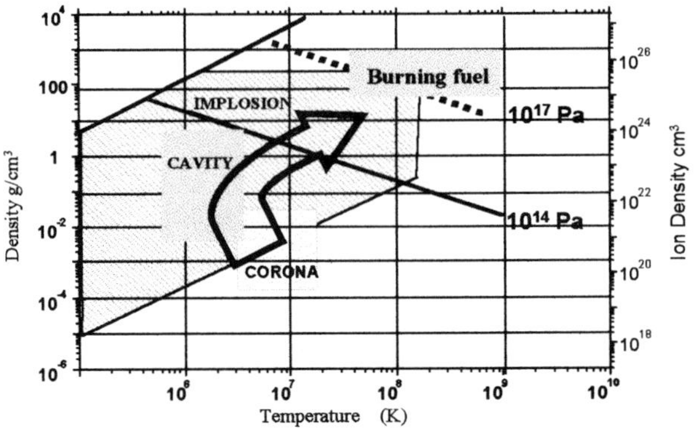

Figure 1. density and temperature of ICF plasmas

As the target gain G scales as

$$G \propto \eta \,.\, (\eta . E_d / \alpha^3)^{0.3} \qquad [1]$$

it is important to maximize the coupling efficiency η. For DD, η depends on rate absorbed energy and hydrodynamical efficiency. For ID η also depends on X-ray conversion efficiency and X-ray capsule coupling.

Each of these terms should be addressed separately. In the case of laser driven experiments, due to the high intensity of the beams propagating through the plasma (10^{14} to 10^{15} W/cm^2), various parametric instabilities may occur[7]. These instabilities are the result of the non-linear coupling between the plasma wave and the laser E.M. wave. As they may strongly affect the efficiency and the quality of the coupling, this problem requires a specific set of diagnostics[8].

To reduce the driver requirements, the DT should be compressed along a low isentrope and remain nearly Fermi degenerate. The fuel equation of state can be then written as:

$$\varepsilon \text{ (Specific energy J/g)} = 3\text{x}10^5 .\alpha. \rho^{2/3} \text{ (g/cm}^3\text{) with } \alpha \geq 1 \qquad [2].$$

As the thermonuclear gain is proportional to α^{-1}, the entropy parameter α has to be carefully controlled. The low entropy compression is realized by a precise shock sequence obtained with an accurate driver pulse shaping (figure 2). The fuel entropy can also be influenced by high energy radiation (photons or particles) generated by the driver plasma interaction and transmitted through the ablator shell.

If the drive flux ϕ around the capsule is not perfectly symmetric, the velocity non-uniformity $\delta V/V$ will result in a non-spherical hot spot. Assuming that the deviation δR_f must be less than $R_f/4$, $\delta V/V$ will be less than $0.25.\ R_f/R_0$, where R_0 is the initial capsule radius. As typical values for R_0/R_f are 30 to 40, $\delta V/V$ should be about 1%. Requirements for $\delta\phi/\phi$ are of the same order.

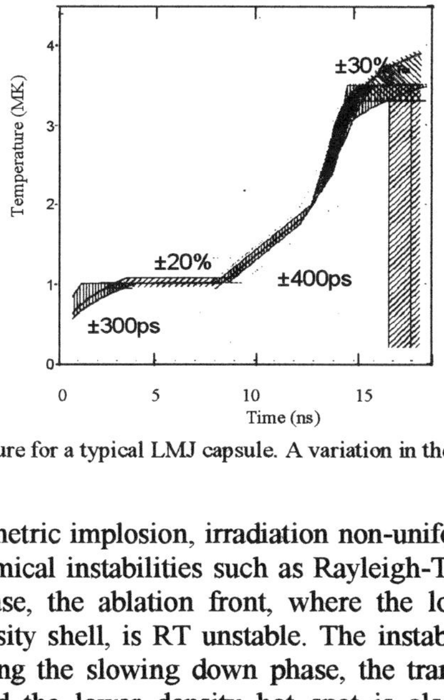

Figure 2 . Drive temperature for a typical LMJ capsule. A variation in the hatched area results in a 25 % yield reduction.

Beside non-symmetric implosion, irradiation non-uniformities and capsule roughness can seed hydrodynamical instabilities such as Rayleigh-Taylor (RT) instability[9]. During the acceleration phase, the ablation front, where the low density evaporated plasma pushes the high-density shell, is RT unstable. The instability growth may lead to shell breakup. Later, during the slowing down phase, the transition zone between the cold, dense main fuel and the lower density hot spot is also RT unstable. The perturbed interface increases the thermal losses for the hot spot to the surrounding fuel, reducing his ability to trigger ignition.

ICF is based on a two zones fuel configuration: a hot spot containing a small fraction (a few %) of the total mass, satisfying ignition conditions to be able to trigger a thermonuclear burn wave in the main fuel compressed on a low isentrope. In the nominal schema, the hot spot results from the driving pulse shaping and from the low-density gas inside the capsule. As several phenomena may affect the hot spot generation, measurements should allow the identification of failure mechanism, if any.

3. EXPERIMENTAL STRATEGY

First experiments on NIF or LMJ will not use cryogenic DT filled capsule. Experimental programs are being constructed to reach the milestones of §2 with increasing complexity and difficulty. The successive stages are:

- ✓ Control the ability to point and synchronize the beams on the target
- ✓ Measure the energy absorption
- ✓ Adjust the cavity parameters for radiation time history and symmetry
- ✓ Check the implosion velocity
- ✓ Determine the fuel conditions at stagnation
- ✓ Measure the burn wave parameters.

Each step requires a specific set of diagnostics.

The measurements required during the first stages, i.e. before ignition experiments, do not need the nominal cryogenic DT capsule and sometimes will even be impossible with such a capsule. These measurements involve surrogate capsules, which mimic or emphasize some aspects of the nominal capsule behavior. When such a capsule is to be used, the diagnostics requirements are established in close connection with capsule requirements. Examples of surrogate capsules are gas filled shell, shell with tracers, low density foam ball...

4. DIAGNOSTICS REQUIREMENTS

4.1 Laser plasma coupling

This item is specific to laser driven experiments and is important for both Direct[10] or Indirect Drive[11] scheme. When the laser beams propagate through the evaporated plasma to the critical density or through the plasma filling the cavity to the wall, parametric instabilities, Stimulated Brillouin Scattering (SBS) and Stimulated Raman Scattering (SRS), in particular, could be significant. Laser-plasma instabilities can scatter a fraction of driver energy out of the plasma, reducing the coupling efficiency. Previous experiments have shown that the time evolution of the backscattered light is generally quite different from the time evolution of the incident power (figure 3). It is important to measure separately the time dependence of SRS and SBS because:

- The time dependence of the pressure driving the capsule implosion is tied to absorbed and not to incident power
- The detailed knowledge of the instability's effects may help to identify an improvement.

Most of the SBS and SRS energy is backscattered toward or near the focusing optics. Some optics is required to collect the reflected light and to inject it into spectrometers coupled to streak camera[12]. The light scattered outside the focusing optics can be measured with an array of filtered fast diodes.

Another prejudicial effect of the plasma instabilities is the generation of suprathermal electrons with temperatures in the range $T_h \sim$ a few 10's to a few 100's keV. These electrons can preheat the DT increasing the fuel entropy (parameter α). The temperature and the number of fast electrons are routinely measured by recording the bremsstrahlung spectra of these electrons as they move through the target material. Measurement are performed with a set of Fast Scintillators coupled to PM tubes and high-pass filters or fluorescence filters.

Thomson scattering (TS) is a classical technique in plasma science, but it was used only recently on large facilities as Nova[13] or Phebus to characterize ICF plasmas (n_e up to 3×10^{21} cm^{-3}, Te, Ti, Z, velocity). TS was also used to analyze the characteristics of electron or ion acoustic wave associated with the parametric plasma instabilities[14]. The possibility to accommodate TS to NIF or LMJ is still to be analyzed (S.H. Glenzer, in this proceedings) as it would require a few kJ probe beam at 0.26 μm wavelength.

4.2 Cavity (hohlraum) physics

This item is specific to Indirect Drive experiments but concerns all drivers. The driver beams deposit their energy at the wall of the cavity or in converters placed inside the cavity. A fraction of absorbed energy is converted to X-rays. These X-rays are confined by the cavity wall, producing a nearly Planckian and nearly uniform radiation environment around the capsule. The radiation field is conveniently expressed as a brightness temperature T_r.

As T_r time history determines the shock sequence required to compress the fuel on a low isentrope, it should be precisely measured (typical values are: $\Delta T_r/T_r$ of the order of 1 %, with less than 100ps time resolution). Many measurements have been already performed to establish the relation between the absorbed driver power and the drive temperature[15].

Broad band X-ray spectrometers are used to infer X-ray intensity escaping from a hole in the cavity wall. They are either based on a transmission grating[16] or on an array of fast diodes[17]. $\Delta T_r/T_r$ of the order of 2 % are routinely achieved. But as the cavity is heated high Z plasma from the wall may obscure the line of sight, reducing the observation hole diameter. To improve the overall precision a companion diagnostic is needed, i.e. a Soft x-ray Imager working at several photon energies (about 300 eV to 1200 eV)[18]. Other techniques exploit the strong dependence on the drive temperature of the radiation wave or of the shock generated in a well-characterized sample.

The radiation field produced in a laser-heated hohlraum is not perfectly Planckian and reflects the wall material atomic structure. The M-band radiation of wall material at a few keV can penetrate deep in shell and increase of fuel entropy. So the amount of energy radiated in the range 2 to 5 keV has to be measured and an absolutely calibrated, time resolved, $\lambda/\Delta\lambda \sim 300$ spectrograph should also be available.

As ICF cavities never reach a complete equilibrium, the radiation environment around the capsule evolves in time. In ID scheme driver beams (and laser entrance holes) positions generate low order ($l < 10$, if the intensity distribution is expressed in terms of spherical harmonic functions Y_{lm}) asymmetries at the capsule surface. The intensity pattern around the capsule can partially be controlled by changing the target geometry or the relative power of the beam cones[19]. Drive symmetry measurements are well essential to optimize the irradiation geometry. One possibility is to image the compressed shell at stagnation[20]. Although quite valuable, this kind of measurement integrates the time evolution of the non-uniformity making the analysis much more difficult. Several time resolving techniques using X-ray imaging and surrogate capsules have been developed[21,22,23]. The comparison of these methods is beyond the scope of the present paper, but is important to notice a lot of work is still to be done. As a matter of fact, for example the NIF specifications on flux asymmetry are 1% for a_2/a_0, and 0.25 % for a_8/a_0. If we assume azimuthal symmetry, the flux is expressed as $\Sigma\ a_l(t)P_l(\Theta)$ (P_l = Legender polynomial). It translate to $a_8 < 2\mu m$, so high spatial resolutions are required.

4.3 Implosion phase

Numerical simulations and models indicate that, for a given driving pressure, the ignition energy scale as[24] $E_{ign} \propto \alpha^{1.88}.V^{-5.89}$ with V= final implosion velocity and α = fuel entropy parameter during the implosion. If the coupling between the capsule and the driver is correct, most of the energy transferred to fuel appears, at the end of the implosion phase, as shell kinetic energy. Therefore, the measurement of the final implosion velocity V is also an essential over-all control of the coupling efficiency. On the other hand, α higher than expected is the signature of fuel heating by suprathermal particles or, more likely, an inappropriate pulse shaping.

Rather than V, the beginning of the stagnation phase can be measured. The end of the acceleration phase is marked by the collapse of the strong shock generated in the DT gas by the shell motion (commonly named "bang time"). At that time, the gas temperature increases sharply generating X-ray and neutrons emissions. Typically, the bang time is measured with an X-ray streak camera and temporal resolution less than 30ps, satisfying NIF/LMJ requirements.

The entropy parameter is not directly accessible. It has been proposed to measure the shell thickness ΔR by X-ray radiography (ΔR is proportional to $\alpha^{-0.6}$). The measurement would require a backlighting source at 6 to 8 keV and a high spatial and temporal resolution X-ray imager (typically less than $2\mu m$ and 10ps).

Concerning the problem of hydrodynamical instabilities during the acceleration phase, although experiments using spherical corrugated shell have already been performed[25], the analysis of instability growth will probably be performed in cylindrical[26] or planar geometry[27]. The measurements will require X-ray radiography capabilities with high spatial resolution (<10 μm) X-ray imaging at a few keV.

4.4 Fuel assembly

The capsule hydrodynamics during the stagnation phase will be analyzed in several situations: surrogate capsules (D_2 filled, low DT mass, tracer...), deliberately under driven cryogenic DT capsule and nominal capsules that fail to ignite. The parameters to be measured are given in Table 1 in conjunction with some proposed diagnostics[28]. Many of these measurements have been carried out on smaller scale capsules (with laser energy up to 40 kJ), to infer the core and pusher final conditions. Some of the diagnostics could meet the requirements for MJ scale capsules with minimal R&D (yield measurements, ions temperature...). Others need significant improvements, for example, the measurement of size and shape of the burning zone. In recent experiments on Omega laser, neutron imaging with penumbral imaging achieved about $20\mu m$ spatial resolution, getting 5 μm resolution requires more work (detector, aperture, data processing). Lastly, some presently used techniques are limited to small-scale capsules, but similar method can be extrapolated. For example fuel areal density has been measured by secondary neutron emission (Reaction in Flight) in D_2 filled balloons, but the measurement is limited to $\rho R < 0.1$ g/cm^2. For MJ DT filled capsules, ρR measurement will be possible via tertiary secondary neutron emission.

Table 1. Measurements for fuel assembly characterization

Parameter	Diagnostic	Range for MJ experiments	Precision
Number of reactions Y	Activation measurements	10^6 to 10^{17}	10 %
Emission time	X-ray or γ detector [29] + Streak camera	10 to 20 ns	100 ps
Emission duration	idem	< 300 ps	10 ps
Hot spot			
Ion Temperature Ti	* Neutron spectrograph / Time of flight * X-ray spectroscopy	4 to 15 keV	5 %
Density ρ	* X-ray spectroscopy [30]: * Tracer Stark broadening	50 to 200 g/cm^3	10 %
Areal density ρR	Reactions in Flight[31, 32]: * Neutron spectrograph/Time of flight * Proton spectrograph Radiochemistry[33]	< 0.5 g/cm^2	10 %
Radius and shape	- Neutron Penumbral Imaging[34] - X-ray imaging (10 to 15 keV)	15 to 100 μm	5 μm
Main Fuel			
Areal Density ρΔR	1/Reactions in Flight: * Neutron spectrograph / Time of flight * Proton spectrograph 2/Radiochemistry 3/X-ray spectroscopy absorption measurement	< 1.5 g/cm^2	10 %
Density ρ		1500 g/cm^3	10 %
Integrity	X-ray imaging[35]		5 μm

4.5 Ignition and burn

The phase devoted to the study of burning capsules requires less measurements than the previous ones. The first reason is that if ignition is achieved, it implies nominal hot spot conditions. The second is that when the main fuel burns, it conceals the hot spot emissions. Finally, if capsules operate not very far from ignition threshold, it could not tolerate measurements inducing some perturbations such as tracers, observation windows ... The core measurement for burning capsules is the burn efficiency by the reactions number, the emission time and duration, the ion temperature and the total areal density with a neutron spectrograph. The related diagnostics are the same than in previous step, only the ranges differ.

5. CONCLUSIONS

The measurements requirements for moderate thermonuclear gain ICF experiments have been analyzed considering the critical yardsticks. Solutions have been proposed to meet most of these requirements and the diagnostics have been successfully tested on existing facilities. R &D are still needed for some of them to reach the specified spatial or temporal resolutions. Some measurements involve specific surrogate capsules or geometry: the transposition to the nominal capsule has to be further analyzed.

It is important to stress that most of these measurements requires absolutely calibrated diagnostics and that they will have to work in a perturbed environment for high yield experiments[36].

References

[1] M.D. Campbell and W.J. Hogan, Plasma Phys. Controlled Fusion **41**, B39 (1999).

[2] M. André, M. Novaro and D. Schirmann, Chocs **13**, 73 (1995).

[3] J.D. Lindl, Phys. Plasmas **2**, 3933 (1995).

[4] S.E. Bodner et al., Phys. Plasmas **5**, 1901 (1998).

[5] M. Tabak et al., Phys. Plasmas **1**, 1626 (1994).; R. Kodama et al., Phys. Plasmas **8**, 2268 (2001).

[6] K.A. Tanaka et al., Phys. Plasmas 7, 2014 (2000).

[7] W.L. Kruer , *The Physics of Laser Plasma Interaction* (Addison-Wesley, New-York, 1988)

[8] M. Pergament (these proceedings)

[9] S. Chandrasekhar, Hydrodynamic and Hydromagnetic Stability (Oxford University Press, London, 1961).

[10] S.P. Regan et al., Phys. Plasmas **6**, 2072 (1999).

[11] S.H. Glenzer et al., Phys. Plasmas 7, 2585 (2000) ; J.C. Fernandez et al., Phys. Plasmas **4**, 1849 (1997).

[12] B.J. MacGowan et al., Phys. Plasmas 3, 2029 (1996).

[13] S.H. Glenzer et al., Phys. Plasmas 6, 2117 (1999).

[14] C. Labaune et al., Phys. Rev. Lett. **75**, 248 (1995) ; S. Depierreux et al., Phys. Rev. Lett. **84**, 2869 (2000).

[15] R. Sigel et al., Phys. Rev. Lett. **65**, 578 (1990) ; H. Nishimura et al., Phys. Rev. A **44**, 8323 (1991) ; R. Kauffman et al., Rev. Sci. Instrum. **66**, 678 (1995) ; C. Bayer et al., 23rd ECLIM, St John's College, Oxford, England (1994) ; C. Decker et al., Phys. Rev. Lett. **79**, 1491 (1997) ; R.J. Leeper et al., Nuclear Fusion, **39**, 1283 (1999).

[16] G.A. Chandler et al., Rev. Sci. Instrum. **70**, 561 (1999).

[17] J-L. Bourgade et al., Rev. Sci. Instrum. **72**, 1173 (2001).

[18] K.L. Baker et al., Phys. Plasmas 7, 681 (2000).

[19] S.W. Haan et al., Phys. Plasmas **2**, 2480 (1995).

[20] A.A. Hauer et al., Phys. Plasmas **2**, 2488 (1995).

[21] N.D.Delamater et al, Phys. Rev.E **53**, 5240 (1996).

[22] P.Amendt et al, Phys. Plasmas **4**, 1862 (1997).

[23] S.M.Pollaine et al, Phys. Plasmas **8**, 2357 (2001).

[24] M.C. Herrmann et al., Phys. Plasmas **8**, 2296 (2001).

[25] C. Cherfils et al., Phys. Rev. Lett. **83**, 5507 (1999).

[26] J.M. Scott et al., Rev. Sci. Instrum. 72, 643 (2001).

[27] B.A. Remington et al., Phys. Fluids B4, 967 (1992) ; J.P. Knauer et al., Phys. Plasmas 7, 338 (2000).

[28] T.J. Murphy et al., Rev. Sci. Instrum.**72**, 773 (2001).

[29] R.R Berggren et al., Rev. Sci. Instrum. **72**, 873 (2001).

[30] B. Yaakobi et al., Phys. Plasmas **4**, 3021 (1997) ; C.J. Keane et al., Phys. Rev. Lett. **72**, 3029 (1994).

[31] E.G. Gamalii et al., JETP Lett. **21**, 70 (1975) ; H. Azechi et al., Laser Part. Beams **9**, 119 (1991).

[32] R.D. Petrasso et al., Phys. Rev. Lett. 77, 2718 (1996)

[33] E.M. Campbell et al., J. Appl. Phys. **51**, 6062 (1980).

[34] G.L. Morgan et al., Rev. Sci. Instrum. **72**, 865 (2001) ; J-P. Garconnet et al., Laser Part. Beams **17**, 563 (1994).

[35] B. Yaakobi et al., Phys. Plasmas 7, 3727 (2000).

[36] J-L. Bourgade (these proceedings).

CORE PERFORMANCE AND MIX IN DIRECT-DRIVE SPHERICAL IMPLOSIONS ON OMEGA

C. Stoeckl,[a] J. A. Delettrez,[a] R. Epstein,[a] V. Yu. Glebov,[a] R. L. Keck,[a] R. L. McCrory,[a] P. W. McKenty,[a] F. J. Marshall,[a] D. D. Meyerhofer,[a,1] P. B. Radha,[a] S. P. Regan,[a] S. Roberts,[a] W. Seka,[a] S. Skupsky,[a] V. A. Smalyuk,[a] C. Sorce,[a] J. M. Soures,[a] R. P. J. Town,[a] B. Yaakobi,[a] J. A. Frenje,[b] C. K. Li,[b] R. D. Petrasso,[b,2] F. H. Seguin,[b] K. Fletcher,[c] S. Padalino,[c] C. Freeman,[c] N. Izumi,[d] R. Lerche,[d] T. W. Phillips,[d] and T. C. Sangster[d]

Abstract

A wide variety of diagnostics have been deployed to study direct-drive spherical implosions of gas-filled, plastic shells on the 60-beam OMEGA laser system at the Laboratory for Laser Energetics. Neutron detectors using plastic scintillators measure primary yield, ion temperature, secondary neutron yield, and burn history. Charged-particle detectors record, on CR-39 samples, the secondary protons in pure D_2 implosions and elastically scattered protons, deuterons, and tritons in DT implosions to measure the shell and fuel areal densities. Primary D^3He protons from plastic shells with a CD layer at the shell–gas interface and a ^3He fill provide a significant signal only if the shell and fuel regions are mixed. Shells with an Ar-doped fill are used to infer the emissivity-averaged core electron temperatures and densities from the measured time-dependent Ar K-shell spectra. The level of nonuniformity in the shell areal density at stagnation is measured using gated x-ray images of Ti-doped shells.

The application of a large number of diagnostics to these implosions significantly improves the understanding of the core conditions and fuel–shell mix.

---

[a] Laboratory for Laser Energetics, University of Rochester, Rochester, NY 14623.
[1] Also Departments of Mechanical Engineering and Physics and Astonomy, University of Rochester
[b] Plasma Science and Fusion Center, MIT
[2] Also visiting scientist, LLE.
[c] SUNY Geneseo, NY
[d] Lawrence Livermore National Laboratory

1. INTRODUCTION

In the direct-drive approach to laser-driven inertial confinement fusion (ICF)[1] a spherical target is symmetrically illuminated by a number of individual laser beams. Target and laser illumination nonuniformities are the primary determinants of implosion performance. Illumination uniformity includes both individual-beam uniformity and on-target beam-to-beam power history differences (power balance), while target nonunformities include those on both the inner and outer capsule surfaces. Nonuniformities of either kind can lead to distortions in the compressed core due to secular growth of low-order ($\ell \leq 10$) modes and shell breakup and mix due to the Rayleigh–Taylor (RT)[2,3] growth of perturbations imprinted by high-order ($\ell > 10$) nonuniformities. Ignition with direct-drive ICF relies on the generation of a "spark" in the compression "hot spot" to begin the nuclear burn process.[4] The hot spot is a high-temperature, low-density region surrounded by a low-temperature, high-density region into which the burn wave propagates, leading to significant energy gain. For an ICF target to ignite, a significant fraction of the hot spot must remain "clean," unperturbed by the evolution of the RT instability. Thus, one expects that the compressed core of a direct-drive ICF implosion will consist a central hot spot, which is mostly compressed fuel, with a region where fuel and shell materials are mixed at larger radii. This region is surrounded by cold shell material. This paper describes techniques used to study the conditions in the compressed target, including those that are more sensitive to one or more of the regions described above.

The effects of Rayleigh–Taylor instabilities on target performance have been studied on direct-drive implosions of gas-filled plastic shells on the OMEGA laser system.[5] These targets are surrogates[6] for cryogenic implosions. The stability of these shells driven with a 1-ns square laser pulse have been shown to be similar to OMEGA cryogenic targets,[7] which are energy scaled from the NIF direct-drive-ignition targets. Plastic shells are advantageous because a wide variety of shell/gas conditions and diagnostics can be applied to study various details of the implosions.[8,9] OMEGA's reproducible laser pulses give confidence to the assumption that the implosion hydrodynamics is unchanged for different target types and fill-gas makeup. Various diagnostics have been used to probe different parts of the compressed target. Primary (single-step) neutron and charged-particle yields are most sensitive to the regions of the core where fuel exists at high ion temperatures. Secondary (two-step) neutron and charged-particle yields probe both central and mix regions, depending on the temperature profile.

2. DIAGNOSTICS

2.1. Neutron and Particle Diagnostics

The primary (N_p) neutron yield is measured using quenched plastic scintillators coupled to fast photomultipliers.[10] For the range of yields recorded, the typical uncertainty in these measurements is 10%. The neutron-averaged ion temperature is inferred from the thermal broadening of the neutron signal recorded on the detector. A convolution $m(t) = g(t) \otimes d(t)$ of a Gaussian $g(t)$ and an exponential decay $d(t)$ as is fitted to the measured signal to improve the accuracy of the ion-temperature measurement.[11] The ion-temperature-measurement uncertainty for these experiments is of the order of ~0.5 keV.

The fusion-reaction-rate history is recorded by the neutron temporal diagnostic (NTD)[12] with a time resolution of 25 ps (FWHM); NTD is based on the fast rise time of a commercially available plastic scintillator. Neutron collisions with the scintillator convert neutron kinetic energy to 350 to 450-nm-wavelength light. The light is collected and transported by an *f*/2 optical system to the photocathode of a fast streak camera, whose output image is recorded by a CCD camera. An optical fiducial system provides absolute timing with an uncertainty of ≤50 ps.

For DT-filled implosions, the fuel areal density is determined from the number of elastically scattered knock-on fuel particles:[13,14] n + T(D) → n' +T′(D′). The shell areal density is inferred from the elastically scattered shell protons: $n + p \rightarrow n' + p'$. The yield of knock-on particles is insensitive to the electron temperature profile.[13,14]

The secondary proton D +D → ^{3}He + n followed by ^{3}He + D → p(12 ~ 17 MeV) + ^{4}He and neutron D + D → T + p followed by T + D → n(12 ~ 17 MeV) + ^{4}He, yields depend on the electron temperature profile in the core and typically provide limits on the ρR_f and the core electron temperature, as well as information about the deviations in temperature and density profiles from 1-D predictions.[15,16] The areal density of the plastic shell, ρR_s, during stagnation can be inferred using the downshift of the secondary protons from D_2 implosions (produced with 12- to 17-MeV energies). The secondary protons are slowed by an amount proportional to ρR_{total}. Using the measured fuel areal density, the shell areal density can be inferred:[15] $\rho R_s = \rho R_{total} - \rho R_{fuel}$.

The core–fuel mix characteristics are inferred in a number of ways. The measured secondary neutron and proton yields from the shell regions can be compared to 1-D simulations. When the yields are significantly different from those predicted, they provide information about fuel–shell mix. An implosion of a plastic shell with a CD layer and a ^{3}He fill provides a significant primary D^{3}He proton signal only if the shell and fuel regions are mixed. This yield depends on the characteristics of the mix, either microscopic (diffusive) or macroscopic, where islands of shell material penetrate the core.

The secondary neutron yield is measured by the multichannel, single-hit detector MEDUSA[17] or by current-mode detectors.[18] The secondary proton and knock-on particle yields are measured with wedged range filters[15] and charged-particle spectrometers (CPS).[19] CR-39 nuclear emulsion is used in both detectors to determine the yield and the energy spectrum.

2.2. X-ray Spectroscopic Diagnostics

In some cases, the D_2 fuel is doped with 0.05 atm of Ar. This level of Ar dopant has only a small effect on the target performance. Time-resolved Ar *K*-shell spectra are recorded with a flat crystal spectrometer coupled to an x-ray streak camera with 25-ps resolution. The emissivity-averaged core electron temperature and density evolutions are inferred from the time-dependent Ar *K*-shell spectral line shapes of the Ar He-β, He-γ,

He-δ, and Ly-β resonant transitions and their associated satellites.[20] The line shapes depend strongly on the electron density and are relatively insensitive to the electron temperature, while the relative intensities are sensitive to variations in both electron temperature and density.[21] Spectral lineouts at different times are compared to a large database of synthetic spectra calculated by following Ref. 22.

2.3. X-ray Imaging

Areal-density modulations in a titanium-doped layer in the CH shell are determined from the intensity ratios of time-resolved, 2-D core images taken above and below the Ti K edge.[23] The 20-μm-thick shells contained an inner layer of 1.9-μm-thick CH doped with Ti (6% by atom). The initial Ti areal density in the shell is about 0.05 mg/cm^2. The emission from Ti-doped CH shells is imaged onto a four-strip framing camera[24,25] using a 6-μm-diam pinhole array, protected by a 200-μm-thick Be filter. The upper two strips of the framing camera are filtered by a 75-μm-thick Ti filter and the lower two strips by a 50-μm-thick Fe filter to image core radiation at photon energies below (~4.8 keV) and above (~6.5 keV) the Ti K edge (4.966 keV), simultaneously. The two x-ray energy channels have spectral bandwidths ($\Delta E/E$) of about 20%. Each image taken with the framing camera had a temporal resolution of ~40 ps and a spatial resolution of ~6 μm. The framing camera output was captured on a CCD camera with a 9-μm pixel size and a magnification of 25.

3. LASER CONDITIONS

An ~23-kJ, 1-ns square pulse delivered by the 60-beam OMEGA laser system[5] was used to drive the implosions described in this work. The beam-to-beam UV energy balance was typically ≤5% rms.[26] When beam overlap on target was included, the on-target nonuniformity due to beam-to-beam energy imbalance was <1% ($\ell \leq 12$). Individual-beam smoothing is accomplished by using smoothing by spectral dispersion (SSD),[27] distributed phase plates (DPP),[28] and polarization smoothing (PS) with birefringent wedges.[29] The two-dimensional (2-D) SSD operates at a single color cycle with 1-THz bandwidth.

4. EXPERIMENTAL RESULTS

This section reports the results of the various diagnostics described above on a series of implosions using targets with (940±30)-μm-diam, (19.5±1)-μm-thick plastic shells. Some shells have CD or titanium-doped (CHTi) internal layers. The targets are filled 15 atm of either He^3, D_2, or DT, sometimes doped with trace amounts of Ar. For the targets with pure He^3, the gas pressure is adjusted so that the hydrodynamics of the implosion, determined mostly by the electron and ion pressure, is the same. These implosions have convergence ratios (initial to final radius of the fuel–shell boundary) calculated by 1-D simulations of 15. The ratio of the measured primary neutron yield to that predicted by 1-D simulations ["yield over clean" (YOC)] for CH shells with D_2 fills is very reproducible around 35±5%.[26] The neutron-averaged ion temperature was measured at 3.5 keV, which is about 0.5 keV higher than the 1-D predictions.

The 1-D predictions of areal density for these shells are ρR_f = 16 mg/cm^2 and ρR_s = 60 mg/cm^2. The total ρR as determined from the D^{3}He secondary protons of D_2-filled shells is found to be 76 mg/cm^2. The ρR_f inferred from the D knock-on yield is 15 mg/cm^2, while ρR_s ~ 61 mg/cm^2 from the knock-on protons, both from DT-filled targets. These measurements show that the sum of fuel and shell areal densities for DT implosions is in good agreement with the total areal density independently inferred from D_2 implosions, close to those predicted from 1-D simulations.

Comparing the fusion-reaction-rate history measured by NTD with 1-D simulations gives valuable information about the accuracy of physical models used in the codes, such as laser absorption and heat transport. In particular, the timing onset of neutron production at shock convergence along with peak neutron production is very sensitive to the value of the flux-limiter.[30] Figure 1 shows a comparison between measured and predicted fusion-reaction-rate histories. A relatively minor change in the flux limiter of about 15%, from 0.06 to 0.07, leads to a significant shift in the burn history, well outside the experimental uncertainty of 50 ps

The emissivity-averaged electron temperature and density history is inferred from spectroscopic measurements of Ar-doped capsules. Figure 2 shows the evolution of the emissivity-averaged electron temperature and density during peak compression. The evolution of the continuum emission at ~3.5 keV is shown for comparison.[29] Note that the peak neutron emission occurs ~170 ps before the peak continuum emission, consistent with 1-D simulations. The measured peak electron temperature of ~2 keV is much lower than the corresponding neutron-averaged ion temperature of 3.5 keV, which indicates that the Ar emission originates from the outer, colder regions of the core and the fuel–shell mix region. The amount of fuel–pusher mixing can be inferred by combining the measurements of the fuel areal density, the core size from x-ray emission (~34 μm),

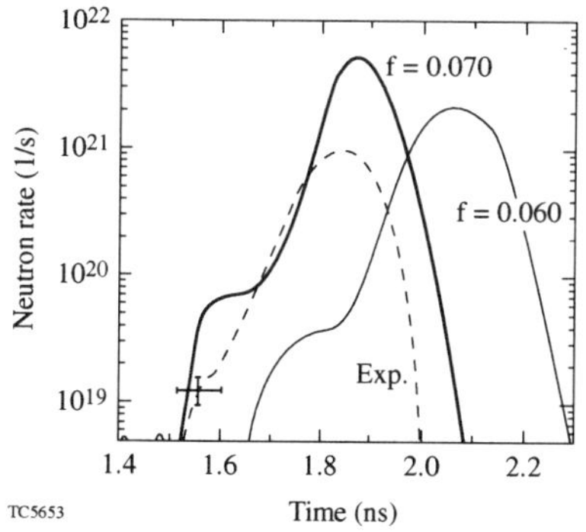

Figure 1. Comparison between measured and predicted fusion-reaction-rate histories for 20-μm shells filled with 15 atm of D_2. A relatively minor change in the flux limiter of about 15% from 0.06 to 0.07 leads to a significant shift in the burn history, well outside the experimental uncertainty of 50 ps.

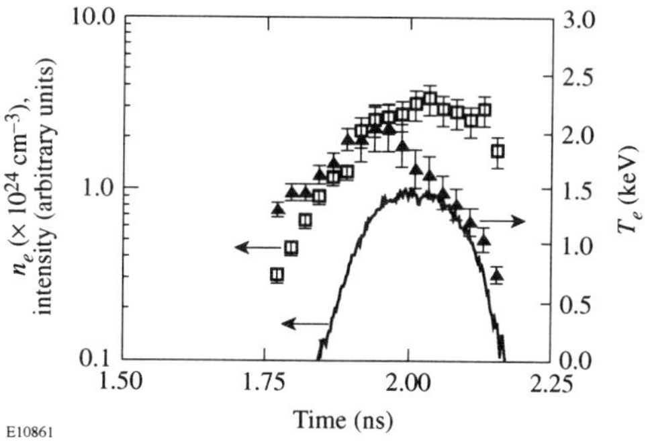

Figure 2. Temporal evolution of the emissivity-averaged core electron temperature and density near peak compression. The temporal evolution of the ~3.5-keV continuum emission is shown for comparison.

and the electron density from spectroscopic analysis. This analysis gives an average D_2 density of 4.4±1.3 g/cm^3 and a CH density of 2.6±0.8 g/cm^3, suggesting that approximately 1/3 of the core density consists of CH from the shell in the region that dominates the Ar emission. The present Ar spectroscopic data, however, provides no information about the spatial profiles in the core.

Figure 3 shows the evolution of areal-density modulations obtained by differential x-ray imaging above and below the Ti *K* edge: 1.92, 1.97, 2.02, and 2.07 ns. The spatial scale of modulations decreases from t = 1.92 ns to t = 2.02 ns (as evident from the motion of the "circular" structure in corresponding images). After that, the shorter-scale modulations start to grow, which can be seen in the image at 2.07 ns. The image size at 2.07 ns is smaller than images at early times because the level of backlighter emission drops late in the implosion. The areal-density modulations are relatively modest (18% at peak compression), indicating a shell behavior close to 1-D. The modulations decrease toward peak compression, probably due to the convergence of the target.

5. MIX MEASUREMENTS

The D^{3}He proton yield from a CH shell with an inner 1-mm-thick CD layer was 1.0±0.2 × 10^7 and dropped by an order of magnitude when the CD layer was offset from the inner surface of the shell by 1 μm. This indicates the presence of fuel–shell mixing during the compression. A model of the mix profiles, described in Refs. 26 and 31, indicates that approximately 50% predicted by 1-D simulations is "clean," with close-to predicted parameters.

6. SUMMARY

A wide variety of neutron, particle, and x-ray diagnostics have been deployed to study the performance of CH-shell implosions on OMEGA. The experimental observations are consistent with the picture that the compressed core at peak neutron production has an inner clean region, whereas a significant amount of shell material is mixed into the outer part of the core. The measured primary neutron yield is ~35% of that predicted from 1-D calculations. The fuel and shell areal densities determined by knock-on particles and the slowing down of secondary protons are close to their predicted values. The areal-density modulations of the shell as measured by differential x-ray imaging are below 20%, which independently indicates a shell behavior close to 1-D.

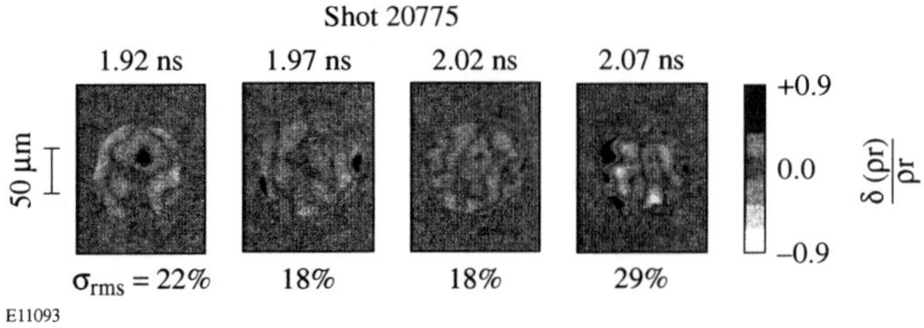

Figure 3. Evolution of areal-density modulations obtained by differential x-ray imaging.

7. ACKNOWLEDGMENTS

This work was supported by the U.S. Department of Energy Office of Inertial Confinement Fusion under Cooperative Agreement No. DE-FC03-92SF19460, the University of Rochester, and the New York State Energy Research and Development Authority. The support of DOE does not constitute an endorsement by DOE of the views expressed in this article.

8. REFERENCES

1. J. Nuckolls, L. Wood, A. Thiessen, and G. Zimmerman, Laser compression of matter to super-high densities: thermonuclear (CTR) applications, *Nature* **239**, 139–142 (1972).
2. J. D. Kilkenny, S. G. Glendinning, S. W. Haan, B. A. Hammel, J. D. Lindl, D. Munro, B. A. Remington, S. V. Weber, J. P. Knauer, and C. P. Verdon, A review of the ablative stabilization of the Rayleigh-Taylor instability in regimes relevant to inertial confinement fusion, *Phys. Plasmas* **1**(5), 1379–1389 (1994).
3. S. E. Bodner, D. G. Colombant, J. H. Gardner, R. H. Lehmberg, S. P. Obenschain, L. Phillips, A. J. Schmitt, J. D. Sethian, R. L. McCrory, W. Seka, C. P. Verdon, J. P. Knauer, B. B. Afeyan, and H. T. Powell, Direct-drive laser fusion: Status and prospects, *Phys. Plasmas* **5**(5), 1901–1918 (1998).
4. J. D. Lindl, *Inertial Confinement Fusion: The Quest for Ignition and Energy Gain Using Indirect Drive* (Springer-Verlag, New York, 1998), Chap. 6, pp. 61–82.
5. T. R. Boehly, D. L. Brown, R. S. Craxton, R. L. Keck, J. P. Knauer, J. H. Kelly, T. J. Kessler, S. A. Kumpan, S. J. Loucks, S. A. Letzring, F. J. Marshall, R. L. McCrory, S. F. B. Morse, W. Seka, J. M. Soures, and C. P. Verdon, Initial performance results of the OMEGA laser system, *Opt. Commun.* **133**, 495–506 (1997).
6. F. J. Marshall, J. A. Delettrez, R. Epstein, V. Yu. Glebov, D. R. Harding, P. W. McKenty, D. D. Meyerhofer, P. B. Radha, W. Seka, S. Skupsky, V. A. Smalyuk, J. M. Soures, C. Stoeckl, R. P. Town, B. Yaakobi, C. K. Li, F. H. Séguin, D. G. Hicks, and R. D. Petrasso, Direct-drive high-convergence-ratio implosion studies on the OMEGA laser system, *Phys. Plasmas* **7**(5), 2108–2113 (2000).
7. V. N. Goncharov, P. McKenty, S. Skupsky, R. Betti, R. L. McCrory, and C. Cherfils-Clérouin, Modeling hydrodynamic instabilities in inertial confinement fusion targets, *Phys. Plasmas* **7**(12), 5118–5139 (2000).
8. M. D. Cable, S. P. Hatchett, J. A. Caird, J. D. Kilkenny, H. N. Kornblum, S. M. Lane, C. Laumann, R. A. Lerche, T. J. Murphy, J. Murray, M. B. Nelson, D. W. Phillion, H. Powell, and D. B. Ress, Indirectly driven, high convergence inertial confinement fusion implosions, *Phys. Rev. Lett.* **73**(17), 2316–2319 (1994).
9. D. K. Bradley, J. A. Delettrez, R. Epstein, R. P. J. Town, C. P. Verdon, B. Yaakobi, S. Regan, F. J. Marshall, T. R. Boehly, J. P. Knauer, D. D. Meyerhofer, V. A. Smalyuk, W. Seka, D. A. Haynes, Jr., M. Gunderson, G. Junkel, C. F. Hooper, Jr., P. M. Bell, T. J. Ognibene, and R. A. Lerche, Measurements of core and pusher conditions in surrogate capsule implosions on the OMEGA laser system, *Phys. Plasmas* **5**(5), 1870–1879 (1998).
10. R. A. Lerche and T. J. Murphy, Geometry compensation for improving speed and efficiency of scintillator-based neutron time-of-flight detectors, *Rev. Sci. Instrum.* **63**(10), 4880–4882 (1992).
11. T. J. Murphy, R. A. Lerche, C. Bennett, and G. Howe, Ion-temperature measurement of indirectly driven implosions using a geometry-compensated neutron time-of-flight detector, *Rev. Sci. Instrum.* **66**(1), 930–932 (1995).
12. R. A. Lerche, D. W. Phillion, and G. L. Tietbohl, 25 ps neutron detector for measuring ICF-target burn history, *Rev. Sci. Instrum.* **66**(1), 933–935 (1995).
13. S. Skupsky and S. Kacenjar, Measuring fuel ρR for inertial fusion experiments using neutron elastic-scattering reactions, *J. Appl. Phys.* **52**(4), 2608–2613 (1981).
14. C. K. Li, D. G. Hicks, F. H. Séguin, J. A. Frenje, K. Green, R. D. Petrasso, D. D. Meyerhofer, J. M. Soures, V. Yu. Glebov, P. B. Radha, S. Skupsky, C. Stoeckl, S. Roberts, and T. C. Sangster, Study of direct-drive, DT gas-filled-plastic capsule implosions using nuclear diagnostics on OMEGA, submitted to *Phys. Plasmas.*

15. F. H. Séguin, C. K. Li, D. G. Hicks, J. A. Frenje, K. M. Green, R. D. Petrasso, J. M. Soures, D. D. Meyerhofer, V. Yu. Glebov, C. Stoeckl, P. B. Radha, S. Roberts, C. Sorce, T. C. Sangster, M. D. Cable, S. Padalino, and K. Fletcher, Using secondary proton spectra to study imploded D_2-filled capsules at the OMEGA laser facility, submitted to *Phys. Plasmas.*
16. H. Azechi, M. D. Cable, and R. O. Stapf, Review of secondary and tertiary reactions, and neutron scattering as diagnostic techniques for inertial confinement fusion targets, *Laser Part. Beams* **9** (1), 119–134 (1991).
17. J. P. Knauer, R. L. Kremens, M. A. Russotto, and S. Tudman, Using cosmic rays to monitor large scintillator arrays, *Rev. Sci. Instrum.* **66**(1), 926–928 (1995).
18. V. Yu. Glebov, D. D. Meyerhofer, C. Stoeckl, and J. D. Zuegel, Secondary-neutron-yield measurements by current-mode detectors, *Rev. Sci. Instrum.* **72**(1), 824–827 (2001).
19. D. G. Hicks, Charged particle spectroscopy: A new window on inertial confinement fusion, Ph.D. thesis, Massachusetts Institute of Technology, 1999.
20. S. P. Regan, J. A. Delettrez, P. A. Jaanimagi, B. Yaakobi, V. A. Smalyuk, F. J. Marshall, D. D. Meyerhofer, W. Seka, D. A. Haynes, Jr., and C. F. Hooper, Characterization of direct-drive-implosion core conditions on OMEGA with time-resolved Ar K-shell spectroscopy, submitted to *Phys. Plasmas* (2001).
21. D. A. Haynes, Jr., C. F. Hooper, Jr., R. C. Mancini, D. K. Bradley, J. Delettrez, R. Epstein, and P. A. Jaanimagi, Spectroscopic analysis of Ar-doped laser-driven implosions, *Rev. Sci. Instrum.* **66**(1), 755–757 (1995).
22. D. A. Haynes, D. T. Garber, C. F. Hooper, R. C. Mancini, Y. T. Lee, D. K. Bradley, J. Delettrez, R. Epstein, and P. A. Jaanimagi, Effects of ion dynamics and opacity on stark-broadened argon line profiles, *Phys. Rev. E* **53**(1), 1042–1050 (1996).
23. V. A. Smalyuk, B. Yaakobi, J. A. Delettrez, F. J. Marshall, D. D. Meyerhofer, S. P. Regan, and R. P. J. Town, Evolution of shell nonuniformities near peak compression of spherical implosion, submitted to *Phys. Rev. Lett.* (2001).
24. V. A. Smalyuk, T. R. Boehly, L. S. Iwan, T. J. Kessler, J. P. Knauer, F. J. Marshall, D. D. Meyerhofer, C. Stoeckl, and B. Yaakobi, Fourier-space image processing for spherical experiments on OMEGA (invited), to be published in *Rev. Sci. Instrum.*
25. C. J. Pawley and A. V. Deniz, Improved measurements of noise and resolution of x-ray framing cameras at 1–2 keV. *Rev. Sci. Instrum.* **71**(3) 1286–1295 (2000).
26. D. D. Meyerhofer, J. Delettrez, R. Epstein, V. Yu. Glebov, V. N. Goncharov, R. L. Keck, R. L. McCrory, P. W. McKenty, F. J. Marshall, P. B. Radha, S. P. Regan, S. Roberts, W. Seka, S. Skupsky, V. A. Smalyuk, C. Sorce, C. Stoeckl, J. M. Soures, R. P. J. Town, B. Yaakobi, J. D. Zuegel, J. A. Frenje, C. K. Li, R. D. Petrasso, F. H. Séguin, K. Fletcher, S. Padalino, M. R. Freeman, N. Izumi, R. A. Lerche, T. W. Phillips, and T. C. Sangster, Core performance and mix in direct-drive spherical implosions with high uniformity, *Phys. Plasmas* **8**(5), 2251–2256 (2001).
27. S. Skupsky, R. W. Short, T. Kessler, R. S. Craxton, S. Letzring, and J. M. Soures, Improved laser-beam uniformity using the angular dispersion of frequency-modulated light, *J. Appl. Phys.* **66**(8), 3456–3462 (1989).
28. T. J. Kessler, Y. Lin, J. J. Armstrong, and B. Velazquez, Phase conversion of lasers with low-loss distributed phase plates, in: *Laser Coherence Control: Technology and Applications*, edited by H. T. Powell and T. J. Kessler (SPIE, Bellingham, WA, 1993), Vol. 1870, pp. 95–104.
29. T. R. Boehly, V. A. Smalyuk, D. D. Meyerhofer, J. P. Knauer, D. K. Bradley, R. S. Craxton, M. J. Guardalben, S. Skupsky, and T. J. Kessler, Reduction of laser imprinting using polarization smoothing on a solid-state fusion laser, *J. Appl. Phys.* **85**(7), 3444–3447 (1999).
30. J. Delettrez, Thermal electron transport in direct-drive laser fusion, *Can. J. Phys.* **64**(8), 932–943 (1986).
31. P. B. Radha, V. Yu. Glebov, F. J. Marshall, D. D. Meyerhofer, S. P. Regan, W. Seka, S. Skupsky, V. A. Smalyuk, J. M. Soures, C. Stoeckl, B. Yaakobi, and R. D. Petrasso, A measurement-based picture of core conditions in OMEGA implosions, *Bull. Am. Phys. Soc.* **45**(7), 164 (2000).

LASER–PLASMA INTERACTION DIAGNOSTICS FOR ICF FUSION RESEARCH

W. Seka,[a] R. S. Craxton,[a] R. L. Keck,[a] J. P. Knauer,[a] D. D. Meyerhofer,[a] S. P. Regan,[a] C. Stoeckl,[a] B. Yaakobi,[a] R. E. Bahr,[a] D. Montgomery,[b] H. Baldis,[c] R. Kirkwood[d]

1. INTRODUCTION

Understanding of the coupling of laser light to the plasma corona is essential for the success of laser-driven inertial confinement fusion (ICF). The most desirable way of imparting laser energy to the target plasma is via inverse bremsstrahlung absorption.[1] In addition, at elevated intensities, the thresholds for various instabilities may be exceeded, channeling laser energy into scattered electromagnetic (EM) waves and various electrostatic waves. As a result, part of the incident laser energy may be unavailable for heating the plasma and the target may be irradiated less uniformly. Furthermore, some of the laser energy may ultimately be transferred to energetic electrons that can preheat the fuel and degrade target performance. Many of these processes, such as stimulated Brillouin scattering (SBS), stimulated Raman scattering (SRS), two-plasmon-decay (TPD) instability, and filamentation, have been studied extensively.[2] Presently, however, the understanding of these processes is not sufficient to predict their importance under realistic ICF plasma conditions. Therefore continued surveying of laser–plasma interaction processes is required to guard against unforeseen surprises. This paper surveys the plasma diagnostics used on the OMEGA laser facility at the University of Rochester's Laboratory for Laser Energetics (LLE).

[a]Laboratory for Laser Energetics, University of Rochester, Rochester, NY 14623.
[b]Los Alamos National Laboratory, Los Alamos, NM.
[c]ILSA, Lawrence Livermore National Laboratory and Dept. of Appl. Sciences, University of California at Davis, CA.
[d]Lawrence Livermore National Laboratory, Livermore, CA.

Advanced Diagnostics for Magnetic and Inertial Fusion, Edited by Stott *et al.*
Kluwer Academic/Plenum Publishers, New York, 2002

2. PLASMA DIAGNOSTICS AND RESULTS

Absorption measurements on OMEGA are obtained from 17 plasma calorimeters placed inside the target chamber that consist of dual, thin (25-μm) Ta foils attached to low-mass thermocouples. A Cu pad with equally low thermal mass serves as reference. One of the Ta foils is covered with a UV-transmitting window that measures the scattered UV energy, while the second foil is open and measures the total energy of the scattered UV light, plasma, and x rays. The plasma plus x-ray energy is obtained by subtracting the UV signal from the open-foil signal. The calorimeters are absolutely calibrated using particle beams and UV sources. The random error is ~5%, and the systematic error is estimated at ~4%. For plastic-shell OMEGA implosion experiments, the absorption fractions range between 70% and 80%, depending on the target and pulse shape used. The measurements generally agree with the predictions from hydrodynamic code simulations.

Parametric interaction processes generally do not impart useful energy to the imploding target. Depending on the process, the drive energy may be reduced, the irradiation uniformity may be decreased, or the fuel may be preheated by energetic electrons. A detailed survey of interaction processes on OMEGA is obtained by monitoring the light scattered back through the focusing lens of two of OMEGA's 60 beams. A schematic layout of the full aperture backscatter stations (FABS) is shown in Fig. 1. A 4% pickoff reflection is focused by a near-normal-incidence concave mirror. A 351-nm high reflector separates the UV and visible energies, which are then measured by two volume-absorbing calorimeters. The UV (SBS) calorimeter is cross-calibrated and the visible calorimeter (SRS) is absolutely calibrated and corrected for all intervening optical components. The estimated error on the energy measurements is 5% to 10%

Another two 4% reflections are used to image the target onto two large-core (435-μm) gradient-index fibers filtered for 351 nm (SBS) and the visible for SRS. The two fibers are routed to two spectrometers, each coupled to a streak camera for time-resolved spectroscopy. Typical SBS and SRS spectra are shown in Fig. 2. The broadband SRS spectra have been corrected for group-velocity dispersion in the fiber. Modal dispersion in the fiber is insignificant. The overall time resolution is 80 ps for SBS and 100 ps for SRS. The corresponding spectral resolutions are 0.04 nm (SBS) and 4 nm (SRS).

Figure 3 shows the results obtained from the two FABS's on OMEGA for SBS when two different beam-smoothing techniques are employed, i.e., 2-D SSD[3] at 1-THz

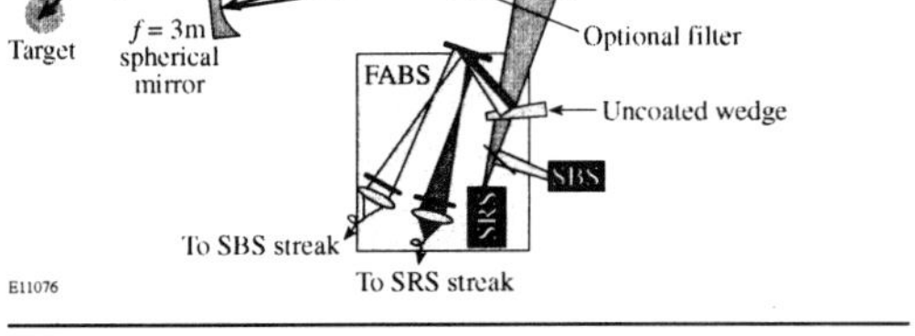

Figure 1. Schematic layout of the full-aperture backscatter station (FABS) on OMEGA, which measures the energy and temporal evolution of the spectrum in the SBS- and SRS-wavelength regions.

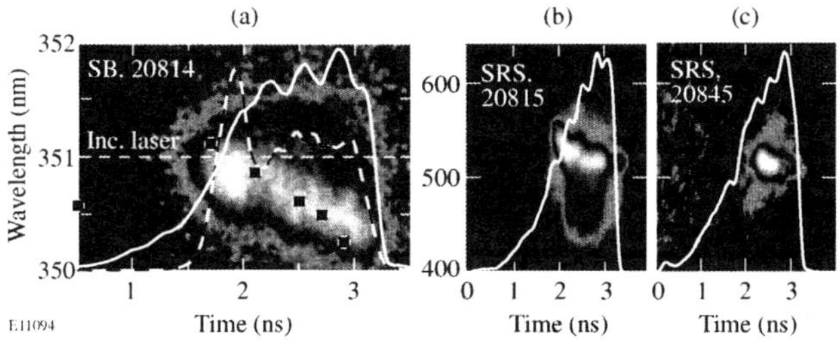

Figure 2. Typical time-resolved SBS (a) and SRS [(b) (c)] spectra obtained on OMEGA long-scale-length interaction experiments at 1-THz SSD beam smoothing. The short-wavelength (at $\lambda \lesssim 500$ nm) SRS emission (b) originates in filaments. Application of polarization smoothing suppresses filamentation and the short-wavelength SRS signal (c). Laser pulse shapes are superposed on all spectra. In (a) the dashed line is the spectrally integrated SBS backscattered power; the squares represent calculated SBS shifts using 2-D hydrodynamic predictions and SBS threshold calculations.

bandwidth with and without polarization smoothing. The two sets of data are clearly separated (○ and ■). Replotting the data for polarization smoothing at half their nominal peak intensity collapses the two sets of data to a single set (● and ■), which leads to the conclusion that SBS primarily originates in the highest-intensity speckles of the intensity distribution.

Parametric interaction processes involving plasma waves are likely to generate energetic electrons.[1] Plasma waves generated by SRS typically give rise to hot-electron temperatures of T_h = 40 to 60 keV, while the TPD plasma waves lead to T_h = 50 and 150 keV. These hot electrons generate x-rays, which are measured with filtered photomultiplier-scintillator detectors.[4] The absolute energy contained in the hot electrons is obtained via cross-calibration using absolutely calibrated x-ray spectroscopy.[5] For ICF studies, the scaling of the temperature and energy content in the fast electrons is important for assessing their possible detrimental impact on target performance. An example of the scaling of T_h with overlapped irradiation intensity is shown in Fig. 4. These data indicate that the TPD instability is susceptible to multibeam-interaction effects contrary to conventional wisdom.

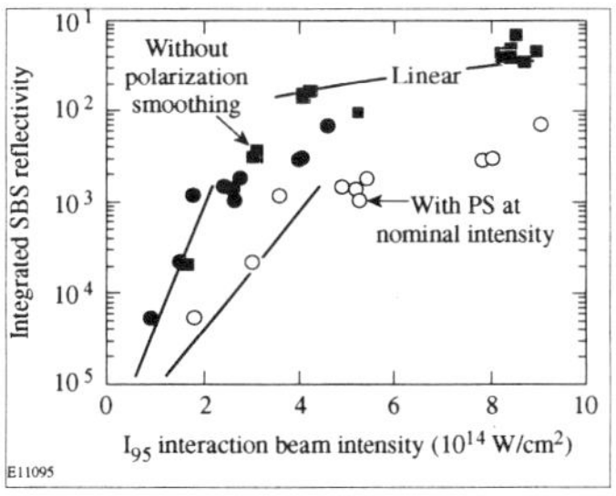

Figure 3. SBS reflectivities obtained in single-beam, long-scale-length, laser–plasma interaction experiments on OMEGA with 1-THz SSD beam smoothing, with (○) and without (■) PS. The ● are data with PS plotted at half their nominal intensities.

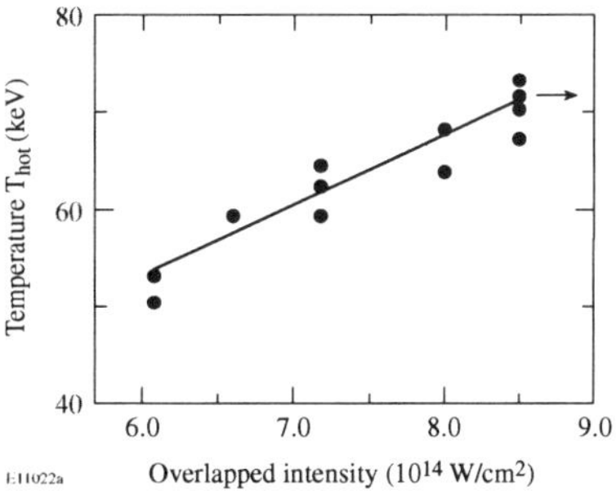

Figure 4. Scaling of hot-electron temperature with overlapped beam intensity inferred from time-resolved, hard-x-ray measurements in the range from 50 to 500 keV. The data were taken with filtered photomultiplier-scintillator detectors on 60-beam OMEGA implosion experiments.

Filamentation in the underdense corona affects the uniformity of target irradiation by expelling plasma from the high-intensity speckles in the light distribution. This process has a low threshold that increases with laser bandwidth when SSD is applied. Filamentation is difficult to measure directly, but there are indirect indicators for it, such as the short-wavelength SRS signal[6] at $\lambda \lesssim$ 500 nm (for λ_{Laser} = 351 m). Interaction experiments on OMEGA have shown clear evidence for suppression of filamentation in experiments with 1-THz SSD [Figs. 2(b) and (c)].

The temperature T_e influences the instability thresholds and gains. Thus, knowing T_e helps to interpret of interaction experiments.[7] In planar-target experiments T_e has been measured using small (~0.1 μm thick and 200 μm in diam) embedded micro-dots of KCl, Ti, or V. The hydrodynamics of the overall plasma are not measurably influenced by the presence of these dots. Time-resolved x-ray spectroscopy then yields electron temperatures from time-varying x-ray line ratios.

3. CONCLUSION

A wide array of plasma diagnostics supports the interaction experiments on the 60-beam OMEGA laser facility. The results obtained from these experiments continue to refine our understanding of the interaction processes and lead to improved target designs for future large laser fusion facilities (NIF in the US and LMJ in France).

4. ACKNOWLEDGEMENT

This work was supported by the U.S. Department of Energy Office of Inertial Confinement Fusion under Cooperative Agreement No. DE-FC03-92SF19460, the University of Rochester, and the New York State Energy Research and Development Authority. The support of DOE does not constitute an endorsement by DOE of the views expressed in this article.

5. REFERENCES

1. W. L. Kruer, *The Physics of Laser–Plasma Interactions, Frontiers in Physics*, Vol. 73, edited by D. Pines (Addison-Wesley, Redwood City, CA, 1988), Chap. 4, pp. 37–43.
2. H. A. Baldis, E. M. Campbell, and W. L. Kruer, Laser plasma interactions, in: *Handbook of Plasma Physics*, edited by A. Rubenchik and S. Witkowski, Physics of Laser Plasma, edited by M. N. Rosenbluth and R. Z. Sagdeev (North-Holland, Amsterdam, 1991), Vol. 3, pp. 361–434.
3. S. Skupsky, R. W. Short, T. Kessler, R. S. Craxton, S. Letzring, and J. M. Soures, Improved laser-beam uniformity using the angular dispersion of frequency-modulated light, *J. Appl. Phys.* **66**(8), 3456–3462 (1989).
4. C. Stoeckl, V. Yu. Glebov, D. D. Meyerhofer, W. Seka, B. Yaakobi, R. P. J. Town, and J. D. Zuegel, Hard x-ray detectors for OMEGA and NIF, *Rev. Sci. Instrum.* **72**(1), 1197–1200 (2001).
5. B. Yaakobi, C. Stoeckl, T. Boehly, D. D. Meyerhofer, and W. Seka, Measurement of preheat due to fast electrons in laser implosions, *Phys. Plasmas* **7**(9), 3714–3720 (2000).
6. R. W. Short and A. Simon, Collisionless damping of localized plasma waves in laser-produced plasmas and application to stimulated raman scattering in filaments, *Phys. Plasmas* **5**(12), 4134–4143 (1998).
7. S. P. Regan, D. K. Bradley, A. V. Chirokikh, R. S. Craxton, D. D. Meyerhofer, W. Seka, R. W. Short, A. Simon, R. P. J. Town, B. Yaakobi, J. J. Caroll III, and R. P. Drake, Laser-plasma interactions in long-scale-length plasmas under direct-drive national ignition facility conditions, *Phys. Plasmas* **6**(5), 2072–2080 (1999).

REQUIREMENTS FOR DIAGNOSTICS IN CONTROLLING ADVANCED TOKAMAK MODES

Y. Kusama*

INTRODUCTION

Advanced tokamak modes are tokamak operation modes having a possibility to establish a concept of an attractive tokamak fusion reactor with a small size, high fusion power density and a capability of steady-state operation. Present tokamaks have realized some advanced tokamak modes such as a high β_p ELMy H-mode [1] and a negative magnetic shear mode [2]. These advanced modes are considered to be candidates for an operation mode in realizing the non-inductive steady-state operation in the International Thermonuclear Experimental Reactor (ITER) [3]. Advanced tokamak modes are accompanied with an internal and/or edge transport barrier forming a steep gradient on the density and the temperature profile. It is commonly understood that simultaneous optimization of the plasma pressure profile and the plasma current density profile, keeping the MHD stability, is required for obtaining and sustaining a high-performance plasma. Moreover, stabilization of MHD modes destabilized in a regime of high plasma pressure and high bootstrap current fraction is important.

Exploration of advanced tokamak modes in present tokamaks has shown that measurements of the plasma current density profile (safety factor profile), the pressure profile (density and temperature profiles of electrons and ions) and fluctuation measurement of MHD modes are indispensable for controlling the advanced tokamak modes. Measurement of plasma rotation profile is also important because transport barrier formation and occurrence of MHD modes relates closely to it. In obtaining and sustaining the advanced tokamak modes, control of the transport barrier is essential. Therefore, measurements must be done with enough spatial resolution to identify the location of the transport barrier and to control the plasma pressure gradient. Other important aspect of the diagnostics for controlling advanced tokamak modes is whether measurements can be utilized in real-time for the feedback control of the plasma.

In this paper, role and issues of the plasma control towards advanced tokamak modes,

* Y. Kusama, Naka Fusion Research Establishment, Japan Atomic Energy Research Institute,Naka-machi, Naka-gun, Ibaraki-ken, 311-0193, Japan.

Advanced Diagnostics for Magnetic and Inertial Fusion, Edited by Stott *et al.*
Kluwer Academic/Plenum Publishers, New York, 2002

key diagnostics and requirements for diagnostics, plasma controls towards advanced tokamak modes in JT-60U, issues in developing diagnostics and feedback control schemes for controlling advanced tokamak modes will be presented.

ROLE AND ISSUES OF PLASMA CONTROLS TOWARDS ADVANCED TOKAMAK MODES

An "Advanced Tokamak Mode" would be defined as a tokamak operation mode, which has a possibility to establish a concept of an attractive tokamak fusion reactor with a small size, high fusion power density and a capability of a steady-state operation. These features are based on high energy confinement, high MHD stability, high bootstrap current fraction and high capability of heat and particle exhaust. A concept of a steady-state tokamak reactor has been developed. For example, following key physics elements are required to be realized simultaneously for the Advanced Steady-State Tokamak Reactor (A-SSTR); H-factor to L-mode ~2.7, a ratio of averaged electron density to the Greenwald density n_e/n_{eGW} ~1.2, normalized bate β_N ~ 3.7, bootstrap current fraction ~ 0.8, non-inductive current fraction ~1, fuel purity ~ 0.85, radiation power to heating power ~0.95 [4]. Role of plasma controls towards advanced tokamak modes is to establish a stable and well-reproducible route to a reactor-grade plasma and sustain it at the highest performance in a steady state.

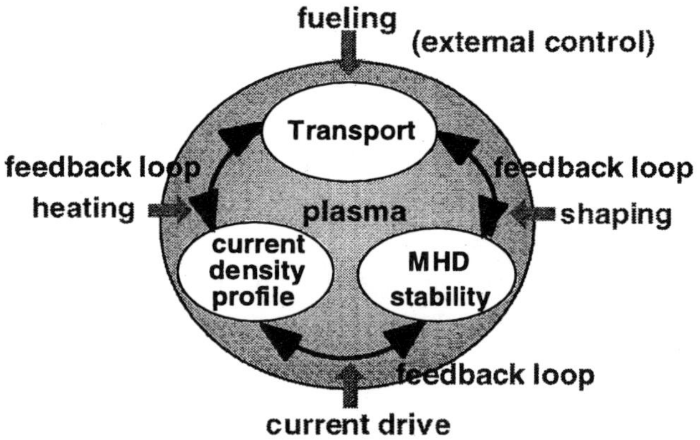

Figure 1. A schematic diagram of mutual interactions between the confinement, the current density profile and the MHD stability in a tokamak plasma and plasma control using methods of heating, current drive, shaping, fueling.

Figure 1 shows a schematic diagram of mutual interactions between the confinement, the current density profile and the MHD stability in a tokamak plasma [5]. A change in the transport characteristics modifies the pressure gradient of the plasma and it affects the bootstrap current. The changed current density profile affects the MHD stability, and the pressure profile modified due to the change in the transport affects the MHD stability directly. In a real plasma, many internal feedback loops act mutually and they determine plasma behaviors. In a case, the internal feedback act in one direction and it leads to a plasma disruption. Therefore, simultaneous optimization of transport, MHD stability and current density profile is required using external controls of heating, current drive, shaping and fueling. Compatibility of a high performance plasma with divertor is also required in a high fusion power plasma.

KEY DIAGNOSTICS AND REQUIREMENTS FOR CONTROL DIAGNOSTICS

A large fraction of the bootstrap current induced by the pressure gradient of the plasma is necessary to establish a concept of an attractive steady-state tokamak reactor. Candidates of the advanced tokamak modes are high β_p ELMy H-mode [1] and negative magnetic shear mode [2] etc. having internal transport barrier and/or edge transport barrier (ITB/ETB). Figure 2 shows profiles of the electron density, the electron temperature, the ion temperature and the safety factor in the negative shear plasma on JT-60U [6]. In a shadowed region density and temperature profiles have a steep gradient due to the ITB formation, which is closely connected to the MHD stability. Therefore, control of ITB/ETB under constraint of the MHD stability is essential in controlling advanced tokamak modes and it requires key diagnostics. The safety factor profile is the most important for producing advanced tokamak modes because it affects significantly the optimization of the plasma confinement, the current drive capability and the MHD stability. The motional Stark effect polarimeter [7] used widely in present tokamaks is a first candidate of the measurement. The Thomson scattering measurement, interferometer, electron cyclotron emission (ECE) measurement and charge-exchange recombination spectroscopy (CXRS) are necessary to measure the electron density and temperature profiles and ion temperature profile, on which a result of the plasma control appears directly. In sustaining the high performance plasma with a large fraction of the bootstrap current (high poroidal beta β_p) and a high toroidal beta β_t, internal structure measurement of MHD modes such as the neoclassical tearing mode (NTM) [8] and the resistive wall mode (RWM) [9] is required to suppress these modes.

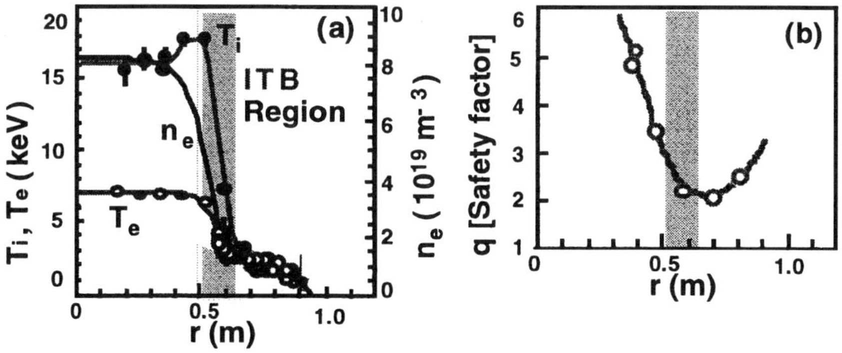

Figure 2. Profiles of (a) the electron density, the electron temperature, the ion temperature and (b) the safety factor in the negative shear plasma on JT-60U.

For control diagnostics, following additional conditions are required, compared to those used for physics understanding because a plasma disrupts at a large stored energy in the worst case if the plasma control is missed.

(1) High reliability and accuracy: Control diagnostics must be used routinely with high accuracy and without being affected by plasma phenomena.

(2) Sufficient time and spatial resolution: Simple scalars and complete spatial profiles must be measured with sufficient time and spatial resolution for physics phenomena to be controlled. For example, a growth rate $\gamma \sim (10\ \mathrm{ms})^{-1}$ for the RWM and an ITB width ~a/10 is an important time and spatial scale to determine the time and the spatial resolution of the diagnostics.

(3) Real-time processing of acquired data: Acquired data must be processed in real-time to establish real-time and active feedback control loops.

PLASMA CONTROL TOWARDS ADVANCED TOKAMAK MODES IN JT-60U

Real-time feedback control schemes in JT-60U

Real-time feedback control schemes of physics quantities presently used or under development for creating and sustaining advanced tokamak modes in JT-60U is listed in Table 1. Used quantities, objectives and applications are summarized in it [10]. In addition to scalars and a single point data of a profile, utilization of a part of the electron temperature profile and the current density profile are being explored.

Quantity	Objectives	Applications
•Averaged electron density (FIR interferometer)	•Optimize target and operation density	•Versatile in a variety of exp.
•Neutron emission rate (fission chamber)	•Control stability, fusion burn	•Establish route toward high Q_{DT}^* in R/S plasmas
•Plasma stored energy (diamagnetic loop)	•Control stability, avoid disruption	•Sustain plasma performance in a steady-state
•Central electron temp. (ECE measurement)	•Optimize target temperature	•Study non-inductive CD, High Te / Ti confinement
•Temperature gradient (ECE measurement)	•Control internal transport barrier	•Sustain bootstrap current for a steady-state operation
•Temperature fluctuations (ECE measurement)	•Suppress neoclassical tearing mode	•Sustain high performance in a steady-state
•Current density profile (MSE polarimeter)	•Sustain minimum q above 2	•Sustain a R/S plasma stably
•Main plasma radiation (bolometers)	•Improve confinement by enhanced radiation	•Explore Radiative Improved mode
•Outer gap distance (magnetics & equilibrium)	•Keep stable RF coupling, investigate wall effect on stability	•Sustain R/S by LHRF heating, improve stability in R/S plasmas
•Divertor pressure (ionization gauge)	•Control MARFE/detachment	•Sustain radiative divertor
•Divertor radiation (bolometers)	•Reduce divertor heat load	•Explore radiative divertor

Table 1. Real-time feedback control schemes of physics quantities in JT-60U.

Feedback control of ∇T_e in a negative shear plasma

Figure 3 shows a feedback control of ∇T_e in a negative shear plasma on JT-60U [11]. This technique has been developed to sustain the ITB by controlling a gradient of the electron temperature at the ITB, aiming at sustaining the bootstrap current constant towards a steady-state operation. The plasma stored energy W_{dia} is controlled to produce internal transport barriers in a early phase. Increases in ∇T_e measured with ECE Fourier transform spectrometers shows the formation of the ITB in the electron system. After the formation of the ITB, a control logic is changed form the W_{dia}-control to the ∇T_e–contorl by the central heating using neutral beams. When the ∇T_e becomes lower than the reference value, the neutral beam power is increased. The ∇T_e is successfully kept almost constant with a similar profile to that at 6.5 s. The ECE Fourier transform spectrometers have been replaced recently by the ECE radiometers in the control loop. It is because electron temperature measurements with the Fourier transform spectrometer is affected by edge localized modes (ELMs) during a scan of the mirror in the spectrometer.

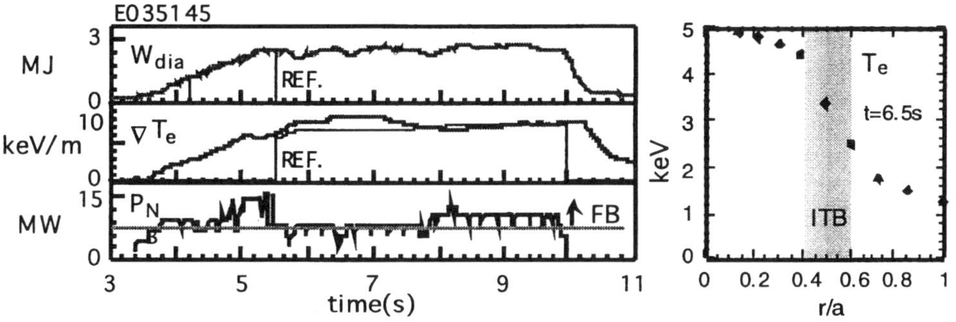

Figure 3. (a) Waveforms of ∇T_e feedback control applied to the ITB control in the negative shear plasma on JT-60U and (b) electron temperature profile at 6.5 s.

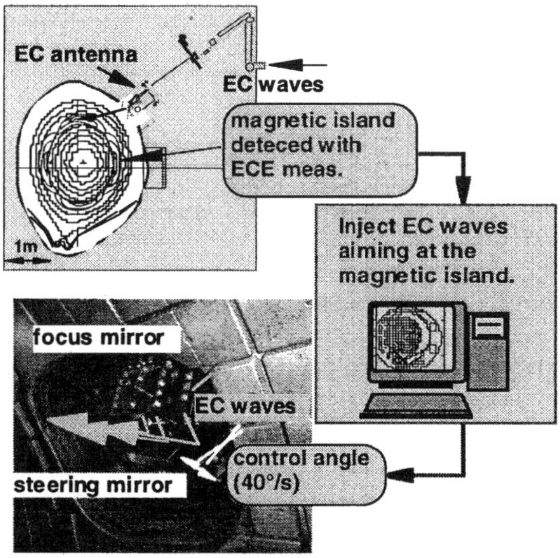

Figure 4. Schematic diagram of a complete loop of the NTM suppression by detection with ECE radiometers and EC wave injection.

Suppression of the NTM with ECE measurements and EC wave injection

Suppression of the NTM is one of important issues in a plasma with high beta and high bootstrap current and has been succeeded with the electron cyclotron (EC) wave injection in ASDEX [12], DIII-D [13] and JT-60U [14]. A complete loop, in which the location of magnetic islands of the NTM is identified by ECE radiometers and EC waves are injected aiming at the magnetic island by controlling the steering mirrors of the EC antenna, is under development in JT-60U. Figure 4 presents the control loop schematically.

Time traces of magnetic fluctuations of the NTM and the peak perturbation of electron temperature are presented on the left and profiles of the electron temperature perturbation are shown on the right in Figure 5. The electron temperature profile is measured with ECE radiometers at every 20 μs. Acquired data of eight channels around the magnetic island are processed in real-time at every 2 ms to obtain a perturbation profile of the electron temperature. When the NTM grows, a clear "notch" indicating the center of the magnetic island is detected in profiles of perturbation level of electron temperature (4-th channel is excluded in the figure because of a large detector noise) [15]. The injection of EC waves was successfully triggered when the peak perturbation exceeded the reference level of 2%. The complete loop will be tested with high power EC wave injection for the current drive to suppress NTMs in a high β_p ELMy H-mode plasma.

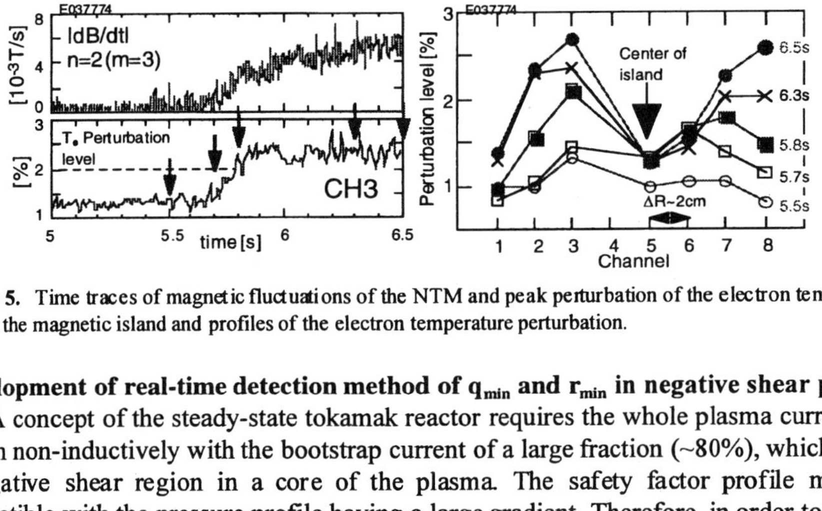

Figure 5. Time traces of magnetic fluctuations of the NTM and peak perturbation of the electron temperature around the magnetic island and profiles of the electron temperature perturbation.

Development of real-time detection method of q_{min} and r_{min} in negative shear plasma

A concept of the steady-state tokamak reactor requires the whole plasma current to be driven non-inductively with the bootstrap current of a large fraction (~80%), which forms a negative shear region in a core of the plasma. The safety factor profile must be compatible with the pressure profile having a large gradient. Therefore, in order to sustain such a plasma continuously, the information on the safety factor profile is indispensable in real-time although q profile can be obtained fairly accurately using the MSE data and magnetic data in a equilibrium code after the discharge is over. In JT-60U, a detection method of minimal value of the safety factor (q_{min}) and the location on minor radius (r_{min}) is under development for the real-time control of the safety factor profile. Figure 6 shows temporal behavior of q_{min} and normalized minor radius r_{min}/a. Filled circles show those evaluated with the equilibrium code using the data from the MSE polarimeter and the magnetic data. Solid waveforms present those obtained only from the MSE polarimeter by identifying the location of measurement points on one typical equilibrium of a similar discharge. Temporal behaviors of values evaluated with two methods agree relatively well even though difference of ~1 for q_{min} and ~0.1 for r_{min}/a (~10 cm) can be seen between them. The difference is considered to be due to a discrepancy in the position and the shape of the real plasma from those referred to map the measurement position. Spikes in solid waveforms arise from the plasma light and the turn-off of the neutral beam used for the MSE polarimeter. It is important to improve the accuracy and the reliability by solving the problems in further developments.

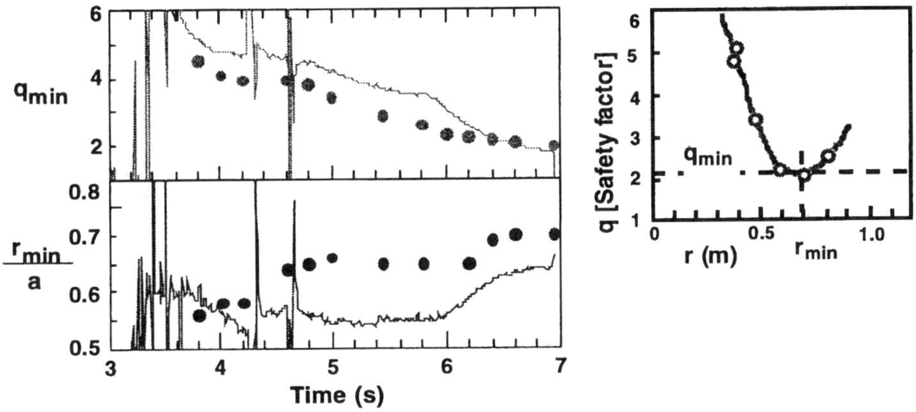

Figure 6. Time traces of q_{min} and r_{min}/a. in the negative shear plasma on JT-60U.

ISSUES IN DEVELOPING DIAGNOSTICS AND FEEDBACK CONTORL SCHEMES FOR CONTROLLING ADVANCED TOKAMAK MODES

Various feedback control schemes have been developed in present tokamaks and have contributed to create and sustain advanced tokamak modes. However, further developments are required in order to establish an operation mode, which can be utilized as a reactor plasma. Following items are issues in developing diagnostics and feedback control schemes for controlling advanced tokamak modes;

(1) Improvement of the reliability of diagnostics to be used as control diagnostics,
(2) Production of diagnostic data available in real-time to control systems,
(3) Need for profile data to be built into plasma control systems for the fine control of advanced tokamak modes. For example, $q(r)$, $n_e(r)$, $T_e(r)$, $v_\phi(r)$, etc,
(4) Establishment of various control algorithm based on physics basis of advanced tokamak modes.

SUMMARY

The most important in creating advanced tokamak modes is to control internal and edge transport barriers, optimizing the confinement, the MHD stability and the current density profile simultaneously. This requires key diagnostics. Measurements of plasma current density profile and MHD mode structure are the most important. High reliability and accuracy, sufficient time and spatial resolution, real-time processing of acquired data are required for control diagnostics. Successful demonstrations of advanced tokamak modes have been shown in present tokamaks. Further improvements are required both in diagnostics and physics understanding for the fine control at the highest performance of advanced tokamak modes.

ACKNOWLEDGEMENT

The author would like to thank Dr. T. Fujita, Dr. T. Oikawa and Dr. A. Isayama for supplying diagnostic and experimental data. The author would like to appreciate Dr. T Fukuda for useful discussions on the plasma control. The author would like to thank members of the ITER Expert Group on Diagnostics for continuous discussions.

REFERENCES

1. Y. Koide, et al., in *Plasma Physics and Controlled Fusion Plasma Research*, Seville, 1994, Vol. 1, 199-210 (IAEA, Vienna, 1994).
2. T. Fujita, et al., Phys. Rev. Lett. **78**, 2377-2380 (1997).
3. Y. Shimomura, et al., Proc. 18th IAEA Fusion Energy Conference, Sorrento, Italy, 2000, IAEA-CN-77/ITER/1.
4. M. Kikuchi, Fusion Technol. **30**, 1631-1635 (1996).
5. K. H. Burrell, Rev. Sci. Instrum. **72** (1), 906-914 (2001).
6. T. Fujita, et al., Nucl. Fusion **39** (11Y), 1627-1636 (1999).
7. F. M. Levinton, et al., Phys. Rev. Lett. **75**, 4417-4420 (1995).
8. Z. Chang, et al., Phys. Rev. Lett. **74**, 4663-1245 (1995).
9. A. Bondeson, D. J. Ward, Phys. Rev. Lett. **72**, 2709-2712 (1994).
10. T. Fukuda, et al., JAERI Review 2000-035, 81-84 (2000).
11. T. Oikawa, JT-60 Team, Nucl. Fusion 40, 1125-1136 (2000).
12. G. Gantenbein, et al., Phys. Rev. Lett. **85**, 1242-1245 (2000).
13. R. Prater, et al., Proc. 18th IAEA Fusion Energy Conference, Sorrento, Italy, 2000, IAEA-CN-77/EX8/1.
14. A. Isayama, et al., Nucl. Fusion **41**, 761-768 (2001).
15. A. Isayama, et al., to be published in Review of JT-60U Experimental Results in 2000.

DIAGNOSTIC NEEDS FOR DIVERTOR AND EDGE PHYSICS

A.W. Leonard*

1. INTRODUCTION

In order to determine the diagnostic needs for divertor and edge physics, it is important to understand the current state of divertor research and the goals driving that work. In future large tokamak reactors, such as that envisioned by ITER-FEAT, the divertor, and edge plasma, must meet several goals and criteria. The divertor must first be able to handle the power that crosses the separatrix into the Scrape-Off-Layer (SOL), by careful design of the target plates and use of plasma radiation to disperse the heat load. The divertor must also sufficiently pump helium that is produced by fusion reactions in order to avoid dilution of the central plasma. The boundary solution must also provide fueling for the main chamber to reach the high density that is required for optimal production of fusion energy. Finally the boundary solution must be designed to contain impurities that are produced at the plasma facing surfaces and prevent them from contaminating the main plasma.

Though a few other outstanding issues remain, some solutions to the individual requirements listed above have been demonstrated in a number of current experimental devices. A large number of divertor and edge diagnostics have been developed to demonstrate and study these solutions. A complete set of power balance measurements are routinely available in today's experiments. This includes IR cameras that measure the surface temperature, and thus infer the power flux, at the divertor target and other plasma facing components.[1] Bolometry can now determine the 2-D profile of radiated power that accounts for a major fraction of power balance.[2] Spectroscopy is routinely used to determine how the radiated power is split between different radiating species.[3,4] Numerous spectroscopic techniques are also used to determine impurity levels and profiles as well as other plasma parameters. Pressure gauges have been developed to operate in magnetic fields to measure the neutral pressure within divertor pumping structures.[5] Even measurements of the divertor target surface erosion have been made to project future divertor component lifetimes.[6]

Sophisticated 2-D modeling codes have also demonstrated a number of these solutions. These codes include most all the physical processes thought to be important and place them within the actual geometry of the experiments they are modeling.

*General Atomics, P.O. Box 85608, San Diego, California 92186-5608

Advanced Diagnostics for Magnetic and Inertial Fusion, Edited by Stott *et al.*
Kluwer Academic/Plenum Publishers, New York, 2002

Qualitative, and some quantitative, agreement has been found between the modeling codes and solutions demonstrated in current experiments.[7] These codes are then used for predicting and designing the operation of future divertors which will be larger and must handle greater power densities. The codes attempt to answer a number of concerns for future divertor design. What level of impurities are needed in the divertor to produce sufficient radiation and reduce the target heat flux to a manageable level? What level of core contamination does this lead to? What upstream separatrix density is attained for a radiative divertor solution?

In order to have confidence in code prediction of future divertor operation a more careful comparison and analysis is needed of code modeling of existing experiments. This paper will describe the diagnostic measurements that are needed in today's experiments in order to build confidence in the results of physical modeling using computer codes.

2. PARALLEL TRANSPORT

The divertor and SOL plasma is essentially a complicated transport problem with many atomic and molecular processes which act as sources and sinks of energy and particles. Energy and particles are deposited in the main plasma by heating and fueling techniques. Both diffuse radially across the plasma until crossing the separatrix. Then parallel transport along the open field lines carries energy and particles towards the divertor. Radial transport broadens the deposition on the divertor targets and can lead to plasma flux to the main chamber walls. At the same time atomic processes also transport energy to the chamber walls through radiation and charge-exchange with neutrals. After the plasma flux recombines in the wall it is released into the plasma volume as a neutral fueling source. Impurities are also created by the plasma wall interaction and may enter the plasma. Modeling of such a system then requires sophisticated computer codes which can account for many coupled plasma processes in a realistic 2-D, and sometimes 3-D geometry. To have confidence in these models it is necessary to compare the experimentally measured processes driving the transport, as well as the resulting sources and sinks, with the modeling predictions.

We start with parallel energy transport which can be described by

$$q_{\parallel} = -\kappa T_e^{5/2}\frac{dT_e}{ds} + nv_{\parallel}\left(\frac{5}{2}\left(T_e + T_D\right) + \frac{1}{2}m_D v_{\parallel}^2\right) \qquad (1)$$

where s is the parallel field line length, κ is the parallel electron thermal conductivity, T_e and T_D are the electron and deuterium ion temperatures respectively, n is the plasma density, m_D is the ion mass and $v_{\parallel}$ is the parallel plasma fluid velocity. Under most parameter regimes the SOL and divertor parallel energy transport is dominated by the first term, electron thermal conduction. To account for heat flux, measurements of T_e in the divertor and SOL have been made for some time by a number of techniques. Langmuir probes mounted in plasma facing surfaces have measured T_e of plasma arriving at the target while insertable Langmuir probes have measured T_e at various upstream locations.[8] Other techniques include line ratios of impurity radiation and Thomson scattering. An example of a 2-D T_e profile of the divertor region is shown in Figure 1. This profile was obtained from a Thomson scattering diagnostic where a steady-state divertor plasma was swept across the diagnostic view locations.[9] These T_e measurements have been found to be in general agreement with parallel transport determined from the power balance measurements described in the introduction.[7] The remaining terms in

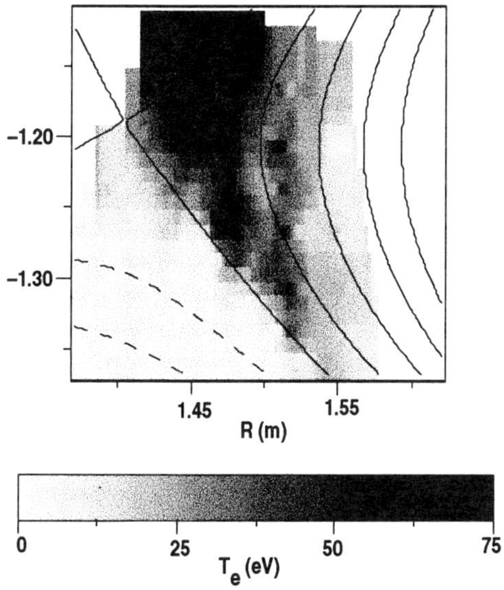

Figure 1. A 2-D profile of the electron temperature in the DIII–D divertor. The profile was taken with a Thomson scattering system by sweeping the divertor configuration across the vertical array of measurement locations.

Eq. (1) represent convection and become important at low T_e near the divertor target and sheath boundary. At the target plate some discrepancy has been found between the measured T_e and expected heat flux.[10] Additional measurements associated with the sheath, including T_D, will be needed to resolve this issue.

The parallel deuterium particle flux is mainly driven by pressure balance and results from particle sources and sinks. Proper modeling of the parallel deuterium flow requires an accurate measurement of the ionization source, or recycling. The neutral particles and impurities that enter the plasma result in most part from plasma flux to material surfaces. Computer modeling should be able to calculate most of these sources if the plasma flux to all surfaces is known. Since it is not, as will be described later, it is important to measure these sources. This is very much a 3-D problem. Most all today's magnetic fusion devices have an outer midplane wall that is very irregular in the toroidal direction. Plasma recycling can and is very much a strong toroidal function. Greater coverage of existing diagnostics, such as vacuum gauges, wall mounted Langmuir probes and D_α measurements would be a good start to address this issue.

Deuterium flux in the poloidal direction can also result from plasma drifts. Radial electric fields near the separatrix of H–mode plasmas, for exapmle, can make a significant contirbution to the total poloidal flow. In principle, most of the plasma drifts can be calculated by today's 2-D modeling codes. In practice, however, obtaining self-consistent solutions with all the drifts terms has proved difficult. Experimental measurements of the poloidal flow profile and the underlying electric fields, or potential, would be a great help in determining where and how they may affect the modeling solutions.

Some useful information on deuterium ion flows have been obtained from Mach probes.[11] These measurements have revealed and/or confirmed basic parameters of SOL and divertor operation. Such measurements, though, are limited to regions, or conditions, that allow probe access. Also, it is very important that both the parallel and perpendicular flow be measured. It is the poloidal flow that is important with the toroidal direction being the one of symmetry. Existing measurements indicate that the contributions to the total poloidal flow from the $E_r \times B$ drift and the parallel flow can be of comparable magnitude. One technique for obtaining simultaneous perpendicular and parallel flows is with a multi-factored Mach probe.[12] Obtaining 2-D profiles, though remains a challenge.

For the impurity ions, the primary forces that drive the parallel flow of an impurity, i, are given by:

$$F_i = -\frac{1}{n_i}\frac{d(n_i T_i)}{ds} + \frac{(v_D - v_i)}{\tau_s} + \alpha_e \frac{dT_e}{ds} + \beta_D \frac{dT_D}{ds} + \ldots \tag{2}$$

where the subscript D refers to the main deuterium ion parameters, τ_s is the impurity ion equilibration time and α_e and β_D are the electron and ion temperature respective gradient force coefficients.[13]

Impurity ion densities, temperatures, and flow velocities have been measured by a number of spectroscopic techniques. Doppler broadening and line shift of impurity emission can be used to obtain ion temperature and flow velocity.[14] A limitation of these techniques is that the measurement is spatially localized to the region where a specific impurity charge state exists in sufficient density. In hotter regions of the divertor or SOL the impurity ion will be ionized to a higher charge state.

The main ion properties, as seen in Eq. (2), are also very important in determining the parallel forces on an impurity ion. Though the impurity ion temperature and velocity can be measured by spectroscopy, the main ion may have very different properties. Because the parallel flow times can be shorter than the equilibration time, an impurity ion species can be created in one region and travel to another before it is heated or cooled to the local deuterium temperature. The flow velocity of the impurity may also be very different as can be seen from Eq. (2) where different forces act on the impurity and main deuterium ions.

One prospect for divertor ion temperature measurements is charge-exchange recombination (CER) spectroscopy. Main ions in the divertor would charge exchange with a diagnostic neutral beam. The resulting deuterium atom would be left in an excited atomic state that can be detected by spectroscopy. This is the same basic technique used for main plasma ion temperature measurements. An example of a main plasma CER diagnostic observing the SOL in DIII–D is shown in Figure 2. In this example the deuterium temperature and flow velocity can be different from impurity species that are often measured. Difficulties with this technique include discrimination of the signal from the high level of background recycling and spatial resolution due to the finite lifetime and flight of the measured neutral deuterium atom. A dedicated diagnostic neutral beam may also be expensive. However an exciting prospect for this diagnostic is the possibility of 2-D profiles using a planar beam with imaging measurement technologies.

Finally, modeling of impurity behavior requires measurement of the sources. Impurities are generated by plasma flux to the walls making distributed measurement of plasma wall fluxes, from Langmuir probes or D_α for example, important to this topic also. Measurements of neutral impurity emission can be a more direct measure of impurity generation, but requires some knowledge of the local plasma parameters. It is usually a more difficult spectroscopic measurement.

Impurities released at a surface due to a plasma flux may also vary depending on the physical or chemical structure of that surface. A number of techniques have been developed to study surfaces that have been exposed to plasma operation. However, many of these techniques require long exposure, or must remain in the machine under many

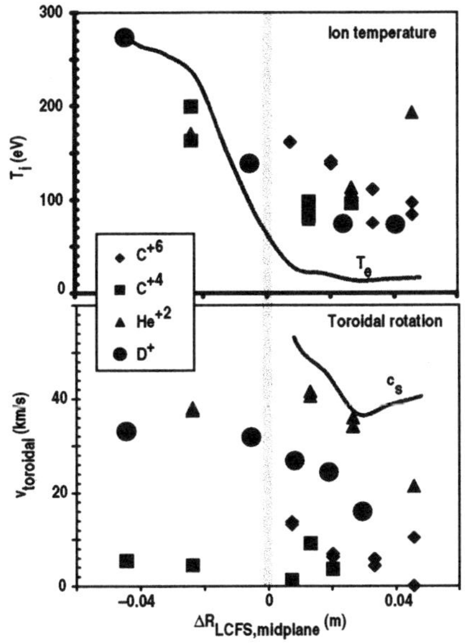

Figure 2. The ion temperature and toroidal rotation profile at the outer midplane as measured by the CER diagnostic on DIII-D for deuterium, He^{+2}, C^{+4} and C^{+6}.

different operational conditions before they can be removed and examined. Development is needed of diagnostics that can measure surface conditions *in situ* and real-time during plasma experiments.

3. PERPENDICULAR TRANSPORT

Perpendicular transport of energy and particles in the divertor, and the SOL in particular, is another issue that needs new measurements. Perpendicular transport is responsible for the width of the heat flux on the divertor target. It also controls the particle flux to all plasma facing components, including the main chamber walls. However, perpendicular transport is not well understood, or even well characterized. To describe perpendicular transport computer codes model the edge plasma with diffusion coefficients that do not vary radially, or poloidally. The coefficients in the codes are varied empirically to match experimental plasma profiles. How these coefficients might scale to future larger tokamaks is unknown. Also, experimental evidence and theoretical considerations suggest that radial transport can vary greatly from inboard to outboard as well as radially. The consequences of this uncertainty are unknown scaling of heat flux widths at the divertor target and unknown particle flux to the chamber walls. The particle flux to the wall is important for understanding main plasma fueling as well as driving SOL flow. The wall plasma flux is also the main source for impurity generation.

A number of diagnostic techniques are used to study SOL and divertor radial transport. Insertable Langmuir probes have measured correlated fluctuating levels of n_e, T_e, and ϕ to obtain localized values of convected and conducted radial fluxes.[15] Other techniques used to observe fluctuation driven transport in the core are now being trained on the SOL. Beam emission spectroscopy (BES) and microwave fluctuation reflectometry are two examples. The largest limitation to all these techniques is their spatial localization, with most having been employed only at the outer midplane. Extending these techniques to get more of a poloidal profile is needed. A 2-D image, encompassing the entire SOL region, would be ideal.

The radial particle flux driven by transport can also be inferred by radial profile measurements of n_e, T_e, and the ionization source profiles.[16] An example of this for DIII–D is shown in Figure 3. A short coming of this technique is that it relies on assumptions of toroidal and poloidal symmetry.

More complete coverage of plasma wall flux is needed. As described earlier the irregular walls in tokamaks today lead to strong toroidal asymmetries. Distributed measurements of plasma to surfaces and/or the resulting neutral density will help define these asymmetries. By knowing the total wall flux profile, coupled with ionization source measurements from spectroscopy, much can be learned about the level of radial transport. These are the same measurements that also provide particle source information for correctly modeling deuterium flow.

4. 2-D PROFILES

The divertor is very much a 2-D system with strong radial and poloidal gradients. The 2-D nature nature of the profiles exists in all the parameters that have been described as important to measure. In the case of the neutral density profile it is even a 3-D problem.

Imaging techniques are ideally suited for this task. Wide angle views of main chamber and divertor recycling have been made in the past. More quantitative work needs to be done with this kind of data in order to turn qualitative pictures into neutral density profiles on a regular basis. Quantitative analysis of visible image data presents some challenges. The emission rates of visible lines are often very strong functions of the plasma density and temperature. To calculate neutral or impurity densities based on this data requires some measurements of the density and temperature profiles. Reflections from other surfaces is also a concern at visible wavelengths. Measurements of lower order transitions $\lambda \leq 150$ nm have less ambiguous interpretation, but imaging these lines is much more difficult because of the technology at shorter wavelengths. Some development in this area is the image of CIV emission taken from DIII-D and shown in Figure 4(a).[17] In this case the detection is made on a phosphor plate in vacuum that is then recorded with a visible camera. Imaging techniques for other parameters would aid in producing 2-D profiles that have been mentioned earlier.

Another difficulty with image data is spatial interpretation, or inversion. While it is easy to discern qualitative features in an image, a detailed quantitative comparison between an image and the results of 2-D modeling is much more difficult. For emission that is expected to have toroidal symmetry some analysis techniques have been developed. On DIII-D and JET,[18] toroidally viewing images have been inverted to produce a 2-D profile on one poloidal plane. An example of a raw and inverted image is shown in Figure 4(b). For signals that may not have toroidal symmetry, such as recycling D_α, or low charge state impurities, other assumptions would be required to invert such images.

A final difficulty with imaging is the sheer size of the data throughput. Though the technology exists for capturing and digitizing image data, more standardized techniques for computational processing and archiving the data are needed.

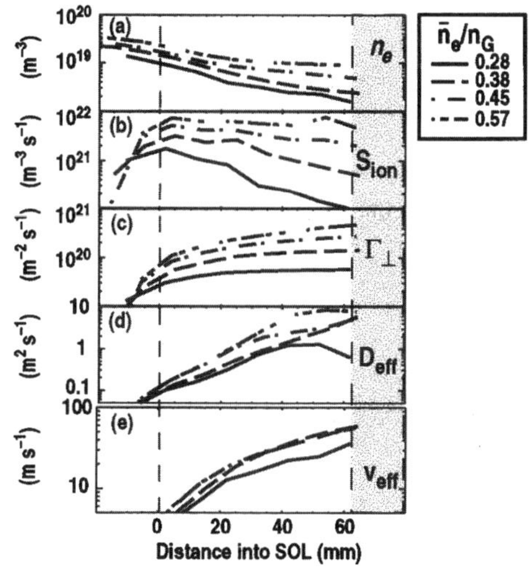

Figure 3. Midplane profiles on DIII–D of (a) electron density from Thomson scattering, (b) ionization source from spectroscopy, (c) the radial flux from particle balance calculations, (d) the effective diffusion coefficient if all flux assumed diffusive and (e) the effective radial fluid velocity if all flux assumed convective.

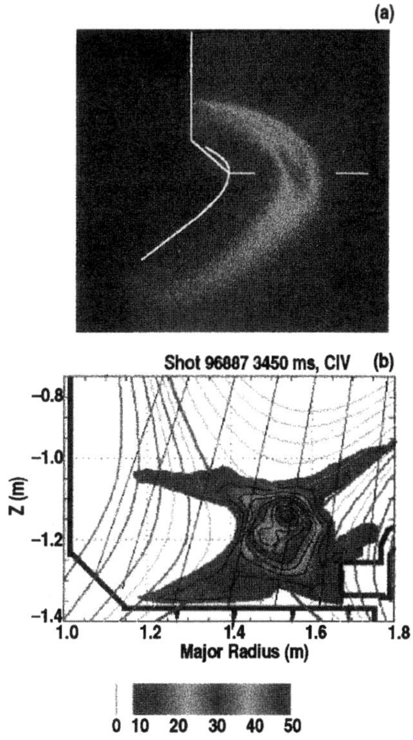

Figure 4. (a) A tangential image of CIV in the lower DIII–D divertor. (b) A inversion of the tangential image onto a poloidal plane.

5. SPECIAL ISSUES

One of the largest concerns in the design of a future fusion device, such as ITER-FEAT, is the issue of tritium retention.[19] Experimentally it has been found on JET that a large fraction of the tritium that has been injected during DT experiments has remained bound up in the walls of the vessel.[20] The level of tritium retention is such that it would limit the operation of a future burning plasma experiment. It appears that most of the retained tritium is trapped in layers of redeposited carbon. Determining the location of carbon erosion as well as its transport in the plasma to the location of redeposition will be required to design a divertor and first wall that minimizes tritium retention. These are all issues that would benefit from additional diagnostic development work as described above.

Another serious issue for the design of ITER-FEAT are the transient heat pulses that arrive at the divertor due to ELMs.[21] ELMs result from a periodic relaxation of the main plasma boundary during H-mode operation. A 1.0 MJm^{-2} pulse of energy could be deposited on the divertor target in as little as 100 μs. Though the heat pulse originates in the main plasma, understanding how the SOL and divertor plasma responds to such a transient in transport is important in designing future large tokamaks. The issues involved in ELM transport are many of those described above. The difficulty arises that measurements of ELM events must be made on a fast time scale. The ELM heat pulse typically lasts 100–500 μs. Measurements of temperature and density in the divertor must be made on a faster time scale in order to study ELM evolution. Another complication is that the ELM can change the parameters, or signal levels, by more than an order of magnitude during the pulse. This requires diagnostics with an extended dynamic range, or measurements must be dedicated to ELM observation only.

6. CONCLUSIONS

In order to advance our understanding of SOL and divertor plasmas, basic measurements of plasma properties are still needed. Two areas stand out in particular. First, measurements of the main ion properties are needed, in particular the temperature and flow velocity profiles. The other is characterization of the fluctuation driven radial transport, and its spatial profile. Much the same as progress has been made in understanding core radial transport, the same can be done for the SOL and divertor. From a diagnostic point of view, though, it is more difficult because it is an inherently 2-D and in some cases 3-D problem. Advances in imaging techniques to obtain 2-D profiles of a number of plasma parameters can also play an important role in advancing divertor physics.

7. ACKNOWLEDGMENT

Work supported by U.S. Department of Energy under Contract DE-AC03-99ER54463.

8. REFERENCES

1. C.J. Lasnier, et al., Nucl. Fusion **38** 1225 (1998).
2. J.C. Fuchs, et al., Proc. 21st EPS Conf. On Control. Fusion and Plasma Physics, Montpellier (France).
3. H. Kubo, et al., Nucl. Fusion **33** 1427 (1993).
4. R.D. Wood, et al., 23rd European Conf. on Control. Fusion and Plasma Physics, Kiev, Ukraine, Vol. 20C, Part II (1996) 763.
5. G. Haas, et al., J. Nucl. Mater. **121-151** (1984).
6. D.G. Whyte, et al., J. Nucl. Mater. 266-269, 67 (1999).
7. G.D. Porter, et al., Phys. Plasmas **3** 1967 (1996).
8. G.F. Mathews, et al., Plas. Phys. Contr. Fusion **36** 1595 (1994).
9. S. L. Allen, et al., J. Nucl. Mater. **241-243** 595 (1997).
10. A.H. Futch, et al., J. Nucl. Mater. **196-198** 860 (1992).
11 J. A. Boedo, et al., J. Nucl. Mater. **266-269** 783 (1999).
12. C.S. MacLachy, et al., Rev. Sci. Instrum. **63** 3923 (1992).
13. J. Neuhauser, et al., Nucl. Fusion **24**, 39 (1984).
14. R. C. Isler, et al., Phys of Plasmas **6** 541 (1999).
15. D. L. Rudakov, et al., Rev. of Sci. Instr. **72** 453 (2001).
16. B. LaBombard, et al., Phys of Plasmas **8** 2107 (2001).
17. D. G. Nilson, et al., Rev. of Sci. Instr. **70** 738 (1999).
18. K. Itami, et al., J. Nucl. Mater. **290-293** 633 (2001).
19. G. Federici, et al., J. Nucl. Mater. **266-269** 14 (1999).
20. A. T. Peacock, et al., J. Nucl. Mater. **266-269** 423 (1999).
21. G. Janeschitz, et al., J. Nucl. Mater. **290-293** 1 (2001).

ITER MAGNETICS: KEY DESIGN ISSUES

George Vayakis and the ITER International Team and Home Teams*

1. INTRODUCTION

The ITER magnetic diagnostic system provides the main electromagnetic parameters of the plasma (eg I_p, V_{loop}); the shape and position of the plasma boundary; key internal parameters (eg β, l_i); the vertical speed of the current centroid; and analysis of the amplitude and mode number of fluctuations from quasi-static to fast (100 kHz). A subset of these measurements is used for plasma current and shape control. The application of magnetics to a burning plasma experiment poses a number of new issues: (a) The high performance cost of any wasted volume or lost shaping capability means that measurement targets (Table 1) have to be very tight; (b) The requirement for long-term (> 20 year) reliability with none or little maintenance means multiple measurement backups are essential and a backup or repair strategy has to be defined and validated in the design stage; (c) The presence of a space-filling conducting blanket in front of the sensors means that eddy currents have significant effects on physics and control - relevant time scales (a few ms); (d) In addition to the standard requirements for insulator breakdown resistance and frequency response and the issues of lifetime under irradiation, sensors and their wiring have to be optimised to fit within the limited volume, to have negligible radiation-induced errors and to dissipate the considerable nuclear heating; (e) the long-pulse targets imply that the sensors –and their wiring– have to provide high signal to noise signals and low drift integrators must be used. This paper summarises the extent to which these issues have been solved. Remaining R&D issues are highlighted.

2. KEY DESIGN ISSUES

Reliability and Redundancy: For the position, shape and speed measurements, there are two independent groups of sensors. The first, mounted on the inside surface of the vessel and in the divertor, has a built-in redundancy of 2, achieved by two independent sector pairs of equilibrium coils, each cancelling n = 1 and, together by appropriate summation, n = 2. These sets are combined with a set of 6-fold toroidal symmetry coils to

* George Vayakis, ITER NAKA JWS, 801-1 Mukouyama, Naka, Ibaraki 311-0193, Japan.

measure independently n = 2 mode activity, and two independent sets of 3-fold toroidal symmetry divertor coils. The second group, mounted between the vessel skins in a more benign environment, can correct any slow drifts of the inner group, albeit with reduced accuracy. For all sensors, reliability is assured by the choice of tough materials, such as mineral insulated cable (MIC), for the coil construction. Of the two in-vessel sets, one is permanent, without connectors to the feedthrough. The second can be replaced (Figure 1), in this way guarding against common-mode failure of the sensor. The divertor coils are replaceable. Other measurements, have similar levels of redundancy.

Table 1. Measurement targets for the magnetics.

MEASURE-MENT	PARAMETER	COND-ITION	RANGE/ COVER AGE	ΔT or ΔF	ΔX or Δk	ACCURACY (2 σ)
1. Plasma Current	I_p	Default	0 – 1 MA	1 ms	Integral	10 kA
			1 – 17.5 MA	1 ms	Integral	1 %
		I_p Quench	17.5 – 0 MA	0.1 ms	Integral	30 % + 10 kA
2. Plasma Position and Shape	Main plasma gaps, Δ_{sep}	$I_p > 2$ MA, full bore	-	10 ms	-	1 cm
		I_p Quench	-	10 ms	-	2 cm
	Divertor channel location (r dir.)	Default	-	10 ms	-	1 cm
		I_p Quench	-	10 ms	-	2 cm
	dZ/dt of current centroid	Default	0 – 5 m/s	1 ms	-	0.05 m/s (noise) + TBD % (absolute)
3. Loop Voltage	V_{loop}	Default	0 – 30 V	1 ms	4 locations	5 mV
		I_p Quench	0 – 500 V	1 ms	4 locations	10 % + 5 mV
4. Plasma Energy	β_p	Default	0.01 – 3	1 ms	Integral	5 % at β_p=1
		I_p Quench	0.01 – 3	1 ms	Integral	~ 30%
8. Locked Modes	B_r(mode) / B_p		$10^{-4} – 10^{-2}$	1 ms	(m,n) = (2,1)	30 %
9. Low (m,n) MHD Modes, Sawteeth, Disruption Precursors	Mode complex amplitude at wall		TBD	DC – 3 kHz	(0,0) < (m,n) < (10,2)	10 %
21. Halo Currents	Poloidal current	In disrupti on	0 – 0.2 I_p	1 ms	9 sectors	20 %
22. Toroidal Magnetic Field	B_T		2 – 5.5 T	1 s	2 locations x 2 methods	0.1 %
27. High Frequency Macro Instabilities (Fishbones, TAEs)	Fishbone–induc ed perturbations in B,T,n		TBD	0.1 –10 kHz	(m,n) =(1,1)	–
	TAE mode – induced perturbations in B,T,n		TBD	30 –300 kHz	n = 10 - 50	–

In-vessel structure eddy current effects. Compared to the target measurement intervals of Table 1, relatively long time scales are predicted for the poloidal (~ 7 ms) and toroidal (~10-20 ms) eddy current modes in the blanket modules. Their effect is a prompt increase of the signal measured by the poloidal field sensors and the in-vessel diamagnetic loop compensation coil, by excluding flux from the module. For reconstructions of the plasma and measurement of stored energy this effect can in principle be compensated after careful modeling. A preliminary study of the effect of the poloidal current eddies on the measurement of the vertical speed has shown that the slight lead term introduced by them is not detrimental to the speed control of the plasma.[1]

Sensor layout optimisation, ie selection of the number and locations of the sensors, is an iterative process. First results[1] indicate that the requirements of Table 1 can be met. Further studies are underway to quantify the degree of redundancy available.

Sensor design optimisation: Selection of the coil parameters was made by a multiple optimisation of coil construction parameters subject to performance constraints, for example frequency response and environmental constraints, such as space and radiation effects. For example, for the poloidal field coils, the coil effective area, NA, has to be above 0.25 turn m^2 in order to ensure sufficient S/N in the wiring and electronics; the coil should not arc ($E < 0.25$ MV / m) for field changes of the order 100 T / s ; the self-resonance frequency of the coil should exceed 10 kHz; Radiation Induced Conductivity (RIC) should affect the signal by less than 0.2 %. In addition, the coil diameter (including case) should be less than ~50 mm to match the blanket cut-out and the length should be less than 100 mm; space for connectors etc must be allowed. Other constraints are an even number of layers (cancel thermal gradient, reduce cross-field contribution); working temperature < 300° C (insulator); and a minimum bend radius of 2.5× the cable diameter.

Radiation-induced Electromotive Force (RIEMF). Nuclear radiation induces a current (~ μA) between the coil core and sheath. In the presence of asymmetries in coil construction, this can lead to a small but significant steady state voltage across coil terminals, despite the low coil impedance (~ Ω), if sufficient asymmetries are present. Scaling studies show that this voltage, when normalised to the desired output, varies little with construction parameters (size, number of turns etc.) once all the other constraints are included. The most recent tests with γ irradiation show that this asymmetric voltage is, at worst, tolerable and could be negligible. Further tests are planned.

Integrator targets. A summary of the integrator requirements for key sensors is shown in Table 2. By comparison, integrators have been made with equivalent drift after 3600 s of 0.1 mVs in the lab,[2] 1.9 mV s in semi-controlled environments[3,4] and 8.6 mVs connected to a tokamak during a pulse,[5] reflecting the difficulty of controlling noise pickup in a realistic machine environment. The target performance appears within reach.

3. OUTSTANDING ISSUES

Issues remain where technical solutions exist but the final design choice requires evaluation of the reliability risks, for example choice of wiring, installation and maintenance options. In-reactor RIEMF tests are planned. Detailed thermo-mechanical studies will be performed on all sensors. Finally, R&D on steady-state sensors has yet to validate a design meeting the ITER requirements; further work in this area is desirable.

4. REFERENCES

1. EU Home Team, Progress Report on the activity performed at CREATE, Report 31 May 1999 (unpublished)
2. K. Kurihara and Y. Kawamata, in: *17th IEEE / NPSS Symposium on Fusion Engineering Vol. 2*, (IEEE, 1997), p. 799.
3. S. Ali-Arshad and L. de Kock, Rev. Sci. Instrum. 64 (9), September 1993, p. 2679.
4. E. Strait et al, Rev Sci. Instrum. 68 (1), January 1997 p. 381.
5. K. Kurihara and Y. Kawamata, private communication.

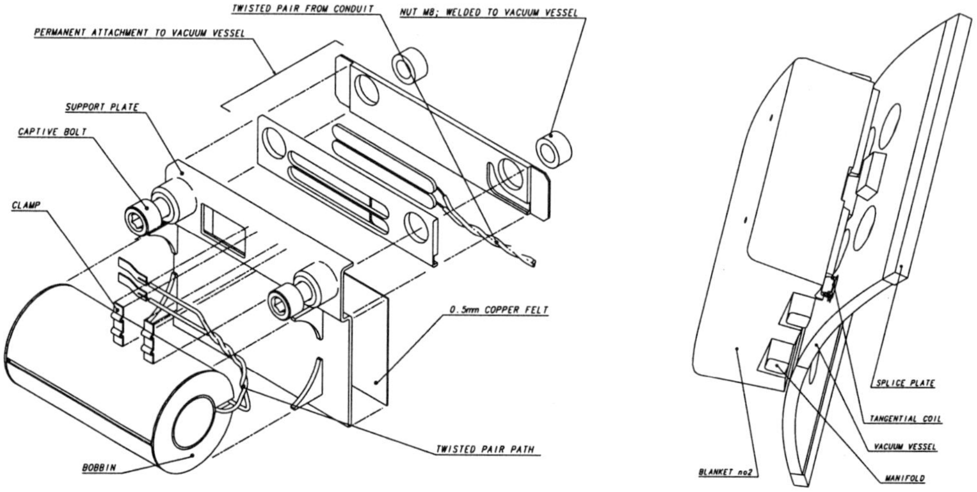

Figure 1. Exploded (left) and mounted (right) views of a replaceable poloidal field sensor.

Table 2. Summary of the integrator (digital and analog) requirements for key sensors

Sensor	**Vessel coils**			**In-vessel**		**External Rogowski**	**Diamagnetic loop**
Parameter	**Tan-gential**	**Nor-mal**	**Partial Saddle Loop (+)**	**Tangential (&)**	**Nor-mal**		**Compens-ation coil**
Peak Voltage (± V)	7.1	3.0	5	27	0.79	20	59
Maximum Flux (± Vs)	2.4	1.0	1.7	0.35	0.27	10	~ 2
Maximum Allowed Flux Error (± mVs)	14	5.8	19	1.9	1.6	40	54
Input offset control required (± μV)	3.9	1.6	5.2	0.52	0.45	11	15
Gain, ± 5V ADC	0.70	1.7	1	0.19	6.4	0.25	0.085
Noise at ADC input (μV/√Hz)*	163	165 (97)*	311	5.8	171 (27)*	170	54
Meas. rate (Hz)	10	10	10	2×10^3	10	10	1×10^3
Minimum effective ADC resolution	13.4	13.5	12.5	14.5	13.4	13.3	11.3
Noise at analog integrator (μV/√Hz)	330	140	440	44	38	940	890
Analog integrator time constant (s)	0.5	0.2	1	0.064	0.054	2	~0.5

(*)where a preamplifier is necessary, the noise at the input of the integrator is shown in brackets.
(+) 1 / 30th of the area of a differential measurement is assumed (~ 1.67 m2). (&) includes divertor coils.

DIAGNOSTICS FOR ENERGETIC CHARGED PARTICLES RELATED TO FAST IGNITOR EXPERIMENTS

For invited volumes

Yoneyoshi Kitagawa, Kunioki Mima, Kazuo Tanaka, Ryosuke Kodama, Yasuhiko Sentoku and Tatsuhiko Yamanaka [1]

1. INTRODUCTION

To demonstrate the feasibility of the fast ignitor scheme,[1,2] a 100 TW Petawatt Module laser (PWM) was illuminated onto an uniformly imploded spherical shell target. The neutron yield is enhanced from below 10^5 up to 10^6 for the first time and agree with those estimated from the simultaneously detected high energy deuteron yield and spectrum.. A few hundred keV deuterons collide with the cold deuterons in core. Without PWM, no neutrons are observed above the detection threshold of 10^5.

A 60 TW GEKKO Module Laser(GMII laser) was illuminated on plastic foils. CR-39 film stacks measured the energy and angular distribution of the protons emitted from the rear surface of the plastic foils. The ring structure of proton emission leads us to that a toroidal magnetic field associated with the hot electrons work on the fast ions and deflect them. From the logarithmic slope of the ring diameter versus the ion energy, the product of the magnetic field $\times$ the length over which it works on the ions is estimated to be up to 1000 MG·μm. The simulation shows that a strong toroidal magnetic field makes the proton's angular distribution to have a similar logarithmic slope.

The 100 TW Petawatt Module Laser (PWM), constructed in 1997,[3,4] is up graded to the world biggest Petawatt Laser (say PW Laser). The PW Laser is to be used for the proof of principle of the scheme and as well for many ultra-intense field reseraches.

2. 100 TW PWM LASER ILLUMINATION ON IMPLODED CD SHELL AND NEUTRON ENHANCEMENT

A target, 500 μm in diameter and 7 μm in thickness, is preimploded by 12 green beams from the GEKKO XII laser. The wavelength is 527 nm and the energy is 2.3 kJ in 2.2 ns Gaussian shape. The 12 beams are focused tangentially onto the target through the random phase plates. At the maximum compression timing, which a fluid code ILESTA-1D[5] predicted to be 2.9 ns after the pulse rise, illuminated is the PWM laser, whose wavelength is 1053 nm, and energy is 90 J on target, respectively. The pulse is the Gaussian of 800fs in FWHM. A parabollic mirror of f-number of 3.5 focused the PWM laser at a point about 200μm from the target center. The critical density point corresponding to the 1μm light is 110 μm from the center. The experimental set up is shown in Fig.1. To detect the D-D neutrons, a plastic(BC-422) scintillator is fully collimated and set at 3.0 m from the target and 40 degrees back to the PWM laser focusing. Figure 2 shows a typical neutron time of flight signal through a numerical noise filter. The spectrum scatters around 2.45 MeV, which shows the feature that the

[1]Yoneyoshi Kitagawa, Kunioki Mima, Kazuo Tanaka, Ryosuke Kodama, Yasuhiko Sentoku and Tatsuhiko Yamanaka, Institute of Laser Engineering, Osaka University,Yamada-oka, 2-6, Suita, Osaka, 565-0871, Japan

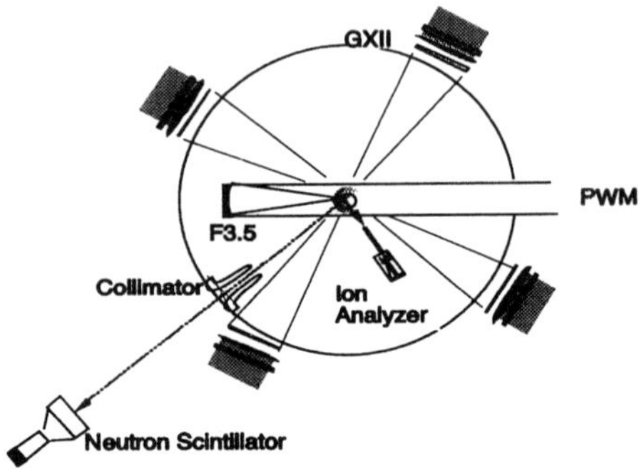

Figure 1. Experimental set for 100 TW PWM laser onto the GXII 12 beam-imploded CD shell target.

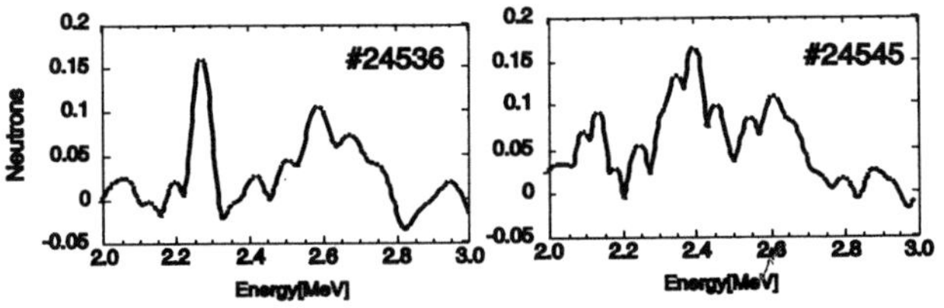

Figure 2. Neutron time of flight signals after a numerical filtering for #24536 (PWM 71 J and GXII 2.2 kJ) and #24545 (86 J and 2.3 kJ) show a beam fusion feature.

energetic deuteron flows collide with dense but cold deuterons in CD plasmas. From the figure, we estimate the averaged spread of the beam to be 25 $\pm$ 15 keV for #24545(Fig. 2). Without PWM laser, we have observed no neutrons above the detection threshold of 10^5. As increasing the PWM laser energy, the neutron yield linealy increases. To simultaneously detect the deuteron emissions, a magnetic analyzer with CR-39 film is set 60 degrees forward the laser. The pinhole of Ta is set at 24 cm and the film at 40 cm from the target. After 1 hour etching in 6N NaOH the pit size on the CR- 39 film becomes 3 $\sim$ 5μ m, which comes from keV to MeV deuterons.[6]

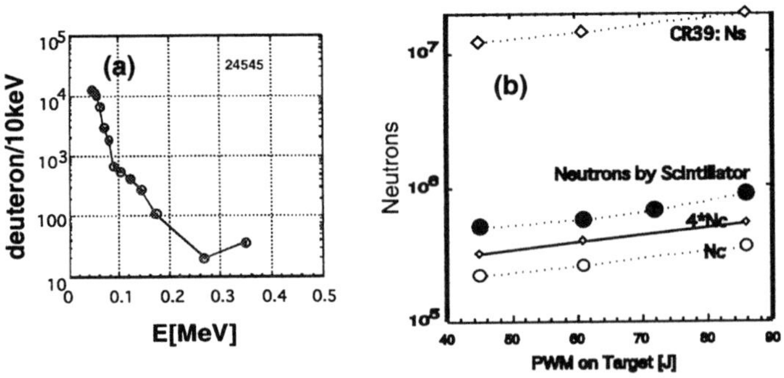

Figure 3. (a) Deuteron spectrum for #24545. (b) Neutron yield comparison between direct scintilaltion data and those estiamted from deuteron emissions for the 1μm critical density N_c, the green light critical $4 * N_c$, and the solid N_s.

Figure 3(a) is the deuteron spectrum for the same shot as Fig. 2. The vertical scale is the particle number/10 keV coming into the detector. The logarithmic temperature is around 20 keV. From the spectrum in Fig. 3(a), we have setimated the neutron yields and compared with those from the plastic schintillator, which is plotted in Fig. 3(b). We set three radial points, at which the laser generates and emits the deuterons in 4π radian. The points are the red light critical density N_c, the green light critical, that is, $4 * N_c$, and the solid N_s. For each case, we estimated the ion stopping power, fusion cross section and the resulting neutron yield. The plots at $4 * N_c$ are close to the scintillator yield , as in Fig. ??. This may lead to that the PWM laser-generated deuteron beams interact with the cold deutrons within the $4\times$ critical density sphere.

3. PROTON REAR EMISSION AND MAGNETIC FIELD GENERATION FROM GMII LASER ILLUMINATED PLASTIC FOILS

Here presented are the observation of proton rear emission and gigagauss scale magnetic fields from 60 TW laser illuminated plastic target. The results are in a good agreement with a two-dimensional PIC simulation.[16] The GEKKO MII CPA laser, used here, produces 25 J at 1.053 μm with 0.45 ps pulse length.[17] The laser pulse, with diameter 140 mm, was focused onto a C_8H_8 target of 3 mm square with an f/3.8 off axial parabolic mirror. 68% of the beam energy was deposited in a spot size of 25 μm in diameter. The target was positioned at 40° to the axis of laser propagation for p-polarization. The intensity on target was $(5.5\pm0.3)\times10^{18}$ W/cm^2. The intensity contrast ratio was 10^3. A one-dimensional hydro code ILESTA-1D[5] predicts a pre formed plasma with a density scale length of 10 μm in the vicinity of the critical density, when the main pulse hits the target. Behind the target, at a distance L of 30 mm and aligned with the normal to the target we placed a 7.5 μm-thick tantalum filter and a stack of four 175 μm thick CR-39 plastic track detectors of 60 mm in diameter, as in Fig. 4.[18–20] Since the tantalum filter stops both carbon ions of less than 10 MeV and protons less than 1 MeV,[21] almost all the tracks on the CR-39 films will be from protons.

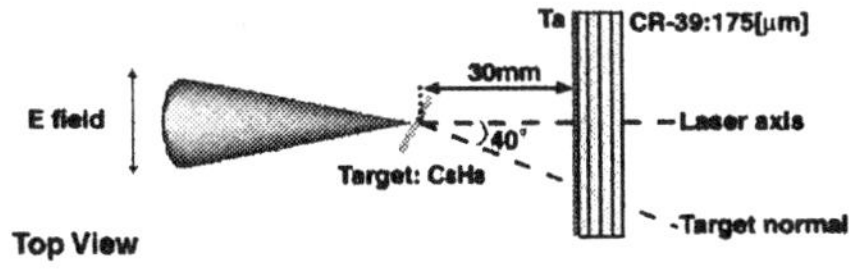

Figure 4. CH target and CR-39 track detector setup. Target is 3-mm square in size and 5 to 100 μm in thickness. CR- 39 film, stack of four layers, covered with Ta foil, is 6 cm in diameter, and is put 3 cm apart from target.

Figures 5 (a) to (d) show the proton tracks on four layered films from 5 μm-thick target after 4 hour etching and Figs. 5 (e) to (h) from 100 μm target. Both the photographs show coaxial ring structures around the target normal axis. Figures 5 (e) and (f) clearly show the ring with a white belt due to much concentration of big pits. For the thickness of 5 and 25 μm, no changes are observed either on the ring structure or emission direction, except that the track number is somewhat reduced for 25 μm. For 100 μm target, large reductions of the track number and ring size are observed. Since a CR-39 film, as well as a Ta foil, is an energy filter, the stacking can analyze the energy of injected ions. The tracks on the first film indicated by (a) and (e) are due to protons of ≥ 1 MeV, the second film due to $>$ 4.8 MeV protons ((b) and (f)), the third $>$ 6.7

MeV ((c) and (g)), and the fourth > 8.2 MeV ((d) and (h)), respectively. The diameter of track pits is again a function of proton energy.[19] So that the energy from the track diameter and the stopping energy on each stack are combined to determine the proton spectrum over the range up to 10 MeV.

The low energy protons will be distributed all over the rear hemisphere. We have summed up the pit number over the film for each energy range and plotted it in Fig. 6 as a function of beam energy. Solid circle is the data for 5 μm target, triangle is for 25, and diamond is for 100. Plots show exponential decay up to 7 or 8 MeV and then a sharp cut off seems to exist around 10 MeV. As the thickness is increased from 5 μm to 25 and 100 , the slope temperature of the spectrum decreases from 3.39 to 3.04 and further to 2.08 MeV, where the slope is given from the exponential curve best fitted to each data line. The yield decreases from 1.8$\times$ to 1.2$\times$ and to 0.8×10^9, respectively. The protons less than 1 MeV are filtered out by the Ta filter. The PIC simulation will discuss the energy slope and cutoff in a later section. Suppose the protons are generated on the front surface and are transported through the solid plastics (CH) to the rear side. They must loose their kinetic energy inside the solid, leading to a change of the spectrum shape. We can assume a spectrum on the front surface being close to the 5 μm-target spectrum (solid circle in Fig. 6). Then, using this spectrum as the initial value, we calculated the output spectrum on the rear surface of a 100 μm CH, which is plotted by the open circle in Fig. 6. As expected, the temperature increases to 3.6 MeV and the plot is between 5 and 25 μm plots, apparently not close to the 100 μm data(the calculated yield is twice), leading us to believe that the protons are generated not on the front side but on the rear side.

The track pits apparently show ring structure on each film, as in Figs. 5. Almost same energy particles concentrate on the film to make ring structure pits, looking like white circular belts. We assume that a toroidal magnetic field, associated with the hot electron current, deflects proton beams as to form ring structure around the hot electron axis. The diameter d of rings on the film gives the deflection angle θ, as $d = 2L' \tan\theta$, where L' is the distance from the target to CR-39 in the target normal direction, namely, $L' = L/\cos 40°$. The θ is plotted in Fig. 7. The vertical error bar corresponds to the ring width, that is, the width of the white circular belt in Figs. 5. Far right 10 MeV data are from the radius

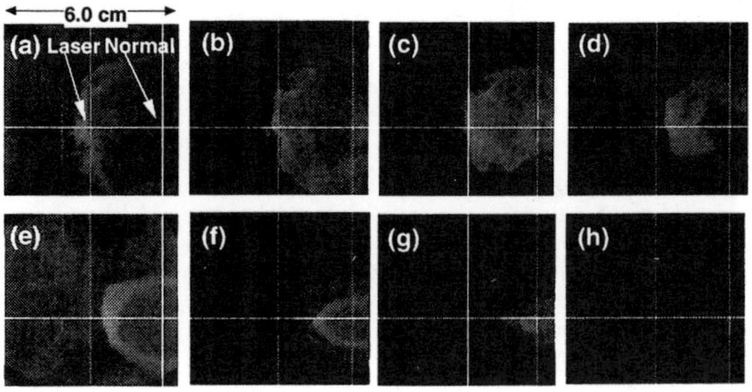

Figure 5. Proton tracks on CR-39 film from 5-μm target (upper (a) to (d)) and 100-μm (lower (e) to (h)): (a) the first film showing the protons of 1 ~4.8 MeV, (b) the 2nd, 4.8 ~ 6.7 MeV, (c) the 3rd, 6.7 ~ 8.2MeV, and (d) the 4th, 8.2 ~ 9.4 MeV. Laser axis is in the center of film and target normal is in the center of track ring. Film aperture is 6 $\times$ 6.5cm. Laser is 5.3×10^{18}W/cm^2 on C_8 H_8 target for 40° p-polarization.

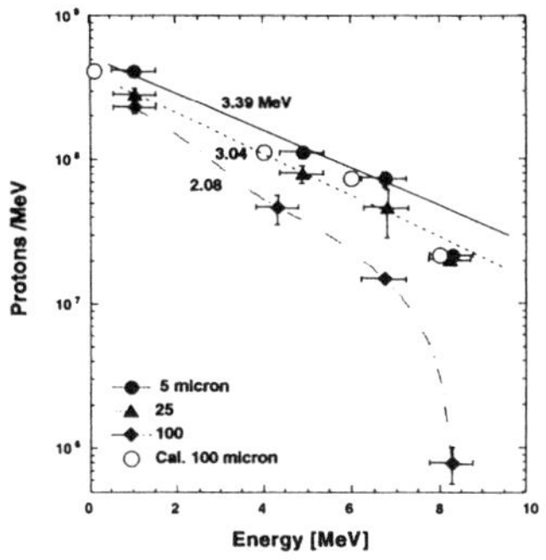

Figure 6. Proton's spectra for the different target thickness ;5 μm (solid circle), 25 μm (triangle) and 100 μm (diamond), respectively. Protons less than 1 MeV are filtered out. Open circle corresponds to 100-μm-thick target, calculated from 5-μm data.

inside the white ring on the final film. A proton gyro radius ρ is given by a magnetic field B_θ and a proton energy $\bar{E}_i$ as

$$\rho[\mu\mathrm{m}] = 1.43 \times 10^3 \bar{E}_i^{1/2}[\mathrm{MeV}]/B_\theta[\mathrm{MG}]. \tag{1}$$

Suppose D is the proton path length in the magnetic field and D is given by $D = \rho\theta$. Also since the proton is accelerated from zero to the final velocity $\bar{E}_i^{1/2}$, when it passes through the magnetic field, we assumed its averaged velocity, $\bar{E}_i^{1/2} \sim E_i^{1/2}/2$. Then

$$\theta[\mathrm{rad}] = \tan^{-1}\left(\frac{d}{2L'}\right) = \frac{B_\theta \cdot D}{1.43 \times 10^3} \cdot \frac{2}{E_i^{1/2}}, \tag{2}$$

where E_i is in the unit of MeV and B_θ in MG, D in μm. The logarithmic slope of the 5 and 25 μm data in Fig.7 equal -1/2 in agreement with that shown in Eq. 2. Figure 7 says that the magnetic fields deflect the protons into a ring structure. Fitting a curve described by Eq. 2 gives the coefficient $B_\theta \cdot D \sim 1000MG\cdot\mu$m for these thick targets. For 100 μm thick, the fitting, though it is somewhat poor, yet yields about 500 MG$\cdot\mu$m. Square plots in the figure are from the two dimensional (2D) particle simulation (PIC) and the slope is also close to -1/2.

The acceleration of ions during the interaction between the target plasma and the ultra intense laser pulse is studied with the 2D PIC simulations. The following are the simulation parameters; the simulation box is 40 μm $\times$10 μm with a fully ionized DH target plasma, whose thickness is 5 μm, placed at 17.2μm. The maximum electron density is 40 times higher than the critical density n_c of 1 μm laser light. In front of the target, there are preformed plasmas, whose scale length near the critical density is about 7 μm and whose density profile is the exponential profile: $0.2 - 1.2n_c$, respectively. This preformed plasma was prospective in the GEKKO MII experiments as discussed in Sec. II. Note here that the target is initially fully ionized and use a DH plasma to simulate the CH one. The acceleration of the fully ionized carbon C^{6+} and the deuteron D^+ is equivalent due to the same charge/mass ratio.

The p-polarized laser irradiates the target normally from the left boundary. The normalized vector potential a_0 is equal to 3.0, which corresponds to the peak intensity $8 \cdot 10^{18}$ W/cm^2. The total pulse duration is about 400 fs. The pulse rises in 50 fs with half-gaussian profile and keeps the intensity during 300 fs, then

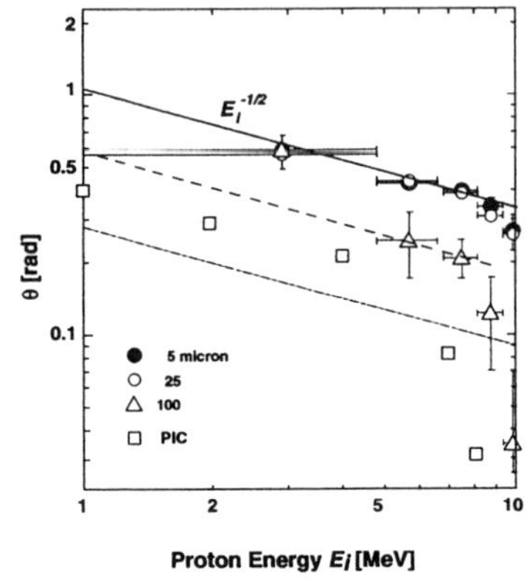

Figure 7. Deflection angle θ of protons versus proton energy. Solid and broken lines from Eq.2 fit to 5 and 100 μm thick target data, respectively. The squares and the dot-broken line are from the 2D PIC simulation.

damps in 50 fs. The transverse profile is also gaussian and its spot size is set to 4μm. The simulation continued 150 fs after the pulse was switched off to see the final proton energy. The total simulation time is 560 fs. Through the simulation, about fifty five percent of the incident laser energy was absorbed in the plasma.

The 2D-PIC simulation shows that the incident laser is reflected at the target surface, then the incident and reflected laser lights are overlapped in the preformed plasma and formed the standing waves. The periodic patterns due to the standing waves are observed in front of the target. Also the very intense positive fields were excited at the target rear side. The maximum E_x, about $0.4 \cdot E_{y0}$, appears at $t = 132$ fs in a few micron scale length. The acceleration energy of this electrostatic field is up to about 5 MeV/μm. Then the E_x expands with expansion of ions. The electrostatic fields at the target rear side have multiple peaks in the case of multiple species plasmas as discussed in Ref.[24] And each peak appears at the expansion front of each ion species. Since the lighter ion is faster than the heavier one, so the first peak appears at the proton expansion front and the second appears at the deuteron front in this simulation.

Quasistatic magnetic fields B_z are caculated both at $t = 198$ fs and 462 fs. In the preformed plasma, bi-pole magnetic fields is driven by the forward accelerated. The intensity is over twenty percent of the laser magnetic field B_{z0}, ~60 MG. At the laser irradiated surface, the magnetic filaments grow from the Weibel instability.[22] The filaments however are localized near the irradiated surface and their intensity is rather weak. Also B_z at the target rear side is still weak, about a few MG, and extended in a several micron at this time (198 fs). Later with plasma expansion, B_z at the target rear side strengthens and extends in the wide region. The time evolution of B_z will be discussed later in this section.

The proton distribution angle in the 2D PIC simulation is smaller than that of the experimental observations. The difference is about factor 2. The evaluated the product of the magnetic field × the length is $B_z \cdot D \sim 0.1 \cdot B_0 \times 8\mu$m= 240MG·$\mu$m. We could attribute this disagreement of the proton distribution angle in the simulation and in the experiments to the difference of the laser pulse duration. The GEKKO MII laser pulse length is about 100 fs longer than that of this 2D PIC simulation (300 fs). The longer pulse increases both the length and intensity of the magnetic fields. Also the transverse electrostatic fields E_y expand the ion angular distribution. These E_y's effects were not taken into account correctly in the simulation due to the small scale and periodic boundaries in the transverse space, especially at the later time.

It is found that the high energy protons (> 7 MeV), which go ahead of the deuteron expansion front, feel less magnetic fields and have different angular dependence, that is the high energy protons distribute in wider angles. This angular divergence is explained by the transverse electrostatic fields. This simulation result implies the protons observed in the experiment of the 5 μm target, are the lower energy parts behind the expansion of the deuteron, since their angular distribution just fits on the scale $E_i^{-1/2}$, as in Fig. 7. If the measurement range of the proton energy is extended over 10 MeV, we must check this and see the energy cut-off.

4. PETAWATT LASER CONSTRUCTION

The PWM is up graded to the world biggest Petawatt Laser (PW Laser). Remaining the main amplifier system as it is, we have up-graded the front end, the compressor and the focusing optics of the PWM. The beam size is enlarged from 20 cm to 50 cm to extract 1 PW or 500 J in 0.5 ps.

We have fixed the parameters according to the database of the PWM laser[4] and the related fast ignitor experiments.[2]

The front end has newly introduced an optically parametric chirped pulse amplifier (OPCP Amplifier) to suppress prepulses. A 40-cm diam. deformable mirror compensates phase distortions on the beam cross section. The final beam size is 50 cm. A pair of 1-m diffraction gratings, set in double path configuration, compresses the pulse length into 500 fs. The PW laser is synchronized to the GEKKO XII beams within 10 ps. As shown in Fig.8, the compression chamber is put on an iron frame at the heigh of 7.16 m from the room floor. To suppress vibrations of optics inside the chamber, the frame is made of a 25 cm ×25 cm squared column.

Figure 8. Compressor, focusing and target chambers.

An off-axial parabola mirror of 3.8 m focal length (the F-number is 7.6) provides the intensity of 2×10^{20} W/cm^2 or more on target. The focusing system is installed in a vacuum chamber of 2.2 m in diameter and 1.18 m in height. The chamber is connected through a gate valve to the horizontal 8 inch port of the target chamber.

5. CONCLUSION

The PWM laser has illuminated the CD shell target, imploded by GXII twelve beams. The laser-generated deuteron beams interact with the cold deutrons within the 4× critical density sphere and enhance ten times the D-D neutron yield.

The energetic protons, emitted from the rear surface of the ultra-intense laser illuminated CH foils. The protons are dragged away from the rear surface, where the hot electrons formed a virtual cathode. For 5 (and 100) μm thick target, 1.8 (and 1.2) $\times 10^9$ protons have an average energy of 3 (and 2) MeV. Coaxial ring structure of protons shows $B_\theta \cdot D$ to be 1000 (500) MG·μm. The 2D PIC simulation is close to the experiment. The energetic protons originate from the rear side acceleration. The magnetic fields grew and extended with the plasma expansion at the target rear side. These toroidal quasistatic magnetic fields, whose intensity is up to several tens megagauss, defocus the proton beams.

The world biggest Petawatt Laser has constructed. The out put is 1 petawatt, which is used also for ultra-intense laser matter interactions above the intensity of 10^{20} W/cm^2.

REFERENCES

1. M. Tabak, J. Hammer, M. E. Glinsky, W. L. Kruer, S. C. Wilks, J. Woodworth, E. M. Campbell, M. D. Perry, and R. J. Mason, **1**, 1626 (1994).
2. K.A. Tanaka *et al.*, Phys. Plasmas **7**, 2014-2022 (2000).
3. Y. Kato *et al.*, Plasma Phys. Controlled Fusion **39**, 1801 (1994).
4. Y. Kitagawa *et al.*, *"Final Report on PWM Laser"*, Institute of Laser Engineering, Osaka University (November 2000).
5. H. Takabe, M. Yamanaka, K. Mima *et al.*, Phys. Fluids **31**, 2884 (1988).
6. M.SADOWSKI *et al.*, Radiation Measurments **28**, 207-210 (1997).
7. A. P. Fews, P. A. Norreys, F. N. Beg, A. R. Bell. A. E. Danger, C. N. Danson, P. Lee, and S. J. Rose, Phys. Rev. Letters **73**, 1801 (1994).
8. E. L. Clark, K. Krushelnick, J. R. Davis *et al.*, Phys. Rev. Letters **84**, 670 (2000).
9. S. P. Hatchett, C. G. Brown, T. E. Cowan *et al.*, Phys. Plasmas **7**, 2076 (2000).
10. S. C. Wilks, A. B. Langdon, T. E. Cowan, M. Roth, M. Singh, S. P. Hatchett, M. H. Key, D. Pennington, A. Mackinnon, and A. Snavely, Phys. Plasmas **8**, 542 (2001).
11. A. Maksimchuk, S. Gu, K. Flippe, D. Umstadter, and V. Yu. Bychenkov, Phys. Rev. Letters **84**, 4108 (2000).
12. A. Puhkov, Z. -M. Sheng, J. Meyer-ter-Vehn *et al.*, Phys. Plasmas **6**, 2847 (1999).
13. E. G. Gamaly, Phys. Fluids **B5**, 3765 (1993).
14. S. Humphries, Jr., *Charged Particle Beams* (John Wiley and Sons, Inc., New York, 1990) Ch. 6 .
15. K. Krushelnick, E. L. Clark, Z. Najimudin *et al.*, Phys. Rev. Letters **83**, 737 (1999).
16. Y. Murakami, Y. Kitagawa, Y. Sentoku *et al.*, to be published in Phys. Plasmas, September, 2001.
17. M.Mori, Y. Kitagawa, R. Kodama *et al.*, Nuclear Instrument Methods in Phys. Research **A410**, 367 (1998).
18. Y. Yamada, Y. Kitagawa, M. Yokoyama, and C. Yamanaka, Phys. Letters **83A**, 9 (1981).
19. A. Szydlowski, M. SAdowski, T. Czyzlwski, M. Jaskofa, and A. Korman, Nuclear Instrument Methods in Phys. Research **B149**, 113 (1999).
20. Y. Kitagawa, K. A. Tanaka, M. Nakai *et al.*, Phys. Rev. Letters **75**, 3130 (1995).
21. J. F. Ziegler, *Stopping and Ranges of Ions in Matter* (Pergamon Press, New York, 1980) Vol. 5.
22. Y. Sentoku, K. Mima, S. Kojima, and H. Ruhl, Phys. Plasmas **7**, 689 (2000).
23. E. Lefebvre and G. Bonnaud, Phys. Review E **55**, 1011 (1997).
24. Y. Sentoku, V. Yu. Bychenkov *et al.* "High energy ion production from short-laser-pulse interaction with high density plasmas" submitted to Phys. Plasmas.
25. Wei Yu, V. Bychenkov, Y. Sentoku, M. Y. Yu, Z. M. Sheng, and K. Mima, Phys. Rev. Letters **85**, 570 (2000).
26. H. Yoshida *et al.*, Rev. Laser Eng. **27**, 291-297 (1999).

ALPHA-PARTICLE MEASUREMENTS NEEDED FOR BURNING PLASMA EXPERIMENTS

Kenneth M. Young*

1. INTRODUCTION

The next major step in magnetic fusion studies will be the construction of a burning plasma (BP) experiment where the goals will be to achieve and understand the plasma behavior with the internal heating provided by the fusion-generated alpha-particles (α-particles). Two devices with these physics goals have been proposed, the International Thermonuclear Experimental Reactor (ITER)[1] and the Fusion Ignition Research Experiment (FIRE)[2]. Extensive conceptual design work for the instrumentation to try to meet the physics demands has been done for these devices, especially ITER[3], and the overall requirements are reviewed in this book[4].

This article provides a new look at the measurements specifically important for understanding the physics aspects of the α-particles. An earlier article addressed a similar topic[5], but two significant events have occurred since then. The first was the completion of physics experiments on JET[6] and TFTR[7] with deuterium-tritium (DT) fueling, and the first chances to study α-physics[8,19], and the second is the realization that relatively compact plasmas, making use of advanced tokamak plasma concepts, are the most probable route to burning plasmas and ultimately a fusion reactor. These plasmas require measurements with excellent spatial and temporal resolutions and a large number of diagnostic signals will feed into the control of the plasmas.

The BP experiment will provide the science basis for the operation of a fusion reactor. Thus it is imperative that the understanding developed should lead to optimization of the plasma performance and being able to use the α-particles in controlling the plasma. The α-particles will become the dominant heating term in any successful BP experiment, and as such will play a key role in the physics of the plasma. Their generation in the core can be expected to significantly affect the density and

* Kenneth M. Young, Princeton Plasma Physics Laboratory, Princeton, NJ 08543, U.S.A.

pressure profiles associated with the confinement regime of choice. The fusion output in the form of these α-particles has got to be fed back into the control of the heating to affect the pressure gradients; it must also be used to assist the current profile, and possible provide current drive; and it must provide momentum to affect the shear flow with its strong influence on the transport properties. These roles then set the purpose for the α-particle measurements, setting up the requirements for specific diagnostics. Table 1 provides a listing of the diagnostic techniques which are being considered for the α-physics measurements in a BP device. They are discussed in the following sections, but some comments indicate where testing on operating devices is very desirable. Many of the diagnostics necessary for these measurements are not fully developed or will be severely constrained in their operation under BP conditions.

2. MEASUREMENTS OF CONFINEMENT AND SLOWING DOWN: EFFECT OF PLASMA HEATING AND PROFILES

The most important measurements to be made for understanding the heating impact of the α-particles are the profile measurements of the electron temperature and ion temperature. The pioneering measurements on TFTR[8] and JET[19] were relatively difficult because of the low-Q and thus small temperature increments. In TFTR averages over a number of shots were required to establish profile differences. The BP plasmas are expected to be created with reversed shear, for which there is little data from TFTR or JET in DT plasmas, so very good measurements of these key parameters, and the safety factor, q(r), will be required.
The best measurement of the source α-particles generated in the fusion reactions will be the 14 MeV neutron flux, with collimation provided to obtain spatial dependence. But it will be necessary to follow the history of the confined α-particles as they lose energy and are distributed across the plasma radius. It is unfortunately not possible to rely on a single measurement technique for this history because of limitations on each of the techniques so far proposed. Collective Thomson scattering may be the best technique, but, so far, has only been successfully demonstrated on the more easily measured thermal plasma[10]. Measurement in the microwave range gives the possibility of relatively good spatial resolution because the scattering angle can be quite large, and the hardware components are mostly available. Unfortunately, there are very limited transmission windows between resonant harmonics across the plasma, which constrains the measurement and causes high background noise levels. At high densities refraction will be an issue. The other possible choice for which sources and detectors are available is a CO_2 laser at 10μ, and a test is planned for studying the fast ions generated by the JT-60U negative-ion-based neutral beam[11]. In this case, though, the scattering angle must be very small (<1°) making spatial resolution and accessibility to the plasma very problematic. A compromise solution would be in the far infra-red (FIR) spectral region, using technologically feasible lasers, but for which developments in the lasers and detectors are necessary.

Neutral particle analyzers have proven capable of obtaining core α-particle

Table 1 Summary of diagnostics for alpha-particle physics

Diagnostics for alpha-particle physics	Comments relevant to a burning plasma (BP) device
Confinement and Slowing Down*	
14 MeV Neutron profile for α-source	Proven technique; needs same spatial resolution as $T_i(r)$.
Collective scattering	Demonstration required (JET, JT-60U); spatial, energy res. concerns.
α-CHERS	Proven technique; needs neutral beam; radiation-effects concerns.
High-energy neutral particle analyzers	Requires low-Z impurity (JET); spatial resolution is an issue.
Knock-on neutron measurement (neutron spectrometer or bubble chamber)	Difficult to get temporal/spatial resolution (JET).
Impact on Plasma Stability*	
Mirnov coils (high frequency)	Proven technique, but tough radiation environment.
Correlation reflectometry	Rapidly evolving technique (JET)
Alfvén antennas	Valuable for JET physics studies. Necessary for BP?
Beam emission spectroscopy	Proven technique; needs neutral beam.
Ion cyclotron emission	Proven measurement; has it any value?
Instabilities affecting redistribution and loss*	
Mirnov coils	Proven technique.
ECE grating polychromator	Proven technique.
Escaping-α diagnostics (IR, Faraday cup, scintillator, diamond)	For energy, source measurement, need scintillator but neutron, thermal issues (JET); diamond demonstration needed (JET).
Alpha-ash and removal*	
Thermal-He CXRS	Proven technique; needs neutral beam.
Divertor He measurement	Proven technique.
Neutron measurement of core n_D, n_T	2.5 MeV neutrons to be measured in larger 14 MeV neutrons (JET).
CXRS measurement of core n_D, n_T	Proven technique; needs neutral beam.
Fast-wave reflectometry of core n_D, n_T	Prototyping needed (DIII-D, JET)

* **Assumes that there are good measurements of $n_e(r)$, $T_e(r)$, $T_i(r)$, $v(r)$, $q(r)$, magnetic equilibria, etc.**

distribution functions in two somewhat different applications. In JET, Gondhalekar et al. made use of the enhanced charge-exchange cross-sections of natural low-Z impurities, Be and C, in the core in unfolding the spectrum of α-particles from the combined D- and α-spectra[9]. There was only one vertical line of sight providing a line-integral measurement, and it is to be hoped that a future JET measurement will have additional sightlines. On TFTR, the novel pellet charge exchange (PCX) concept, using a neutral particle analyzer in conjunction with a low-Z lithium pellet radially-injected into the plasma with horizontal observation was applied[13]. The pellet flight time provided the spatial resolution for the α-particle distribution function and showed redistribution of trapped α-particles in reversed-shear discharges. In high-density plasmas with neutral beam heating, the failure of the pellet to penetrate limits its applicability. A spectroscopic technique making use of the charge exchange of the α-particles with neutral beam particles, the so-called α-CHERS (charge exchange recombination spectroscopy), was also used on TFTR[8]. The high-energy tail of the 468.6 nm He^+ line is spectrally resolved to determine the slowing-down spectrum of the α-particles below about four-times the neutral beam energy (i.e. up to about 700 keV in the TFTR case). One result was that the anomalous diffusivity of the α-particles was less than 0.03 m^2/s. This measurement of the tail requires very precise knowledge of the visible bremsstrahlung background and other noise sources. The principal of operation of this technique is the same as that used for measurement of the He-ash or the ion temperature and plasma rotation, though the instrument requires a higher optical throughput. All these instruments depend on a neutral beam source, with energy in the range 100 - 150 keV/amu. Particles with this beam energy do not penetrate well into the proposed burning plasmas and development of a short-pulse intense diagnostic neutral beam has been proposed[13].

The high-energy tail of the neutron spectrum contains a "knock-on" component created by the α-particles colliding with the fusing atoms and imparting additional center-of-mass velocity, leading to higher energy neutrons. The spectrum of this very small tail, tied to the α-energy spsectrum, approximately 10^{-3} of the peak, has been measured on JET by a neutron spectrometer, averaging over many DT shots[14]. It is difficult to see how such a spectrometer could provide sufficient counts for both adequate temporal or spatial resolution. Hence a development of a set of pulsed bubble chamber detectors, with very sharp low-energy cut-offs, the lowest at about 15 MeV, so that only the "knock-on" tail's spectrum is measured, has been proposed[15]. The detectors could be installed in collimator tubes of a neutron camera for a BP device.

Measurement of escaping α-particles, while important for the issue of α-particle confinement is described in section 4.

3. IMPACT ON PLASMA STABILITY; MACROSTABILITY AND COLLECTIVE EFFECTS

Fast ions and α-particles have very significant impacts on the plasma in that they can stabilize MHD modes and destabilize a large variety of high-frequency (> 100 kHz) modes of the Alfvén eigenmode (AE) type. The behavior of low-frequency MHD modes can be relatively easily followed by the Mirnov coils. The normal pressure diagnostics must be monitored closely to observe any build-up of the pressure.

High-frequency Mirnov coils give a clear picture of the presence of the various AE instabilities, being particularly good at following the evolution in frequency (chirping, etc.) which have provided the necessary clues for mode identification. In BP reversed-shear plasmas a build-up of an avalanche of these instabilities to cover a large fraction of the plasma cross-section, providing a clear rapid transport path, is a concern. Correlation reflectometry on TFTR has been effective in spatially localizing modes and providing a resource for theoretical analysis and prediction[16]. Since its sensitivity to small amplitude density fluctuations is much greater, it can detect fluctuations not visible on the Mirnov coils. For a BP plasma, it is essential that there be coverage across the full plasma cross-section, so that a combination of O- and X-mode correlation reflectometry is required.

Another opportunity for studying AE modes has been taken at JET where a set of external coils has been used to destabilize stable AE modes, hence allowing for study of the growth of the modes and tying behavior closely to theoretical projection[17]. It is not clear that there is value to extending this type of experiment to a BP where the internal drives are likely to be very strong.

Another very clear instability observed on TFTR and JET occurs at harmonics of the ion cyclotron frequency at the outside, low-B_T, plasma edge, caused by trapped α-particles interacting with the low plasma density[8,9,18]. The interactions are so complex that it is doubtful whether any meaningful information about the α-particle population will be obtained in a BP at high magnetic field.

4. IMPACT OF INSTABILITIES ON THE ALPHA-PARTICLE BEHAVIOR; REDISTRIBUTION AND LOSS

Good measurements of the MHD and the higher frequency MHD modes are required and internal T_e fluctuation measurements are particularly important. In TFTR DT experiments, the possibility of investigating the redistribution of the α-particles was provided by the two diagnostics, PCX and α-CHERS[8]. Sawteeth were shown to redistribute the trapped and non-thermal α-particles and TAE modes redistributed trapped α-particles. The key to the measurements was the good spatial resolution of the instruments.

The limit to redistribution is loss, which must be measured effectively with an array of detectors. Surface temperature measurements by infra-red imaging gives good spatial information about the overall energy loss of fast ions. But localized detectors are necessary for obtaining information about the energy and source of the α-particles. An array of such detectors in the lower-outside quadrant (depending on the toroidal field direction) of the first wall can provide this information. On TFTR, an array of thin-scintillator detectors behind shielded apertures gave very specific data on the pitch angle and gyroradius of the lost particles[8]. In quiescent plasmas, the losses could be assigned to first orbit losses and toroidal field ripple induced loss. In discharges with active MHD, and with the total light being fed to a photomultiplier, quantitative measurements of the losses were obtained for sawteeth, fishbones, locked modes, tearing modes, the thermal quench prior to major plasma current disruptions, high-frequency kinetic ballooning modes and ICRF heating. For application in a BP device, improved scintillators which can tolerate high temperatures and radiation-hard optics must be developed. An alternative technique using Faraday cup detectors has been proposed and installed on JET[19]. These detectors are easier to locate and install, requiring only electrical wiring, but they only provide information about the energy of the α-particles passing through the aperture.

5. MEASUREMENT FOR STUDIES OF TRANSPORT OF FAST IONS AND α-ASH REMOVAL

The measurement of helium in the core of the plasma has been carried out spectroscopically in many tokamaks to study the diffusion of this impurity[7]. In a BP this α-ash builds up as the α-particles thermalize and for a reactor could stop the burning process. Hence it is critical to measure it both in the plasma core and in the efflux through the divertor. In the core, charge-exchange spectroscopy, using the full 468.6 nm He^+ spectral line, is the only available technique. Again, there is a requirement for a neutral beam. In the divertor, residual gas analysis and Penning gauges can be used for quantifying the helium flow from the plasma.

One expects that there will be specific experiments aimed at manipulating the α-particles, to be able to enhance the α-ash loss[7] or to channel the high-energy particles[7] toward the goals of controlling plasma behavior. Suggestions for affecting the current drive and the plasma rotation and shear have been made and have to be studied much more deeply than in preliminary TFTR and JET experiments[7,6]. The diagnostics required for these physics studies will be the standard set of profile diagnostics, and magnetic measurements for studies of current drive.

An associated aspect of controlling the α-particle behavior is the fueling of the hydrogen isotopes, deuterium and tritium. The measurement of the core densities of these ions can be done spectroscopically, with a neutral beam to provide the source of particles for the neutralization. There are proposals that the relative densities could be obtained by neutron spectroscopy, even though the DD neutron intensity will be many orders of magnitude smaller than the DT intensity, and it is hoped to test this idea on JET. An alternative technique which uses simple coils to carry out fast-wave reflectometry was prepared for test on DIII-D and is awaiting test[13].

6. SUMMARY AND RECOMMENDATIONS FOR MEAUREMENTS FOR α-PHYSICS IN A BP DEVICE

We have briefly described four areas of physics study to be carried out in burning plasma experiments for understanding the behavior of the heating particles, emphasizing the measurements needed and examples of diagnostics for carrying out those measurements. The set of diagnostics must be aimed at achieving the mission of the BP device, the ultimate goal of which must be establishing the optimal operating conditions for the more-engineering-oriented next step. For the latter, access to the plasma for diagnostics will necessarily be more restricted and plasma control, rather than physics understanding, will be the prime motivator for the instrumentation.

The set of diagnostics to enable thorough α-physics studies that can be considered now for implementation on a BP device is shown in table 1. It is essential that there be very high quality diagnostics for measuring the profiles of the main plasma parameters and for measuring the fluctuations over a wide range of frequencies. Many of the techniques with the best spatial resolution are dependent on a neutral beam, which is a major challenge for high-density and large tokamak plasmas. Good spatial and temporal resolution neutron measurements will be key, particularly for confinement studies, since they will define the α-source. Good measurements of the fluctuations, externally with many Mirnov coils and internally with many spatial locations of ECE and reflectometry to provide local temperature and density variations are essential for defining the instability modes.

Finally a set of diagnostics for the α-particles themselves must be installed on a BP device. Physics studies of confinement and redistribution demand techniques with very good spatial resolution, such as was available with α-CHERS and PCX on TFTR. Both of these diagnostics are difficult to implement on a BP device, the former needing a neutral beam and the latter a high-speed lithium pellet. Collective Thomson scattering techniques and neutral particle analysis dependent on natural levels of low-Z impurities must be developed further, but it is difficult to see how they can provide sufficiently good spatial resolution to resolve some of the physics issues. An array of lost-α detectors at the wall to give energy and source information about the escaped particles will enhance the physics understanding. All these techniques should, if possible, be applied in a second JET DT campaign to determine the value of installing them on a BP device.

7. ACKNOWLEDGEMENTS

I am very grateful for many discussions with and guidance from Raffi Nazikian. This work was supported by U.S. DOE Contract No. DE-AC02-76CH0307.

8. REFERENCES

1. R. Aymar, The ITER Joint Central Team, ITER-FEAT – "The future international burning plasma experiment overview", in *Proceedings of 18th IAEA Fusion Energy Conference, Sorrento* (IAEA, Vienna, 2001) paper IAEA-CN-77/OV/1, http://www.iaea.org/programmes/ripc/physics/fec2000/html/fec2000.htm.
2. D.M. Meade. S.C. Jardin, C.E. Kessel, M.A. Ulrickson, J.H. Schultz, P.H. Rutherford, et al., Mission and design of the Fusion Ignition Research Experiment (FIRE), in *Proceedings of 18th*

IAEA Fusion Energy Conference, Sorrento (IAEA, Vienna, 2001) paper IAEA-CN-77/FTP2/16, http://www.iaea.org/programmes/ripc/physics/fec2000/html/fec2000.htm.

3. ITER Physics Expert Group on Diagnostics, ITER Physics Basis Editors, ITER physics basis, Chapter 7: Measurement of plasma parameters, *Nucl. Fusion* **39** (12), 2541 – 2575 (1999).
4. A.E. Costley, in: *Advanced Diagnostics for Magnetic and Inertial Fusion*, edited by P.E. Stott and G. Gorini (Plenum Press, New York, 2001) this volume.
5. S.J. Zweben, S. Putvinski, M.P. Petrov, G. Sadler, K. Tobita and K.M. Young, in: *Diagnostics for Experimental Thermomnuclear Fusion Reactors,* edited by P.E. Stott, G. Gorini and E. Sindoni (Plenum Press, New York, 1996) pp 467 – 476.
6. M. Keilhacker, A. Gibson, C. Gormezano, P.J. Lomas, P.R. Thomas, M.L. Watkins et al., High fusion performance from deuterium plasmas in JET, *Nucl. Fusion*, **39** (2), 209 – 234 (1999).
7. R.J. Hawryluk, Results from deuterium-tritium tokamak confinement experiments, *Rev. Mod. Phys.,* **70** (2), 537 – 587 (1998).
8. S.J. Zweben, R.V. Budny, D.S. Darrow, S.S. Medley, R. Nazikian, B.C. Stratton et al., *Nucl. Fusion* **40** (1), 91 – 149 (2000).
9. JET Team (prepared by P.R. Thomas), Alpha particle studies during JET DT experiments, *Nucl. Fusion*, **39** (11Y), 1619 – 1625 (1999).
10. H. Bindslev in: *Advanced Diagnostics for Magnetic and Inertial Fusion*, edited by P.E. Stott and G. Gorini (Plenum Press, New York, 2001) this volume.
11. T. Kondoh, S. Lee, D.P. Hutchinson and R.K. Richards, Collective Thomson scattering using a pulsed CO_2 laser in JT-60U, *Rev. Sci. Instrum.*, **72** (1), 1143 - 1146 (2001).
12. R.K. Fisher, H.H. Duong, J.M. McChesney, P.B. Parks, S.S. Medley, R.V. Budny et al., Alpha diagnostics using pellet charge exchange: results in the Tokamak Fusion Test Reactor and prospects for ITER, *Rev. Sci. Instrum.*, **68** (1), 336 - 339.
13. A.E. Costley, D.J. Campbell, S. Kasai, K.M. Young amd V. Zaveriaev, ITER R&D: Auxiliary Systems: Plasma Diagnostics, *Fus. Engg. & Des.*, **55** (2-3), 331 – 346 (2001).
14. J. Källne, L. Ballabio, J. Frenje, S. Conroy, G. Ericsson, M. Tardocchi, E. Traneus and G. Gorini, Observation of the alpha particle "knock-on" neutron emission in magnetically confined DT fusion plasmas, *Phys. Rev. Lett.*, **85** (6) 1246 - 1249 (2000)
15. R.K. Fisher, P.B. Parks, J. Liptac, D.C. Ingram, C.E. Brient, S.S. Medley et al., Measured response of bubble neutron detectors and prospects for alpha knock-on diagnostics, *Rev. Sci.. Instrum.,* 72 (1), 796 - 800 (2000).
16. R. Nazikian and E. Mazzucato, Reflectometer measurements of density fluctuations in tokamak plasmas, *Rev. Sci. Instrum.*, **66** (1), 392 - 398 (1995).
17. D. Testa and A. Fasoli, The effect of plasma shaping on the damping of low n Alfvén eigenmodes in JET tokamak plasmas, *Nucl. Fusion,* **41** (7), 809 – 812 (2001).
18. N.N. Gorelenkov and C.Z. Cheng, Alfvén cyclotron instability and ion cyclotron emission, *Nucl. Fusion*, **35** (12), 1743 - 1752 (1995).
19. F.E. Cecil, B. Roy, C. Sutton and N. Wasinger, A thin film device as a low energy, high flux charged particle spectrometer, *Rev. Sci Instrum.,* **72** (1), 792 - 795 (2001).

DIAGNOSTICS BASED ON HELIUM NEUTRAL BEAMS IN ITER

M G O'Mullane, H Anderson, Y Andrew, M Brix, C Giroud, A G Meigs, M Proschek, H P Summers, K-D Zastrow and contributors to the EFDA-JET workprogramme *

1. INTRODUCTION

Reducing the activation of the vessel and its components during the early phase of ITER operation is important. One scheme of minimising the activation is to use helium as the heating neutral beam species. This gives rise to a concern about how effective diagnostic spectroscopy will be with such beams. The diagnostic use of deuterium heating beams is well established yielding, for example, ion temperature profiles, plasma rotation, magnetic fields, impurity concentration, Z_{eff} and beam penetration measurements. The experience with helium beams is more limited and there are greater uncertainties in the fundamental atomic data. A helium neutral beam has certain potential advantages: a reduced beam halo, no fractional energy components, greater penetration than deuterium beams and possibly more efficient charge exchange donation in certain energy regions.

At JET a series of experiments have been undertaken with helium beams to address these issues. A number of these plasmas have very high concentrations of He, up to 100%, which is also of relevance for future ITER operations. Beam emission and charge exchange spectra were detectable and useful. It is shown that the diagnostic capability is high matching that of deuterium but with some marked differences.

2. JET EXPERIMENTS AND IMPLICATIONS FOR ITER

The JET experiments have used both a He-doped deuterium beam and a full helium beam. The doped D_2/He system was developed[1] to make fast He beams routinely available by using the heating beam as a parasitic He source. The beam emission is viewed by a twelve chord spectrometer whose lines of sight are almost parallel to the beam at their intersections, located between the boundary and the plasma centre. By sweeping the plasma, a much higher spatial resolution is achieved together with an in situ relative calibration between lines of sight. A

*M G O'Mullane, H Anderson and H P Summers, Department of Physics and Applied Physics, University of Strathclyde, Glasgow, UK. Y Andrew, A G Meigs and K-D Zastrow, Euratom/UKAEA Fusion association, Culham Science Centre, Abingdon, UK. M Brix, Institut für Plasmaphysik, Forschungszentrum Jülich GmbH, EURATOM Association, D-52425 Jülich, Germany. C Giroud, Association EURATOM-CEA sur la fusion, CEA Cadarache, F-13108, St Paul lez Durance, France. M Proschek, Institut für Allgemeine Physik, TU-Wien, Austria, (Association EURATOM-ÖAW).

Advanced Diagnostics for Magnetic and Inertial Fusion, Edited by Stott *et al.*
Kluwer Academic/Plenum Publishers, New York, 2002

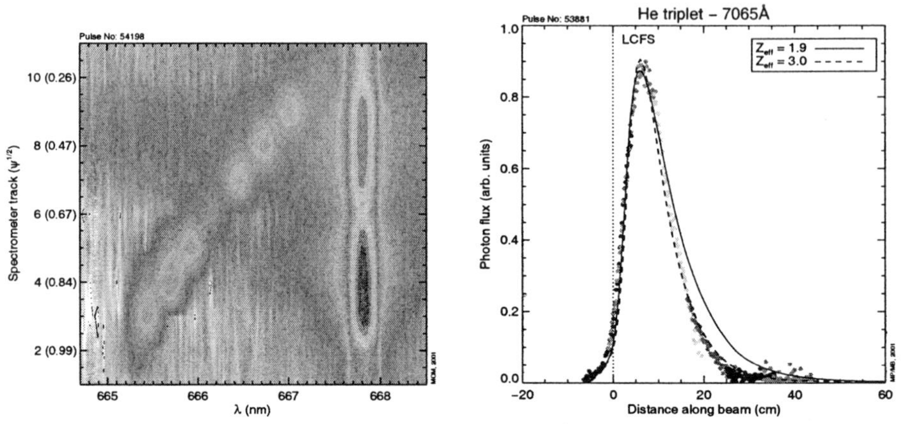

Figure 1: Beam emission characteristics of HeI: (left) the singlet, 667.8nm is present across the plasma radius and is well separated from the edge emission by the Doppler shift. The triplet, 706.5nm (right), is not measurable beyond 300mm along the beam. Note that line averaged Z_{eff} measured from the bremsstrahlung is 1.9.

high clearance L-mode plasma was selected for all experiments to give ELM free plasmas. The plasma was swept by $\pm$6cm with a ramp rate of $\sim$ 5cm/s. The volume varied by $\sim$ 2% and the line averaged density was constant to within 5%.

2.1. Phenomenology

The two electron nature of helium produces a rich spectral phenomenology from both HeI and HeII emission, from both beam and plasma. This has no equivalent in a deuterium beam but initial studies show it to have special diagnostic potential. Note that at the JET beam energies there is a negligible motional (linear) Stark effect on the $n = 3$ He levels. Although this simplifies the spectral analysis, there is no direct MSE diagnostic. The immediate analysis focuses on beam emission (BME) and impurity charge exchange (CXS) which are the equivalents of these for deuterium beams and on the role of metastables.

2.2. Beam Emission

Monoenergetic, essentially Stark unperturbed beam emission from helium is much simpler than that from of deuterium. The Doppler beam emission spectra is well separated from the 'cold' edge HeI emission at most radii, and is not contaminated by impurities. Figure 1 shows the principal characteristics of the singlet and triplet emission. Weak collisonal coupling between the two metastables of helium, $1s^2\,^1S$ and $1s2s\,^3S$, means that there are effectively two beams in the plasma attenuating with different decay lengths. The true excited populations have formation and relaxation lengths of $\lesssim$ 3cm and can be treated as quasi-static relative to the unrelaxed beam. The singlet beam is observed to penetrate to the centre of the plasma while the triplet side stops at $\sim$ 30cm along the beam. Initial modelling of the triplet[2,3] indicates a larger Z_{eff} than measured by bremsstrahlung (without allowance for 'hollow' profiles).

The edge triplet BME reflects the decay of an initial metastable entry fraction in the beam rather than a collisonally regenerated population in the plasma. This fraction can be estimated directly from the edge singlet BME as seen in figure 2. Full collisional-radiative BME coeffi-

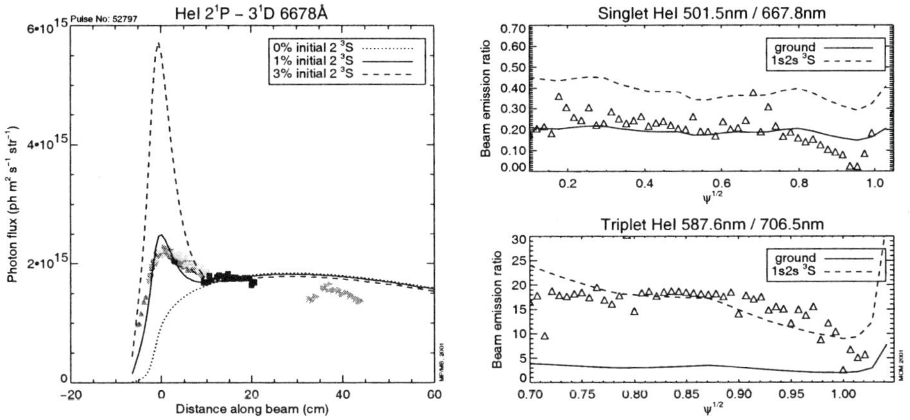

Figure 2: Composition of He beams: (left) modelled [2,3] 667.8nm emission at three initial metastable fractions. (right) measured ratios (triangles) of singlet and triplet beam emission showing their principal driving source and inter alia validation of the BME calculations. Note that the $\psi^{1/2}$ scales are different reflecting the region of valid data across the plasma.

cients have been calculated. Ratios are plotted in the figure to cancel any uncertainities in the viewed volume.

When scaled to ITER conditions the triplet BME can be used to measure the rapidly varying n_e edge profile and ITB shear region. These clear triplet and singlet BME signals can be used to diagnose beam deposition, plasma density and Z_{eff} more clearly than the deuterium BME. The precision of the deduction is limited by the quality of the ion impact excitation cross sections for higher Z impurities ($Z \geq 3$).

2.3. Charge Exchange

Core impurity measurements rely on modelling the re-distribution and cascade emission following state selective single charge transfer from the beams into high $n\ell$ states of plasma impurities. In comparison with the D beams it is notable that the CXS emission is stronger towards the edge, where the He metastable is an efficient donor, and weaker at the core where the tightly bound ground state is the donor. The HeII $n = 4 - 3$ charge exchange emission at 468.5nm is present, as previously recorded[4]. The spectral analysis is complicated since it requires modelling of the helium plume, which is appreciable. It is nonetheless tractable and is a favourable method of measuring core helium concentrations.

Neon was actively introduced (up to 4% concentration) to investigate the CXS diagnostic capability of higher Z impurities with He beams launched into He plasmas. Figure 3 shows that one of the significant effects of a helium plasma is to completely suppress carbon in the confined plasma, as seen by the absence of the CVI CXS emission (the suppression is also confirmed by VUV spectroscopy). The NeX $(11 - 10)$ CXS signal is evident and comparable to that in D beam observations. Results from the different viewing lines show that, again, the most effective CXS is in the edge region where the triplet is the primary donor. In the practical diagnostic procedure with D beams, the CVI CXS emission is used to determine T_i and the weaker NeX signal yields the concentration. Its absence in the He beam/He plasma case means that the HeII CXS signal must provide the ion temperature. Hence the methodology will be different but CXS with He beams is a feasible diagnostic. For a primarily D plasma CVI CXS driven by the He beam is expected to be useful. Confirmation in the experiment was not

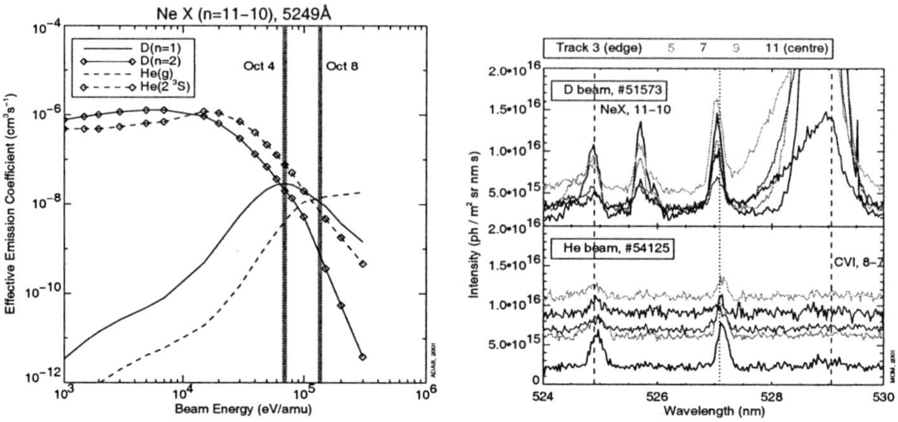

Figure 3: Charge exchange emission: (left) metastable beam donor influence on the effective emission coefficients from fully ionised Ne X — the coefficient is given relative to the donor population. (right) Progression of NeX 524.8nm ($n = 11 - 10$) CXS emission across the plasma in a deuterium and helium plasma.

possible since dilution of the helium plasma by introducing CD_4 was prohibited.

3. CONCLUSIONS

Experiments at JET confirm that the combined beam emission and charge exchange analysis can provide a significant diagnostic for ITER plasmas. Using helium beams and He plasmas to minimise vessel activation does not imply a major impairment to diagnostic capability but reduction techniques must be altered. The most marked difference from D beams is the clear BME signals. Over the extent of the He triplet, BME measurements yield edge parameters and conventional CXS analysis gives ion temperatures and concentrations. Deeper into the plasma the combination of BME and CXS signals can be used to deduce a collisionality profile.

Continuing work will address the limitations in the precision of interpretation due to the present ion impact excitation data and improvements will be incorporated into the Atomic Data and Analysis Structure (ADAS)[5]. This provides modelling of effective stopping and emission data for helium beams at all energies for ITER.

4. REFERENCES

1. H D Falter et al, Helium doped hydrogen or deuterium beam as cost effective and simple tool for plasma spectroscopy, *Rev. Sci. Instr.*, **71**, 3723–3727, (2000)
2. M Proschek et al, Measurement of line emission profiles of fast helium beams and their application to plasma diagnostics, *28th EPS conference on controlled fusion and plasma physics*, paper 1.100, (2001)
3. M Brix, Ph.D Thesis, Internal report Forschungszentrum Jülich, Jül-3686, (1999)
4. U Gerstel et al, Quantitative simulation of non-thermal charge-exchange spectra during helium neutral beam injection, *Plasma Phys. Control. Fusion*, **39**, 737–756, (1997)
5. H P Summers, The ADAS User Manual — version 2.2 (2000), *http://adas.phys.strath.ac.uk*

This work has been conducted under the European Fusion Development Agreement and is partly funded by Euratom and the UK Department of Trade and Industry.

ADVANCED CONCEPTS IN FAST IGNITION AND THE RELEVANT DIAGNOSTICS

Hideaki Habara, Kathryn Lancaster, Chris Reason, Chris Aldis, Bill Lester, T. Craig Sangster* and Peter Norreys

1. INTRODUCTION

Recent developments in high power lasers have allowed the study of new regimes of laser-matter interactions relevant to astrophysics[1], nuclear physics[2], and fusion energy research[3, 4]. In the context of inertial fusion energy research, the 'fast ignitor' (FI) concept[5] was proposed in order to relax the strict symmetry requirements for the laser irradiation of the spherical target and to reduce the drive energy needed to achieve high densities and the formation of the spark. However, other problems have gradually become clear with this scheme. These are relevant to the temporal and spatial stability of the propagation of the additional heating source, for example the high-energy electron beam, due to losses and deflection[6] of the ultra-intense laser pulse in the surrounding plasma and the transport through a considerable length of a plasma[7]. Here we describe a new compression geometry that eliminates these problems utilizing the cone-guided compression scheme[8], or the so-called 'advanced fast ignitor'[9]. The attraction of this scheme is that it makes easy to access to fuel core without the temporal and spatial uncertainty. Using this scheme, colleagues from Japan and UK reported the first experimental results of fast heating of the compressed super-solid density matter using an ultra-intense short pulse laser and showed that efficient compression and heating are both possible simultaneously[9, 10]. This new approach provides a route to efficient fusion energy production.

It is also important for fast ignitor scheme to understand how to accelerate the electrons and ions using short-pulse intense laser pulses in the overdense plasma and also how the particles propagate into the super-critical plasma from the viewpoint of an additional heating source. Recently there have been many reports in both simulation and experiment that demonstrate a number of different acceleration mechanisms. Some of these suggest the acceleration strongly depend upon the laser and the plasma conditions such as the laser polarization, intensity and the density scale-length[11]. In particular, recent simulations predict that the energy of the accelerated ions suddenly increases with higher laser intensity and that the ion acceleration direction is changed from the longitudinal (or laser) direction to the transverse direction as plasma density scale-length is increased[12]. Therefore, measurement of the accelerated ion momentum distribution helps us to understand the laser-plasma interactions under various controlled conditions.

Hideaki Habara, Kathryn Lancaster, Chris J. Reason, Chris Aldis, Bill Lester and Peter A. Norreys, Rutherford Appleton Laboratory, Chilton, Didcot, Oxon, OX11 0QX, UK.
*T. Craig Sangster, LLE, Univ. of Rochester, 250 E, River Road, Rochester, NY, 14623-1299, USA.

Advanced Diagnostics for Magnetic and Inertial Fusion, Edited by Stott *et al.*
Kluwer Academic/Plenum Publishers, New York, 2002

Neutron spectroscopy is one of the most fascinating methods to investigate the accelerated ion distribution processes generated by ultra-intense laser-plasma interactions (using beam-fusion neutron spectroscopy) [13, 14]. For this purpose, we are re-constructing the multi-channel neutron spectrometer, LaNSA[15, 16]. The system records the timing when the neutrons are detected at each scintillator in the array. These timings are converted into neutron spectrum via the time-of-flight (TOF) method. Neutron energy spectra taken at several different directions can then be used to determine the momentum distribution of the accelerated ions by taking into account the Doppler shifts of neutron spectra from 2.45MeV for each viewing angle. In this paper, we discuss the re-commissioning of the LaNSA system and the future experiments that are planned for the RAL PW laser system.

2. NOVEL GEOMETRY OF FAST IGNITION SCHEME

The first question to address in guided compression is the effect of the high Z walls on the stability of the foil during the compression phase. To study this interesting stability problem, an experiment was conducted on the Nd:glass laser VULCAN[17] at the Rutherford Appleton Laboratory. The laser can be supplied sub-ns long pulse laser and sub-ps short pulse laser simultaneously. Figure 1 shows the first experimental layout. The short pulse beam was focused onto an aluminum plane with Bi coating to backlight the plasma into the v-groove of copper block. The short pulse beam was focused onto an aluminum plane with Bi coating to backlight the plasma into the v-groove of copper block.

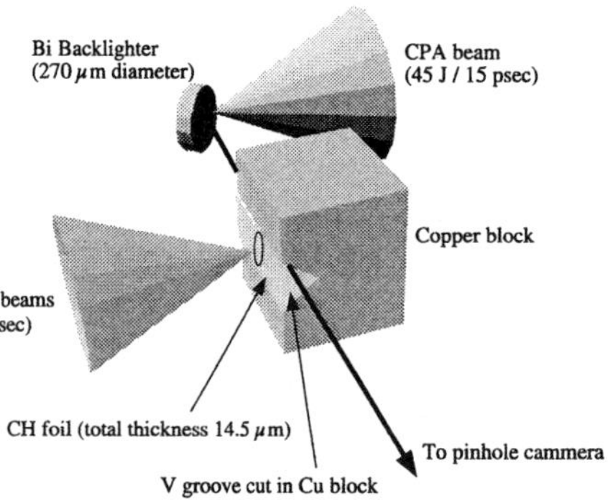

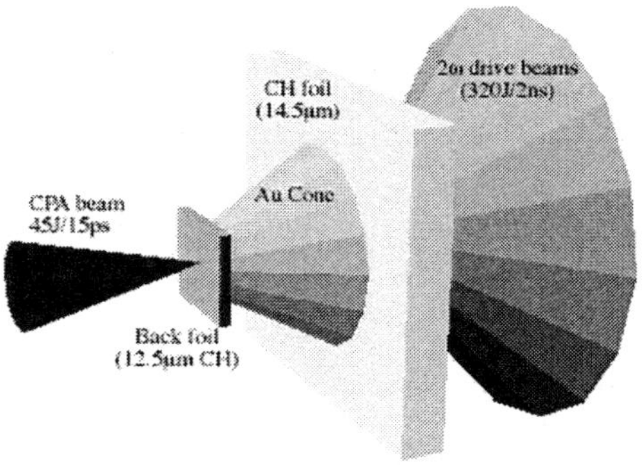

Left: **Figure 1**. Schematic illustration of the back-lighting experiment. Right: **Figure 2**.Schmatic illustration of conical implosion experiment.

The short pulse beam was focused onto an aluminum plane with Bi coating to backlight the plasma into the v-groove of copper block. The incident angle of the beam is 30 degree at peak intensity of 5×10^{14} W/cm^2. The 2ω drive beams were focused 500μm on the foil from the edge of the Cu block. The backlighting images were recorded with x-ray pinhole camera. In the results, we obtained an image of the guided plasma taken 1.0ns from the leading edge of the 2ns/5×10^{14}W/cm^2 pulse. The foil image shows some inner curvature at the edges of the v-groove without exhibition of the growth of any hydrodynamics instability at the foil surface.

A fully integrated conical compression experiment was conducted to determine the efficiency of fast electron heating. In this second experiment, the six 2ω nsec-duration beams irradiated a 12.5μm flat polyethylene CH foil, which was covered with a 2.0 μm of deuterated polystyrene CD foil as shown in Fig.2. A 100 μm diameter, 100nm thick signature layer of Ti and a 500nm thick layer of chlorinated paralyene were deposited between the CH and CD foils, with the CD layer facing the apex of the Au cone. The distance from the entrance hole to the apex was 250 μm and the Au cone material was 20 μm thick. A 20 μm hole was machined in apex of the cone, so that the target could be precisely positioned and the fast electrons transport was through the compressed CH/CD material only. A 12.5 μm thick, 270 μm diameter polyethylene foil was placed over the apex of the cone in order to control the scale-length of the short pulse interaction process. The p-polarized 45J/15ps pulse was focused onto this foil in a 20μm spot, giving a peak intensity on target of 10^{18}W/cm^2.

A significant increase in x-ray emission that peaked at 1.7ns after the beginning of the main pulse, corresponding to the onset of stagnation, was observed with the time resolved x-ray spectrometer. The time-integrated spectra indicated an upper bound to the heated plasma electron temperature of 1keV. Neutrons generated through DD thermal reactions were only observed at the same time of the peak of the X-ray emission, of which yield was 2×10^4 neutrons. Assuming the ρR of the plasma is 10mg/cm^2 from the simulations of the experiment, the plasma temperature will be estimated at 500eV, which agree with the result from X-ray spectroscopy. These results indicate that about 7% of the short pulse laser energy was coupled to the plasma to produce this temperature. While this coupling efficiency appeared somewhat low, it could be increased by a more suitable match of the range of the fast electrons to the achieved ρR. These experiments at RAL suggested that the high densities predicted by 2D computer simulations for the cone insert geometry could be achieved in practice.

The cone-guided compression experiments on heating of super-solid density plasmas were performed on the GEKKO XIII laser at the University of Osaka. This laser has 12 beams for nanosecond pulses with a maximum energy of 15kJ at 0.53μm wavelength and a synchronized sub-ps pulse beam with a power of 100TW and pulse energy of 60J[18]. The schematic diagram for the experiment is shown in Figure 3. Long pulse lasers irradiate the spherical target excepted with the area where the cone is inserted. At the proper timing of the implosion, ultra-intense laser is focused on the imploded core guided the propagation path free from the disturbance by an expansion and implosion of the plasma. The distance between the cone apex and center of the shell is 50μm, which is enough short to access to the core fuel with reduction of the sensitivity to electron beam propagation instabilities and losses. The cone walls are on a radius from the shell centre in order to minimize the disruptions to the spherical symmetry at the implosion and achieve high density. The laser energy of the drive pulses was 1.2kJ to measure the core heating taking account of the internal energy of the compressed core (about 5% of drive energy) to be comparable to the short pulse laser energy. The imploded core was observed at the center of the shell, which is estimated 40-45μm from X-ray self-emission image of the compressed plasma.

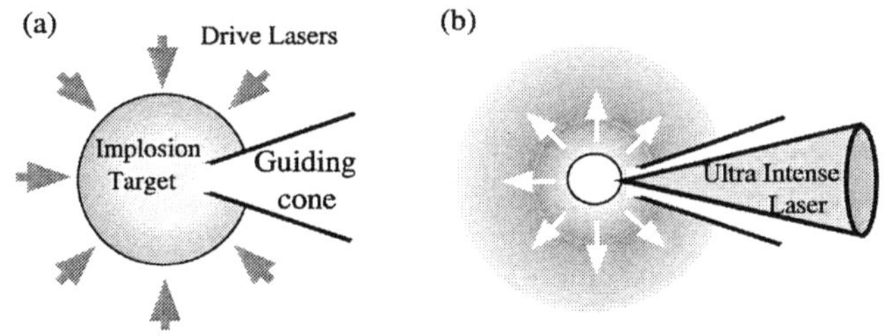

Figure 3. Schematic illustration of the cone-guided implosion. (a)Drive lasers irradiate the spherical target with a cone. (b) Ignition via additional heating by the fast electrons generated by the ultra-intense laser, which is propagated into the cone instead of the plasma channel.

The compressed core area is observed without short pulse injection, resulting in elliptical shape of $40 \times 25 \mu m^2$. The sequences of images of the core emission with and without the heating pulse were also taken with an X-ray flaming camera. The sequence when the short pulse is injected after the maximum compression of the shell clearly indicates the re-heating of the core by the short pulse after the brightness due to the compression is decreased. The size of the emission region at the peak of compression is about 30μm and increases to 50μm due to the injection of the heating pulse. These results coincide with the independent measurement of the heating region using the Al block.

Heating of the highly compressed plasma was measured through neutron spectroscopy for 2.45MeV thermonuclear DD neutrons. The neutron energy spectra were obtained using time-of-flight scintillator/photomultiplier detectors from two different angles[19]. Peak at 2.45MeV energy on the spectra is observed when the heating pulse was injected at maximum compression timing corresponding to a thermonuclear neutron yield of $2 \pm 1 \times 10^5$ neutrons. This neutron yield was more than 10 times above the numbers ($9 \pm 1 \times 10^3$) without or delayed injection of the short pulse. The efficiency of the energy coupling of the energetic electrons to the highly compressed plasma was estimated from the neutron yield and the heated volume inferred from the x-ray images. The heated mass is estimated from the density (50-70g/cc), the volume estimated by X-ray emission ($25 \times 40 \times 40 \mu m^3$). In order to increase the neutron yield by a factor of 10, a temperature at the region has to increase about 120eV (+/-20%) from the initial temperature regions of several hundred eV, which corresponds to 12-16J of the energetic electrons derived from the ultra-intense laser-matter interaction experiment using a plane target. The total coupling efficiency of the short pulse laser to the core could therefore be 20-27% with the estimation errors of 8% from the uncertainness of the temperature and the heated volume. This is a very encouraging result for the first demonstration of short pulse laser heating. In future experiments, we plan to increase the short pulse laser energy to see if the conversion efficiency remains high using the PW lasers at RAL and Osaka University. We also plan to measure the growth of any Kelvin-Helmholtz instability that develops on the edge of the cone that could mix high Z material into the compressed core plasma by placing suitable H-like and He-like emitting layer there and measuring them by X-ray spectroscopy.

3. NEUTRON SPECTROSCOPY IN ULTLA-INTENSE LASER-MATTER INTERACTIONS

We are also planning multi-channel neutron spectroscopy to measure the momentum distribution of accelerated ions inside the target. One possible way of measuring the ion momentum distribution is by direct observation of ions with plastic nuclear track detectors such as a CR-39. However, the ion motion is likely to be strongly affected by a large electric and/or magnetic fields in the plasma. Neutron spectroscopy, on the other hand, has the advantage that neither the target potential nor the self-generated magnetic fields can affect the motion of neutrons generated in the target. Unfortunately, the measurement of the neutron spectrum from just one direction does not give us the three-dimensional ion momentum distribution. It is necessary to measure the neutron spectra from at least two different directions to obtain the ion momentum distribution in the target.

Figure 4 (a) shows an example of neutron spectra generated by beam-fusion nuclear reactions from three different viewing angles[14]. The s-polarized, 1μm laser light was obliquely incident onto a CD 5μm target from 40 degree from target normal at an intensity of 10^{19}W/cm^2. The solid lines in all figures show the observed neutron spectra. Figure 4(a) shows experimental spectra at three observation angles: 90° or parallel to the target surface (solid line), 56° (dashed line) and 1° (dotted line) to the target normal, respectively. The peak of the spectra from the front side of the target shifts to lower energy (2.2MeV at 56°, 1.9MeV at 1°) from 2.45 MeV given by D-D thermal neutron energy while there is almost no shift in the spectra along the target surface direction (solid line). These shifts suggest that ions are accelerated into target inside taking account of the momentum conservation, which is similar to a Doppler effect on the neutron spectra. To investigate the precise momentum distribution of the ions, we performed 3D Monte-Carlo simulations to compare the calculated spectra with the experimental results. Figure 4(b) shows the well fitted calculated spectra at the same viewing angle as the experiments, which indicates that the ions are accelerated into the target, exactly opposite to the target normal direction. The ion distribution is shown in Fig.4(c) as a contour plot projected on the x-y plane from the 3-D distribution. The distribution collimated the target rear direction is given as a momentum ratio of $P_x:P_y:P_z=2.3:1:1$. The ion energy to the x-direction corresponding to the direction into the target is 330keV whereas the energy to the y and z-direction is about 70keV.

The neutron detector that we are re-constructing for the next generation experiments is the multi-channel spectrometer, LaNSA[15, 16], which originally consisted of 960 channels of scintillator-photo-multiplier detectors. In order to measure the neutron angular distribution, these modules will be divided into three separate units, each of which has 240 channels. The neutron detectors consist of an array of Thorn-EMI9902KB05 photo-multipliers coupled to BC505 liquid scintillator. The electrical signals are delivered to a Time-Digital-Converter (TDC), which records the arrival time of the signal via a discriminator. These timings are collected and then converted into a neutron spectrum through the time-of-flight (TOF) method. The data acquisition of the system is based on PC to control all CAMAC and Fastbus modules.

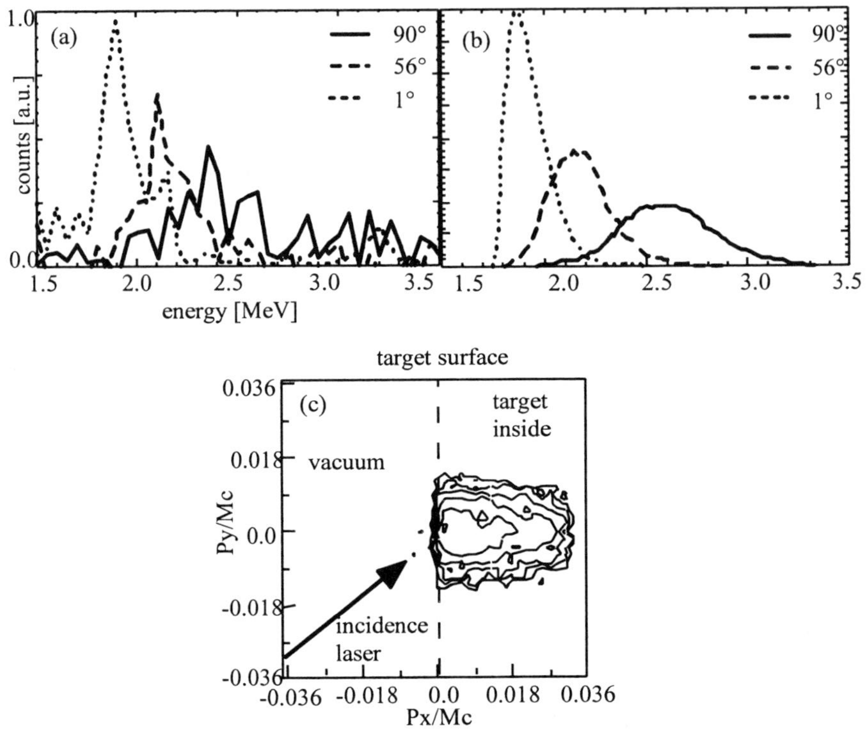

Figure 4. An example of (a) neutron spectra, (b) well-fitted calculated spectra and (c) the well-fitted ion momentum distribution[14].

These three modules will be positioned in target area PetaWatt (TAP) and target area west (TAW) of the VULCAN laser facility at the Rutherford Appleton Laboratory. Figure 5 shows an schematic image of the neutron array for one spectrometer in TAP together with an overview of the setup for PW laser and the neutron spectrometers. An image of chamber and the module at TAW is also shown in Fig.6. The neutron detector modules are represented by yellow box and the neutron flow is given by green gimlet area in both images. The distance between the two modules and the chamber center in TAP is about 12m and 5m. The distance of the array in TAW is about 8m from the chamber center.

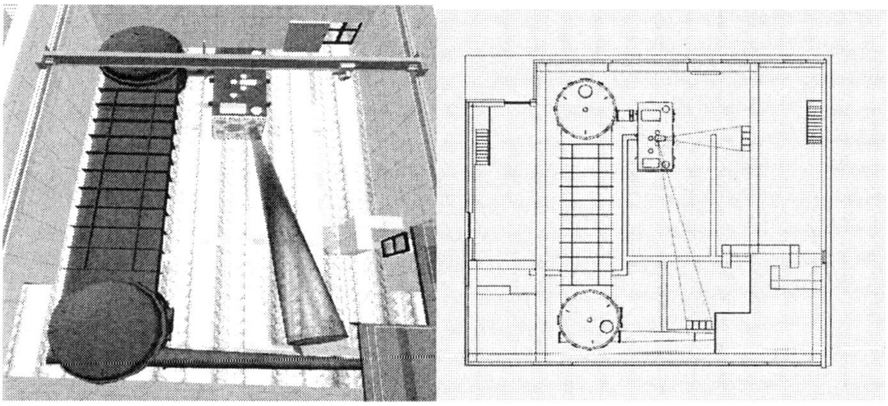

Figure 5. Image for one module at PW laser chamber room (TAP) and overview of PW laser and the neutron spectrometers.

The energy resolution for each module as a function of distance from the target is shown in Fig.7 for different neutron energies. The shorter the distance between the spectrometer and the focal position, the worse the energy resolution of the neutron spectrum becomes. For example, at 5m distance, the module has 300keV of energy resolution for 2.45MeV neutrons, whereas there is only 800keV resolution for 14MeV neutrons. On the other hand, at 12m distances, the energy resolution of the module becomes much smaller (100keV for 2.45MeV neutron and 350keV for 14MeV neutron). Therefore, the module that is nearest the target will be set up so that it views an angle that is perpendicular to the expected ion acceleration direction. By adjusting the target rotation and the incident laser direction, the energy resolution of both modules can remain at a comparably low level. On the other hand, the dynamic range to detect the neutrons can be set by adjusting the voltage of the photo-multipliers and the discriminator threshold.

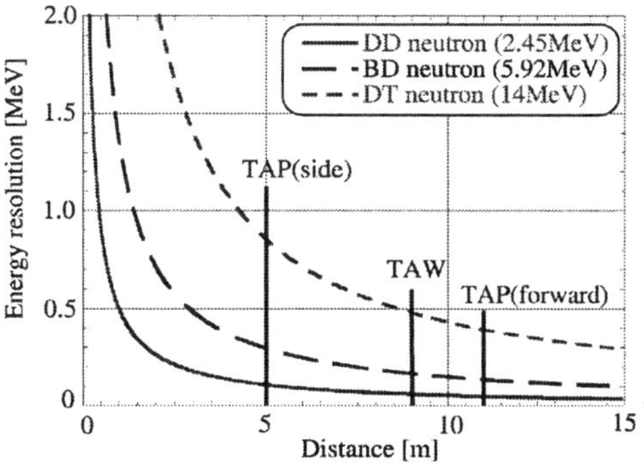

Left: **Figure 6**. Image of neutron detection at TAW. Right: **Figure 7**. Energy resolution for different neutron energy as a function of the distance between the spectrometer and laser focal position.

Using these neutron spectrometers, we plan experiments to measure the ion acceleration distribution for incident intensities up to 10^{21} Wcm^{-2}. It will be possible to change plasma density to determine the effect of scale-length on the ion acceleration mechanisms. Measurement of neutron spectra generated by a number of different nuclear fusion reactions has other interesting applications. In particular, B-D and Li-D reactions generate higher energy neutrons and have higher energy cross sections than those of the DD reaction - this may lead to the simultaneous observation of the high energy part of the ion momentum distribution.

ACKNOWLEDGEMENTS

The work is supported by the UK Engineering and Physical Science Research Council. The LaNSA array is on loan from the U.S. Department of Energy to the UK Ministry of Defense.

REFERENCES

1. B. A. Remington, D. Arnet, R. P. Drake, and H. Takabe, Modeling astrophysical phenomena in the laboratory with intense lasers, *Science* **284**, 1488-1493.
2. E. P. Liang, S. C. Wilks, and M. Tabak, Pair production by ultraintense lasers, *Phys. Rev. Lett.* **81**, 4887-4890 (1998).
3. J. Nuckolls, L. Wood, Z. Thiessen, and G. Zimmerman, Laser compression of matter to super-high densities, *Nature* **239**, 139-142 (1972).
4. J. Lindl, R. L. McCrory, and E. M. Campbell, Progress toward ignition and burn propagation in inertial confinement fusion, *Phys. Today* **45**, 32-40 (1992).
5. M. Tabak et al., Ignition and high gain with ultra powerful lasers, *Phys. Plasmas* **1**, 1626-1634 (1994).
6. B. J. Duda, R. G. Hemker, K. C. Tzeng, and W. B. Mori, A long-wavelength hosing instability in laser-plasma interactions. *Phys. Rev. Lett.* **83**: 1978-1981 (1999).
7. S. Hain, F. Cornolti, and H. Opower, Hydrodynamic models and schemes for fast ignition. *Laser and Part. Beams* **17**, 245-263 (1999).
8. S. Hatchett and M. Tabak, presented at the anomalous absorption conference, ocean city, May 2000, (unpublished).
9. P. A. Norreys et al., Experimental studies of the advanced fast ignitor scheme, *Phys. Plasmas* **7**, 3721-3726 (2000).
10. R. Kodama et al., Fast heating of ultrahigh-density plasma as a step towards laser fusion ignition, *Nature* **412**, 798-802 (2001).
11. M. I. K. Santala et al., Effect of the plasma density scale length on the direction of fast electrons in relativistic laser-solid interactions, *Phys. Rev. Lett.* **84**, 1459-1462 (2000).
12. C. Toupin, E. Levebvre, and G. Bonnaud, Neutron emission from a deuterated solid target irradiated by an ultraintense laser pulse, *Phys. of Plasmas* **8**, 1011-1021 (2001).
13. P. A. Norreys et al., Neutron production from picosecond laser irradiation of deuterated targets at intensities 10^{19} Wcm^{-2}, *Plasma Phys. Control. Fusion* **40**, 175 (1998)
14. H. Habara et al., in: *High-Power Lasers in Energy Engineering*, edited by K. Mima, G. L. Kulcinski, and W. Hogan, (Proceedings of SPIE Vol. 3886, 2000), pp. 513-520.
15. M. D. Cable, S. P. Hatchett, and M. B. Nelson, Neutron spectroscopy with a large neutron time-of-flight detector array (LaNSA), *Rev. Sci. Instrum.* **63**, 4823-4827 (1992).
16. M. B. Nelson and M. D. Cable, LaNSA:A large neutron scintillator array for neutron spectroscopy at NOVA, *Rev. Sci. Instrum.* **63**, 4874-4876 (1992).
17. C. N. Danson et al., Well characterized 10^{19} Wcm^{-2} operation on VULCAN – an ultra-high power Nd:glass laser, *J. Mod. Opt.* **45**, 1653-1669 (1998).
18. Y. Kato et al., Fast ignition and related plasma physics issues with high-intensity lasers. *Plasma Phys. Contril. Fusion* **39**, 145-151 (1997).
19. N. Izumi et al., A highly efficient neutron time-of-flight detector for inertial confinement fusion experiments, *Rev. Sci. Instrum.* **70**, 1221-1224 (1999).

RECENT EXPERIENCE AND FUTURE PLANS FOR JET DIAGNOSTICS

J. Sanchez[1] and contributors to the EFDA-JET work programme

1. INTRODUCTION

The operation of JET under the EFDA agreement has provided experience in two important areas concerning the future organisation of supra-national Fusion Facilities.

The scientific exploitation of the device during 2000 and 2001 has been done by scientists from the different European laboratories, with a significant role of the British UKAEA Association as Operator of the facility. The integration of the operation team with the visiting scientists has been quite successful, not only in the general areas of Tokamak Physics but also in the particular area of Diagnostics, which required a very close collaboration of both groups.

In addition to the operation of JET, a complex program to enhance the JET facilities (JET Enhanced Performance) is being prepared and could be implemented during the period 2001-2006.

The present paper will briefly discuss the aspects of the 2000-01 operation and enhancements (with emphasis on the diagnostics systems) and give an overall description of longer term enhancement project, focusing on the diagnostics. Given the broad spectrum of the activities described, involving the work of many persons, deep technical details will not be discussed but the author will try to provide a comprehensive list of references when available.

2. OPERATION OF JET DIAGNOSTICS DURING 2000-01

Under the EFDA agreement, JET operates as a "Users Facility". The Operator team takes care of all the basic elements of the machine operation, including heating and diagnostics systems, whereas the European Associations (including the UKAEA) determine the scientific programme of JET and provide the staff for the experiments and analysis, organised into "Task Forces".

The exploitation of the diagnostics system follows the same complementary organisation: Basic operation of existing diagnostics is carried by the Operator (up to

[1] Asociacion EURATOM/CIEMAT, 28040 Madrid Spain

the data acquisition, including calibration in many cases). The role of the visiting (or UKAEA) scientists, under Task Force Diagnostics (TFD) was:

To collaborate with Operator personnel in the operation of diagnostics.

To propose and perform diagnostic oriented experiments.

To provide validation and scientific analysis of the data.

To propose and construct new diagnostics or diagnostics enhancements (interface by Operator Team).

In parallel with the above explained philosophy, one of the main goals when preparing the campaigns was the formation of integrated teams, trying to avoid any mismatch between the members of the Operator Team (with an extensive JET experience) and the visitors (more than 60, in many cases visiting JET for the first time). The result was quite encouraging: thanks to the effort and enthusiasm from both sides the integration was excellent, leading to a successful operation of the diagnostics.

One of the advantages of the broad participation was that, for every system, we had the opportunity to get the visit of the best European (and US) experts. A number of systems like Li beam (edge density profile studies) [1] , IR cameras (heat loads to divertor during ELMs) [2] and Doppler spectroscopy for poloidal rotation studies [3], among others, benefited from this fact. Even if the general recommendation for visitors interested on diagnostics exploitation was to stay for full experimental campaigns, short visits were welcomed as a possible seed for a future deeper involvement of the particular experts in the JET programme.

Diagnostics oriented experiments, requiring devoted machine time, were grouped in four areas: q profile measurements by MSE (leading among other results to the streaking observation of a core region whit zero plasma current on ITB discharges) [4], Electric field measurements by MSE (trying to use the comparison of full and half energy beam components)[5], High Z ablation experiments (identification of high Z impurity lines as erosion markers in a hypothetical high Z ITER) and He beam spectroscopy experiments (by the use of a He doping system in one of the Heating Neutral Beams) [6]

Validation activities are organised by the Data Validation Coordination Committee, which runs monthly meetings. Many of the problems analysed there triggered working groups or actions involving TFD members. In particular we can mention the effort on: ECE-Thomson comparison (both in the aspect of geometrical interpretation of Te profiles as in the possible effects of non-maxwellian electron distributions), comparison of edge profiles from different diagnostics [7], alignment of CXRS data, comparison of Zeff obtained from different diagnostics (Visible Bremsstrahlung and CXRS), magnetic reconstruction issues (iron core, zero current at plasma centre) and the work done on validation of interferometry data (fringe corrections).

Finally, an ambitious programme of diagnostics enhancements has been prepared. Most of them lie within the JET EP project, described below, but also a number of shorter term enhancements have been developed during 2000-01 and will be available for the 2002 campaigns: the new microwave reflectometer for correlation studies of the density turbulence (US collaboration), the upgrade of the MSE system for operation with a new dedicated beam at 130 MeV (US collaboration), the development of new Langmuir probe heads for turbulence studies, the installation of a Quartz Microbalance [8] for deposition studies, the installation of high efficiency detectors for the edge LIDAR system and a number of refurbishments and maintenance actions undertaken by the Operator, i.e. the review and replacement of magnetic sensors.

3. THE JET EP PROJECT

The JET Enhanced Performance Project (JET EP) is a proposal for extending the JET operation beyond 2002, with a series of significant enhancements. The main physics objective is the support of ITER, trying to improve the subsystem design and the operation capabilities of the machine.

In order to accomplish the goals, the capabilities of the machine might be upgraded with additional heating systems (7.5 MW NBI, to be commissioned in 2002, 5 MW ECRH at 113 GHz and a new ITER-Like ICRH antenna, able to operate in ELMy H mode), a new Divertor, able to handle the extra power, (as well as high triangularity configurations, which presently would have the strike points outside the divertor plates) and a comprehensive set of new or upgraded diagnostics, which will be discussed below.

The present schedule considers commissioning of JET EP enhancements during 2004, though the detailed strategy (schedule and project content) might change in accordance with the overall planning of the European Fusion Programme for the period 2003-06 (6th Framework Programme).

3.1. Diagnostics for JET EP: Generation of the Proposal

In order to prepare a proposal on the diagnostics enhancements for JET EP a working group (Diagnostics Working Group, DWG) was formed. This group involved both diagnosticians and general plasma physicists from the different Task Forces, from the Culham and Garching Close Support Units and members of the ITER Diagnostics Expert Group.

The reference elements for the DWG were the Physics programme of JET EP (with a strong ITER-driven component) and the ITER needs in terms of diagnostics physics and technology. Proof of principle diagnostics would be considered only if ITER relevant.

A summary list of the urgent Physics developments required for ITER is given below (extracted from the document on European Domestic Assessment of the ITER FEAT design)

- High density operation, (high confinement), safe Divertor operation, n_e peaking
- β-limit in ELMy H-mode: Neoclasical Tearing Modes.
- ELMs control (for aceptable lifetime of plasma facing components)
- Disruption mitigation, halo currents, runaway electrons
- T retention: codeposition and removal
- Advanced scenarios, sustainability, influence of energetic particles
- Heating technology/physics
- Diagnostics technology/physics

On the other side, the framework constraints included, in addition to the usual procurement budget and manpower ceilings, the specific condition of JET as a consolidated Tritium-compatible device. This required reinforced attention to the interface impact: port allocations, knock on relocations of systems, in vessel work (tritiated environment), availability of in vessel wiring, remote handling compatibility, waste disposal and impact on the shutdown duration.

Within the terms of reference, the DWG analyzed the possible enhancements taken into account some facts to be weighted against the resources demanded:

- Fraction of the programme which would benefit from the enhancement
- Commissioning date: when could the programme start using the enhancement ?
- Technical risks (to be kept to a minimum for a device like JET)
- Uniqueness of JET as diagnostic bed, for ITER driven developments
- Availability of interested experts within the European programme
- Existence of non EU interested parties able to contribute equipment and expertise.

3.2. Proposal on Diagnostics for JET EP

More than 80 possible enhancement actions were considered, those proposals were provided by external sources (individual scientists, Associations, CSU,s ITER...) or generated by the DWG itself in order to give response to identified needs. Within the given constraints, only a fraction of the proposed actions could be recommended for implementation, whereas many proposals of significant scientific value had to be left out. The proposed enhancements are commented below.

3.2.1. Systems for Overall Support of the JET Physics Programme

3.2.1a. High Resolution Thomson Scattering. Working with the traditional 90° geometry this system would complement the existing LIDAR (core and edge) systems, providing coverage for the edge and ITB regions (in practical terms the whole outer radius of the machine) with 10 Hz repetition rate and 15 mm spatial resolution. The physics aims of the system would be the studies of ITBs (which show sharp n_e and T_e gradients) and H-mode pedestal/ELMs (a primary concern for the operation of ITER), in addition to an overall benefit to every JET experiment.

The reference proposal has been produced after a deep analysis by a group of Thomson Scattering experts. The analysis work and the details of the final proposal are described elsewhere on this publication [9,10].

3.2.1b. Upgrade Of The Heterodyne and new Michelson ECE Spectrometers. The existing ECE heterodyne system will be upgraded with 48 new channels, this will improve the coverage of the whole plasma radius under a great variety of configurations. (20-40mm, 10 μs resolution)

The new Michelson system (100mm, 5 ms resolution) will provide the coverage of the full ECE spectrum, allowing comparison of the different harmonics. This system might substitute the aged existing Michelson and would be useful as calibration standard. The system will also play an important role on high density DT discharges (1st harmonic under cutoff), where detailed time evolution of the central T_e is required to assess the alpha heating.

3.2.1c. Real Time Profile Analysis. The sustainability of ITBs needs the control on real time of the plasma profiles (being of capital importance the current profile control).

An ambitious programme has been already started for the integration of profile diagnostics in a real time analysis system (which will be used as the "sensor" part of a feedback loop, to be completed by "actuators" as gas injection or heating systems) [11,12]. During 2001 and 2002 data from a number of systems (ECE, polarimetry, FIR interferometry, CXRS, MSE and magnetics) will be included in the real time VME server. The project will require some moderate effort on hardware procurements and interfacing but will need a substantial effort to generate the fast algorithms (diagnostic specific) necessary for the real time analysis.

3.2.1d. Li Beam for Edge Density Measurements. The aim of this project will be the improvement of reliability and performance for the 50 kV Li beam, which already operated during the 2000-01 campaigns. The work to be done will include the development of a new source for the beam and the improvement of the analysis software. This diagnostic will play an important complementary role on the characterization of the edge density profile (relevant for ELM and pedestal physics).

3.2.1e. He Doped Beam for Edge n_e and T_e Measurements. This (modest cost) activity would be the continuation of the experiments performed during 2000-01 on the use of he doped heating beams as edge diagnostic. Preliminary results [6,13] show good prospects for n_e measurement, whereas the sensitivity of the method for T_e is smaller. The proposed upgrade involves the installation of He doping systems in both NBI boxes and the development of remote operation tools to make the diagnostic routinely available. Analysis will progress on the actual feasibility of this diagnostic, mainly in the weaker (presently) T_e side, which unfortunately can be hardly obtained from other diagnostics.

3.2.1f. Fast Data Acquisition System. The main fast Data acquisition system in JET is "CATS", a very sophisticated transputer-based system, able to operate with intelligent triggers, which has been of great help during the last ten years. Nevertheless, this system is difficult to upgrade and its performance has been surpassed by the simpler "crude force" modern systems, equipped with large memory and fast ADCs.

The aim of the new fast DAQ would be to provide full discharge coverage for the systems requiring high sampling rate (up to 2 Ms/s, 12-14 bit). Presently these systems are limited on the number of channels and must operate with "time windows" leading to loss of the relevant data if the events are not properly synchronised.

The project will use a Lynux based system, controlled by PCs with digitising boards able both to store the data locally (64 Ms) or to channel it into the PC memory for continuous sampling. Isolated inputs will be used for higher quality signal recording..

3.2.1g.Vertical Bolometer Camera. It will substitute the, presently unreliable, vertical bolometer. It will have a high density of channels viewing the Divertor region. This camera, together with the existing horizontal ones (to be relocated) and those located in the Divertor (to be refurbished) will complete a basic bolometry system to ensure the global radiation measurements, radial profiles and main plasma/divertor radiation ratio.

3.2.1h. Current Profile and E Field Diagnostics. The successful results of the existing MSE system will be complemented in two areas, the first one (already consolidated as a project) would be the upgrade of the poloidal rotation CXRS system for the measurement of radial electric fields, the second one (still under assessment) would be the development of a Li beam-based spectrometer for edge current studies. This system would use either traditional polarimetry of the Zeeman split lines or a new technique[14] based on line intensity ratio measurements. The measurement is considered of capital interest for edge stability studies, linked to ELM control.

3.2.1i. Core CXRS Upgrade. The final decision on upgrade of this key diagnostic is presently being refined. The main targets will be: improvement of s/n and time resolution, installation of a He dedicated spectrometer (determination of He^3 concentrations on minority heating experiments) and, possibly, improved resiliency to NBI unavailability (by having redundant systems looking to beams from both injectors).

3.2.1j. Magnetics Upgrade. The control of some of the proposed plasma shapes, as well as the improved configuration control which will be required for detailed analysis of edge stability (among other elements), recommend to refurbish and upgrade the aged system of magnetic sensors. A detailed study, with cost/performance optimisation of the possible upgrade, is underway.

3.2.1k. Divertor Diagnostics. The implementation of the new (Mark III High Power) Divertor will require the reinstallation of several diagnostics and will provide an opportunity for the refurbishment of several systems, in some cases with minor upgrades. The main systems considered are: Langmuir probes (90+30), Thermocouples (42), Bolometers (7 Cameras+ upper view KB5), Pressure Gauges (2), Quartz microbalances (2), Long term samples (mech. indexes, marker films, sticking boxes...), Thermal He beam, μw antennas for interferometry/reflectometry, Equilibrium and Halo Coils.

3.2.2. Specialised Systems

3.2.2a.TAE Antennas. TAEs are actively excited and the damping rate measured with magnetic sensors (also reflectometry, ECE) to the determine their stability properties. The proposal includes the installation of 8 antennas, able to excite the ITER relevant high n (10-15) modes. The system can indirectly determine q values and isotope ratio n_D/n_T (linked to the Alfven frequency).

3.2.2b. Time of Flight Neutron Spectrometer. To be used for 2.5 MeV DD neutrons, this system will be designed for high count rate [15], useful on high performance and ICRH discharges. Line integrated fuel Ti and flow rotation (if tangential view) will be provided.

3.2.2c. Vertical VUV spectrometer. It will substitute the existing vertical VUV spectrometer equipped with an unreliable vibrating mirror. Resolution on space, wavelength and time will be gained by using a CCD detector.

3.2.3 ITER Relevant Systems (urgent measurement or technology development)

3.2.3a. Halo current sensors. Very important to determine the possible effect of disruptions on ITER (forces). Detailed design of coils and locations is underway.

3.2.3b. Microwave Access. For Reflectometry and ECE, using corrugated waveguides (38 mm diam) for broadband operation (the reference solution for these diagnostics on ITER). They will be complemented with quasioptical launchers able to focuse the beams.

3.2.3c. Infrared Viewing system . The main goal will be the detection of thermal loads to the first wall, as possible side applications: alpha particle losses (geometrical distribution) and synchrotron radiation from runaways (disruptions) are also considered. The system will include zoom capabilities and a fast camera will be used for 100μs resolution.

3.2.3d. Tritium retention diagnostics. Very urgent ITER need. In addition to the already programmed long term samples and quartz microbalances, shuttered samples, spectroscopy systems and other novel techniques will be discussed for further implementation.

3.2.3e. Mirror Tests for ITER. Technology task to assess the effects of the plasma erosion and redeposition on plasma facing mirrors for ITER. Samples will be placed on the JET Divertor, with different orientations and, possibly, shutters to select type of plasma.

3.2.4. DT Diagnostics

JET has already a comprehensive set of DT diagnostics, tested during the DTE1 campaign. For a possible DT campaign on JET EP, a few complementary systems are considered.

3.2.4a. Lost Alpha Detectors. Either with Faraday cup or scintillator box technology, should be able to determine the energy of the escaping alphas. Also an important test for ITER.

3.2.4b. Magnetic Proton Recoil Spectrometer Upgrade. This would improve background noise and extend to 2.5 MeV the successful existing 14 MeV MPR system [16]. With this upgraded capabilities the system will aim to the analysis of the alpha knock on spectrum (linked to the pressure of fast alpha particles)[17] and the comparison of DD to DT peak intensity (for isotope ratio measurement).

3.2.4c. Neutral Particle Analyzer Upgrade. The upgraded high energy NPA system could be useful for alpha particle distribution measurements (also a new detector able to discriminate alphas from fast deuterons is included in the design) with a rough spatial resolution (achieved by the implementation of three channels along the plasma radius).
One of the goals of the system would be the analysis of alpha particle redistribution by MHD phenomena.

4. ACKNOWLEDGEMENTS

The author is deeply indebted to many collaborators: TFD and Operator team members, components of the JET EP Diagnostic Working Group and proponents of the more than 80 proposals received, many of them with a substantial background work behind.
This work has been conducted under the European Fusion Development Agreement

5. REFERENCES

1. M. Brix, A. Korotkov, M. Lehnen, P. Morgan et al., Determination of edge density profiles in jet using a 50 kV Lithium beam, 28th *EPS Conference, Madeira (2001) P1 101*

2. T. Eich, A. Herrmann, V. Riccardo, P. Andrew et al., Analysis of power deposition in JET MKII-GB divertor by ir thermography, 28th *EPS Conference, Madeira (2001) P5 010*

3. F. Sattin, Y. Andrew, C. Giroud, N. Hawkes et al., First core poloidal flow velocity measuremants in JET, 28th *EPS Conference, Madeira (2001) P1 096*

4. N. C. Hawkes, B.C. Stratton, T. Tala, C.D. Challis, et al., Observation of zero current density in the core of JET discharges with lower hybrid heating and current drive, *Phys. Rev. Letters **87**, 115001 (2001)*

5. S. Reyes Cortes, N. C. Hawkes, P. Lotte, C. Giroud et al., Techniques for measuring the plasma radial electric field using the motional Stark effect diagnostic at JET, *Submitted for publication to the Czechoslovak Journal of Physics*

6. M. Proschek, M. Brix, H.D. Falter, H. Anderson et al, Line emission of fast helium beams for fusion plasma diagnostics, 28th *EPS Conference, Madeira (2001) P1 100*

7. M. N. A. Beurskens, E. Giovannozzi, J. Gunn, Y Andrew et al., Analysis of plasma edge profiles at JET, 28th *EPS Conference, Madeira (2001) P3 084*

8. G.F. Neill, J.P. Coad, H.G. Esser and D.J. Wilson, High temperature quartz microbalance for the measurement of deposition inside an experimental fusion device, *HITEN Conference, Oslo, 5-8 June (2001)*

9. C. Gowers, C.J. Barth, R. Behn, M. Beurskens et al., High resolution Thomson Scattering on JET- an assessment of the feasibility, *These proceedings.*

10. P. Nielsen and R. Pasqualotto, High resolution Thomson scattering for JET, *These proceedings.*

11. D. Mazon, X. Litaudon, M. Riva, G. Tresset, et al., Real-time plasma control of internal transport barriers in JET, 28th *EPS Conference, Madeira (2001) P2 0 11*

12. G.Tresset,D.Moreau,X.Litaudon,C.D.Challis et al. Characterization of internal transport barriers in jet and simulations of Control Algorithms, 28 th *EPS Conference, Madeira (2001) P2 0 14*

13. M.G. O'Mullane, Y. Andrew, M. Brix, C. Giroud, et al., Diagnostics based on Helium neutral beams in ITER, *These proceedings.*

14. A.A. Korotkov, K.McCormick, P. Morgan, J. Scheweinzer et al., Line ratio method for poloidal magnetic field measurement using Li multiplet ($2\,^2S$-$2\,^2P$) emission, *These proceedings.*

15. A. Hjalmarsson, S. Conroy, G. Ericsson, G. Gorini, et al., The 2.5 MeV neutron time-of-flight spectrometer for optimized count rate (TOFOR), *These proceedings*

16. G. Ericson, S. Conroy, G. Gorini, H. Henriksson et al., The neutron magnetic proton recoil spectrometer upgrade (MPRu), *These proceedings.*

17. J. Kallne, E. Ballabio, J. Frenje, S. Conroy et al., Observation of alpha particle "knock-on" neutron emission from magnetically confined DT plasmas, Phys. Rev. Letters **85**, 1246 (2000)

STRATEGY OF THE TORE SUPRA DIAGNOSTIC SET FOR STEADY-STATE HIGH POWER OPERATION

C. Laviron*

1. INTRODUCTION

In most tokamaks as well as in other fusion devices, the set of necessary and useful diagnostics is mainly determined from the plasma parameters to be measured. For next step devices, steady-state high power operation is foreseen, and the diagnostics will increase their role for other missions, such as machine operation or safety control.

High power operation induces possible high thermal and neutron fluxes on diagnostic components close to the plasma. Having high power together with steady-state operation adds a new constraint: not only the thermal flux, but the integrated thermal load on components has to be considered. Such constraints have been examined for ITER [1] for its design, but they have to be applied now to TORE SUPRA which is operational.

TORE SUPRA is a tokamak with superconducting magnetic coils which has originally been designed for long pulse operation. The initial nominal value was 30 seconds when it started in 1988. Nevertheless, this duration has been exceeded when it reached a record pulse of 120 seconds with a controlled total energy of 280 MJ injected and extracted[2]. TORE SUPRA is now facing a major upgrade, with the installation of new internal components[3]. The design is such that the machine would be able to handle a power up to 25 MW for a pulse length of 1000 seconds.

As TORE SUPRA is a machine without tritium, the neutron load is quite low, and there is no need for remote handling. The main consequence of high power operation for long durations is that the total integrated energy has to be considered. A power of 25 MW for 1000 s would give a total energy of 25 GJ in the machine, which has to be injected, controlled and extracted. Compared to the present record of 280 MJ, this is two decades more ambitious in terms of thermal load.

Within this context of high power density for a long time, TORE SUPRA diagnostics have been reassessed, and the priorities were redefined.

---

* Clément LAVIRON, Association Euratom-CEA, Département de Recherches sur la Fusion Contrôlée, CEA/Cadarache, 13108 Saint Paul lez Durance Cedex, France

2. CONSTRAINTS

Steady-state high power operation of fusion plasmas implies severe constraints on diagnostics. Compared to shorter high power pulses, thermal load on components for long durations has to be accounted not only in terms of maximum power but also of integrated energy. Numerical simulations have been made for windows, mirrors and mechanical supports : by example, for a total radiated power of 10 MW, the thermal flux could reach 32 kW/m^2 on a window facing the plasma at a distance of 80 cm, which cannot be sustained for 1000 seconds without damage. Solutions have been found with intermediate actively cooled mirrors and/or tubes protecting some windows resulting to a reasonable temperature increase (less than 200°C after 1000 s). A test bed facility has been developed to validate these solutions, and drastic control procedures are also made before installation in the vessel, as any single water leak would be very dramatic for the experimental campaign.

Another important point to be considered is the data management system, taking into account the fact that the duration of the discharge is long at human scale, that many kinds of feedback loops have been developed on TORE SUPRA to ensure a good control of the plasma, and that the data has to be accessed in real time by the physicists. For all these requirements, a new data management system has been developed[4], with direct exchange of information at low level between diagnostics, together with real time transfer, real time processing and real time availability and display.

3. TORE SUPRA DIAGNOSTICS

Within this objective of steady-state high power operation, Tore Supra diagnostics have been classified in four groups, depending of their main purpose.

The first group corresponds to diagnostics which are essential to make and sustain a plasma. Without them, no plasma would be possible, and the machine useless. In this group, two important plasma diagnostics have been placed: magnetics and interferometry. Strictly speaking, it would be possible to make a plasma without these diagnostics, but it would be uncontrolled and at low performance (far from our "steady-state high power operation" objective). In addition, some tools (not strictly plasma diagnostics) necessary for the pilots are joined to this group, such as a CCD camera and a mass spectrometer.

Magnetic measurements are necessary for plasma control in terms of shape, position and plasma current. They also provide additional information (stored energy, MHD, ...), which could be classified within an another group of diagnostics (see below). On Tore Supra, the magnetic measurement system has been entirely renewed[5] for an optimized adaptation to the new geometry and with enough redundancy to allow failure of some sensors..

For interferometry, only the central vertical chord is considered as essential, as it is used for the density control feedback. In steady-state operation, a good particle control is essential because the pumping conditions and the recycling on the plasma facing components can vary strongly during the pulse (due to wall saturation and components temperature change). This diagnostic has been completely rebuilt and Tore Supra is equipped with five vertical chords[6], coupled with polarimetry measurements.

The second group of diagnostics include all those necessary to ensure a safe operation, without creating any strong damage to the machine. If the essential systems listed in the first group are more important, because needed to control a plasma, this second group is strictly necessary for a relevant experimental programme: it is not possible to aim at performing plasmas without the safety diagnostics. Five of them have been placed in this category: neutron monitoring, infra-red thermography of limiters, infra-red monitoring of neutralisers, ripple loss measurements and calorimetry.

Taken as a safety diagnostic, neutron monitoring controls the neutron flux for human and material protection. In addition, it is also used for physics purposes such as ion temperature.

One of the main plasma facing components of Tore Supra in its new configuration is the toroidal pump limiter (TPL), with the ability to stand 15 MW of convected power for 1000 seconds. The safety of this TPL relies on a set of infrared cameras viewing the plasma through endoscopes[7]. This diagnostic is a good example of all constraints : optical components very close to the plasma are maintained at 20°C through a water cooled internal system, while external components evacuate high thermal fluxes through a second layer of water cooling system (200°C, 30 bars). It has to be robust and allow a feedback action on the plasma (shape, position heating). This diagnostic is also used to monitor the additional heating antennas (ICRH and LH) and their thermal protections.

The neutralisation part of this TPL is in a very tight zone, and cannot be observed by IR cameras. A separate diagnostic, based on IR fibers and bichromatic pyrometers[8], is developed to monitor these neutralisers which can locally handle a flux of 20 MW/m^2.

Due to its superconducting toroidal coils which are very compact, the toroidal field encounters a noticeable ripple at the plasma edge. This ripple induces particle losses in the vertical ports which can be high for some plasma conditions. Ripple losses are measured by the intermediate of an array of twelve graphite collectors in a vertical port, each collector being related to a known vertical chord across the plasma by the drift trajectory of the localized particles[9].

Calorimetry, with about 150 temperature and 50 flow measurements, ensures the safety of all actively cooled components. It is also used for analysis of plasma power and energy balance, and is therefore considered as a plasma diagnostic.

Being used for control and safety, all diagnostics of these two groups need to be in the feedback loop system.

The third group of diagnostics include those necessary for a basic monitoring of the plasma. They are not necessary for safely making a plasma, but the plasma would have no interest without measurement of these parameters. In this group, UV spectroscopy, Zeff, Hα, Thomson scattering and bolometry (2 cameras) have been selected. With the combination of the first two groups, electron temperature and density profiles are obtained, with information on radiation, impurities and plasma equilibrium. This provides the basis for energy balance and other standard studies.

Finally, the fourth group of diagnostics include those who give additional and useful information, but which are not essential by themselves. They are linked to specific physics programmes, or used to complete and improve the knowledge of important parameters measured by diagnostics listed above. The choice of the diagnostics of this group obviously depends on the scientific needs, but with some weighting in priority from other considerations such as what already exists, the budget and manpower for renewing, the technical constraints due to steady-state high power, etc.

After the long shutdown from the end of 1999 to this summer 2001, many if not all diagnostics have been subjected to strong modifications, and only part of them are operational for the 2001 campaign: ECE radiometer, X-crystal spectroscopy (mainly for ion temperature measurement), soft X-ray tomography, hard X-ray tomography, reflectometry, polarimetry, barometry, Langmuir probes (fixed in many places and pop-up in the TPL), reciprocating probe, bolometry (4 additional cameras), Zeff (full profile), neutral particle analyser. Other diagnostics are due for 2002 and 2003 campaigns.

4. CONCLUSION

TORE SUPRA is now facing a very important challenge, with the objective of very long steady-state high power discharges. The associated severe environmental constraints are taken into account, and in particular the thermal load on diagnostic components. In addition, extension in the feedback system and the possibility to access the data during the pulse are implemented.

Therefore, diagnostics have been revisited, and priorities redefined, ordered in different groups: essential diagnostics needed to make and sustain a plasma, safety systems to avoid creating any strong damage, basic diagnostics for a minimal plasma monitoring, and others dedicated to the programmatic purpose.

REFERENCES

1. A. Costley, Diagnostic requirements and issues for next step burning plasma experiments, these proceedings.
2. X Litaudon et al, Stationary regimes of improved confinement in Tore Supra, *Plasma Phys. Control. Fusion* **38**, A251–A268(1996)
3. P. Garin, CIEL in Tore Supra: how to master power and particles on very long discharges, *J. Plasma Fusion Res. SERIES* **3**, 10-15 (2000).
4. B. Guillerminet, M. Le Luyer and B. Rothan, Tore Supra data acquisition : A system for long duration discharges, Fusion Engineering and Design, **43**, 259-264 (1999)
5. P. Moreau, P. Defrasne, E. Joffrin and F. Saint-Laurent, New magnetic diagnostic for the CIEL project on Tore Supra, *Proc. of 21st Symposium On Fusion Technology* (2000)
6. C. Gil, D. Elbeze and C. Portafaix, Development of the IR interfero-polarimeter for long pulse operation at Tore Supra, *Proc. of 21st Symposium On Fusion Technology* (2000)
7. D. Guilhem et al, Infrared surface temperature measurement for long pulse ; real time feed- back control in an activelcooled machine, *Rev. Sci. Instrum*,**70**(1), 427-430 (1999)
8. R. Reichle et al, Thermography of target plates with near-infrared optical fibres at Tore Supra, *Journal of Nuclear Materials*,.**290-293**, 701-705 (2001)
9. V. Basiuk et al, Profile measurements of localized fast electrons and ions in Tore Supra, *Proc. of Int. Conf. on Plasma Physics, Innsbruck*, 16C(I), 175-178 (1992)

IMAGING OF PLASMAS USING PROTON BEAMS GENERATED BY ULTRA-INTENSE LASER PULSES

M.Borghesi, D.H.Campbell, R.J.Clarke, M.Galimberti, L.A.Gizzi, M.Haines, A.J.MacKinnon, A.Schiavi, O.Willi*

INTRODUCTION

During the interaction of ultraintense laser pulses with plasmas and solid targets, a considerable fraction of the laser energy is deposited into highly energetic charged particles. One of the most exciting results recently obtained in this area of research is the observation of very energetic beams of protons, generated during the interaction of ultraintense short pulses with solid targets. In a number of experiments, performed with different laser systems and in different interaction conditions, protons with energies up to several tens of MeV have been detected behind thin foils irradiated with high intensity pulses[1]. In these experiments it was seen that the particle beams are directed along the normal to the back surface of the target, and have a small angular aperture at the highest energies. As proton beams are observed even using target which nominally do not contain hydrogen, protons are thought to originate from hydro-carbon impurities located on the target surfaces[2] or from bulk contamination of the target. Proposed theoretical models indicate that the protons gain their energy from the enormous electric field (~MV/micron) set up by laser accelerated fast electrons via space-charge at the back target surface[3]. As protons are easier to accelerate than heavier ions, protons are the dominant component of the ion emission.

The particular properties of the proton beams (small source size, high degree of collimation, short duration, energy dependence on the target characteristics, ease of synchronization with laser pulses) make them very suitable for application as a probe in dense plasmas investigations. Beside the standard radiographic use, i.e. the detection of aerial density variations[4], proton beams can also be used to detect electric and magnetic

* M.Borghesi, Department of Pure and Applied Physics, the Queen's University of Belfast, Belfast (UK); D.H.Campbell, A.Schiavi, M.G.Haines, Imperial College of Science, Technology and Medicine, London (UK), O.Willi, Institut für Laser- und Plasmaphysik, Heinrich-Heine-Universität, Düsseldorf (Germany), L.A.Gizzi, M.Galimberti, IFAM-CNR, Pisa (Italy), A.J.MacKinnon, Lawrence Livermore National Laboratory, Livermore (CA, USA)

Advanced Diagnostics for Magnetic and Inertial Fusion, Edited by Stott *et al.*
Kluwer Academic/Plenum Publishers, New York, 2002

fields generated during the interaction of intense laser pulses with plasmas. In this context, proton imaging can be developed as a diagnostic with great potential for Inertial Confinement Fusion (ICF) studies. For example, the possibility of accessing directly electric field distributions in dense plasmas may shed new light on issues as hydrodynamic and electromagnetic instabilities highly detrimental for ICF. The technique has great potential for access to the complex and yet unexplored electric and magnetic field distributions in indirect drive target assemblies. In a Fast Ignitor context, it can contribute to the study of the large transient electric and magnetic fields generated in high-intensity laser-matter interactions. The magnitude and characteristics of these fields are currently the subject of much conjecture in the short pulse, high-intensity plasma physics community. In this paper the results of an experimental campaign carried out at the Central Laser Facility, Rutherford Appleton Laboratory will be presented and discussed, in which the proton source has been applied for the first time as a particle probe in a range of laser-plasma interaction conditions[5].

EXPERIMENTAL SET-UP AND DIAGNOSTIC ARRANGEMENT

The VULCAN Nd:glass laser operating in the Chirped Pulse Amplification mode (CPA) was used in the experiment. The main targets (used for the production of the proton beams) were Al foils, 1-2 mm wide and 3 - 25 μm thick. The 1.054 μm CPA interaction pulse, 1 ps in duration, with energy up to 100 J was focused by an F/3.5 off-axis parabola (OAP) onto the centre of the main target. The incidence was about 15° off normal. There was the possibility, exploited in the final part of the experimental campaign, of splitting the main CPA pulse in two separate high-intensity pulses, and to focus them employing two separate F/3.5 Off-Axis Parabolas. The focal spot varied between 8 and 10 μm in diameter at full width at half maximum (FWHM), containing 30-40 % of the energy, and giving intensities up to $5\text{-}7 \times 10^{19}$ W/cm^2. Heating pulses with duration 0.5-1 ns were also available, and were used to provide, in some parts of the experiment, intensities on target in the range $10^{13}\text{-}10^{15}$ W/cm^2. A low-energy fraction of the main CPA pulse was frequency quadrupled and used as a transverse optical probe pulse. A summary description of the beam arrangements used in the various experimental tests performed will be given at the beginning of each experimental section.

The detector employed in the whole campaign consisted of a stack of several layers of radiochromic (RC) film. The film consists of 270 μm thick plastic containing a double layer of organic dye, which reacts to ionizing radiation[6]. The equivalent dose of energetic protons stopped in the film can be measured from the changes in optical density undergone by the film, yielding information on the number and energy of the protons. By using them in a stack, each layer of film acts as a filter for the following layers. Since protons deposit energy mainly in the Bragg peak at the end of their range, each RC film layer spectrally selects a narrow range of proton energies. 25-50 μm Al filters were placed in front of the first layer of film giving a minimum detectable proton energy of about 3 MeV.

The laser-produced proton beam was characterized in view of its application as a particle beam probe. The protons were produced by focusing the CPA pulse onto the Al foil, and detected by placing the RC film stack at the back of the target, typically at a distance of about 2 cm. It was found that the proton beams have high brightness, typically with 10^{12} protons with energy above 3 MeV per shot (at a laser irradiance of the order of 5 10^{19} Wcm^{-2}). As observed in previous experiments, the beams were highly directional, propagating along the normal to the back surface of the target with small angular divergence (about 15° for 10 MeV protons). The response of each RCF layer to high-energy protons has been determined using Monte Carlo modelling of the deposited energy by the particles[7]. Each film shown responds to ions within an energy band of less than 1MeV. For I ~5-7 10^{19} W/cm^2 a strong proton signal was observed out to the layer corresponding to proton energy just above 20MeV. The energy deposited within each film layer can be extracted from the absolutely calibrated film. By doing this for each film layer and fitting an exponential energy spectrum, an estimate of the proton energy spectrum and total energy could be extracted from the film data[7]. The mean proton energy of the exponential fit is 3.75MeV ± 0.3MeV with a total energy of 2.2J in an equivalent Maxwellian with a temperature of 2.5MeV. During the experiments the source size of the proton beam was estimated using a penumbral edge method, setting an upper limit of 15-20 μm diameter for the source of 10 MeV protons (there is debate if this is the real or just the apparent source size).

The main reasons making laser-produced proton beams particularly interesting for probing applications are: the small source size, leading to high spatial resolution in imaging applications; the short pulse duration, measured to be a few ps in our conditions; the low divergence and high brightness of the beam; the long stopping range of 10-30 MeV protons, i.e. the possibility of penetrating reasonably thick objects. In addition, high spatial and spectral resolution is guaranteed by an appropriate choice of detector (e.g. RCF stacks), and the proton beam characteristics can be tuned by suitably changing the laser and target parameters. The temporal resolution obtainable with these beams is far beyond the possibilities of beams provided by conventional accelerators. Further, as it will be seen in the following sections, the fact that the beams are not monochromatic is far from being a disadvantage for the proposed applications. Finally their ease of use and synchronization is a clear advantage in laser-plasma and Inertial Confinement-Fusion related experiments.

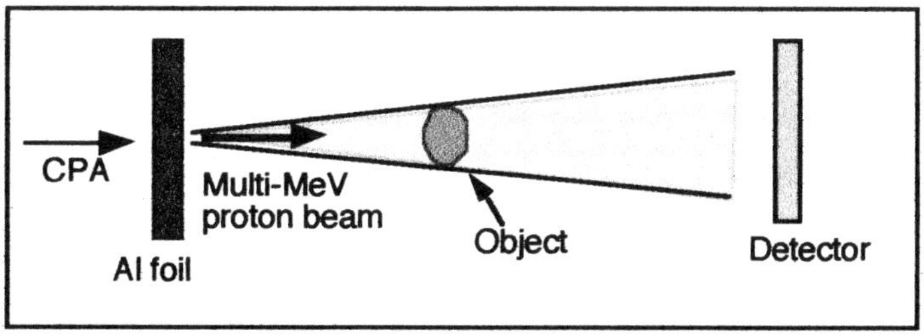

Figure 1 - Schematic arrangement for proton imaging.

Thanks to the small size of the source, we were able to implement a scheme of point projection imaging, with the object placed between the proton source (i.e. the back side of the primary target) and the detector (see fig. 1). The magnification of the image was determined by the ratio of the detector-to-object and object-to-source distances

IMAGING OF THIN CHARGED OBJECTS

When thick solid obstacles were placed in the beam a shadow of the object was observed in the proton beam, due to collisional stopping in the target. However, a shadow was observed even when objects with a thickness much smaller than the proton penetration depth were used, e.g. meshes formed by 10 μm Cu wires with 25 μm spacing, or by 25 μm Cu wires with 100 μm spacing. The meshes were parallel to the main target, i.e. perpendicular to the proton beam. The distance between the proton source and the mesh was 1 mm, while the first of the radiochromic film layers was positioned at 22 mm from the source.

A shadow of the mesh was observed on the radiochromic film, as shown in fig. 2(a). The image shown in the picture has been obtained with 15 MeV protons and the 25 μm spacing mesh. Fig.2(b) shows the optical density lineout across the shadow of a mesh with 100 μm spacing, obtained with 20 MeV protons. The optical density modulation observed is of the order of 0.2. It should be reminded that the stopping range of 20 MeV protons in Cu is about 800 μm, and the collisional energy loss for such energetic protons in 25 μm of Cu is just $\Delta E/E \sim 2\ 10^{-4}$. Such a variation is too small to produce any detectable effect on our diagnostic. Under these conditions scattering is also unimportant, therefore the shadow observed must be due to some different effect. Our present explanation is that the mesh charges up positively, due, for example, to hot electrons preceding the proton beam and expelling cold electrons from the target.

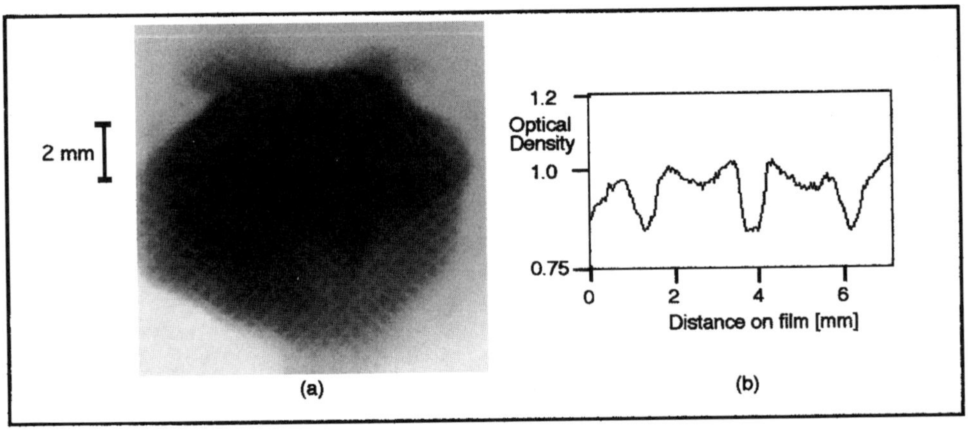

Figure 2 - (a) Image of a mesh with 25 μm spacing (10 μm Cu wires) obtained with 15 MeV protons; (b) Lineout across image of a mesh with 100 μm spacing (25 μm Cu wires) obtained with 20 MeV protons.

The deflection of the protons due to the electric field in the proximity of the wires is then responsible for the creation of the mesh shadow on the detector, where pile-up of the deflected protons is observed beside the regions of minimum transmission. An estimate of the linear charge λ on the mesh can be obtained via particle tracing, giving $\lambda \sim 10^{-6}$ C/m.

IMAGING OF LASER-PRODUCED PLASMAS

The sensitivity of the proton beam to electric fields, demonstrated in the measurements reported in the previous section, makes proton imaging a valid and unique diagnostic for investigating electric field distributions inside plasmas. The diagnostic potential of this application was tested in part of our experimental campaign. A preformed plasma was created by irradiating a secondary target with 600 ps, 0.527 μm heating pulses, prior to the CPA pulse interaction with the primary target. Both *face-on* (proton propagation axis parallel to the main plasma axis) and *side-on* (proton propagation axis perpendicular to the main plasma axis) geometry was used.

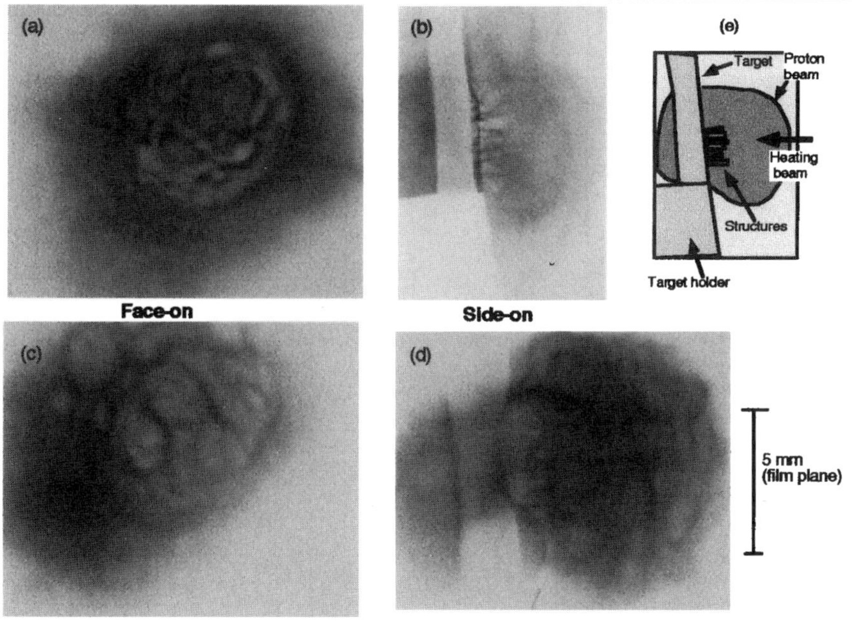

Figure 3 - Face-on and side-on proton images of plasmas produced by laser irradiation of 25 μm foils: (a),(b): Heating focal spot : 100 μm diameter with no beam smoothing, irradiance: I ~ 5 10^{14} W/cm^2; (c), (d) Heating focal spot: 250 μm diameter with beam smoothing (Phase Zone Plates) I~ 10^{14} W/cm^2. (e) Schematic showing the main characteristics of the side-on images.

The proton beam was produced as usual from a 25 μm Al foil, and the secondary target also consisted of a 25 μm Al foil. The heating irradiance on target varied between 5 10^{13} and 10^{15} W/cm2, with typical delays between the plasma formation and the CPA interaction with the primary target in the range 200-800 ps. When no preformed plasma was produced onto the secondary Al foil, the proton beam was, as expected, unperturbed, except for some charging-up-induced deflection at the edge of the foil. When the secondary foil was laser-heated the situation changed, and, under certain conditions, the presence of the plasma perturbed significantly the intensity distribution of the proton beam. Most importantly the perturbations imposed on the intensity profile of the proton beam appeared to depend strongly on the plasma conditions.

As a matter of fact, changing the irradiation parameters led to striking differences in the modulation pattern superimposed onto the proton beam. An example of this is given in fig. 3, where the RC film images pertaining to different irradiation conditions are shown. The structures observed differ markedly between the case in which no beam smoothing is applied to the beam, and the case in which Phase Zone Plates (PZP)[8], producing an ordered hot spot structure in the focal spot profile, are employed. In particular side-on images indicate clearly the presence of chaotic structures in the former case, while a regular pattern is observed in the latter. In all cases these structures were only observed within the duration of the heating pulse, and disappeared at later times. We believe that these structures are caused by electric fields present inside the plasma, which are strictly related to the laser intensity distribution. Detailed analysis and modelling is undergoing.

IMAGING OF ULTRAINTENSE INTERACTIONS

In part of the experimental campaign, proton imaging was applied to the study of ultraintense laser-plasma interactions. These measurements took advantage of the unique possibility of employing two temporally and spatially independent high-power CPA pulses. The second pulse (CPA_2) was split off the main pulse (CPA_1) before amplification. Due to set-up constrains, the energy content of each pulse was about 20 J, focusable to intensities in the range 0.5-1.0 10^{19} W/cm^2. The temporal separation of the two pulses could be varied shot-to-shot. The two beams were focused from different directions with separate f/3.5 optics. The CPA_2 pulse was used to produce a proton beam from an Al foil, and the proton beam was used to probe, transversely, the interaction of the CPA_1 pulse with solid targets.

It should be noted here that, when probing an object undergoing fast changes, the fact that the proton beams are not monochromatic is actually an advantage, as it can be exploited to obtain, in a single shot, information on the temporal history of the object to be probed. In fact, if the object to be probed is situated at a finite distance from the source, protons with different energies will reach the object at different times. By employing a detector that performs spectral selection, the information on the temporal evolution of the target can be retained. In particular, proton imaging was employed to observe the fast evolution of the electric fields following the irradiation of solid targets (wires, foils and micro-balloons) with the CPA_1 pulse.

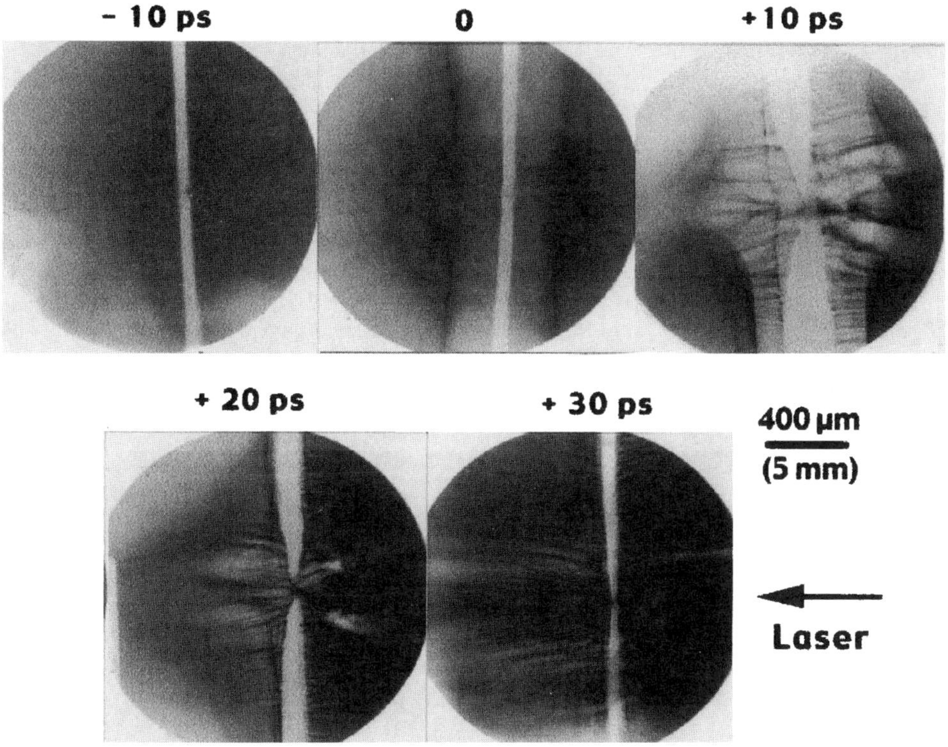

Figure 4 - Temporal evolution of proton images of the interaction zone for different interaction-probing delays. The target was a 50 μm Ta wire and the energy of the protons employed was 6-7 MeV. The delay between CPA_1 interaction and probing is indicated above each of the pictures. Spatial scales for both the target and the image plane (between bracket) are indicated.

Particularly dramatic effects were observed probing the interaction of the CPA_1 pulse with small diameter metal wires. Fig.4 shows proton images of the interaction with a 50 μm Ta wire. The RCF layers shown (second layers of the RCF stacks, corresponding to proton energies of 6-7 MeV) have been obtained in separate shots for different CPA_1-CPA_2 delays. When the proton probe arrived on target before the CPA_2 interaction pulse (e.g., t=-10 ps, first image in fig.4), the shadow of the Ta wire, thick enough to slow down the protons can be seen, with some small effect visible in the interaction region due to preplasma present ahead of the interaction. However when the probe is coincident with the interaction a dramatic effect is observed, with the protons being deflected away from the wire surface, which charges up due to hot electron expulsion. The charge is seen to decay in a few tens of picoseconds, as filamentary structures (horizontal striations) are seen to appear. The appearance of these structures is consistent with the growth, at the surface of the target, of an electromagnetic heat-flow instability, such as for example the electro-thermal instability[9].

The instability causes filamentation of the hot current, and as the filaments are magnetized, the effect is imprinted on the proton beam via the radial electric fields present inside the filaments. The onset of pronounced symmetrical horizontal features above and below the interaction axis is also detected. Such structures may be related to the radial expansion of shocks following the interaction, likely presenting non-neutral regions at their leading front.

CONCLUSION

Proton imaging is a diagnostic with enormous potential for the investigation of fundamental plasma physics problems which were impossible to explore up to now. By using this diagnostic, for the first time the measurement of transient electric fields in dense plasmas has been obtained, determining their evolution on a picosecond scale with micrometric spatial resolution. The data is of great relevance to Inertial Confinement Fusion both in the conventional and Fast Ignitor approach. Detailed analysis and modelling is presently undergoing.

ACKNOWLEDGEMENT

The work was supported by an ESPRC grant. We acknowledge the invaluable contribution to the work provided by the Central Laser Facility staff at the Rutherford Appleton Laboratory, and fruitful discussions with Dr. H.Ruhl (General Atomics).

REFERENCES

1. E.L Clark *et al*, Measurements of energetic proton transport through magnetized plasma from intense laser interactions with solids, Phys. Rev. Lett. **84**, 670 (2000); R.D.Snavely *et al*, Intense high-energy proton beams from Petawatt-laser irradiation of solids, Phys. Rev. Lett., **85**, 2945 (2000)
2. S.J.Gitomer *et al*, Fast ions and hot electrons in the laser-plasma interaction, Phys. Fluids **29**, 2679 (1984)
3. S. Wilks *et al*, Energetic proton generation in ultra-intense laser--solid interactions, Phys.Plasmas **8**, 542 (2001); A.Pukhov, Three-dimensional simulations of ion acceleration from a foil irradiated by a short-pulse laser, Phys. Rev. Lett, **86**, 3562 (2001); H.Ruhl *et al*, Computer simulation of the three-dimensional regime of proton acceleration in the interaction of laser radiation with a thin spherical target, Plasma Phys. Rep. **27**, 363 (2001)
4. J.A. Cookson, Radiography with protons, Naturwissenschaften, **61** 184 (1974)
5. M.Borghesi *et al*, Proton imaging: a diagnostic for Inertial Confinement Fusion/Fast Ignitor studies, to be published in Plasma Phys. Control. Fusion (2001)
6. W.L.C. McLaughlin *et al*, Sensitometry of the response of a new radiochromic film dosimeter to gamma radiations and electron beams, Nucl. Instr. Methods Phys. Res. A, **302**, 165 (1991)
7. A.J.Mackinnon *et al*, Effect of plasma scale length on multi-MeV proton production by intense laser pulses, Phys. Rev. Lett., **86**, 1769 (2001)
8. T.H.Bett *et al*, Binary phase zone-plate arrays for laser-beam spatial-intensity distribution conversion, Appl. Opt., **34**, 4025 (1995)
9. M.G.Haines, Thermal instability and magnetic field generated by large heat-flow in a plasma, especially under laser-fusion conditions, Phys. Rev. Lett. **47**, 918 (1981)

NUCLEAR DIAGNOSTICS OF ICF

N. Izumi, R.A. Lerche, M.J. Moran, T.W. Phillips, T.C. Sangster, G.J. Schmid, M.A. Stoyer, L. Disdier, J. L. Bourgade, A. Rouyer, R. K. Fisher, R. R. Berggren, S. E. Caldwell, J. R. Faulkner, J. M. Mack, J. A. Oertel, C. S. Young, V. Yu. Glebov, P. A. Jaanimagi, D. D. Meyerhofer, J. M. Soures, C. Stoeckl, J. A. Frenje, C. K. Li, and R. D. Petrasso*

1. INTRODUCTON

In inertial confinement fusion (ICF), a high temperature and high density plasma is produced by the spherical implosion of a small capsule[1]. A spherical target capsule is irradiated uniformly by a laser beam (direct irradiation) or x-rays from a high Z enclosure (hohlraum) that is irradiated by laser or ion beams (indirect irradiation). Then high-pressure ablation of the surface causes the fuel to be accelerated inward. Thermonuclear fusion reactions begin in the center region of the capsule as it is heated to sufficient temperature (10 keV) by the converging shocks (hot spot formation). During the stagnation of the imploded shell, the fuel in the shell region is compressed to high density ($\sim 10^3$ times solid density in fuel region). When these conditions are established, energy released by the initial nuclear reactions in center "hot-spot" region can heat up the cold "fuel" region and cause ignition.

We are developing advanced nuclear diagnostics for imploding plasmas of the ignition campaign on the National Ignition Facility (NIF). The NIF is a 1.8MJ, 192-beam glass laser system that is under construction at Lawrence Livermore National

* N. Izumi, R.A. Lerche, M.J. Moran, T.W. Phillips, G.J. Schmid, M.A. Stoyer, Lawrence Livermore National Laboratory, P. O. Box 808 Livermore CA, 94550; L. Disdier, J. L. Bourgade, A. Rouyer, Commissariat à l'Energie Atomique, Bruyères-le-Châtel, France; R. K. Fisher, General Atomics, San Diego, CA 92121; R. R. Berggren, S. E. Caldwell, J. R. Faulkner, J. M. Mack, J. A. Oertel, C. S. Young, Los Alamos National Laboratory, Los Alamos, New Mexico 87545; T.C. Sangster, V. Yu. Glebov, P. A. Jaanimagi, D. D. Meyerhofer, J. M. Soures, C. Stoeckl, Laboratory for Laser Energetics, University of Rochester, Rochester, New York 14623; J. A. Frenje, C. K. Li, and R. D. Petrasso, Plasma Science and Fusion Center, Massachusetts Institute of Technology, Cambridge, Massachusetts 02139

Advanced Diagnostics for Magnetic and Inertial Fusion, Edited by Stott *et al.*
Kluwer Academic/Plenum Publishers, New York, 2002

Laboratory[1]. One objective of the NIF is to demonstrate ignition and gain in an inertial confinement fusion plasma.

Extreme physical conditions characterize the imploded plasmas on the NIF. First, the thickness of the plasma, expressed by areal density (plasma density times radius), is large, up to ~1 g/cm^2. Highly penetrating probes such as energetic neutrons, hard x-rays, or γ rays are required to see deep inside the plasma. Second, the implosion time is quite short. The implosion process takes ~20 ns and the duration of the fusion reaction is on the order of 100 picoseconds. To observe the time history of the nuclear reactions, time resolution better than 10 ps is required. Third, the size of the imploded plasma is quite small (~100 μm). To see the shape of burning region, a spatial resolution of ~ 5μm is required for imaging systems. Fourth, the diagnostics operate in a harsh background. In implosion experiments, strong bursts of electromagnetic pulses, x-rays, neutrons, and neutron-induced radioactivity are produced. Therefore the diagnostics have to be designed to survive in these backgrounds. In addition, to prevent materials ablated from diagnostic components close to the target from being deposited on the laser optics, these components will be excluded from a zone around the target with a radius in the range of 0.5 m to 5m. This exclusion zone has a large impact on diagnostic design.

2. BASIC NUCLEAR REACTIONS IN ICF PLASMAS

This section reviews the fundamental nuclear reaction in the fuel plasma. For ignition experiments, the target capsules are cryogenic and consist of an ablator layer, a solid deuterium-tritium (DT) fuel, and a central DT gas region.

2.1. Primary reactions in DT capsule

In the DT filled fuel, the primary reactions that are important for diagnostics* are[1]

$$\begin{cases} D+T \rightarrow \alpha(3.5\ MeV) + n(14.1\ MeV) \\ D+T \rightarrow \gamma(16.7 MeV) + {}^5He. \end{cases} \qquad (1)$$

Due to its large cross section, the t(d,n)α reaction is dominant in DT filled capsule. The spectral width of the DT neutrons, (full width at half maximum) ΔE (keV) is given[2] as a function of ion temperature T_i (keV), by $\Delta E = 177 \times T_i^{0.5}$. This relation is used for the ion temperature measurement. However for a neutron emission history measurement, the neutron velocity distribution due to the thermal broadening degrades the temporal resolution at exclusion zone distances. Instead of the neutrons, the γ ray branch is more useful for a burn history measurement. The branching ratio of this reaction is 5×10^{-5} (ref. 3). A burn history measurement using 16.7-MeV γ-rays will be discussed.

2.2. Tertiary reaction in DT fuel

---

* d(d,n)^{3}He, d(d,p)t, t(t,2n)α are also primary reactions in DT filled target.

Tertiary reactions in DT filled capsules will be used for an areal density measurement[4, 5]. It is also possible to use tertiary protons for an areal density measurement via ^3He seeding in DT fuel. A small fraction of primary neutrons recoil from ions in the fuel,

$$\begin{cases} n(14.1MeV) + D \rightarrow n' + D^* (< 12.5MeV) \\ n(14.1MeV) + T \rightarrow n' + T^* (< 10.6MeV) \\ With\, ^3He\ seeding, \quad n(14.1MeV) + {}^3He \rightarrow n' + {}^3He^* (< 10.6MeV). \end{cases} \tag{2}$$

The probability of these ion recoils is proportional to the areal density of the target ions in the capsule. (Areal density measurement using scattered DT primary neutrons is also being developed[6, 7].) These recoiling energetic ions cause tertiary reactions,

$$\begin{cases} D^* \left(< 12.5\ MeV\right) + T \rightarrow \alpha + n\left(12 - 30\ MeV\right) \\ T^* (< 10.6MeV) + D \rightarrow \alpha + n\left(9.2 - 28.1\ MeV\right). \end{cases} \tag{3}$$

With ^3He seeded in the initial DT fuel, the ion recoils produce tertiary protons by

$$\begin{cases} D^* \left(< 12.5\ MeV\right) + {}^3He \rightarrow \alpha + p\left(12.5 - 30.8\ MeV\right) \\ {}^3He^* \left(< 10.6\ MeV\right) + D \rightarrow \alpha + p\left(9.7 - 28.9\ MeV\right). \end{cases} \tag{4}$$

When the areal density of the fuel is small and the energy loss of the recoiling energetic ions (D, T, ^3He) is negligible, the production ratio of tertiary neutrons to primary neutrons is proportional to the square of the areal density. As the energy loss becomes significant, the tertiary neutron production increases more slowly with areal density.

2.3. Primary neutron in D_2 fuel

For various pre-ignition experiments such as implosion symmetry studies, deuterium filled capsules will be used. In deuterium filled fuel, the primary reactions are,

$$\begin{cases} D + D \rightarrow n(2.45\ MeV) + {}^3He(0.82\ MeV) \\ D + D \rightarrow p(3.02\ MeV) + T(1.01\ MeV). \end{cases} \tag{5}$$

The expression in this case for thermal broadening of the neutron spectrum ΔE(keV) is $\Delta E = 82.5 \times T_i^{0.5}$ (ref. 2). This relation is again useful for ion temperature measurement.

2.4. Secondary reaction in D_2 fuel

Energetic ^3He and T cause secondary reactions in the fuel[8],

$$\begin{cases} {}^3He^*(< 1.01\ MeV) + D \rightarrow \alpha + p(12.5 - 17.4\ MeV) \\ T^*(< 0.82\ MeV) + D \rightarrow \alpha + n(11.8 - 17.1\ MeV) \end{cases} \quad (6)$$

When the energy loss of the energetic particles (^3He and T) is negligible, the ratio of secondary to primary reaction product is proportional to the fuel areal density. Since the cross-sections for these reactions are functions of particle energy, the yield ratio of secondaries to primaries diverges from proportionality with increasing particle energy loss[5]. For the high areal densities expected of NIF capsules (~ 1g/cm^2), these particles are stopped in the fuel and the secondary to primary ratio is a measure of the ion range.

2.5. Scattered primary neutrons from D_2 filled capsule

The 2.45 MeV primary neutrons (Eq. 5) lose energy by elastic scattering in the target. The areal density $\rho_D R$ of the deuterium plasma can be determined from the intensity ratio of lower energy (0.27-0.6 MeV) to primary neutrons (2.3-2.7 MeV). The areal density of deuterons can be determined from

$$\rho_D R \sim 6.0 g \cdot cm^{-2} \times \frac{Y_n\left(0.27 - 0.6\ MeV\right)}{Y_n\left(2.3 - 2.7\ MeV\right)}. \quad (7)$$

This relation (Eq. 7) saturates at higher areal densities. Allowing 15% divergence from linear relation, this technique is applicable up to $\rho_D R$ = 3 g/cm^2.

2.6. Scattered secondary neutrons from D_2 filled capsule

Secondary neutrons lose energy via d(n,elastic) and d(n,2n)p reactions in the target. The areal density $\rho_D R$ of the deuterium plasma can be determined from the ratio of lower energy scattered neutrons (4-10 MeV) to the secondaries (12-17 MeV) as follows[9],

$$\rho_D R \sim 9.2 \cdot g \cdot cm^{-2} \times \frac{Y_n\left(4 - 10\ MeV\right)}{Y_n\left(12 - 17\ MeV\right)}. \quad (8)$$

This technique is applicable for a full scale NIF DD surrogate capsule.

3.CORE DIAGNOSTICS

Core diagnostics are defined as a group of basic diagnostic instruments, which have been used at NOVA, OMEGA and other large laser fusion facilities. The applicability and reliability of these diagnostics have already confirmed experimentally and require minimal R&D for application to NIF. In this category[10] the following diagnostics are included: (1) Primary and secondary neutron yield measurements with nuclear activation, (2) Neutron yield, ion temperature, and nuclear emission time measurements with current mode scintillation detectors, (3) Areal density measurement via secondary and tertiary neutrons using a single-hit neutron detector array, and (4) Areal density measurements with a target activation technique using debris collection.

4. ADVANCED NUCLEAR DIAGNOSTICS

The expected increase of nuclear reaction yields on the NIF will make it possible to utilize nuclear reactions with low cross sections for advanced diagnostics. In this section, recent progresses on the development of these advanced diagnostics are reported.

4.1 Areal density measurement

The areal density of the imploded plasma is a crucial parameter that characterizes the performance of the implosion. The areal density of the hotspot $\rho R_{HS} > 0.3$ g/cm^2 is a necessary condition for ignition. The burn efficiency is proportional to the areal density of the fuel region. As discussed in a previous section, the diagnostic methods of areal density are based on the observation of the reactions of thermonuclear products with fuel ions. For DT capsules, neutrons or protons produced by tertiary reactions, or scattered primary neutrons will be used for areal density measurements. For D_2 filled fuel, observations of the scattering by fuel ions of secondary or primary neutrons can be employed. To observe these neutrons, the detector must have a large dynamic range and operate in this harsh background environment. In a time-of-flight system, a strong burst of direct primary neutrons hits the detector before the scattered neutrons arrive at the detector. In order to detect the faint scattered signal after the primary burst, the detector must be able to recover quickly. The neutrons scattered by "non-target materials" (mainly vacuum chamber wall) cause background. There are two different ways to reduce the background due to "non-target" scattering. (1) Using a fast response detector within the target chamber and completing all measurements before the primary neutrons hit the chamber wall. (2) Using a sensitive detector outside the chamber and reducing the background with collimation. With these concepts, we are developing both the CVD diamond detector and the lithium-glass scintillation fiber detector.

4.1.1. CVD diamond detector for short path operation.

At present we are developing a large area diamond detector to be used in current mode as a neutron TOF detector for a scattered secondary neutron measurement[11]. This device consists of a large polycrystalline diamond disc, which is fabricated by the chemical vapor deposition (CVD) technique. Fast neutrons incident on the detector cause ^{12}C(n,elastic), ^{12}C(n,n'γ), and ^{12}C(n, n')3α reactions in the diamond and excite hole-electron pairs. The charge from these pairs is extracted through ohmic contact electrodes. The advantages of a CVD diamond detector are follows: (1) Low noise even at room temperature due to its wide band gap ~5.5eV, (2) Fast time response (~ few hundred ps) due to the short lifetimes of the carriers, (3) Radiation hardness due to its strong bonding high-density lattice. Due to its fast time response, a neutron time-of-flight system with a CVD diamond detector can be operated with a short flight path. Thus the detector can complete the TOF measurement before the primary neutrons hit the chamber walls and before the neutron-induced background increases. We have fabricated the CVD diamond detector (using 1-mm thick 10-mm diameter Type-IIa wafer) and performed preliminary tests at the Omega laser fusion facility[12]. The detector shows a clear response to 14 MeV neutrons. A detailed impulse response for neutrons and the detection efficiency at 4-10 MeV are being investigated.

4.1.2. Lithium Glass SCIFI detector

Counting scattered DD primary neutrons is an alternative method for measuring the areal density of a D_2 filled capsule. We are developing a novel lithium-glass scintillation fiber detector for scattered primary neutron detection[13]. Since the scattered neutrons arrive at a detector after the primary neutrons, the detector needs a huge dynamic range ($> 10^4$) and enough signal-output for low energy neutrons (down to 272 keV). A lithium glass scintillation fiber (LG-SCIFI) is expected to serve this purpose. The LG-SCIFI is an optical fiber faceplate made of lithium scintillation glass. The detector will be located 6m away from the target. Since each neutron is detected using the exothermic reaction ^{6}Li(n,t)α with Q = 4.8 MeV, the lithium glass produces a large light output (~7500 photons) for a single neutron. Approximately 3% of the scintillation light (~225 photons/neutron) is guided to the end of a scintillation fiber. The scintillation light is amplified by a gated image intensifier. The gated image intensifier is operated in pulsed mode so that only the signals in the TOF window of the scattered primary neutrons are amplified and recorded by a charge-coupled device. Signals from scattered neutrons are counted as individual events in the gated image. The detector works as a 10^5 channel, scintillation detector array. With such a multi-channel counting mode detector array, the number of primary neutron hits for each channel is reduced and the interference due to "after glow" from the primary neutron burst is reduced. Using Monte Carlo simulations, we evaluated the background using a polyethylene collimator reducing the background due to non-target scattering. The next step is a proof of principle experiment using a neutron source.

4.2. Gas Sampling System

Nuclear activation measurements with a gas sampling system could play an important diagnostic role because of its unique features. A layer of the target is doped with a tracer material and its activation products form the residual gas in the target chamber after the implosion. The specific reaction product in the gas sample is analyzed by a mass spectroscopy or the detection of γ-ray activity. The sensitivity for this system is quite high. The minimum detection limits of specific nuclear products are expected to be one in 10^4 for mass spectroscopy and 1 in 10^5 for a gamma ray counting system. This technique is completely free from the backgrounds due to prompt burst of ionizing radiations from implosion. Although the information obtained is time-integrated, by coupling with a layer-selective tracer-doping technique, the activation technique allows the design of spatially resolved experiments. For example, doping ^{38}Ar initially in the DT gas region of the capsule and detecting ^{37}Ar produced by the ^{38}Ar(n,2n)^{37}Ar reaction is planned for measuring the areal density of the hot spot region. On the other hand, doping the ^{79}Br in the shell region of DT filled capsule and detecting the ^{79}Kr produced by ^{79}Br(d*,2n)^{79}Kr via knock on deuterons and ^{79}Br(p,n)^{79}Kr via primary DD protons is being designed to measure pusher areal density. Proof of principle experiments are planned at the OMEGA laser facility at the Laboratory for Laser Engineering in Rochester NY in FY02.

4.3 Penumbral Neutron Imaging

Neutron imaging gives crucial information on compression dynamics and asymmetries of implosions. A penumbral imaging system is being developed by scientist from Commissariat à l'Energie Atomique (CEA) for neutron imaging. Penumbral imaging is conceptually similar to pinhole imaging, except that the aperture is larger than the size of neutron source. The coded image observed at the detector plane is a convolution of the source image with an aperture function. The coded image can be reconverted to a measure of the source image using recently developed deconvolution methods[14]. The advantage of the penumbral system is higher statistical accuracy due to the large solid angle of the aperture. Since the NIF has an exclusion zone of about 1m, the distance from target to aperture l_1 >1m is required. A distance from the aperture to the detector $l_2 < 25$ m is preferable due to the finite size of the building. This results in an imaging system with relatively low magnification $l_2/l_1<25$. Simulations of the NIF target suggest that the diameter of the neutron image will be less than 25 μm and to see the asymmetry of the neutron producing implosion, a spatial resolution of at least 5 μm is required[10]. Therefore, spatial resolution better than 125 μm is required at the detector plane. Led by scientists of General Atomics, we are collaborating on tests of bubble detectors for high-resolution neutron imaging. The bubble detector is an elastic polymer that contains a freon-like superheated liquid droplet. When a neutron strikes the fluorine or chlorine in the droplet, the droplet vaporizes and grows rapidly to 50 μm in diameter. The spatial resolution of the detector itself is determined by the bubble formation process and readout technique (~30-μm resolution). We performed proof-of-principle experiments on OMEGA. Since the efficiency of the detector is limited to 10^{-5}, the resolution of the decoded image is limited by the statistics of the bubbles. Low yield pre-ignition experiments at NIF will require a liquid bubble chamber that has larger detection efficiency. With a liquid bubble chamber, 5 μm spatial resolution with a signal to noise ratio of 10 should make measurements possible for neutron yields as low as $\sim 10^{15}$.

4.4 Gas Cherenkov Gamma Ray Detector

The time history of nuclear reactions during hot spot formation and fuel burn provides crucial information in diagnosing the performance of ICF implosions designed to ignite. Since the time-of-flight spreading of the primary neutrons traversing the exclusion zone limits the temporal resolution of neutron TOF, the use of 16.7-MeV γ-rays produced by the DT reaction is proposed. To observe the 16.7-MeV γ-ray signal from a DT filled capsule, a gas Cherenkov detector has been fabricated by LANL[15]. The γ-rays produced in the target are converted to relativistic electrons by Compton scattering on a beryllium converter. The relativistic electrons pass through a high-pressure CO_2 gas chamber and produce Cherenkov radiation. The Cherenkov light in the visible and near ultraviolet region is collected by Cassegrain mirror optics and detected by a photo multiplier tube (PMT). Most of the γ-ray backgrounds are produced by (n,γ) reactions (mostly below 10 MeV) in the laser target material. The tunable detection threshold of the gas Cherenkov system reduces these backgrounds. When the CO_2 pressure is about 2 atmospheres, the Cherenkov radiation in the gas chamber has a steep threshold at 12 MeV. Thus the detector is insensitive to the background γ-ray signal below 12 MeV. The detection response was calibrated using an electron linear accelerator and the expected threshold behavior was confirmed experimentally. The system has been used at OMEGA on a high-yield experiment and a strong γ-ray signal was observed. In this case the time resolution of the system was limited by the time response of the PMT (~250 ps). The PMT has been replaced by a streak camera and integrated tests are underway on an electron linac.

5. ACKNOWLEDGMENT

This work was performed under the auspices of the U.S. Department of Energy by the University of California, Lawrence Livermore National Laboratory under contract No. W-7405-Eng-48.

6. REFERENCES

1. J. Lindl, Development of the indirect-drive approach to inertial confinement fusion and the target physics basis for ignition and gain, Phys. Plasmas **2**, 3933-4024 (1995)
2. H. Brysk, Fusion neutron energies and spectra, Plasma Phys. 15,611-617, (1973)
3. F. E. Cecil and F. J. Wilkinson, III, Measurement of the ground-state gamma-Ray branching ratio of the *dt* reaction at low energies, Phys. Rev. Lett. 53,767-770, (1984)
4. D. R. Welch, H. Kislev, and G. H. Miley, Tertiary fusion neutron diagnostic for density-radius product and stability of inertial confinement fusion, Rev. Sci. Instrum. 59, 610-615, (1988); R. D. Petrasso, C. K. Li, M. D. Cable, *et al.*, Measuring implosion symmetry and core conditions in the National Ignition Facility, Phys. Rev. Lett. 77, 2718-2721, (1996); S. Cremer, C. P. Verdon, R. D. Petrasso, Tertiary proton diagnostics in future inertial confinement fusion experiments, Phys. Plasmas 5, 4009-4014, (1998)
5. H. Azechi, M. D. Cable, and R. O. Stapf, Review of secondary and tertiary reactions, and neutron scattering as diagnostic techniques for inertial confinement fusion targets, Laser Part. Beams 9, 119-134, (1991)
6. S. M. Lane, M. D. Cable, W. R. Graves, *et al.*, Advanced Neutron Measurements, Lawrence Livermore National Laboratory Report, No. UCRL-50021-84, 1985, p.5-61
7. J. A. Frenje, K. M. Green, D. G. Hicks, *et al.*, A neutron spectrometer for precise measurements of DT neutrons from 10 to 18 MeV at OMEGA and the National Ignition Facility, Rev. Sci. Instrum. 72, 854-858, (2001)
8. T. E. Blue and D. B. Harris, The ratio of D-T to D-D reactions as a measure of the fuel density-radius product in initially tritium-free inertial confinement fusion-targets, Nucl. Sci. Eng. 77, 463-469 (1981)
9. Stephen P. Hatchett (private communication, Aug. 17, 2000)
10. Thomas J. Murphy, Cris W. Barnes *et al.*, Nuclear diagnostics for the National Ignition facility, Rev. Sci. Instrum. **72**, 773-779 (2001)
11. G. J. Schmid, V. Yu. Glebov, A. V. Friensehner, *et al.*, CVD diamond detectors for current mode neutron time-of-flight spectroscopy at OMEGA/NIF, To be published in the Proceedings of the SPIE, vol. 4510A-10 (2001)
12. T. R. Boehly, D. L. Brown, R. S. Craxton, R. L. Keck, *et al.*, Initial performance results of the OMEGA laser system, Opt. Comm., 133, 495-506 (1997)
13. N. Izumi, R. A. Lerche, T. W. Phillips *et al.*, Development of lower energy neutron spectroscopy for areal density measurement in implosion experiment at NIF and OMEGA, To be published in the Proceedings of the SPIE, vol. 4510A-29 (2001)
14. R. A. Lerche, D. Ress, R. J. Ellis, *et al.*, Neutron penumbral imaging of laser fusion targets, Laser and Part. Beams **9**, 99(1991)
15. R. R. Berggren, S. E. Caldwell, J. R. Faulkner, Jr., R. A. Lerche, *et al.*, Gamma-ray-based fusion measurements, Rev. Sci. Instrum. **72**, 873-876 (2001)

NEUTRON EMISSION SPECTROSCOPY FOR MAGNETIC CONFINEMENT EXPERIMENTS

Jan Källne*

INTRODUCTION

The purpose of neutron diagnostics (ND) for magnetically confined plasmas of extended volumes, such as those of tokamaks, is to measure the flux and the energy distribution (spectrum) of the neutron emission for a representative selection[1)] of volume elements. ND should determine the total and local neutron yield rates of the plasma, $Y_n(t)$ and $y_n(x,y)$, as well as provide information for the deduction of the velocity distribution of the ions responsible for $y_n(x,y)$. Unfortunately, local ND measurements are impossible so three surrogates are used, namely, (i) calibrated flux measurements for information on Y_n; (ii) collimated flux measurements for $y_n(x,y)$ and $Y_n(t)$; (iii) spectral measurements[2)] of collimated fluxes for information on the fuel ionic state of the plasma as well as on fast ions of other species besides collective ion motion (rotation).

Calibrated flux measurements are successfully performed with fission chambers over a large dynamic range of Y_n. However, the calibration reference, for the flux-yield conversion, is problematic as is its constancy over time. The reference is based on neutron transport calculations and modeling with empirical support from *in situ* neutron source measurements [1] besides complementary information from neutron activation of samples close to the plasma; the sample response is time invariant and serves as verifier of the time constancy of the fission detectors. A second calibration route is afforded by neutron cameras (see below).

A collimated flux measurement selects the neutron emission from a plasma volume (V) along the sight-line λ representing the neutron brightness $B_\lambda=\int y_n(x,y)\cdot dv$ for V_λ. A linear array of collimators forms a camera whose sight-lines should intersect a large segment of the poloidal plane so as to provide a neutron brightness picture of the plasma,

* Department of Neutron Research (INF), Uppsala University, EURATOM-VR Association, Uppsala, Sweden

1) Because of toroidal symmetry, the area coordinates (x,y) of the poloidal plane can be used.

2) The direction of observation is an important measurement parameter because of possible anisotropicities in the energy distribution of the neutron emission arising from ion velocity distributions that are anisotropic relative to the magnetic field.

Advanced Diagnostics for Magnetic and Inertial Fusion, Edited by Stott *et al.*
Kluwer Academic/Plenum Publishers, New York, 2002

B_{λ_k}, k=1-k^{max} [3]) . Two orthogonal neutron pictures can, in principle, be used to determine the emissivity distribution $y_n(x,y)$ through tomographic reconstruction while the total yield, Y_n, can be obtained by summation of the B_λ-data. The diagnostic performance depends on the front and back ends of cameras in different ways. The front end defines the neutron optics and the related machine interface space; the latter is a cost to be paid for increased diagnostic benefit. The back end houses the detectors that record the flux of signal neutrons, with suitable and stable detection efficiency, while they also have to handle predictably the background of γ-rays and other radiation besides the admixture of scattered neutrons[4]). The latter is an intrinsic problem of B_λ-measurements and need to be further assessed by theoretical neutron-transport studies and verified empirically with spectrometer, which should ideally be installed in some camera channels. The development of neutron cameras should aim at increased accuracy, resolution, sensitivity and radial extension of brightness measurements.

Neutron spectrometers use a collimated sight-line which is often chosen to pass through the plasma core to give maximum neutron flux. Radial sight-lines can be accommodated in the horizontal plane, or vertically, while those at different angles to the reference direction of the tangent to the toroidal magnetic field ($\underline{B}_T$) are usually in the horizontal plane. Tangential, perpendicular and intermediate viewing directions can be selected in order to change the sensitivity of neutron emission observations to anisotropicities in the fuel ion motion relative to the direction $\underline{B} \approx \underline{B}_T$[5]). This neutron emission spectroscopy (NES) diagnostic is the one least developed relative to its information potential and, yet, is the one that has seen the most instrumental and functional development since its first exploitation at the very beginning of fusion research in the 50's. The deuterium-tritium experiment (DTE1) of 1997 at JET [2] meant a dramatic leap forward through the combination of high neutron flux from the plasma and a new instrument with high count rate (C_n) capability with matching high accuracy and stability of calibration. There is a want of yet higher fluxes (to boost C_n into the several MHz-range) and enhanced instrumental sensitivity. This would make still finer and weaker details in the neutron spectrum accessible for diagnostic exploitation. A new 2.5-MeV spectrometer could facilitate needed NES-diagnosis of D-plasmas as it is now available for DT.

This paper discusses NES diagnostics as seen from its lines of development marked by the mile stones already passed to the present achievements, as illustrated by the JET results, from which the forward role of NES is projected to the planned next-step fusion experiments such as ITER

---

3) Here, k is an index that increases with plasma radius.

4) The two main sources are the plasma wall within the collimator view and the collimator wall; whose admixed flux is suppressed by pulse height discrimination and the rest calculationed.

5) Ion motion anisotropy is a feature that mostly manifests itself in the energy distribution of the neutron emission and to a lesser extent in the brightness distribution observed with the camera.

FUSION NEUTRON SPECTROMETRY SINCE THE 50'S

A cloud chamber was used in 1958 to measure the spectrum of the neutron emission from D-plasmas of ZETA [3]. The spectrum was observed to be Doppler shifted depending on the plasma current direction, in consistence with an ordered center-of-mass velocity of the deuteron pairs producing the neutrons (d+d->τ+n of E_n=2.5 MeV). It was concluded that the fusion reactions were not due to thermal but driven ions thus disproving that thermonuclear reactions had be attained.

The first proof of thermonuclear reactions was obtained with a ^{3}He ionization chamber (see Table 1) at the PLT tokamak in 1979 [4]. Neutron spectra were recorded for D-plasmas injected with neutral beams (NB) of H or D atoms which allowed separation of neutrons from thermal and driven ion reactions. A few years later, the thermal Doppler broadening (Γ_D) of the neutron emission was measured for Omically (Ω) heated D-plasmas produced in the Alcator C tokamak [5]; ion temperature (T_i) was derived from $\Gamma_D=k\cdot\sqrt{T_i}$ with k≈83 for dd-reactions (units in keV) compared to k=177 for14-MeV neutrons from d+t->α+n. This proof-of-principle experiment required summing many discharges which, a decade later at JET, would take only a fraction of a discharge thanks to higher flux. Besides T_i the spectral amplitude (A) could also be measured for a well collimated sight-line which allowed n_d to be determined [6]; since A is proportional to the density-reactivity product $n_d^2\cdot\rho(T_i)$, and, hence, the fuel dilution factor, $(n_e-n_d)/n_e$. This proved unambiguously that the dilution was much higher than had been surmised.

To improve NES data, a neutron time-of-flight (TOF) spectrometer was developed in the 80's which could operate at a significantly higher count rate (C_n of several kHz) and thus acquire better statistics [7]. This made it possible to detect contributions to the fusion power due to reactions between thermal ions and those involving supra-thermal ions caused by injected power (P_{in}) through neutral beams (NB) or radio-frequency waves (RF). This was the beginning of NES as a probe of the fuel ion population while the data quality was short of the limit for quantitative plasma information. This changed with the development of the magnetic proton recoil (MPR) technique

The MPR was designed specifically to attain maximum performance for ignited plasmas [8] and was put to the test for the sub-ignited conditions reached in JET's DTE1 of 1997. This fourth generation NES diagnostic meant drastically improved data quality (up 10^3 by C_n) which, e.g., permitted plasma information to be extracted from the shape of the neutron emission and not only the width.[9]. This resulted in unique information on the fuel-ionic state of plasmas subjected to NB and RF injection relative to the thermal equilibrium plasmas of Ohmic heating. Similarly, the effects of fast fusion ions, viz. tritons from d+d->t+p in D-plasmas and α's from d+t->α+n in DT, were detected for the first time as the spectral components of triton burnup and α-particle knock-on neutron emission (TBN[6]) and AKN[7])) [10]. Plasma rotation was also measured with the MPR.

---

6) Fast (1-MeV) tritons are created in d+d->t+p reactions which, during their slowing-down with a certain probability (about 1 % , typically), suffer nuclear burn-up through t+d->α+n; the triton burnup neutron (TBN) emission is used to study the fast ion behavior in D-plasmas that partly mimics that of α-particles in DT

7) Fast α's create a high velocity component in the deuteron and triton populations through knock-on scattering; these fast d's and t's give rise alpha-particle knock-on neutron (AKN) emission. NES records the

These achievements indicate the development of the next step NES diagnostics towards (i) further increase in sensitivity through proton detectors with better background handling capability and use of orthogonal sight-lines and (ii) development of a high-performance 2.5-MeV spectrometer (TOFOR, see below) for NES diagnosis of D plasmas.

INSTRUMENTS TESTED AT JET

JET has had an extensive NES-program [11] since its start in 1983 in which instruments of different types have been tested and a few were put to diagnostic use (see Table1). Some were commercially (or otherwise) available (nos. 1,2,3,4), while others were specifically developed for the NES application (nos. 3, 5, 6, 10, 11, and 12) of which no. 12 is a NES prototype for ITER. Two (nos. 7 and 13) are the TOFOR and MPRu development projects (12, 13) intended for the forward enhanced program at JET (JET-EP). The latter represent the leading techniques for fusion neutrons of 2.5- or 14-MeV. TOFOR will increase the C_n-capability by two orders of magnitude relative to that of the present generation TOF spectrometers (no. 6) and raise the quality of NES

Table 1. Summary of instruments tested as neutron spectrometers at JET for 2.5- and/or 14-MeV neutrons; also listed are their NES diagnostic status (used, proposed or tested), size, and the 7 type of nuclear reaction/scattering processes underlying the neutron detection. Reference spectrometers are checked in shaded boxes.

Instrument		E_n[MeV]		NES	Size	Type
No	Description	2.5	14	usage		
1	Nuclear emulsion track recorder	√	√	tested	S	S1
2	NE213 liquid scintillator	√	√	tested	S	S1
3	Spherical hydrogen ionization chamber	√		tested	M	S1
4	^{3}He ionization chamber.	√		used	M	R1
5	Forward proton-recoil pulse-height spect.	√		tested	M	S2
6	Neutron time-of-flight (TOF) spect.	√		used	L	S3
7	Optimized rate TOF spectrometer (TOFOR)	√		prop.	L	S3
8	Silicon neutron reaction detector		√	tested	S	R2
9	Diamond neutron reaction detector		√	tested	S	R3
10	Associated-particle neutron TOF spect.		√	(used)	L	S3
11	Annular proton-recoil pulse-height spect.		√	tested	M	S2
12	Magnetic proton recoil (MPR) spect.		√	used	L	S2
13	MPR upgrade spectrometer (MPRu)	√	√	prop.	L	S2

S1 n+$\underline{p}$→$\underline{p}$+n. **S2** n+p→$\underline{p}(\underline{\theta}_{np})$+n. **S3** n+p→p+$\underline{n}(\underline{\theta})$. **S4** n+p→$\underline{p}$+$\underline{n}$. **R1** n+^{3}He→$\underline{p}$+$\underline{t}$. **R2** n+^{28}Si→$\underline{\alpha}$+^{25}Mg. **R3** n+^{12}C→$\underline{\alpha}$+^{9}Be. Underlined particles/angles are measured.

---

presence of confined α's, monitors their confinement/slowing-down behavior (as well as that of fast d's and t's) as function of time, and provides a measure of the α pressure in the plasma.

diagnosis of D-plasmas to the level of that attained with MPR for DT-plasmas. The MPRu would be the only spectrometer capable of measuring both 2.5- and 14-MeV neutrons in either DT or D; for D-plasmas, the MPRu will provide high accuracy for prolonged data accumulation, thus complementing TOFOR.

The MPR- and TOF-spectrometers are the reference instruments for DT and D plasmas with regard to both performance and size. Concerning size, there are small instruments which laterally do no extend beyond the dimensions of a typical collimator bore and some larger, medium sized ones of equally limited NES diagnostic capability. Diagnostic performance does not increase gradually with size but is coupled to certain minimum dimensions as set by the nature of neutron detection. Small size is not a benefit in this connection as lack of performance cannot be compensated for (cf. ref. [11]).

GROSS FEATURES OF THE NEUTRON FLUX SPECTRUM

The collimated flux of direct neutrons of d+d->τ+n and d+t->α+n is restricted to energies around 2.5 MeV and 14 MeV (Fig. 1). Scattered neutrons of these sources form continuous spectra which can embarrass certain NES measurements. Camera detectors with modest capability for pulse height discrimination cannot separate scattered (energy degraded) and direct neutrons. Another example concerns measurement of the peak of 2.5-MeV dd-neutrons from DT plasmas which is superimposed on the continuum of scattered dt-neutrons [14]. The scattered spectrum is presently only known from neutron transport predictions which need data or verification of the results and their improvement. Generic tokamak data on the scattered spectrum is non-existent which would be remedied with MPRu measurements at JET-EP.

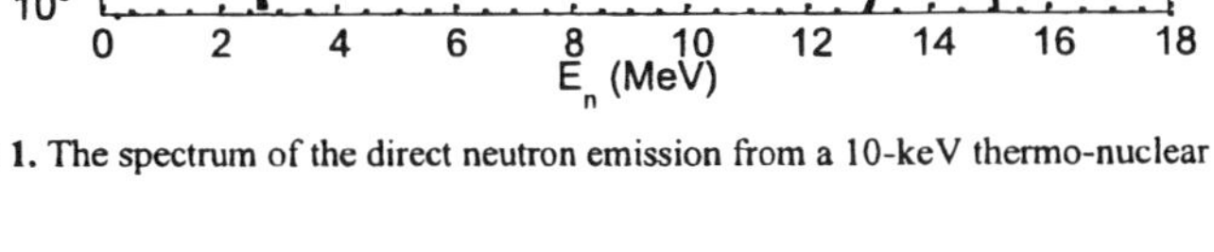

Figure 1. The spectrum of the direct neutron emission from a 10-keV thermo-nuclear DT plasma.

THERMAL AND DRIVEN REACTION CONTRIBUTIONS TO P_F.

The yield rate, $Y_n(t)$, and hence the fusion power, P_f, is routinely measured providing a gauge and compass for the progress in fusion research. Presently studied plasmas are subjected to strong auxiliary heating (P_{aux}≈P_{in}) in the form of NB and RF injection and the ratio $Q=P_f/P_{in}$ is the quantitative progress parameter. The equivalent quality parameter can be defined as $\eta_{TN}=P_{TN}/(P_{TN}+P_{ST})$ where P_{TN} is the thermonuclear reaction contribution to P_f and P_{ST} is the remaining one due to reactions involving driven (supra-thermal) ions. Both Q and η can be predicted based on models describing the underlying plasma conditions which can be well assessed and understood only if empirical η-results are available for comparison. It has been demonstrated that NES can provide this information for DT plasmas and it could also be provided for D-plasmas with the new TOFOR spectrometer. Studies of sub-ignited plasmas will include transitions regions of η-values affected by different P_α/P_{aux}-ratios as schematically shown in Fig. 2. This transition region would be approached with JET-EP and would indeed be the operating regime of ITER-FEAT [13]. NES is the prime source of information on η in these fusion plasma studies and can provide the empirical basis to test our understanding of sub-ignited and similar plasmas while it can also guide in the work to optimize the fusion quality of the plasma, i.e., conditions for maximum P_{TN}.

ANALYSIS OF NES DATA AND INTERPRETATION

A phenomenological model, based on heuristic ion physics assumptions, has been successfully used to describe the salient features of measured neutron spectra [9]. The model uses a limited number of components and is parameterized so that the predicted spectrum can be fitted to the data to give quantitative results in terms of spectral amplitudes and shape parameters (e.g., T_i). The model contains a T-component of Gaussian shape from reactions between thermal ions and an epithermal (E) component from super-thermal ions interacting with the thermal ones (normally being the bulk of the population). There is a third high-energy (H) component involving ions of energies above the critical limit (typically above ≈100 keV) where the slowing down is due to electrons

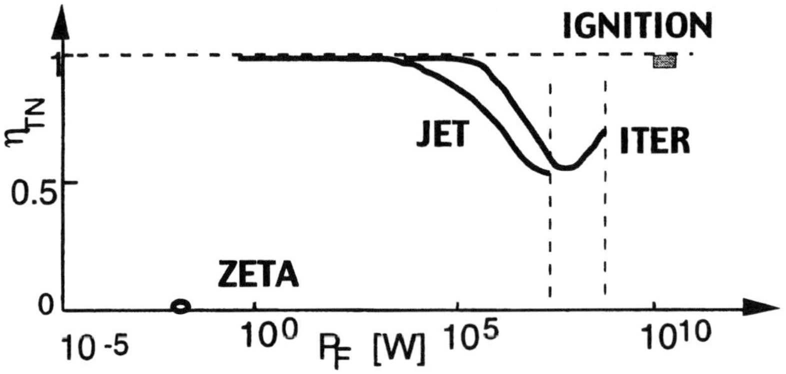

Figure 2. Schematic depiction of the variation of the thermo-nuclear quality factor as function of fusion power for JET and ITER operating with pre-ignition DT plasmas as compared to the first attempt to see thermo-nuclear fusion in the ZETA machine.

with little pitch angle scattering. The high-energy ions can be anisotropic leaving a specific imprint in the H-component which is not present in the E-component. The spectrum is fitted by varying the amplitude of the spectral component and the temperature of the underlying ion velocity distributions whenever a Maxwellian approximation is motivated; the neutron energy is also allowed to vary with a Doppler shift common for all components. The AKN and TBN features are calculated based on the use of computedvelocity distributions to describe the source and slowing-down of the fast ions responsible. It is noteworthy that the neutron energy is calculated to an absolute accuracy of better than 10^{-5} compared to an experimental accuracy, presently in the range 10^{-3} to 10^{-4}. The energy difference between calculated and measured spectra constitute a Doppler shift which is the basis for NES determination of plasma rotation; the present sensitivity level is about ±5 km/s with a potential of improvement to ±1 km/s with the next generation of spectrometers. Generally, the results obtained have been used for systematic studies of the neutron emission response to specific changes made to the plasma as well as for establishing the types of plasma phenomena that show up in NES diagnostic data.

PLASMA RESPONSE TO P_Ω, P_{RF} AND P_{NB}

A tritium discharge (JET#41759) with minority H- and D-admixtures (10 %), (HD)T, can be taken as example reults from the NES heating response studies. [16] The base plasma was formed with P_Ω of the order MW to which was applied a 2-s pulse of 8 MW ICRF-power tuned to the principal cyclotron resonance of hydrogen with 2ω–coupling also to deuterons. There were three 0.2-s NB-blips as well , namely, one during the period of P_ω only and two during the RF-pulse, thus, forming the periods of the heating combinations P_ω [1], $P_\omega+P_{NB}$ [2], $P_\Omega+P_{RF}$ [3] and $P_\Omega+P_{RF}+P_{NB}$ [4]. The plasma of these periods changed markedly as reflected in the NES data which were analyzed to give plasma results of the type described above such as on temperature. Thus, the measured T_i was increased from 1.1±1.6 to 12.9±1.4 keV when P_{NB} was added to P_ω [1→2], and to 8.4±0.5 keV for $P_\omega+P_{RF}$ [1→3]. P_{RF} created also a H-component of deuterons with T_H=300±50 keV; the spectrum recorded for period [3] is shown in Fig. 3. The combined

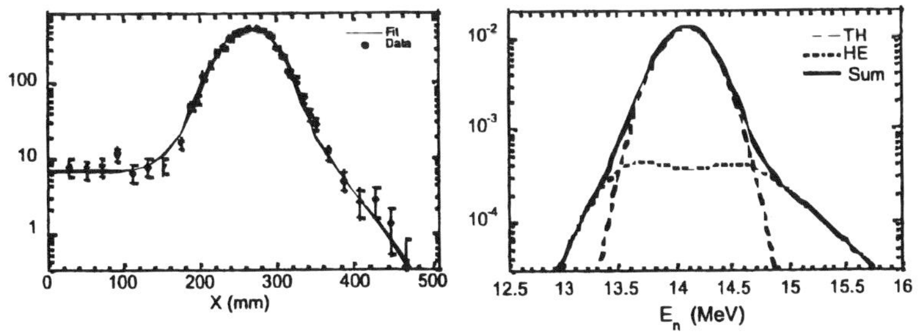

Figure 3: The spectrum recorded for the neutron emission from the period with RF-heating (left panel) and the extracted T and H components of the neutron spectrum in the right panel

injection of P_{NB} and P_{RF} [3→4] was found to raise T from 8 to 18±0.9 keV for the bulk deuterons but without affect on the tail within (T_H=300±120 keV). The results show the prompt and effective temperature impact of P_{NB}-injection and its boost to the effectiveness of P_{RF} in raising the bulk ion temperature.

As an example of derived information we consider the particle number (n) and energy ($w \propto n \cdot T$) densities where, during the RF-pulse, the deuteron density is the sum of the partial densities, $n=n_T+n_H$. The n_T and n_H values can be determined from the measured spectral amplitudes, A_T and A_H, and the reactivities, ρ_T and ρ_H, from the measured temperatures T_T and T_H, i.e., $n_T=n/[A_T/\rho_T+A_T/\rho_T]\cdot(A_{TH}/\rho_T)$ and $n_H=n-n_T$. The result is $n_T=0.98\cdot n$ and $n_H=0.02\cdot n$ for the measured ratio $A_T/A_H=10.1$ and the derived ratio $\rho_T/\rho_H \approx 0.17$. The partial deuteron kinetic energies are estimated to $w_T=0.57\cdot w_{RF}$ and $w_H=0.42\cdot w_{RF}$ of the total for the RF-heated plasma [3]. For the combination RF+NB [4], it is found that $A_T/A_H=24.0$ and $\rho_T/\rho_H \approx 1$. The partial kinetic energies are $w_T=0.59\cdot w^*$ and $w_H=0.61\cdot w^*$ where $w^*=w_{NB}+w_{RF/NB}$, and $w_{RF/NB}$ is due to the RF+NB synergy. The realtive population energies during the different auxiliary heating scenarios are thus estimated to $w_{RF/NB}:w_{RF}:w_{NB}=130:110:100$ while $w_\Omega:w_{NB}=10:100$ for the P_Ω and P_{NB} periods. The synergetic effect of NB during the RF pulse is estimated to be about 20 %.

The discussion above provides an illustration of the ion kinetic information that NES can provide for different heating scenarios including synergy effects. Results are still inclusive of the line integration effects that need to be considered.

FROM LINE INTEGRATED DATA TO LOCAL INFORMATION

NES measurements receive neutron emission along a sight line over which the plasma conditions can vary depending on popoidal locus. The neutron brightness profile is measured with cameras which provide the weight factors for calculating the relative contributions along the sight-line through the plasma. This can now be done qualitatively with a new computer code (YAPAN [17]) which will make it possible to assess profile effects in NES-data and ascribe a radial parameter value to measured quantities. Moreover, the profile calculations will be important for predicting NES signals of anisotropic plasma effects and their dependence of the sight-line angle relative to the B-direction. Preliminary attempts to consider profile effects have been done in comparison with CXRS data on T_i-profile [18]. There has also been an early study on $T_i(r)$-effects in measured spectra which showed that the data for thermal plasmas contain extractable information on the temperature density distribution, $n(T_i)$. With the YAPAN-code and the high quality MPR-data it would be possible, for the first time, to include profile effects in the analysis of NES data and make interpolations with radial parameter assignment to the extracted plasma information.

FUEL DENSITY MEASUREMENTS.

From simultaneous measurement of the spectra of d+d->t+n and d+t->α+n one can potentially determine the densities[5)] n_d and n_t of the mixed fuel DT plasmas This is

[5)] The ratio n_d/n_t would be available with higher accuracy.

done from measurements of the amplitudes A_2 and A_{14} besides T using $A_2 \propto n_d^2 \cdot \rho(T)$ and $A_{14} \propto n_d \cdot n_t \cdot \rho(T)$ where the reactivity function $\rho(T)$ is known. [19]. The A_2-signal (S) is superimposed on a background (B) of scattered neutron with a predicted S/B-ratio of about 1. This S/B-level does not in itself prevent this density diagnosis but it lowers the quality compared to the case S/B>>1 which, however, can be compensated for by increasing the statistics. On the other hand, the tolerance for ambient radiation background is very low because of the weak A_2-signal so that only a neutron spectrometers of the MPR-type are feasible and then only in its new MPRu version. An n_d/n_t-diagnostic is essential on ITER thus motivating the proposal to develop MPRu for JET-EP, where it would be used specifically to test the background handling capability besides providing the first generic data base on the scattered neutron spectrum for tokamaks.

The above derivation assumes that the plasma conditions are thermal but ITER-FEAT will operate in the sub-ignited plasma regime ($Q \leq 10$) where 2/3 or less of P_f is from thermal ion reactions ($\eta \leq 0.7$). The remaining driven contributions will increase the Doppler broadening of the neutron emission thus increasing apparent temperatures to $T_2 = T + \Delta T_2$ and $T_{14} = T + \Delta T_{14}$ so that $A_2 \propto n_d^2 . \rho(T+\Delta T)$ and $A_{14} \propto n_d \cdot n_t \cdot \rho(T+\Delta T)$ if $\Delta T = = \Delta T_2 = \Delta T_{14}$. The boost in A from ΔT can be estimated from the $\rho(T)$-function so the non-thermal plasma conditions of ITER-FEAT should not preclude the use of NES as a density diagnostic. It is interesting to note here that NES provides direct information on the separate fusion reactivity components related to the Q and η parameters which one ultimately wishes to maximize for given $T_i \approx T_{eff}$ where n_d/n_t is an inclusive factor.

CONCLUSIONS

It was noted that neutron emission spectroscopy (NES) has been used for over four decades of magnetic fusion research to answer questions pertaining to the fusion quality of plasmas and to gauge progress towards the ignition condition. The JET deuterium-tritium (DT) experiments of 1997 meant record high fusion power and a breakthrough for NES where both projeceted and new diagnostic functions were demonstrated. The present JET and its planned enhanced program (JET-EP) will offer and require extended use of NES diagnostics in the further exploration of pre-ignition plasmas of yet higher Q and η. This will be exploited with new instrumentation incuding the strategically important development of a high performance spectrometer for D-plasmas. The NES program at JET is the main (presently only) base of developing NES diagnostics for the next step experiments, viz. ITER, operating in the pre-ignition regime where fusion quality issues should be central. Why is then ITER presently without a NES program?

ACKNOWLEDGEMENT

This work has done under auspices of the European Fusion Development Agreement (EFDA).

REFERENCES

1 G Sadler, et al, Calibration issues for neutron diagnostics, in *Diagnostics for experimental thermonuclear reactors 2,* edited by P.E. Stott, G. Gorini, P Parndoni and E. Sindoni (Plenum press, New York, 1998), pp. 501-510.

2 M Keilhacker and the JET team, Nuclear Fusion 39, 209 (1999).

3 B Rose, AE Taylor and E Wood, Nature **181**, 1630 (1958).

4 J Strachan, et al, Nature 279, 626, (1979).

5 W Fisher, S Cien, D Gwinn and R Parker, Nucl. Inst. Meth. **219,** 179-91 (1984).

6 N Jarvis, G Gorini, J Källne, V Merlo, Sadler, v Belle, Nucl. Fusion **27,** 1755 (1987).

7 T Elevant,, et al Nucl. Inst. Meth. **A306,** 331(1991).

8 J Källne, L Ballabio, S Conroy, G Ericsson, J Frenje, G Gorini, M Tardocchi and E Traneus, Rev. Sci. Instr. 70, 1181 (1999).

9 M Tardocchi, in *Neutron emission spectroscopy studies of fusion plasmas of deuterium-tritium in tokamaks,* Ph.D. thesis, Uppsala University , Sweden, 2000,; and to appear in Nuclear Fusion.

10 J Källne, L Ballabio, J Frenje, S Conroy, G Ericsson, M Tardocchi and E Traneus, Phys. Rev. Lett, **86,** 1246-49 (2000).

11 N Jarvis, *Neutron spectrometry at JET (1983-1999),* JET report **P(00)03** (2000), to appear in Nucl Inst and Methods.

12 A Hjalmmarsson , contribution to these proceedings.

13 G Ericsson, contribution to these proceedings

14 P Antozzi, G Gorini, J Källne, and E Ramström, Nucl. Instr. Meth. **A386,** 457(1996)

13 A Costley, contribution to these proceedings .

16 H Henriksson, S. Conroy, G. Ericsson, G. Gorini, A. Hjalmarsson, J. Källne and M. Tardocchi *Neutron emission from JET DT plasmas with RF heating on minority hydrogen*' Uppsala Univ. Neutron Physics Report, UU-NF 01/#7(2001); submitted to Plasma Physics and Controlled Fusion.

17 S Conroy, private communication.

18 H. Heniksson, *The MPR neutron spectrometer at JET and implications for future applications*, Diploma thesis, Uppsala Univ. Neutron Physics Report, UU-NF 99/#1 (1999).

PROJECTED NEUTRON EMISSION SPECTROSCOPY DIAGNOSIS OF JET DEUTERIUM PLASMAS

G. Gorini, S. Conroy, G. Ericsson, L. Giacomelli, H. Henriksson, A. Hjalmarsson, J. Källne, M. Tardocchi*

The DT experimental campaign on JET (1997) represented a major step forward for neutron emission spectroscopy (NES) diagnostics thanks to the high count rate measurements obtained with the magnetic proton recoil (MPR) neutron spectrometer in high fusion yield plasmas[1]. NES measurements were made on JET DT plasmas for different heating conditions such as NB injection or ICRH using, e.g., the (D)T minority scheme[2]. A multiple-component model has been developed for analysis of DT neutron spectra that is based on a simplified description of fuel ion velocity distributions for different heating conditions[3,4]. The same model is used here for projections of NES spectra from JET deuterium plasmas using the DT results as input. This is done on the basis of a system of two complementary NES instruments, the TOFOR (Time of Flight - Optimized Rate)[5] and MPRu (Magnetic Proton Recoil Upgrade) spectrometers that are presently proposed for installation on JET[6]. Both spectrometers would record neutron spectra in D plasmas but for different viewing lines. TOFOR would have a "vertical" view, i.e. at an angle of 90° relative to the toroidal magnetic field. This is the same viewing line used for some of the previous dd NES measurements on JET[7]. MPRu would have a "tangential" view, i.e. horizontal (on the equatorial plane) at an angle of 47° relative to the toroidal magnetic field and opposite to the beam injection direction. This is the same viewing line used for the 1997 DT measurements using the MPR spectrometer.

Projections of dd neutron spectra under the assumptions above are used here for an assessment of the performance and requirements of NES in JET D-plasmas. The model neutron spectrum used for projections is the sum of two components. The first component (High Energy, HE) is due to energetic deuterons. These are strongly anisotropic and have different distributions for NBI and ICRH heating. In the ICRH case,

* G. Gorini, L. Giacomelli, M. Tardocchi, INFM, Physics Department, Milano-Bicocca University, and Plasma Physics Institute, EURATOM-ENEA-CNR Association, Milan, Italy; S. Conroy, G. Ericsson, H. Henriksson, A. Hjalmarsson, J. Källne, Department of Neutron Research (INF), Uppsala University, EURATOM-VR Association, Uppsala, Sweden

a Maxwellian energy distribution is assumed but restricted to the pitch angle range 90°±10° ("RF" model distribution); its temperature (T_{HE}) is a free parameter in the fit. In the NBI case, a uniform energy distribution is assumed for 75 keV<E_d<150 keV in the pitch angle range 60°±15° in the "counter" direction relative to the tangential NES view ("half box" model distribution). The second component (Bulk, B) is due to thermal deuterons plus what we refer to as "epithermal" deuterons. These are close to thermal, isotropic conditions, as is the case, for instance, of NBI deuterons after they are slowed down to a fraction of their initial energy. A single Maxwellian (temperature T_B, free parameter) is assumed to model both thermal and epithermal deuterons.

NES spectra fitted to measured DT spectra for reference JET discharges are shown in Fig. 1-2. The model parameters (both input and output) are shown in Table 1. To be noted is the double hump shape of the HE component. HE components dominate on the high-energy side of the spectrum (above 14.4 and 14.6 MeV in the ICRH and NBI case, respectively); their detectability on the low energy side is hampered by background. From the intensity of the spectral components the relative density of the Bulk and High Energy deuterons is determined. The deuteron distribution is thus completely specified but for a global scale factor.

The deuteron velocity distributions specified by the DT spectra are used as input to project NES spectra for dd reactions using a Monte Carlo code. Strictly speaking, these are simulations of dd NES spectra from the same DT plasmas. However they can be assumed to represent certain features of dd spectra in a more general sense. A two-component deuteron distribution yields three dd NES spectral components: i) Bulk-Bulk (B), ii) Bulk-HE (HE), iii) HE-HE (fast, F). These are reminiscent of the usual "thermal", "beam-plasma", and "beam-beam" components. Note however, that the Bulk component here includes an epithermal part accounting for a large fraction of "beam" deuterons. Hence the Fast component is always very small. Also, Bulk deuterons can have different effective temperatures depending on the reaction considered.

The projection results have different features depending on the type of heating. In the ICRH case, the projected spectra indicate that the energy bite required to detect ICRH dd NES spectra at 10^{-3} intensity would be 1.6-4.2 MeV for a 90° view. This is a very broad energy bite that will require special efforts to be met by NES diagnostics. A further requirement relates to the intensity of the HE component in D plasmas. As 2nd harmonic heating of (H)D-plasmas is not nearly as efficient as fundamental (D)T heating, much weaker HE deuteron components can be expected in D plasmas. High count rate capability and good background discrimination will be required for diagnosing energetic deuterons from ICRH in D plasmas.

Table 1. NES spectra parameters for the reference JET DT plasma discharges

discharge (#)	heating type	A_B (%)	A_{HE} (%)	c^d_B (%)	c^d_{HE} (%)	T_B (keV)	T_{HE} (keV)
42769	ICRH	63±0.8	37±0.6	17.3	0.7	8.0±0.2	109.4±3.7
43134	NBI	73±2.3	27±1.6	83.6	6.4	15.4±0.7	

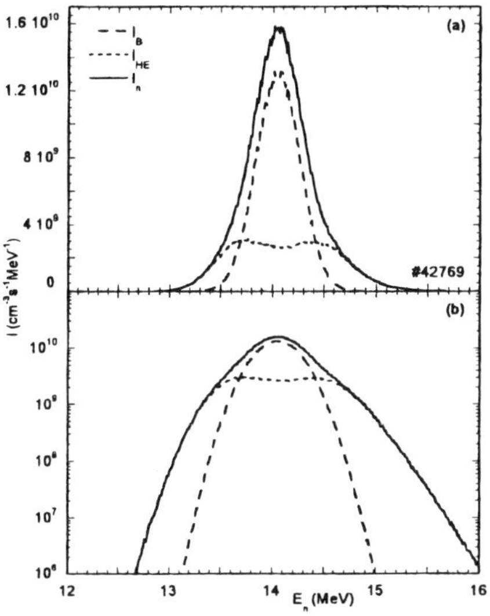

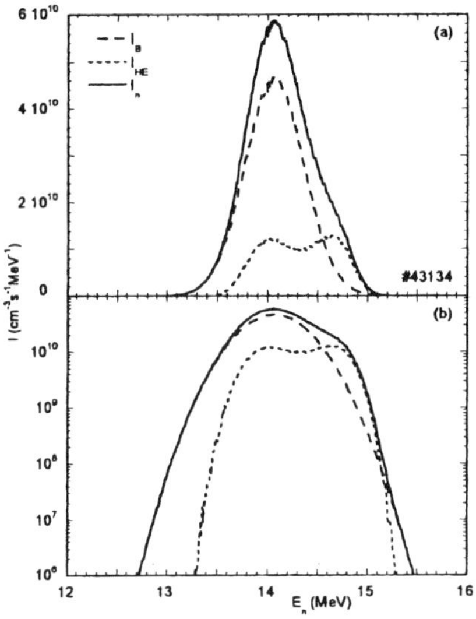

Figure 1. NES spectra fitted to the NES data of JET discharge # 42769 (ICRH discharge) plotted on (a) linear and (b) log scale. The full line (I_n) is the sum of Bulk (I_B) and High Energy (I_{HE}) components, see text.
Figure 2. Same as Fig. 1 but for JET discharge # 43134 (NBI discharge).

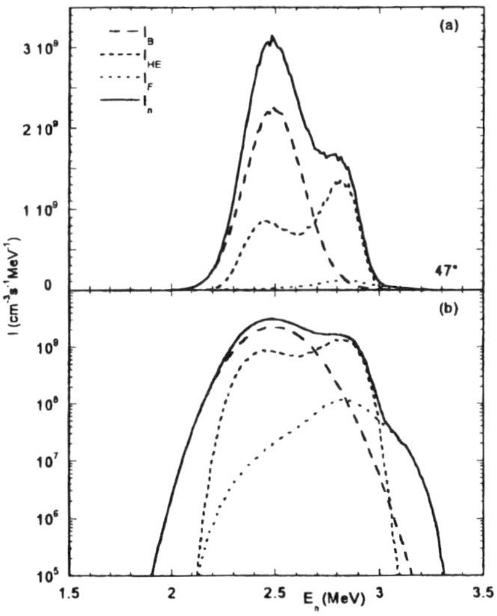

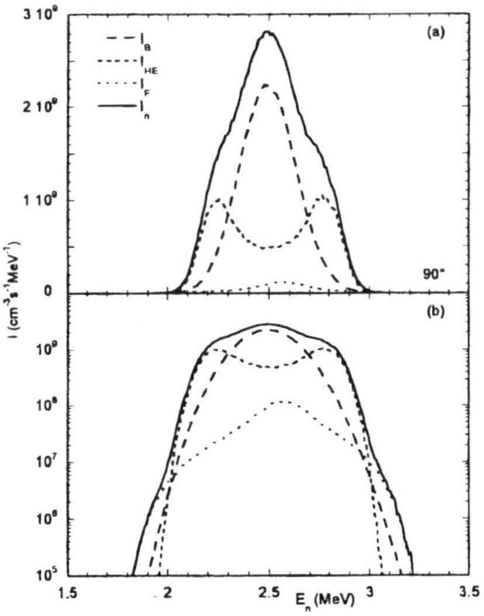

Figure 3. NES spectra for the same conditions as Fig.2 (NBI discharge) but for the dd reaction, plotted on (a) linear and (b) log scale. The full line (I_n) is the sum of Bulk (I_B), High Energy (I_{HE}) and Fast (I_F) components.
Figure 4. Same as Fig.3 but for a 90° viewing angle.

In the case of NBI heating, HE neutrons give rise to a marked asymmetry in the NES dd spectrum at 47° view (Fig.3). The HE component is enhanced over the DT case and dominates above 2.6 MeV. Note that it is the Fast component (at the 1% level) that dominates above 3 MeV. This is a distinct feature that could be detected if statistics and energy bite allow for. The energy bite required to detect NBI dd NES spectra at 10^{-3} intensity would be 2.0-3.1 MeV, i.e. not as demanding as in the ICRH case.

At 90° view (Fig.4), the strong asymmetry of the HE component is lost and the overall shape is triangular. The Fast component is not detectable in practice (it is out of range) and also the HE component is not very well separated from Bulk. The strong anisotropy of the HE component can be used to enhance its detectability by combining multiple NES measurements from different viewing lines, if these are available.

These projection are in qualitative agreement with previous NES measurements in D-plasmas[7]: in particular, observations with 90° view and 80 keV NBI are in qualitative agreement with the simulation of Fig.4.

In conclusion, preliminary results indicate that, for the same fast deuteron content in D and DT plasmas, the high energy component in the 2.5-MeV neutron spectrum is enhanced over that of the 14-MeV one reflecting the energy dependencies of the $d+d \rightarrow {}^3He+n$ and $d+t \rightarrow \alpha+n$ cross sections. This would make the task of detecting energetic deuterons relatively easier in D-plasmas compared to DT. Since the energetic components are anisotropic, much better separation can be achieved by a multiple viewing-line system; e.g., a tangential view is required in order to unambiguously detect the neutron emission from beam deuterons. The main requirements for NES diagnostics in D-plasmas are a large neutron energy bite (≈factor 2) and high signal/background ratio (s/b>>100). If these requirements can be met, a NES system installed on JET would provide unique information on energetic deuterons from ICRH, NBI or combinations of heating methods (including synergetic effects).

In the case of NB heating, a two-component NES model ("bulk", B, and "high energy", HE) has proven to work well for DT. The same model was assumed here for NES simulations in D-plasmas. A conversion from our empirical B/HE ratio into the conventional ratio of thermal and beam-plasma components, as simulated, for instance, by the TRANSP code, is presently being investigated.

This work has been conducted under the auspices of the European Fusion Development Agreement.

REFERENCES

1. J. Källne, these Proceedings.
2. M. Tardocchi et al., *Observation of energetic ions in Ion Cyclotron Heated DT plasmas using neutron emission spectrometry*, paper in preparation.
3. M. Tardocchi et al., *Neutron spectrometry study of RF heated (D)T plasmas*, subm. to Nucl. Fusion (2001).
4. H. Henriksson et al., Res. Sci. Instr. 72 (2001) 832
5. A Hjalmarsson et al, these proceedings.
6. G. Ericsson et al, these proceedings.
7. T. Elevant et al., Nucl. Instrum. Methods A 306 (1991) 331.

PROPOSED MAGNETIC PROTON RECOIL NEUTRON SPECTROMETER UPGRADE (MPRU)

G.Ericsson, S.Conroy, G.Gorini, H.Henriksson, A.Hjalmarsson, J.Källne, M.Tardocchi*

The Magnetic Proton Recoil (MPR) neutron spectrometer was installed at the Joint European Torus (JET) in 1996. The present prototype instrument is designed for 14-MeV neutrons. Its capability as a neutron emission spectroscopy (NES) diagnostic was demonstrated during the Deuterium-Tritium Experimental (DTE1) campaign in 1997 and it has since been continuously used during subsequent D-plasma operation. In this paper we describe an upgrade of the spectrometer's focal plane, recoil-proton detector and radiation shield. The MPRu is intended to make it possible fully to exploit the MPR capabilities in the forward JET experimental programme and to demonstrate the potential role of NES diagnostics in the next step fusion experiments such as planned with ITER.

The MPR instrument, its data acquisition, monitoring and control systems, as well as the calibration and data reduction procedures, have been described elsewhere[1]. The use of fast plastic scintillators as detectors, in the form of a 37-element hodoscope, and simple signal-handling electronics, has ensured that the high count-rate capability of the instrument is well matched to the high neutron flux offered by the position close to the plasma. This made it possible to reach count rates up to 611 kHz during the world record pulse with P_{fus}=16.1 MW (with a neutron rate of $Y_n = 6\cdot10^{18}$ n/s) in 1997. High count rate can be used for time resolved measurements and is a prerequisite to reach the high sensitivity needed to observe weak components in the neutron spectrum. Thus, high sensitivity is attained by a combination of high count rate, providing sufficient counting statistics, and efficient background suppression, i.e., a high signal to background (S:B) ratio. With the present MPR, it has been possible to perform detailed studies of auxiliary (ICRH or NB) heated plasmas[2] and to make the first observation of the so-called Alpha Knock-on Neutron (AKN) emission[3], the latter being the imprint of the energetic alpha particles in the neutron spectrum. Additional information contained in the neutron emission can be accessed if the instrumental sensitivity is further increased.

* G.Ericsson, S.Conroy, H.Henriksson, A.Hjalmarsson, J.Källne, Dept of Neutron Research, Uppsala University, EURATOM-VR Association, Uppsala, Sweden, G.Gorini M.Tardocchi, INFM, Physics Dept, Milano-Bicocca University, and Plasma Physics Institute, EURATOM-ENEA-CNR Association, Milan, Italy.

To fully exploit the potential of the MPR technique, the measurement sensitivity should be increased, which is the purpose of the MPRu project. This will be done by reducing the interference from the ambient background as well as extending the measurement range down to neutron energies of about 1.5 MeV. This will make it possible to cover the full range of fusion neutron energies, including the 2.5- and 14-MeV peaks of dd- and dt-reactions, as well as the extended spectrum of back-scattered neutrons. The most challenging objective is to measure the 2.5-MeV neutron emission from DT-plasmas. To this end, a new recoil-proton detector will be developed with high background immunity over the full energy range. In addition, the radiation shield surrounding the instrument will be reinforced.

The present concrete radiation shield has a plasma-facing wall that is 1.5 m thick; other walls are at least 0.7 m. The shielding design is based on MCNP calculations, which includes a detailed model of the MPR and the JET structures. The calculations agree within a factor of 2 with experimental results[4], which gives confidence in the model used and provides a good basis for assessing the effect of adding shielding in selected places. We have thus identified two regions where quite limited amounts of extra shielding material would reduce the background at the detector by a factor of 10. The rest of the increased sensitivity relies on detector improvements.

For many investigations of the 14-MeV neutron emission, from both D and DT plasmas, the main limitation up to now has been the counting statistics. This was the case for the studies of the Triton Burn-up Neutron (TBN) emission from pure D-plasmas[5] as well as for ohmic DT plasmas. Similarly, counting statistics did not always permit time-resolved studies of auxiliary heated discharges. In the AKN studies the signal was so weak that the measurement limit was set by background. Future experiments with JET-EP (enhanced performance) could reach neutron rates higher than those of DTE1, which would permit an extension of the AKN studies if the detector background were to be reduced; the quality of other studies could be improved as well.

The sensitivity of the instrument is a function of high count-rate capability and efficient background rejection. To be specific: a detector channel in the central part of the neutron spectrum (close to 14 MeV) registers approximately 10^4 signal events per 10^{18} produced plasma neutrons which represents a comfortably high ratio of S:B $\approx 10^3$:1 even with the use of a fairly relaxed pulse-height discrimination setting. The background intensity is rather constant over the hodoscope so the S:B ratio follows the signal intensity which decreases rapidly for the off-center channels, for instance, towards higher energy. However, the background interference can in this case be reduced by applying tighter pulse-height discrimination, which, together with a background correction step in the off-line data analysis, leaves a remnant of about 0.1 events for $Y_n=10^{18}$, or, S:B $\approx 10^5$:1.

The MPR spectrometer can be set to measure neutrons in the energy range E_n = 0 - 19 MeV by changing the magnetic field. The flux detection efficiency, ε, depends on the desired energy resolution and is $\varepsilon=5\cdot10^{-5}$ cm^2 for14-MeV neutrons at a resolution of 2.5% (FWHM), corresponding to the Doppler broadening at T_i=4 keV. For 2.5-MeV neutrons, ε is $2\cdot10^{-5}$ cm^2 at 5 % resolution. For the same resolution, the 2.5-MeV count rate is lower by a factor of about 8 due to efficiency and another factor 100 due to the reaction rate differences between DT and D plasmas. Still, an attempt was made in 1999 to observe the 2.5-MeV neutrons from D-plasmas with the prototype MPR in order to explore the S:B conditions[6]. A 2.5-MeV spectrum could indeed be measured from which a Doppler broadening was determined corresponding to an ion temperature of T_i=11 keV, in line with expectation. The relative background was determined to be S:B≈1:10 which indicates that the background must be reduced a factor $>10^3$ to obtain data of desired quality.

Analysis of the recorded pulse height spectra showed that the background was mainly due to electrons coming from γ-rays undergoing Compton scattering in the structures surrounding the detector. Most electrons that hit a detector pass straight through and produce a peaked pulse height distribution, which happens to be similar to that of stopped 2.5-MeV protons. The electron distribution is determined by the scintillator thickness, which was chosen to match the range of 14-MeV protons. However, to stop 2.5-MeV protons takes only a twentieth of the present thickness of 2.8 mm so a thickness reduction would practically eliminate the electron background. Combined with shielding reinforcements and improved signal selectivity, as discussed below, the factor of 10^3 reduction in background should be achieved. Moreover, it is estimated that with a thinner detector, background due to 2.5-MeV (dd) and 14-MeV (TBN) neutrons, as well as γ-rays from neutron capture, should contribute at the $<10^{-3}$ level in the relevant energy region. Thus, the anticipated reduction of the Compton electron component should in itself result in measurements with the desired data quality.

To measure the 2.5-MeV neutron emission from DT plasmas is a greater challenge than that from D-plasmas because of the much higher relative background. First, the signal of 2.5-MeV neutrons is about 100 times weaker than the 14-MeV signal. Second, experience from DTE1 shows that the background is about a factor 10^4 stronger in the 2.5-MeV region. Thus, to obtain data of the same quality as in the present 14-MeV measurements, an S:B improvement of more than six orders of magnitude might be necessary. With MPRu, the improved measurement situation should allow for exploratory measurements with S:B ratios of the order 1:10, similar to the pilot study performed with the original detector in pure D plasmas. Additionally, important studies of the scattered neutron component could be made in a wide energy range, $1.5 < E_n < 19$ MeV.

Reducing the background to obtain S:B ratios of 100 or better probably requires the design of a dedicated 2.5-MeV spectrometer. Such an instrument could be much more compact than the present MPR, which would allow for both more efficient passive background shielding and the use of smaller detector dimensions, thus further reducing the active volume susceptible to background. Improvements in detector design could also contribute. The studies made possible with the MPRu should contribute significantly to the knowledge base needed to design such a spectrometer, and pave the way for assessing the feasibility of using a dual-arm MPR spectrometer as a fuel ion density diagnostic.

We will investigate the possibility to replace the present monolayer scintillators of the hodoscope with laminated ones in a so-called phoswich arrangement[7]. The use of phoswich detectors would permit the successful 14-MeV measurements to continue, and at the same time open the field for measurements of the 2.5-MeV neutron emission.

Each phoswich element consists of a thin front scintillator of about 200 μm thickness to stop protons of up to 3 MeV. To this is attached a 2.6-mm thick back scintillator, stopping 14-MeV protons. The two layers respond differently to ionizing radiation in terms of pulse shape, with a fast response from the thin layer and a slow response from the thick. This property makes it possible to distinguish the contribution from each layer even when they are connected to the same photomultiplier tube, as would be the case here (Figure 1).

Recoil protons from 2.5-MeV neutrons are distinguished by requiring a signal (of known light yield) from the front layer, while the second layer in this case acts as a veto. This should reject most background induced by heavy charged particles of energies above say 3 MeV and practically eliminate electrons as discussed earlier. For 14-MeV measurements, a pulse from both layers is required. Here it is important to make efficient use of the quite limited light yield from the front layer and careful design of the scintillator readout arrangement will be necessary.

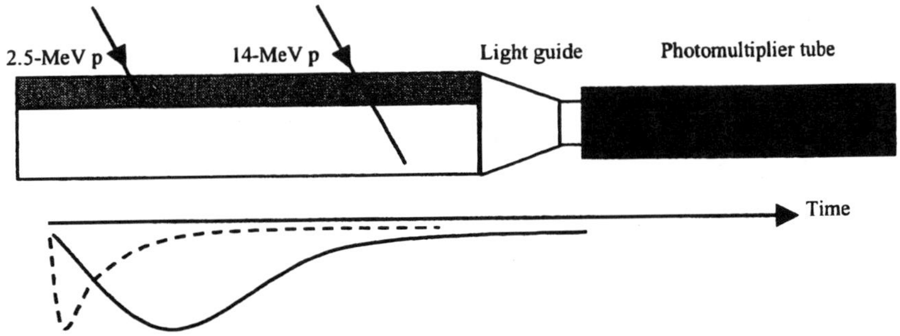

Figure 1. Schematic side view of a phoswich scintillator connected via a light guide to a single photomultiplier tube (top, not to scale); signal shapes from the front (dashed) and back (solid) scintillator layers (bottom).

To fully exploit the potential of neutron emission spectroscopy (NES) in the forward JET programme (and in extension, on ITER), we propose an upgrade of the Magnetic Proton Recoil neutron spectrometer (MPRu) with respect to its detector and radiation shield. The upgrade aims at increasing the sensitivity of the instrument by *reducing the background* in the measurements of 14-MeV neutrons from both DT and pure D plasmas, as well as *extending* the range of high-quality measurements to include 2.5-MeV neutrons from pure D plasmas. MPRu measurements would also provide an experimental basis for determining the feasibility of using the MPR technique for studies of the 2.5-MeV neutron emission in DT plasmas, which, if successful, would establish the MPR as a fuel density diagnostic.

REFERENCES

1 G.Ericsson et al., Neutron Emission Spectroscopy at JET – Results from the magnetic proton recoil spectrometer, Rev. Sci. Instr. 72 (2001) 759

2 M.Tardocchi et al., Time resolved neutron emission spectroscopy of DT plasmas with ion cyclotron resonance heating, Rev. Sci. Instr. 72 (2001) 836

3 J.Källne et al., Observation of the Alpha "Knock-On" Neutron Emission from Magnetically Confined DT Fusion Plasmas, Phys. Rev. Lett. 85 (2000) 1246

4 J.Frenje et al., Response Functions of the MPR Focal Plane Detector, Uppsala Univ. Neutron Physics Rep., UU-NF 97#6, 1997.

5 L.Ballabio et al., Measurement and interpretation of the spectrum of the triton burnup neutron emission from deuterium tokamak plasmas, Nuclear Fusion 40 (2000) 21

6 A.Hjalmarsson, Undergraduate Thesis, Uppsala Univ. Neutron Physics Report, UU-NF 99#9, 1999.

7 D.Fox et al., Calibration of plastic phoswich detectors for charged particle detection, Nucl. Instr. Meth. A374 (1996) 63

THE 2.5-MeV NEUTRON TIME OF FLIGHT SPECTROMETER FOR OPTIMIZED COUNT RATE (TOFOR)

A. Hjalmarsson*, S. Conroy*, G. Ericsson*, L. Giacomelli*, G. Gorini†‡, H. Henriksson*, J. Källne* and M. Tardocchi†

1 INTRODUCTION

Neutron emission spectroscopy (NES) diagnostic can give valuable information of both thermal and auxiliary heated tokamak plasmas. The information can be achieved given that the neutron yield rate (Y_n) is high and that the NES diagnostic can operate at high count rates. This has been demonstrated with the Magnetic Proton Recoil (MPR) neutron spectrometer[1] at the Joint European Torus (JET). The MPR has reached count rates of 0.6 MHz measuring the neutron emission from $d + t \rightarrow \alpha + n$ reactions in deuterium-tritium (DT) plasmas. To achieve the same NES performance in pure D-plasmas, the efficiency of the instrument must be a factor of 100 higher due to the lower Y_n value of the $d + d \rightarrow {}^3He + n$ reaction so correspondingly higher efficiency is needed. This can be achieved with the time-of-flight technique but it has a problem in the count rate C_n capability so the design must be optimized to approach its intrinsic upper limit. This project is about the time-of-flight optimized for rate (TOFOR), which is a NES diagnostic designed for high count rates to give high quality data for D-plasmas comparable with those obtained with the MPR for DT-plasmas.

The C_n limitation is due to the fact that it is a coincidence measurement based on neutron events in two different detectors. The coincidence can be of two types, true and accidental, and the accidental fraction increases with C_n so the system suffers paralysis. So it is essential to control and minimize the fraction of accidental.

The principle design and fundamental specifications of the TOFOR concept has been presented earlier[2] with an estimated count rate of up to 0.5 MHz. The results obtained with the MPR spectrometer[3] can be used as a basis to predict what NES diagnostics can provide for D-plasmas based on measurements of the 2.5-MeV neutron emitted form $d + d \rightarrow {}^3He + n$ reactions.

*Dept. of Neutron Research, Uppsala University, S-751 20 Uppsala, Sweden, EURATOM-VR Association.
†INFM, Physics Department, Milano-Bicocca University, Milan, Italy.
‡Plasma Physics Institute, EURATOM-ENEA-CNR Association, Milan, Italy.

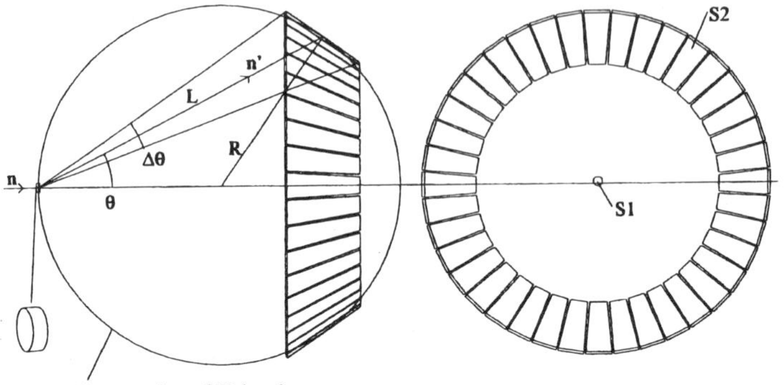

Figure 1: Front and side view of the TOFOR. TOFOR consists of one scattering scintillator (S1) in the neutron beam and 40 scintillators (S2) for detection of scattered neutrons. The S2 scintillators are placed on the sphere of constant time-of-flight as shown to the left; S1 is also shown in magnification.

2 PRINCIPLE OF TOFOR

The TOFOR spectrometer makes use of a CH based plastic scattering scintillator (S1) placed in a collimated neutron flux (beam) and a ring shaped array of scintillators (S2) in a special configuration (Fig. 1). Recoil protons in S1 from $n + p_H \rightarrow n' + p$ reactions are detected while the scattered neutrons (at angle $\theta \pm \Delta\theta/2$) are detected in S2. The recoil-proton events in S1 and S2 give the flight time, t, of the scattered neutron. The radius of the time-of-flight sphere R, gives the flight path $L^2 = 4R^2 cos^2(\theta \pm \Delta\theta/2)$. This gives, together with the measured t, the scattered neutron energy $E_{n'} = 2mR^2 cos^2(\theta \pm \Delta\theta/2)/t^2$, where m is the neutron mass. The incoming neutron energy (E_n) is determined by (non-relativistic) kinematics, $E_n = E_{n'}/cos^2(\theta \pm \Delta\theta/2)$, so that $E_n = 2mR^2/t^2$ is independent of scattering angle.

Due to the finite length of the S2 scintillators, the interaction of neutrons will not always be on the constant time-of-flight sphere. This gives a spread in the time spectrum, which, in turn, implies a broadening in the neutron energy spectrum, i.e, incident energy resolution, $\Delta E/E$.

3 NEUTRON TRANSPORT CALCULATIONS

The neutron transport calculations have been done with the Geant4 code as a means to simulate and predict the performance of TOFOR. The first objective for the calculations was to predict thc TOFOR response to mono-energetic 2.5-MeV neutrons. The geometry and dimensions for the TOFOR scattering scintillator, S1, is a 12.5-mm cylinder with a diameter of 36 mm. The second scintillator, S2, has a trapezoidal shape with sides 100 mm and 65 mm, length 350 mm and thickness 12.5 mm (Fig. 1). In the calculations, the neutrons flux is assumed to be perpendicular to the S1 surface and randomly distributed. The flight time for the scattered neutrons was calculated to determine the incoming neutron energy using the expression above. Moreover, the energy resolution $\Delta E/E$ was predicted compared

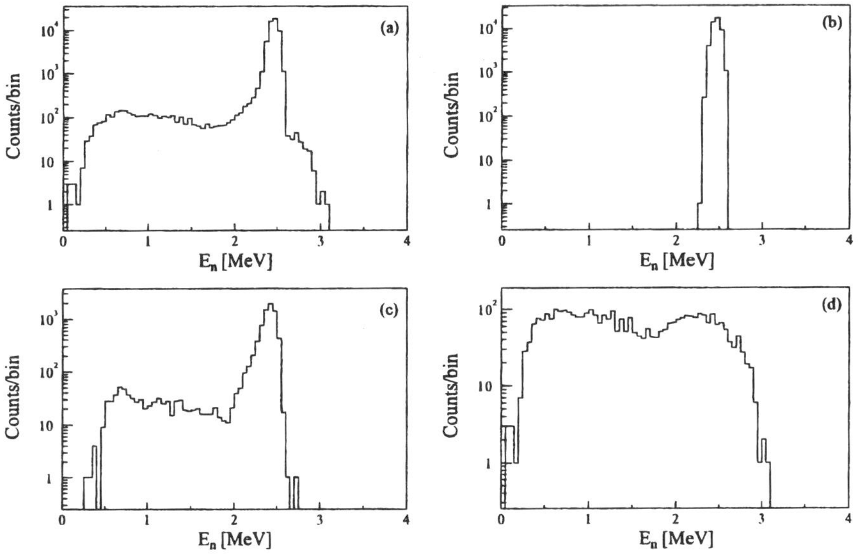

Figure 2: The predicted response of the TOFOR neutron spectrometer to 2.5-MeV mono-energetic neutrons. The total response (a) is made up of several scattering/reaction components such as, single scattering in S1 and S2 (b), single scattering in S1 and multiple scattering in S2 (c) and multiple scattering in both S1 and S2 (d).

with the requirement of $\Delta E/E \leq 5\%$ (FWHM); this value was chosen as a reference as it corresponds to the Doppler broadening of neutrons emitted from low temperature plasmas ($T_i = 2.3$ keV).

4 RESULTS AND DISCUSSION

As shown in Fig. 2, the predicted TOFOR neutron response consists of several different reaction components. The total spectrum is shown in Fig. 2a and its contributions in Fig. 2b-d. For instance single elastic neutron scattering on protons in both S1 and S2, give rise to a Gaussian shaped response in Fig. 2b and single scattering in S1 and multiple scattering in S2 gives Fig. 2c. The low energy tail arises from cross-talk between scintillators. The cross-talk is a consequence of neutron scattering on carbon in one scintillator and then entering a neighboring scintillator where it scatters on a proton. Due to the low energy transfer, when neutrons scatter on carbon, carbon events will not be registered. Neutrons that multiple scatter in both S1 and S2 give rise to the spectrum shown in Fig. 2d.

A Gaussian fit to the histogram in Fig. 2 can be used to estimate the energy resolution. This gives the result $\Delta E/E = 4.6$ % (FWHM) corresponding to $T_i = 1.9$ keV and $\Delta E/E = 4.9$ ($T_i = 2.1$ keV) for Figs. 2b and a respectively . These FWHM-values are only rough estimates of the TOFOR performance as it is the detailed spectrum that will be used as response function to extract plasma information from the measurements; the high energy tail enhancement is too small to impede spectral shape analysis. Finally, the neutron flux detection efficiency was determined to $\epsilon = 0.05\ cm^2$. This means that it would require a neutron flux of about $10^7\ n/cm^2s$ to exploit the estimated count rate capability of 0.5 MHz. Such plasma conditions would be offered at JET.

5 CONCLUSIONS

The time-of-flight neutron spectrometer for optimized rate (TOFOR) has been described in terms of its predicted performance based on neutron transport simulation. The results confirm that the efficiency, energy resolution and count rate are as conceptually anticipated. We therefore conclude that TOFOR should make neutron emission spectroscopy (NES) diagnostic available to D-plasma studies at JET and provide information with a quality approaching that of the MPR neutrons spectrometer already demonstrated for DT-plasmas. The next step in the TOFOR project is experimental studies of the scintillation detectors making up the system to provide input to refined computations of the detailed response function of the final optimized design.

6 ACKNOWLEDGMENTS

This work was performed under the auspices of the Association EURATOM-VR and was supported financially by the Swedish Research Council and EURATOM. Two of the authors (AH, HH) also acknowledges the support from the AIM research school sponsored by the Swedish Foundation for Strategic Research.

REFERENCES

1. J. Frenje, Ph.D thesis, Comprehensive Summaries of Uppsala Dissertations from the Faculty of Science and Technology No. 335 (1998).
2. G. Gorini *et al.*, Rev.Sci.Instum, **63** 4548 (1992).
3. G. Ericsson *et al.*, Rev.Sci.Instum, **72** 759 (2001).
4. G. Gorini *et al.*, KLUWER ACADEMIC/PLENUM PUBLISHERS (these proceedings)

MEASUREMENT OF HIGH ENERGY PROTON TEMPERATURE IN LHD

M. Sasao, A.V.Krasilnikov, M. Isobe, T. Seki, K. Saito, T. Saida, S. Murakami, and the LHD-Experimental Group*

INTRODUCTION

Measurement of energetic ion spectra has been one of the key diagnostics in fusion plasma experiment. Among various methods, passive measurement of fast neutrals escaping from magnetically confined plasma is widely used[1]. In order to derive ion spectra, possible all neutralization processes and spatial distribution of each particle spices responsible for these processes should be carefully considered and the measured spectra should be modified by sum of these energy-dependent charge exchange probabilities. In a burning plasma experiment, the ion spices are hydrogen isotopes, helium isotopes and impurities, and $H^0(D^0/T^0)$, He^0, He^+, A^{q+} ($q < Z$) are responsible for neutralization of protons. Helium concentration will be high in a D^3He plasma and will be low in a DT plasma[2].

In Large Helical Device (LHD), most of ICRF sustained, or ICRF assisted discharges[3] are produced by He+H admixture. A code to calculate H^0, He^0, He^+ distribution, including 6 charge exchange processes, electron/ion impact ionization and recombination processes in plasma of cylindrical geometry has recently been developed (ACHEN-Code)[2]. Using these results and charge-exchange cross sections, proton energy spectra and energetic tail temperatures (T_{eff}) of two vertical veiwing chords of LHD are derived from the neutral spectra measured by Natural Diamond Detectors (NDD)[4].

With the help of ACHEN-Code, not only the T_{eff}, but also the information on the neutral influx can be obtained from the calibrated NDD counting rates, and that on the neutral source temperature can be obtianed from the comparison of T_{eff} of the two vertical chord in poloidal cross section.

---

* M. Sasao, M. Isobe, T. Seki, K. Saito, T. Saida,, S. Murakami and LHD-Experimental Group, National Institute for Fusion Science, Toki, Gifu-5509-5292, Japan; A.V.Krasilnikov, Troitsk Institute for Innovating and Fusion Research (TRINITI), Troitsk, Russia.

NEUTRAL FLUX AND SPECTRUM OF HIGH ENERGY ATOMS

The relation of measured neutral flux, $\Gamma_{meas}(E_z)$ to the local proton distribution function $F(E_z,z)$ can be expressed as[1]

$$\Gamma_{meas}(E_z) = \eta_g \int s(z)\, dz\, P(E,z)\, A(E,z)\, F(E_z,z)\ , \tag{1}$$

where $P(E,z)$: the neutralization efficiency, $A(E,z)$: the attenuation factor*, η_g : the geometrical efficiency, and $s(z)\,dz$: the local observation volume. The neutral efficiency is the summation of contributions from all possible atoms/ions that can transfer one electron to a proton.

$$P(E,z) = \Sigma_i <\sigma^i v>_E n_i(z)\ . \tag{2}$$

Here we consider only H^0, He^0, and He^+, because the impurity concentration is low in LHD. In the following section the calculation of density distribution of these species is discussed, but the ratio of $n_i(z)$'s do not steeply depend on z and we can approximate as the following,

$$P(E,z) \approx n_0(z_R)[\Sigma_i <\sigma^i v>_E n_i(z)/n_0(z)] = n_0(z_R)\, g(E), \tag{3}$$

and z_R is a representative point on the viewing line, such as the resonance point of the ICRF wave. We also assume a uniform proton energy spectrum:

$$F(E_z,z_R) \approx n_p(z_R) f_p(E). \tag{4}$$

Then, E-dependent factor can be separated from the z-dependence factor, as

$$\Gamma_{meas}(E_z) \approx \eta_g\, s(0)\, g(E) f_p(E) \int n_p(z)\, n_0(z)\, dz. \tag{5}$$

In the present analysis procedure, we first calculate the z-dependence factor $\int n_p(z)\, n_0(z)\, dz$, make the velocity(energy) dependent correction, and fit into an exponential line, assuming

$$\begin{aligned} f_p(E)dE &\approx f_p(v)\, v_\perp dv_\perp \\ &\approx v_\perp dv_\perp\, [(m/2\pi)(T_{II}\, T_\perp)^{-1/2} \exp\{-(mv_{II}{}^2/2T_{II})-(mv_\perp{}^2/2T_\perp)\}] \\ &\approx v_\perp dv_\perp\, [(m/2\pi)^{1/2}(T_\perp)^{-1/2} \exp\{-(mv_\perp{}^2/2T_\perp)\}]\ , \end{aligned} \tag{6}$$

where T_{II} is assumed to be the ion thermal temperature and $E \approx mv_\perp{}^2/2$ [2] .

NUETRAL INFLUX CALCULATION

In order to estimate the H^0, He^0, and He^+ distribution, a new code[ACHEN-Code] , including 6 charge exchange processes, electron/ion impact ionization and recombination processes in a plasma of cylindrical geometry has recently been developed. Here all particle spices are grouped into two, a bulk plasma component and a slow component which originates

---

* The attenuation effect is generally smaller in an He plasma by a factor of 4 -10 than in a H/D plasma.

outside the LCFS (Last Closed magnetic flux Surface) with a density of n^0 coming into a plasma column. He^0 penetration depends on its initial velocity across the LCMS, which is determined by the temperature of neutrals, T^0. The He^+ ion density also depends on T^0, because they are mostly produced through the ionization of slow He^0 atoms. Production of He^+ though the recombination of a bulk He^{++} ion becomes dominant when the density is higher than $3\ x10^{19}/m^3$.

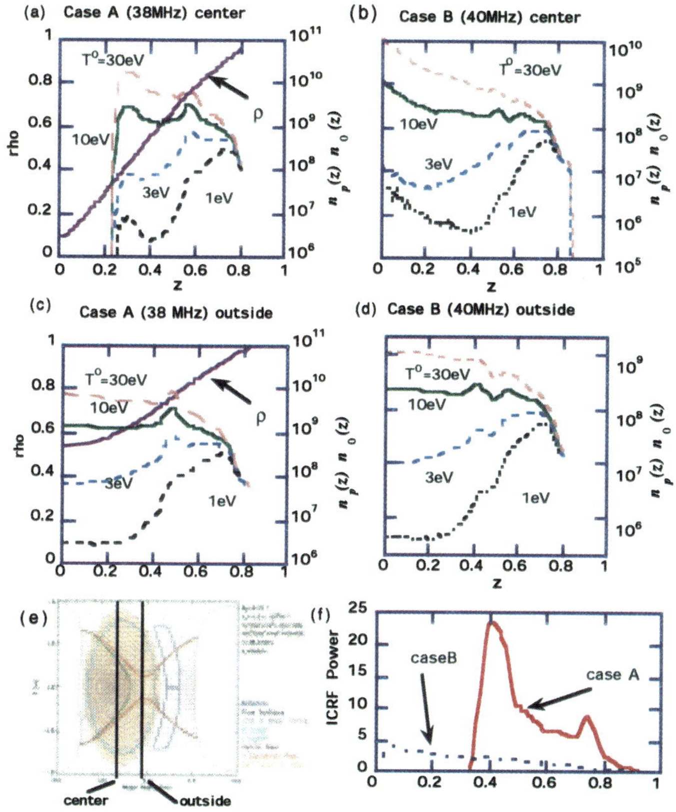

Figure 1. Comparison of two observation lines in ρ and $n_p(z)\ n_0\ (z)$ for two cases of ICRF power deposition (f), 38.47 MHz(caseA) and 40.47 MHz(caseB). (a), (b) are for Case A and (c), (d) are for Case B. (a), (c) are for the central observation line, and (b), (d) are for the outside observation line, as shown in (e) with the resonace surface structures.

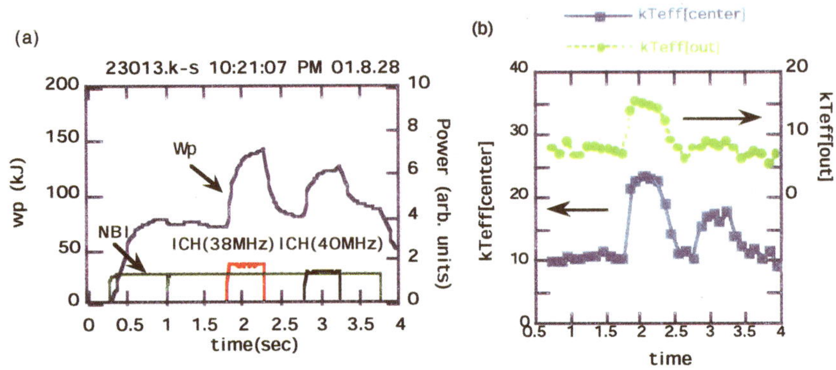

Figure 2 Time evolution of Wp (a) and T_{eff} of central and outside observation chords(b) for ICRF assisted NBI discharge. The changes of T_{eff} are different for two observation lines when the resonance structure changes.

Comparison of two observation lines in ρ and $n_p(z)$ $n_0(z)$ are shown in Fig. 1 for two cases of ICRF power deposition, 38.47 MHz (case A) and 40.47 MHz (case B). Here we assume that the high-energy particle distribution function is proportional to that of ICRF power deposition. When T^0 is less than 10 eV, both center and outside chords are not sensitive to the central region, and hence there would not be difference in the T_{eff}.

EXPERIMENTAL RESULTS

Figure 2 shows one of example of ICRF assisted NBI discharges. ICRF of Two different scheme was imposed, from 1.8 – 2.3 sec (case A), and 2.8 – 3.3 sec (case B). For case A, both chords shows increase in T_{eff}, but T_{eff} increases only in the center chord for case B, This happens when the T^0 is higher than 30 eV and the measured spectrum is composed by a substantial contribution from the central region.

If we assume that protons in the ICRF-coupled plasma volume, which is 14% of the total[3,4], are accelerated by ICRF, the information on the neutral influx can be obtained from the calibrated NDD counting rates. Detail study is now in progress.

REFERENCES

1. TESTA, D., GONDHALEKAR, A., Nucl. Fusion, 40 (2000) 975
2. Sasao M. et.al., Proceedings of the 27th European Physical Society Conference on Controlled Fusion and Plasma Physics (June 2000, Budapest), pp.456-459.
M. Sasao, to be published in NIFS-report.
3. Watari, T., et.al., Nucl. Fusion 41, 325
4. A. V. Krasilnikov, M. Sasao et al., *STUDY OF ACCELERATION AND CONFINEMENT OF HIGH ENERGY PROTONS DURING ICRF AND NBI HEATING IN LHD*, submitted to Nuclear Fusion.
5. M.Sasao, S.Murakami, M.Isobe, A.V.Krasilnikov, et.al. Presented at 18 IAEA Fusion Energy Conference (Sorento, October 2000), IAEA-CN-77/EX9/1

ESCAPING ENERGETIC ION MEASUREMENT ON LARGE SCALE MAGNETIC FUSION DEVICES

T. Saida, M. Sasao[*], M. Isobe[*], D.S. Darrow[*1],
T. Nishitani[*2], A. Ebisawa[*3]

INTRODUCTION

For escaping energetic ion measurements on magnetic fusion devices it is very important to get information on spatial distributions (poloidally and toroidally), pitch angle and energy distributions of escaping energetic ions. This paper presents two kinds of optical measurements of escaping energetic ions. The first is a scintilator-type probe which can resolve simultaneously both pitch angle and gyroradius of lost ions. The second is ceramic scintilator plates fixed on the first wall directly to survey spatial distribution of lost ions. There are difficulties in escaping energetic ion measurements in large scale magnetic fusion devices. For example, signal transportation is severe because loss points of fast ions are away from a port of a fusion device, especially on recent superconducting coils (SC) devices. We have studied these issues on Large Helical Device (LHD) and the signal was successfully transported by approaching an optical fiber close to a scintilator. Plans of lost alpha particle measurement on International Thermonuclear Experimental Reactor (ITER) using these optical techniques are also discussed in this paper.

LOCAL MEASUREMENT OF LOST ION

Here we introduce a scintilator-type probe. A scintilator-type probe has been developed to measure lost alpha particles by Zweben et al. and has been used on a

---

Department of Fusion Science, The Graduate University for Advanced Studies, 322-6, Oroshi, Toki, 509-5292, Japan
[*]National Institute for Fusion Science, 322-6, Oroshi, Toki, 509-5292, Japan
[*1]Princeton Plasma Physics Laboratory, Princeton, NJ, 08543,USA
[*2]Fusion Neutronics Laboratory, JAERI, Tokai, Naka, 319-1195, Japan
[*3]Diagnostics Group, Physics Unit, ITER EDA, Naka Joint Work Site, JAERI, Naka, 311-0193, Japan

Advanced Diagnostics for Magnetic and Inertial Fusion, Edited by Stott *et al.*
Kluwer Academic/Plenum Publishers, New York, 2002

variety of experimental fusion devices, for example Tokamak Fusion Test Reactor (TFTR)[1] and JAERI Fusion Torus-2M (JFT-2M) in tokamaks and Compact Helical System (CHS)[2], Wendelstein7-AS (W7-AS)[3] in stellarators. Detection principals of a scintilator-type probe are following. The probe has a piece of scintilator plate in a metal tube and a finite size aperture on its side. Escaping energetic ions gyrating by magnetic fields pass through the aperture and generate scintilation lights. The emitted scintilation lights are detected by a charge coupled device (CCD) camera linked with an image intensifier and by some photomultipliers (PMTs) through optical fibers, which are installed outside a vacuum chamber. The two-dimensional image distribution detected by CCD camera is verified with grid information on pitch angle and gyroradius of lost ions calculated with a Monte Carlo code. Time evolution of lost ion signal are measured by PMTs. Fast ion loss process can be studied from the probe position and pitch angle and energy distributions of lost ions with help of Lorentz-orbit-code calculation.

We report here preliminary experimental results of a scintilator-type probe on LHD during the 4th campaign. LHD is a Heliotron type stellarator device of pole number l=2, toroidal period m=10 and has a major radius 3.9m. In LHD discharge gas is helium or hydrogen and neutral beam's are H^0's produced from H^-. An entrance edge of the optical fiber is fixed on a vacuum seal window which is 40 cm away from the scintilator. Optical lenses are mounted in the gap space sealed in vacuum for image focusing. A Scintilator probe is installed downside of an equatorial plane because ions are drifted downward by average grad-B drift on LHD. The probe tip is installed about 7cm away from diverter legs during the 4th campaign, but we will install it by diverter legs in the next campaign.

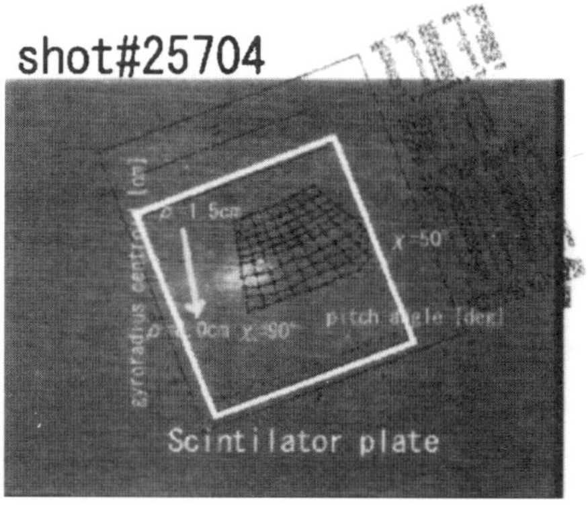

Figure 1. Scintilation light spot detected by a CCD camera. Both pitch angle and gyroradius of escaping energetic ions are estimated simultaneously

Scintilation lights spot detected by a CCD camera is shown in figure 1. By comparing with Monte-Carlo calculation of gyroradius and pitch angle grids it is estimated that lost ions have about 150keV energy and 95 degrees pitch angle. The energy is consistent with the energy of Neutral Beam (NB).

Time trace of one of PMTs signals is shown in Figure 2. When NBI is terminated, fast ion loss signal is decaying with some time constant. PMTs signals don't increase when H-alpha lights are strong due to recombination of electrons and protons at 0.2 sec after the NBI termination. From fast ion orbit calculated using the Lorentz orbit code it is expected that fast particles escaping beyond most outer magnetic surface by charge exchanges are re-ionized and detected.

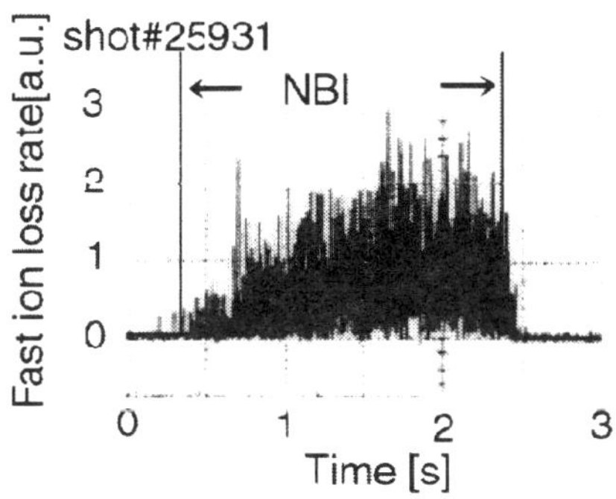

Figure 2. Time trace of PMTs signal. The figure stands for time relation with NBI and PMTs signal.

We study the feasibility to measure escaping alpha particles on ITER by a scintilator-type probe. In ITER a ceramic scintilator[4] with heat and damage resistance will be adopted and installed in a gap between two blanket modules near the first wall to prevent it from strong radiation and heat by burning plasma and from neutron irradiation (Figure 3). The optical signal should be transported by mirror system and optical image fibers to one of port windows.

SPATIAL DISTRIBUTION MEASUREMENT OF ESCAPING ION

There are two kinds orbits of escaping energetic ions on tokamak-type configuration; ripple trapped particles and toroidal banana particles. Konovalov calculated alpha particle losses and showed that loss distributions of alpha particles on the first wall in ITER is localized around from 210 to 240 degrees poloidally[5]. Particles affected by pitch angle scattering or toroidal ripple diffusion or by MHD activities and so on may belong to either orbit. We propose that several tens of ceramic scintilator plates with heat and damage resistance can be installed on the first wall directly. The measurement principal is following. Scintilation lights emitted by energetic alpha particles are selectively observed with optical filters through a plasma by a wide viewing system through mirrors on opposite side (Figure 3). Tobita et al. measured temperature distributions of the first wall on JAERI Tokamak-60 Upgrade (JT-60U) by infrared cameras[6]. Observed temperature rise spots are agreement with calculated hot spots by lost fast ions. There are following advantages in a ceramics scintilator. A ceramic scintilator is selectively sensitive to alpha particle losses because out-going efficiency of scintilation lights is good due to short range of alpha particles. Scintilation lights are opaque if particles have long range. Influences by radiations are negligible on ceramics scintilators. However there are some issues to be studied. For example, whether the plasma light background in the wavelength region of the scintilation lights is low enough or not, whether scintilation lights emitted by neutron, gamma-ray, ultraviolet radiation from plasma and so on can be discriminated from alpha particle signals, whether linearity of scintilation efficiency is satisfied up to expected S/N ratio level, whether ceramics and mirrors have sufficiently heat-resistance and damage-resistance. We will plan to examine these problems by using an accelerator and LHD.

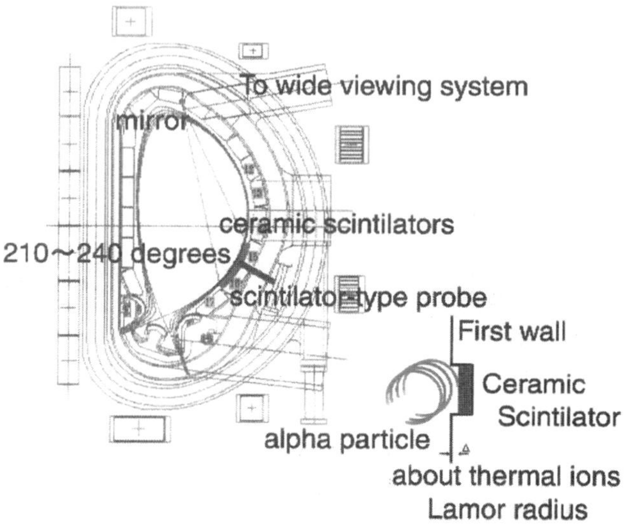

Figure 3. A layout of scintilator-type probe and ceramic scintilator plates on a poloidal cross section of ITER.

SUMMARY

Escaping energetic ions were measured by a scintilator-type probe on LHD during the 4th campaign. We came to settlements of a problem about a measurement method on SC devices in large scale fusion devices. A scintilator-type probe tip will be inserted by diverter legs on LHD in the next campaign and these results and experiences will be useful for designing the measurement system on ITER. Spatial distributions measurement of escaping energetic ions using ceramic scintilators was proposed. We will study the feasibility of this concept in detail by an accelerator and LHD.

REFERENCES

1. D.S.Darrow, et al. Nucl. Fusion. **36** (1996) 1
2. M.Isobe, et al. Rev. Sci. Instrum. **70** (1999) 827
3. A Werner, et al. Rev. Sci. Instrum. **72** (2001) 780
4. C.Greskovich, et al. Annu. Rev. Mater. Sci. **27** (1997) 69
5. S.V. Konovalov, 'RIPPLE AMPLITUDE AND ALPHA PARTICLE RIPPLE LOSS IN ITER FEAT', (Subtask 1, ITER, January 31, 2001)
6. K.Tobita, et al. Nucl. Fusion. **35** (1995) 1585

FAST ION LOSS DIAGNOSTIC FOR THE WENDELSTEIN 7-X STELLARATOR

A. Werner[1], D.S. Darrow[2], R. Kuduk[3] and A. Weller[1]

INTRODUCTION

An important issue of future reactor-like fusion devices is the confinement of fusion born alpha particles required for a burning plasma. A fast ion loss diagnostic, like proposed here, could be an adequate diagnostic tool to study the fast ion confinement properties of W7-X and, thus, could confirm at least one of the 7 optimization criteria given in Ref. [1]. The objectives of such measurements can be classified into 3 categories: (i) The confinement of fusion alphas directly depends on the magnetic field structure. Generally, the loss cone of the configuration should be as narrow as possible to avoid first orbit losses, otherwise alpha particle and auxiliary heating efficiency may be deteriorated. Although alpha particle confinement is expected to be sufficient in a HELIAS reactor (R = 20 m, a = 2 m), alpha particle confinement studies should be performed in W7-X (R = 5 m, a = 0.5 m) with 60 keV protons as a substitute. First orbit losses of ions injected by neutral beam heating (NBI) are caused in cases where the particle sources reside close to the lost/confined boundary in phase space or even are located inside the loss cone. Ion cyclotron heating (ICH) drives mainly perpendicular momentum of the ions, thus a transport into the loss cone is likely. (ii) Even in magnetic fields with perfect ion confinement losses may occur when ions are affected by fluctuating electric and magnetic fields. Shear Alfvén modes as an example are accompanied by poloidal and radial electric field oscillations that change the ion orbit trajectory, provided that particle and wave propagation are roughly in resonance. In this case, a subensemble of fast ions suffer from radial diffusion. Such modes can be excited by inverse Landau-damping with either fast ions of auxiliary heating or fusion alphas. In contrast to this, fusion alphas may be expelled from the plasma by fast ion driven modes, which could endanger reactor operation. However, if specific modes are excited in a controlled way, ion-wave interaction may be beneficial for helium ash removal. In this

[1]Max-Planck Institut fuer Plasmaphysik, , EURATOM Association, Boltzmannstr. 2, D-85748 Garching, Germany

[2]Princeton Plasma Physics Laboratory Princeton, New Jersey 08543

[3]Institute of Physics, M. Curie-Sklodowska University, Lublin, Poland

Advanced Diagnostics for Magnetic and Inertial Fusion, Edited by Stott *et al.*
Kluwer Academic/Plenum Publishers, New York, 2002

case, fusion alphas have to pass a resonance layer at a certain velocity during their slow-down at which enhanced radial transport is induced. (iii) Knowledge about fast ion losses is also essential to improve plasma performance. Strong fluxes of suprathermal ions are an additional drive mechanism for radial electric fields. Consequently, the lost/confined boundary may changes in a positive way and plasma performance can be improved, at least for thermal ions. The fast ion loss diagnostic provides experimental evidence whether such a drive is caused by fast or thermal ions. Another important issue is the influence of impurities. Enhanced impurity concentration increases pitch angle scattering of fast ions. This leads to an enhanced diffusion in phase space, and subsequently, to increased ion loss rates. Strong ion outward fluxes are not desirable, because they cause strong local heat loads on the first wall, which are not acceptable in a steady state plasma operation. Additionally, new impurity sources are generated by sputtering or by ion impact induced desorption. Thereby, the impurity concentration is increased and pitch angle scattering gets further enhanced leading to a non-stationary behavior in all plasma parameter.

DETECTOR CONCEPTS

The purpose of this diagnostic twofold: the measurement of fast ion losses related to auxiliary heating and the detection of fusion products. Fast ion losses will be measured at some locations outside the plasma resolved in gyro radius, being equivalent to the ion kinetic energy, and in pitch angle. Up to now, 4 different detector types have been considered for utilizing them at W7-X: (i) Nuclear track detectors as exposure samples provide absolute loss rates by counting the nuclear tracks on the sample [2]. However, no intrinsic energy and pitch angle resolution is given and no online data evaluation is possible, because the samples have to be removed from the vessel for counting the tracks. Such a system is only reasonable if absolute loss rates with high accuracy are required for performing cross checks with other ion loss detectors. (ii) Thin foil Faraday collectors have their merits in the rather simple setup and their resistiveness against gamma and neutron radiation [3]. These detectors, consisting of a stack of thin foils, need only one current measurement per foil and could be easily integrated into the first wall. However, no energy resolved measurement are possible by evaluation of foil current ratios. Only energy discrimination can be obtained, which is adjusted by the foil thickness. Such detectors are valuable, if ion loss measurements required at many locations in order to estimate global ion fluxes. Up to now, no real proof could be given that these devices are exclusively sensitive to fast ions, which, of course, is a precondition for employing them at W7-X. (iii) Semiconductor based detectors (surface barrier diodes) are appropriate for the detection of fusion products. They provide an intrinsic energy resolution by pulse height analysis which implies that they have to be used in counting mode. Pitch angle resolution may be achieved in combination with an entrance aperture either by diode segmentation or by using position sensitive detectors (charge separation depending on ion strike point location). Using silicon or germanium based diodes in the harsh plasma environment seems to be critical due to damage by neutron fluxes and the necessity to keep the detector at low temperatures. Such problems can be circumvented by utilizing natural diamond detectors, which are already used as neutral particle detectors at other fusion experiments [4]. (iv)Scintillator based detectors, as used in the TFTR alpha loss diagnostic [5], provide energy and pitch angle resolved loss distributions. However, the energy resolution decreases rapidly at large gyroradii. Absolute loss rates can only be obtained if the scintillator plate is used as a Faraday collector simultaneously, since the scintillator efficiency depends on the ion energy and, additionally, on temperature. Scintillators like ZnS can be used at moderate high temperatures, so that operation of these detectors should be compliant with W7-X operating conditions.

In a first stage, only ion loss measurements related to auxiliary heating are proposed, because, as mentioned above, NBI-ions may serve as a substitute for alpha particles and fusion products arc not expected in the very beginning of W7-X operation. Thus, scintillator based detectors are proposed for this stage, since experience already exists from successful operation at W7-AS [6]. The properties of the detector are basically the same, i.e. sensitive to ions having E_{ion} = 5...500 keV and pitch angle χ = 15...85°. Like the W7-AS detector, two staggered apertures systems in one casing with a scintillator plate divided in two domains should allow the measurement of co- and counter-going ions simultaneously.

In a second stage, measurements of alpha particle losses could be envisaged. Since alpha particle loss rates are expected to be lower than fast ion loss rates generated by NBI or ICH, devices based on natural diamond detectors are the choice for this stage of diagnostic development. Natural diamond detector are available with energy resolution $\Delta E/E < 10\%$ constant with E, which is clearly better than scintillator probes at large gyro radii ($\Delta E/E < 30\%$).

FINDING SUITABLE DETECTOR POSITIONS

As mentioned above, a fast ion loss diagnostic should be capable of resolving fluxes in energy and pitch angle. Previous experiments at W7-AS have shown that the distribution in pitch angle gives valuable information about the physical processes concerning slowing down and radial diffusion due to MHD activities. As a precondition, loss detectors have to be placed at the plasma edge in a way that losses of passing and trapped ions can be observed. The term trapped ion losses denotes ions, which are bouncing in local magnetic mirrors and drifting radially outwards by the net ∇B drift velocity, whereas passing ion losses denotes toroidally circulating ions whose displaced orbits, due to ∇B drift, intersect plasma facing vessel components. It is envisaged to choose positions where passing and trapped ions can be observed in both co and counter direction.

First Monte-Carlo slowing down simulations [7] reveal main ion fluxes across the separatrix in regions close to the divertor structure. The spatial distribution of ion strike points on the vessel surface exhibits strongest ion loss fluxes on the target plates. However, a detector implementation into target tiles is not compliant with the expected peak heat loads. Suitable locations can be found in the baffle structure, where significant loss fluxes exist either. Additionally, localized fluxes occur due to coil ripple on the first wall. At such locations only trapped ions strike the wall and the valuable information of pitch angle distribution is suppressed.

The ion strike point pattern found on the divertor structure depends strongly on the magnetic field configuration (W7-X has 7 different standard vacuum field configurations). Slowing down calculations for each configuration at vanishing plasma pressure, where the strongest loss fluxes are expected, reveal that the baffle plates at the ends of the divertor structure are most suited for ion loss detection in all configurations. It is envisaged to equip at least one bottom and one top divertor with three detector heads each.

From the considerations above it follows that the ion loss detectors must be integrated into the baffle plates. Due to technical limitations concerning heat loads, the detector heads cannot be mounted on movable shaft. Instead, they have to be flush-mounted in order to avoid interference with divertor operation and to be close as possible to the plasma edge. Thus, one has to find locations on the baffle surface providing intersection free orbits towards the entrance aperture in both co and counter direction (in principle a watershed on the baffle surface for ion orbits). Detailed numerical investigations are being performed in order to find such locations. This is done with collisionless real orbit calculations, in which ion trajectories are started reversed in time tangentially to the baffle surface at the location of interest. The ion trajectories length is calculated as function of pitch angle and energy providing a map, which easily shows where ion orbit scrape-off occurs predominantly.

PROBE DESIGN

The probe head of the scintillator based probe can be entirely adopted from the design used at W7-AS. The feasibility of integration into graphite tiles can be found as an example in TFTR [5]. The diagnostic mainly consists of the probe head (approximately 80 mm in diameter), which is embedded in the tile with a recess in front of the entrance apertures allowing ions to enter the probe. In order to construct such a recess, a Monte-Carlo simulation of real ion orbits into the probe is used, which also provides a spatial mapping of gyro radius and pitch angle on the scintillator plate. Behind the divertor some space is required for imaging the scintillator image onto a fiber guide or a system consisting of lenses and mirrors, with which the light is guided out of the vessel. It is then connected to a slow-scan CCD camera and a set of photomultipliers for sufficient spatial and temporal resolution, respectively. The compliance of this setup of with the divertor structure is now under investigation.

REFERENCES

1. Grieger G *et al.*, Phys. Fluids **B 4** 208, (1992)
2. Sadowski M *et al.,* Radiation Measurements **31** 185, (1999)
3. Jarvis O N *et al.,* JET-Report **P(99)** 54, (1999)
4. Isobe M *et al.,* Rev. Sci. Instrum. **72** 61, (2001)
5. Darrow D S, Rev. Sci. Instrum. **66** 476, (1995)
6. Werner A *et al.*, Rev. Sci. Instrum. **72** 780, (2001)
7. Strumberger E, Nucl. Fusion **40** 1697, (2000)

GAMMA-RAYS: MEASUREMENTS AND ANALYSIS AT JET

V. G. Kiptily[1], O. N. Jarvis[1], L. Bertalot[2], S. W. Conroy[3], S. Popovichev[1], M. J. Mantsinen[4] and contributors to the EFDA-JET workprogramme

1. INTRODUCTION

Optimisation of auxiliary plasma heating by means of ion cyclotron radio-frequency (ICRF) and neutral beam injection (NBI) as envisaged for the future fusion reactors is one of the main priorities in present research at today's tokamaks. Therefore, investigation of the production of fast ions during heating and of the subsequent fast ion behaviour in magnetically confined plasmas, together with an evaluation of the resulting bulk ion heating efficiency, are of essential importance for fusion reactor development.

Gamma-ray diagnostics, based on the measurement of the gamma-ray emission from nuclear reactions between fast ions and the main plasma impurities, is a valuable technique for studying the fast particle energy distributions. Gamma-ray spectrometry provides information on the energy distribution, and the measurement of emission profiles supplies information on the spatial distribution of the reaction sites.

Since 1987, the γ–ray emission from JET plasmas has been systematically monitored and used successfully in the analysis of heating effects during ICRF and NBI heating in the JET tokamak. The classical character of the fast ion slowing down behaviour has been demonstrated and estimates have been obtained of the fast particle confinement time[1]. The study of sawtooth crashes has demonstrated dramatic spatial redistribution of fast particles and other effects[2].

In recent JET experiments to study the ITER-relevant ICRH scenarios (^{3}He)D and (^{3}He)^{4}He, γ–ray measurements provided information on the fast ion population, with the effective temperature of the energetic tail ions being deduced with the help of a γ–ray spectrum simulation code, GAMMOD. In this paper, the main γ–ray results are presented

---

[1] Euratom/UKAEA Fusion Association, Culham Science Centre, Abingdon, Oxfordshire OX14 3DB UK
[2] Ass. Euratom/ENEA/CNR sulla Fusione, Centro Ricerche Energia ENEA-Frascati 00044 Frascati, Rome
[3] INF, Uppsala University, Euratom-VR Association, Box 535, 75 121 Uppsala, Sweden
[4] Helsinki University of Technology, Association Euratom-Tekes, P.O.Box 2200, FIN-02015 HUT, Finland

Advanced Diagnostics for Magnetic and Inertial Fusion, Edited by Stott *et al.*
Kluwer Academic/Plenum Publishers, New York, 2002

and the capabilities of gamma diagnostics are discussed in the light of the ITER-project programme.

2. EXPERIMENTS AND RESULTS

Gamma-ray energy spectra are recorded with a calibrated bismuth germanate (BGO) scintillation detector (75 mm x 75 mm). This detector is located in a well-shielded bunker and views the plasma tangentially. The detector line-of-sight lies in a horizontal plane about 30 cm below the typical plasma axis. The γ–rays are recorded in the energy range 1-28 MeV with an energy resolution of about 4% for the 10-MeV γ–line.

Gamma-ray emission profiles have been obtained using the JET neutron profile monitor[3]. This monitor consists of two cameras, a vertical camera with 9 line-of-sights and a horizontal camera with 10 line-of-sights. Due to finite collimation, the poloidal viewing extent of each channel at the centre of the plasma is about 10 cm. The radiation detectors are NE213 liquid scintillators. Separation between γ and neutron events is possible by exploiting the differing pulse shape characteristics exhibited by the detectors for the two forms of radiation. The 19 pulse shape discrimination (PSD) electronics modules were set up so as to restrict the detection of gamma rays to the energy range from 1.8 to 6 MeV.

Recent γ–ray measurements demonstrated the first direct evidence for ICRF-induced pinch of *^3He*-minority ions based on profile data[4]. Toroidally asymmetric ICRF waves were used in this experiment. Depending on the toroidal direction in which the launched wave propagates (co-current or counter-current for $+90^O$ or -90° phasing), the resulting spatial pinch of the trapped ions is either inwards (shot 54239) or outwards (shot 54243).

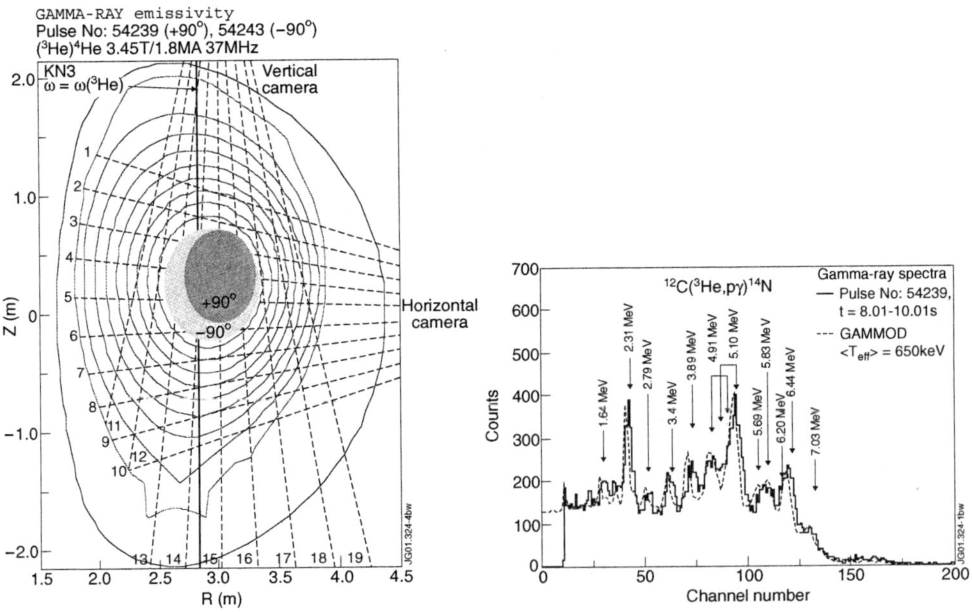

Figure 1. Left: comparison of regions radiating more than 50% of γ–rays from reaction $^{12}C(^3He,p\gamma)^{14}N$ in shots 54239 and 54243. Right: experimental and calculated spectra for shot 54239.

Fig.1 (left) shows the effect of different spatial distributions of the γ–radiation due to nuclear reactions between fast ^3He-ions and carbon, which is the main impurity in JET. The disparity is due to the differing character of the fast ^{3}He ion orbits for two ICRF phasings, as predicted by theory[4].

The γ–ray spectra recorded by the BGO-spectrometer permit an assessment of the energy distribution of the ^3He-ions because the reaction excitation function exhibits a threshold nature.

A dedicated code for γ–ray spectrum modelling, GAMMOD, has been used to deduce an effective temperature of the fast ions-tail. The program is based on basic reaction cross-section data for 16 nuclear reactions, includes 88 γ–ray transitions in the final nuclei and applies calculated response functions for the gamma spectrometer. Experimental data for the time-averaged plasma temperature, the applied NBI power, fuel and impurity densities are used as input parameters. A Maxwellian energy distribution is chosen to describe the line-of-sight averaged ICRF-driven ions' tail. Energy distributions of the fusion products have been approximated by the classical distribution function for steady-state plasmas. The analysis of recorded gamma-ray spectra using GAMMOD gives the effective tail temperatures, the fast ion particle concentrations and the contribution to the neutron yield made by the fast particles.

Experimental and modelled γ–ray spectra are compared in Fig.1 (right) for discharge # 54239, with +90°-phasing. Spectrum analysis indicates that the γ–radiation is mainly produced by the nuclear reaction $^{12}C(^3He,p\gamma)^{14}N$ and the effective temperature of the ICRF accelerated ^3He ions is $\langle T_{eff}\rangle$=650 ±150 keV. In the case of discharge # 54243 with the opposite phasing (-90°) $\langle T_{eff}\rangle$ is roughly the same, but the total γ–radiation intensity is halved.

In a special JET experiment[4], during third harmonic heating of ^4He plasma in the presence of ^4He-beams (120 keV and/or 70 keV), γ–radiation due to reaction $^9Be(\alpha,n\gamma)^{12}C$ was observed for the first time. This reaction has been proposed previously as a dedicated diagnostic for fusion alpha-particles[5,6]. Gamma-ray spectrum analysis shows the acceleration of ^4He ions up to energies in excess 2 MeV. An example of measured and calculated spectra is presented in Fig.2 (left) for discharge # 54172. Observation of strong γ–ray emission from $^{12}C(d,p\gamma)^{13}C$ shows that some ICRF power was also absorbed at the third harmonic D-resonance. It is possible that the D-tail is enhanced by knock-on collisions between thermal deuterium and fast ^4He ions[4]. The results of the spectrum modelling are $\langle T_{eff}\rangle_{4He}$=550 ±150 keV and $\langle T_{eff}\rangle_D$=300 ±100 keV. This calculation is based on an assumption that ratio of impurity concentrations

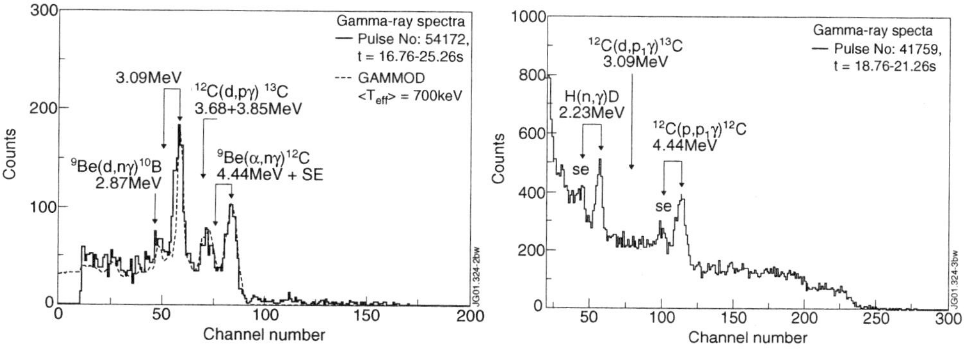

Figure 2. Left: experimental and calculated spectra for shot 54172. Right: experimental γ–ray spectrum recorded during shot 41759; *se* - single escaped peak.

Be/C is in the range from 0.15 to 0.5.

Gamma-ray modelling facilitates the interpretation of the neutron yield produced by the fast particles during low neutron yield discharges. In ^{4}He-discharges, as in shot #54172, the fast ^{4}He-ion component is about 10 times stronger than the D-component; nevertheless, the accelerated deuterons provide more than 90% of the d-d neutron yield.

In the deuterium-tritium JET campaign, DTE1, the γ–ray emission due to the reaction $^{12}C(p,p'\gamma)^{12}C$ indicates the presence of ICRF-accelerated protons with energy exceeding 5 MeV. For tritium discharge # 41759 the effective tail temperature of ICRF-driven H-ions is estimated to be in the range 0.4 to 0.6 MeV. This allows the identification of the threshold reaction $T(p,n)^{3}He$ as the source of 40% excess neutron emission[7]. Since there is no indication of gamma rays due the reaction $^{12}C(d,p\gamma)^{13}C$ in the measured γ–spectrum (Fig.2, right), the surprisingly high neutron yield in this discharge cannot be due to the second-harmonic-accelerated D-ions.

3. CONCLUSIONS

Experiments at JET have demonstrated the capability of the γ–ray diagnostics for studies of fast ion behaviour in deuterium, hydrogen, helium and tritium plasmas. Gamma-ray spectrometry and profile measurements provide information on the energy and spatial particle distribution of the fast ions and the efficiency of ICRF heating of bulk plasma. Further development of γ–ray spectrum modelling and the extension of the nuclear reaction data-base is essential. New possibilities of γ–ray diagnostics for fast confined alpha-particle investigation have been shown in a ^{4}He-plasma experiment. The nuclear reaction $^{9}Be(\alpha,n\gamma)^{12}C$ could be useful not only for DT-plasma[8] but also in the possible He-phase of ITER. In this case, high-energy resolution spectrometers could provide additional information due to the possibility of measuring the Doppler-broadening of the γ–lines[6].

This work has been conducted under the European Fusion Development Agreement and is partly funded by Euratom and the UK Department of Trade and Industry

REFERENCES

1. G. J. Sadler, S. W. Conroy, O. N. Jarvis, P. van Belle, J. M. Adams, M. A. Hone, Investigation of Fast Particle Behaviour in Joint European Torus Plasmas Using Nuclear Techniques, Fusion Technology **18**, 556-572 (1990).
2. O. N. Jarvis, J.M. Adams, P. J. A. Howarth, et al., Gamma Ray Emission Profile Measurements from JET ICRF-Heated Discharges, Nuclear Fusion **36** (11), 1513-1530 (1996).
3. J. M. Adams, O. N. Jarvis, G. J. Sadler, D. B. Syme, N. Watkins, The JET Neutron Profile Monitor, Nucl. Instrum. Methods Phys. Res., **A239**, 277-290 (1993).
4. M.J. Mantsinen et al., ICRF heating scenarios in JET with emphasis on He plasmas for the non-activated phase of ITER, 14th Topical Conference on Radio Frequency Power in Plasmas, Oxnard, 7-9 May 2001.
5. V.G.Kiptily, Capabilities of Gamma Spectroscopy for Fast Alpha-particle Diagnostics, Fusion Technology **18**(4), 583-590 (1990).
6. V.G. Kiptily, A.V. Matjukov, A.S. Mishin et al., In-beam Study of the Reaction $^{9}Be(\alpha,n\gamma)^{12}C$ and Alpha-particle Diagnostics, Fusion Technology **22**(4), 454-460 (1992).
7. M.J. Mantsinen, O.N. Jarvis, V.G. Kiptily et al., First Observation of p-T Fusion in JET Tritium Plasmas with ICRF Heating of Protons, Nuclear Fusion (in press).
8. V.G. Kiptily, T. Kondoh, V.O. Naidenov et al., in: *Diagnostics for Experimental Thermonuclear Fusion Reactors 2*, edited by P.Stott, G.Gorini, E.Sindoni (Plenum Press, New York&London, 1998) pp.511-519.

Spatial Distribution Measurement of High Energy Particle using Time-Of-Flight Neutral Particle Energy Analyzer in Large Helical Device

T.Ozaki, P.Goncharov, S.Murakami, S.Sudo, M.Shoji, M.Emoto, H.Funaba, M.Goto, K.Ida, H.Idei, K.Ikeda, O.Kanko, K.Kawahata, T.Kobuchi, S.Kubo, R.Kumazawa, S.Morita, O.Motojima, S.Muto, T.Mutoh, Y.Nagayama, Y.Nakamura, H.Nakanishi, K.Narihara, N.Noda, S.Ohdachi, N.Ohyabu, Y.Oka, M.Osakabe, B.Peterson, K.Saito, S.Sakakibara, M.Sasao, K.Sato, T.Seki, T.Shimozuma, Y.Takeiri, K.Tanaka, H.Torii, K.Tsumori, T.Watanabe, T.Watari, V.Zanza, G.Bracco, A.Sibio, B.Tilia and G1/G2 Group

INTRODUCTION

The time-of-flight neutral particle measurement system[1] has been installed on the 10-O port at the 2nd campaign. It has the capability of two-dimensional measurement with a larger viewing angle by using the pivot on the rectangular port, which is set inside of the 10-O mother port. The movable stage with the vertical scan mechanism rotates on the fan shaped stainless rails around the pivot. The sight lines of -2 to 32 degrees against the 10-O port surface makes the angle between the sight line and the magnetic axis of the standard configuration (3.6 m) of 95 to 35 in the 4th campaign. Therefore the analyzer can observe high-energy particles with these pitch angles. The higher scan speed of 0.17 degree/second than that in the last campaign (0.05 degree/s) can be obtained in this campaign. Precise spatial distribution of the high-energy particles can be expected because it enables the continuous scanning in the long discharge.

* T.Ozaki, S.Murakami, S.Sudo, M.Shoji, M.Emoto, H.Funaba, M.Goto, K.Ida, H.Idei, K.Ikeda, O.Kanko, K.Kawahata, S.Kubo, R.Kumazawa, S.Morita, O.Motojima, S.Muto, T.Mutoh, Y.Nagayama, Y.Nakamura, H.Nakanishi, K.Narihara, N.Noda, S.Ohdachi, N.Ohyabu, Y.Oka, M.Osakabe, B.Peterson, K.Saito, S.Sakakibara, M.Sasao, T.Seki, T.Shimozuma, Y.Takeiri, K.Tanaka, K.Tsumori, T.Watanabe, T.Watari, and G1/G2 Group, National Institute for Fusion Science, Toki, Japan. P.Goncharov, T.Kobuchi, Graduate University for Advanced Studies, Hayama, Japan. K.Sato, H.Torii, Department of Energy Engineering and Science, Nagoya University, Japan. V.Zanza, G.Bracco, A.Sibio, B.Tilia, ENEA, Frascati, Italy.

THE NEUTRAL PARTICLE MEASUREMENT SYSTEM

The time-of-flight type NPA has the capability of a high S/N ratio against the radiation including soft/hard X-rays from plasmas[1]. The analyzer and the movable stage are installed at the mid-plane of LHD on the 10-O port. There are two NBIs (NBI#1 and #2) at neighboring ports to 10-O. In particular, the beam path of NBI#1 crosses the sight line of the analyzer. Therefore a large amount of particles from the plasma center can be expected, which is suitable for the high-energy distribution measurement. The scanning is performed with the pivot in front of the 10-O port by using a stepping motor, which is remotely controlled from -2 degrees to +31 degrees (0 degrees indicates the direction perpendicular to the 10-O port flange surface). The angles are equal to the pitch angle (the angle between the magnetic axis and the sight line) of 100 degrees and 40 degrees, respectively. The Web camera monitors the angle.

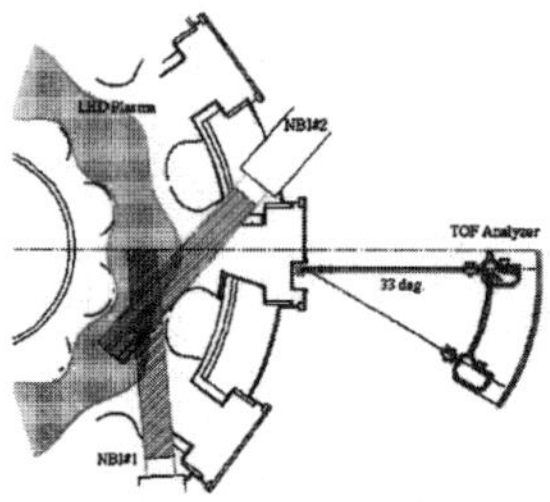

Figure 1. The apparatus of the neutral particle analyzer and LHD.

RESULTS IN NBI LONG DISCHARGE

The spatial resolved energy spectra can be observed during a long discharge of NBI plasma by continuously scanning the neutral particle analyzer[2]. Figure 2(a) shows the time evaluated (=angular distributed) 3 dimensional spectrum obtained by overlapping of three NBI plasma discharges. The pitch angle means the angle between the sight line and the central magnetic axis. In these discharges, the plasmas are initiated by the ECH heating, after that NBI#2 (Co-injection) sustains the plasma during 40 – 60 seconds. The scanned pitch angle is from 44 degrees to 74 degrees. The electron density measured by the interferometer is kept to be almost $2x10^{19}$ cm^{-3} during discharge except the plasma initiation phase. At the initiation phase by ECH plasma, the high-energy particle flux is enhanced as the density is too low and the background neutral density is high. The injected hydrogen neutral beam energy of NBI#2 is only 130 keV because the original ion source polarity is negative. The high-energy beam over the injection beam energy cannot be observed, as there is no ICRF heating. Figure 2(b) shows the spectra at some typical pitch angles. The shape of spectra is almost similar from 44 degrees to 53 degrees. However the spectra from 55 degrees are strongly varied. It reflects the injection pitch angle of the beam according to the simulation (53 degrees at Rax=3.75 m in simulation). The beam keeps the pitch angle at incidence until the beam energy becomes to the energy, which the pitch angle scattering is occurred by the energy loss due to the electron collision. The low flux region can be observed around 10-15 keV which is 15 time so the electron temperature. The energy region may be equal to the energy at which the pitch angle scattering is occurred. At the energy, the particle is scattered by the collision with the plasma ion and some of particles may run away from the plasma because they have a possibility to enter the loss cone (the ripple diffusion). Figure 2(c) shows the simulation result. According to the simulation, the loss cone can be expected at the 10 keV with the small angular dependence. The depth of the loss cone is deep at the small pitch angle. The hollow in the spectrum may be concluded to be the loss cone as the tendency is almost agreed with the experimental result.

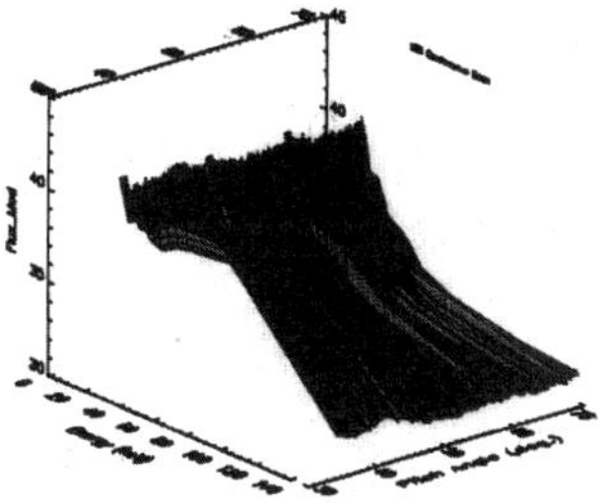

Figure 2(a). The spatial-resolved neutral particle energy spectrum in NBI long discharge.

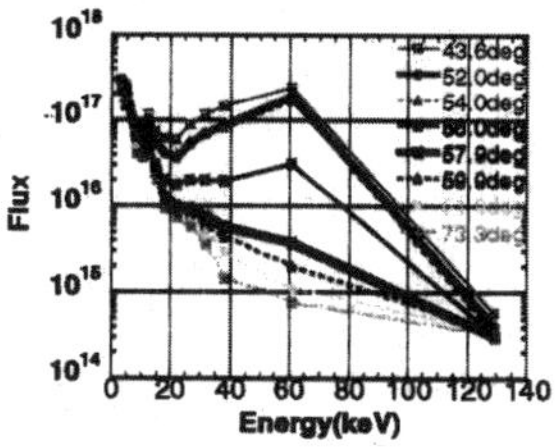

Figure 2(b). Energy spectrum of each pitch angle obtained by Fig. 2(a).

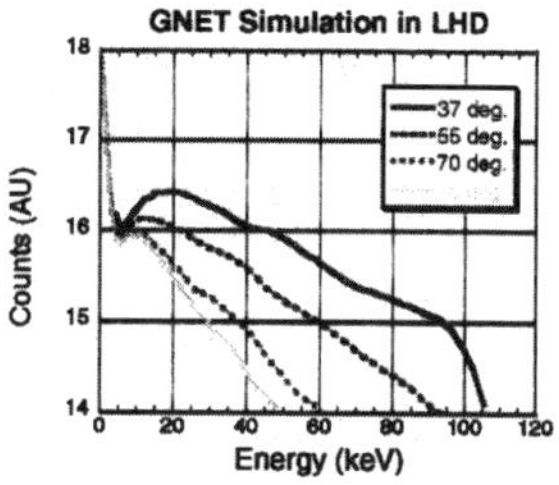

Figure 2(c). Energy spectrum of each pitch angle obtained from GENET simulation.

RESULTS IN ICH PLASMA

High-energy particles over 250 keV can be observed from ICH in NBI plasma. The pitch angle distribution can be obtained by using the same scanning method as the scanning in NBI plasma. Figure 4(a) shows the spectra during the ICH phase. The complicated energy spectrum can be obtained when the particle is integrated along the sight line because the high-energy particle has the pitch angle dependence. It is easy to compare with the predicted simulation result when the spectra are translated on to the velocity phase space. (Fig. 3(b)). The Vperp and V// can be obtained from the pitch angle and the energy in Fig. 3(b)[3]. The equatorial line shows the logarithm of the flux. The "butter-fly" shape structure (peak angle of 63 degree), which is well known in the theory, can be obtained. The difference of the sight line length (flux) is not so effective against the structure because the equatorial line is plotted in logarithmic scale. The pitch angle of particles accelerated by ICH is oblique rather than perpendicular against the magnetic axis. The fact shows the particles accelerated by ICH are not bulk ion rather than the NBI particles.

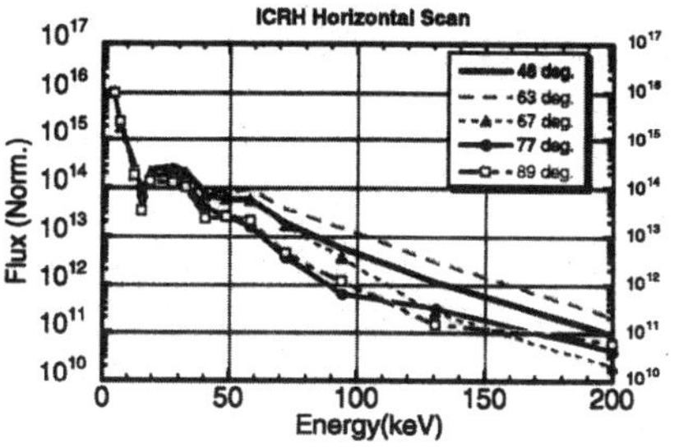

Figure 3(a). The energy spectra of horizontal scan in ICH plasma.

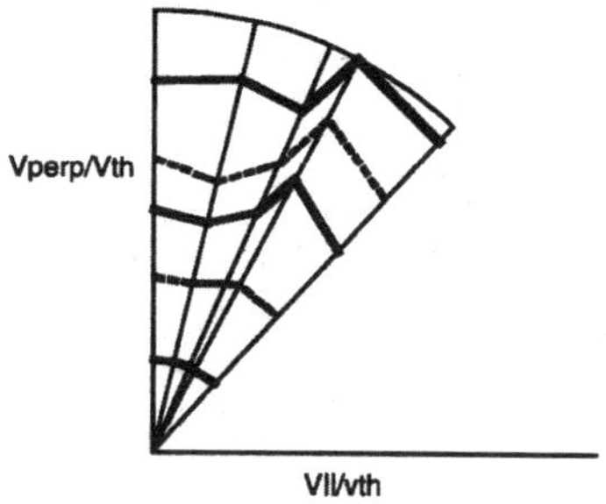

Figure 3(b). Flux plotting on the velocity phase space. The maximum flux is observed at the pitch angle of 63 degree.

RESULTS IN ICH LONG DISCHARGE

The energy spectra are measured when the analyzer is continuously scanned in the long discharge. 120 seconds on ICH plasma can be succeeded in the 4th campaign of LHD. The merit in the diagnostics of the long discharge is the capability of the accumulation of data and the spatial scanning under the almost same experimental condition. In the experiment, the analyzer is scanned from 60 to 90 degrees during three-60 seconds discharges (Figure 4). The experimental condition is as follows; the magnetic axis, the magnetic field, the stored energy and the density are 3.6m, 2.75T, 100kJ and $1x10^{-19}$ m^{-3}, respectively. ICH of the power of 1.0MW is applied from two antennas at 3.5U and 3.5L. It makes the resonance region around a/2 at the lower magnetic field side. The large amount particle at the high-energy region can be detected when the sight line crosses the resonance area over 65 degree. The large amount particle at the energy range of lower than 6 keV is observed between 70 to 80 degrees. It is the trapped particle accelerated by ICH near the plasma edge. It is not observed in the NBI plasma. The result is in agreement with the simulation.

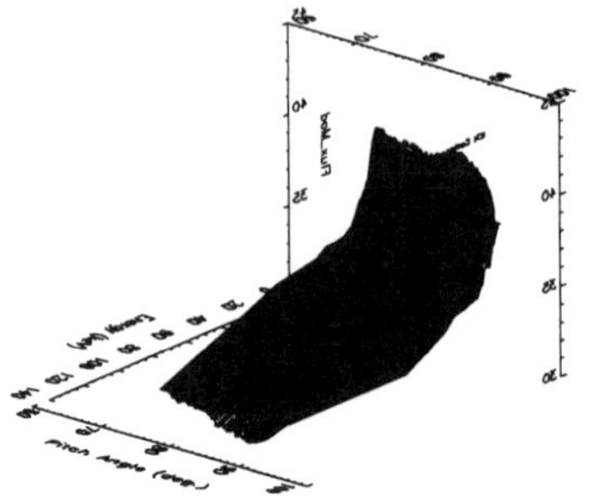

Figure 4. The spatial-resolved neutral particle energy spectrum in ICH long discharge.

CONCLUSION

The time-of-flight type neutral particle analyzer system can be provided the pitch angle distribution in NBI and ICRF heating plasmas. The hollow in the energy spectrum of NBI plasma may be the loss cone according to the comparison with the simulation. The spatial distribution of high-energy particles in ICH plasma can be also obtained. The butterfly structure of the distribution function based on the helically trapped particle can be observed.can be observed in NBI plasma.

REFERENCES

1. T. Ozaki, V. Zanza, G. Bracco, A. Moleti, B. Tilia, A. Sibio, S. Sudo, H. Nakanishi, M. Kojima, M. Shoji and G1/G2 experimental group, High Energy Neutral Particle Measurement System in Large Helical Device, *Rev. of Sci. Instrum.*, **71**(7), 2698-2703 (2000) .
2. T. Ozaki, S. Sudo, et al., Preliminary Results of the Neutral Particle Measurements in Large Helical, *Plasma Phys. and* Fusion Research., SERIES, Vol.3, 444-448 (2000) .
3. K.Saito, R.Kumazawa, T.Mutoh, T.Seki, T.Watari, T.Yamamoto, Y.Torii, N.Takeuchi, Y.Zhoa, A.Fukuyama, F.Shimpo, G.Nomura, M.Yokota, K.Kato, M.Sasao, M.Isobe, T.Ozaki, et al., STUDY OF HIGH-ENERGY IONS CREATED BY ICRF IN LHD, *Plasma Physics and Controlled Fusion* (to be submitted).

A NEW DIAGNOSTIC TO ACHIEVE NEUTRON SPECTROSCOPY IN A HIGH GAMMA BACKGROUND IN ICF EXPERIMENTS

M. Houry, L. Disdier and A. Fedotoff *

INTRODUCTION

The burn of deuterium-deuterium (DD) or deuterium-tritium (DT) fuel performed on inertial confinement fusion experiments provides neutrons, charged particles and photons in various energy. Studies on these particle emissions are of interest since they carry crucial information on experimental conditions required to achieve ignition. Neutron emissions give access to characteristics like the fusion yield, burn and bang times, fuel temperature, core dynamic and areal densities ($<\rho.R>$) of the capsule and the fuel. Each of these measurements requires specific and challenging diagnostic.

In the present work, we propose a novel spectrometer concept to determine the fuel-areal density ($<\rho.R>$). The method is based on the spectroscopy of secondary and tertiary neutrons produced in DD and DT capsules respectively [1]. Expected yield of secondary and tertiary neutrons (i), measurement of the high gamma background induced by primary neutron interactions in the environment (ii) and the performances of the proposed diagnostic (iii), are discussed.

SECONDARY AND TERTIARY NEUTRON SPECTROSCOPY AS A MEASURE OF $<\rho.R>_{FUEL}$

The DD or DT fusion reactions at the fuel temperature lead to primary neutrons (2.45 and 14.1 MeV respectively) and charged particles emission. Then, in-flight fusion between these products and fuel leads to high energy neutron emissions :

In a DD capsule, tritium ions produced by the DD fusion interact with the deuterium fuel. Energy spectrum and yield of these so-called secondary neutrons are a direct measurement of the DD fuel-areal density $<\rho.R>$.

In the case of a DT capsule, tritium or deuterium ions are scattered by primary neutrons; recoils ions then interact with atoms at rest in the fuel. This in-flight fusion is a third order process and when the range of the accelerated ion is higher than the $<\rho.R>$ the number of tertiary neutrons is proportional to $<\rho.R>^2$.

• Commissariat à l'énergie atomique, CEA/DAM-Ile-de-France, F-91680Bruyères-le-châtel ; michael.houry@cea.fr

However in the limit when ions are slowed down in the fuel, energy spectrum and yield of secondary and tertiary neutrons depend on the <ρ.R> and on the electron temperature of the DT fuel [2].

Experimental studies performed at Laser OMEGA (Rochester) have shown fuel-areal density[3] lower than 20 mg/cm^2. Based on this value, the slowing down of accelerated ions is assumed negligible and the ratio of secondary and tertiary neutrons to the number of primary neutron is about $10^{-1}.\langle\rho.R\rangle$ (DD capsule) and $5.10^{-3}.\langle\rho.R\rangle^2$ (DT capsule*). Hence, the flux is estimated to $2.5.10^3$ secondary neutrons/cm^2 and 1.6 tertiary neutrons/cm^2 at 10 meters and for 10^{13} primary neutron yield.

These values are dimensioning parameters for the neutron spectrometer. For example, dynamic and detection area must be consistent with the relative low tertiary neutron yield.

GAMMA BACKGROUND INDUCED BY NEUTRONS

The number of delivered primary neutrons is considerable ; on the OMEGA facility the 14.1 MeV neutron flux reaches 10^{14} in 4π. A part of these neutrons interacts with material surrounding the target (reaction chamber, supports, ground, walls, ...), leading to photon and charged particle emissions. Primary neutron induced background is not negligible and it is important to quantify the dose since it influences the diagnostic design : for large photon doses, some instruments need shielding or cannot run. Hence the expected yields of secondary and tertiary neutrons are low, and a spectrometer using a photon sensitive detector could be blinded.

Background calculations for the LIL facility

Calculations using the TRIPOLI code[4] have been performed for the LIL facility[5]. To obtain reliable results, the defined experimental area geometry includes target chamber, walls, grounds, diagnostics insertion systems, stairs, ...[6].

In Fig.1.1, the neutron and photon flux/cm^2 (at 10 meters from the target) are displayed as a function of time. The calculation has been scaled to 10^{13} primary neutrons of 14.1 MeV. About 70 ns after the DT fusion, the interaction between neutrons and the target chamber skin creates an intensive flux of photons which decreases in time ; signatures of hardware located in the hall are seen. After a time of flight of about 195 ns, the primary neutrons reach the diagnostic. The following neutrons are mainly scattered neutrons.

The mean energy of produced photons is 1.2 MeV. The photon dose is calculated in a 80 ns width time window starting on the first photon pulse. The measured dose is 0.4 $mRad_{air}$ at 8 meters.

The number of photons arriving within the same time window as tertiary neutrons is about 1.8 10^5 by cm^2 at 10 meters from the target and for 10^{13} primary neutron yield. Hence the photon flux is much higher than the tertiary neutron flux (see above).

- Due to the energy overlap between tertiary neutrons (≈ 9 to 30 MeV) and primary neutrons (≈ 14 MeV), only tertiary neutrons with energy between 16 and 30 MeV are considered.

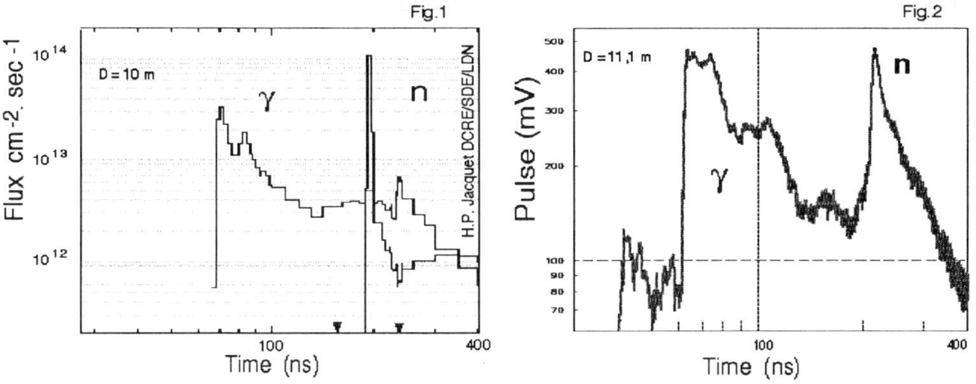

Figure 1. The TRIPOLI calculation displayed in Fig1.1, exhibits photon and neutron flux as a function of time of flight for the LIL facility (at 10 meters and for 10^{13} 14 MeV neutrons in 4π. Fig.1.2 shows the pulse measured with a detector located at 11.1 meters of the target in the OMEGA experimental hall.

Background measurements at OMEGA

Measurements of gamma background have been performed at OMEGA using a calibrated detector made of a plastic scintillator and a fast PM. Fig.1.2 exhibits the output signal measured as a function of time.

In Fig1.2, the main gamma rays pulse comes from the target chamber skin. The signal observed before is created by insertion mechanisms inside the chamber. The measured signal after the 14 MeV pulse include both contribution of gamma rays and scattered neutrons. The measured dose is about 0.1 $mRad_{air}$ and compared well with results of the calculation presented above.

The photon flux arriving within the same time window as tertiary neutrons is about 4.4 10^4 by cm^2 at 10 meters and for 10^{13} primary neutron yield.

EXPERIMENTAL SETUP : THE DEMIN SPECTROMETER

Theoretical and experimental studies show that secondary and tertiary neutron yields are low compared with the photon flux. Lead shielding is insufficient and induces perturbations in neutron measurements. Hence, we propose a novel diagnostic able to provide neutron spectroscopy with a relative photon transparency.

The instrument is based on a recoil ion gas detector called "DEMIN", Detector MICROMEGAS[7] for Neutrons (Fig.2). A neutron-to-proton (*np*) conversion foil is set in front of the gas detector ; with a plastic foil of 2 mm thick, the *np* conversion efficiency is about 2.10^{-3} and the probability of a photon striking an electron is 10^{-3} for the same thickness. Protons and electrons respectively produced by neutrons and photons enter then in the drift stage (0.5 mm thick) of the gas detector. Since the gas ionization depends on the charged particle mass, a proton creates more electron-hole (e^-h) than an electron; for example, 1 MeV protons provide ≈ 100 e^-h, whereas 1 MeV electrons are below the ionisation threshold (in 0.5 mm of helium at 1 bar). The detection efficiency of a charged particle in the gas chamber is close 1.

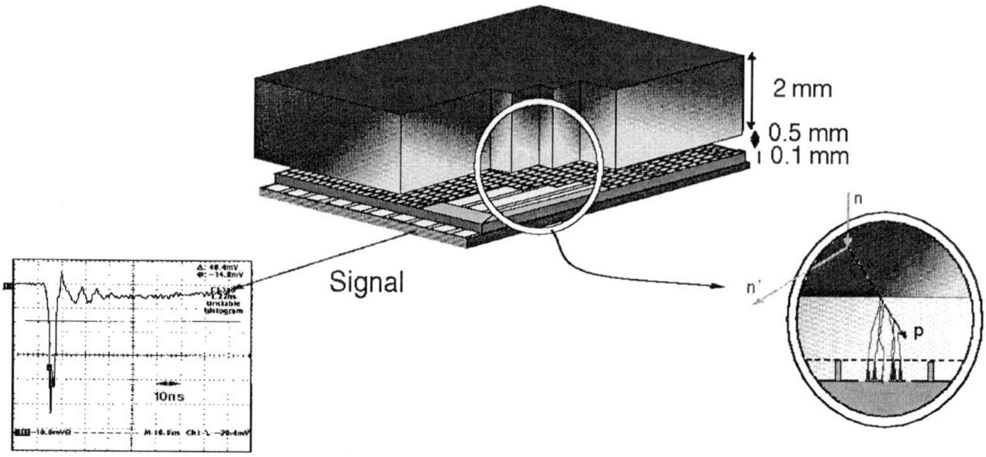

Figure 2. DEMIN experimental set-up.

In the second stage of the gas detector(<100 μm), high electric field induces avalanche phenomena with a gain of 10^3 to 10^5 (depending on gas nature, electric field, ...). The signal is then collected on parallel strips. The signal rise time is about 1 ns, and exhibits a width of 5 ns with 0.5 mm of isobutene gas (a pulse sample is shown in Fig.2).

CONCLUSION

The extensive photon background induced by neutrons in ICF experiment will perturb secondary and tertiary neutrons measurements. On account of this background a novel neutron diagnostic has been designed for the fuel-areal density measurement. The method is based on the spectroscopy of the secondary (DD capsule) or tertiary (DT capsule) neutrons with a relative photon transparency. The association of a 2 mm converter foil with a thin and high-speed gas chamber exhibits adequate detection efficiency. The short rise time (1ns) provides a good resolution for neutron energy spectroscopy by time of flight measurements. The dynamic is large because of the versatile characteristics of the detector: easy to move, modular area.

REFERENCES

1. H. Azechi, M.D. Cable, and R.O. Stapf, *Laser and Particle Beams*, 9 (1991) 119.
2. C. Li and R.D. Petrasso, *Phys. Plasmas*, 2 (1995) 2460.
3. LLE Review, *Quarterly Report*, Vol. 86, (2001), 54. http://www.lle.rochester.edu/.
5. M. André and F. Jequier, *proceeding of the 4th Symposium on Current Trends in International Fusion Research*, 12-16 March 2001, Washington-US.
6. H.-P. Jacquet, private communication.
7. Y. Giomataris et al., *Nucl. Instr. And Meth.* A 376, (1996) 29.
8. D.C. Wilson, W.C. Mead, L. Disdier, M. Houry, J.-L. Bourgade, and T.J. Murphy, to be published in *Nuc. Instrum. Method* A.

NATURAL DIAMOND DETECTOR BASED SPECTROMETERS FOR MAGNETIC AND INERTIAL CONFINEMENT FUSION

Anatoli V.Krasilnikov, Andrey G.Alekseyev, Vladimir N.Amosov, Yuri A.Kaschuck, Igor N.Rastyagaev*

1. NDD BASED DT NEUTRON SPECTROMETRY AND FLUX MONITOR SYSTEM FOR ITER-FEAT

ITER-FEAT radial and vertical neutron cameras with included spectrometers should provide measurements of: total neutron yield, emission profiles and ion temperature in the ranges 3×10^{16}–3×10^{20} n/s, ~4×10^{14}–4×10^{18} n/s/m^3, and ~0.5–50 keV, respectively, with required accuracy, temporal and spatial resolution. So neutron cameras with incorporated spectrometers will provide essential part of ITER-FEAT requirements to neutron diagnostic and could be discussed as basic neutron diagnostic complexes of ITER-FEAT.

Compact natural diamond detector (NDD)[1,2] based neutron spectrometers and flux monitors were proposed for the application in ITER[3,4]. Due to small size they could be installed inside collimators of the radial and vertical neutron cameras (see Figure 1) in front of other detectors without noticeable disturbances of neutron flux.

After the tests at DT neutron generators SNEG-13, FNG, FNS[5] several NDDs have provided DT neutron spectrometry and flux monitoring on TFTR[6], JT-60U[7] JET[8] and escaping charged fusion product spectrometry inside Tore-Supra vacuum chamber[9]. Energy resolution and sensitivity for DT neutron spectrometry of used NDDs reached 1.2-3% and 1.7×10^{-5} counts/(n/cm^2). Useful for flux monitoring sensitivities reached value $2\div3\times10^{-4}$ counts/(n/cm^2). This provided neutron flux monitoring with useful count rate up to $6.7*10^5$ cps at total event rates up to $8.3*10^7$cps during JET DTE1 experiments. Major issue of NDD effective application as DT neutron flux monitors in ITER-FEAT is relatively low sensitivity. Total detector sensitivity could be increased by increase the number of sensitive elements inside detector and by installation of hydrogen contained radiators in front of the diamond sensitive elements. In such an arrangement eachNDD will measure neutrons and protons recoiled from radiator. Detector blocks consisting of one NDD and one polyethylene radiator of different thickness were manufactured and

* All authors from Troitsk Institute for Innovating and Fusion Research TRINITI, Troitsk, Moscow region, 142092, Russia, e-mai: anatoli@triniti.ru

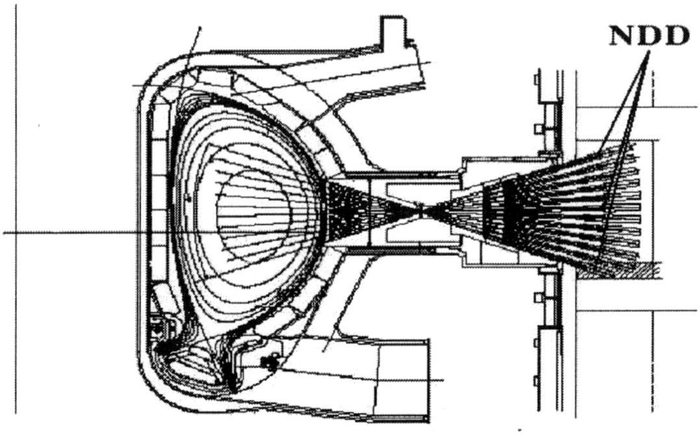

Figure 1. Arrangement of the array of compact NDD based high resolution neutron spectrometer located inside radial neutron camera collimators.

studied in DT neutron flux of NGM-17[10] neutron generator. Detector block pulse height spectra essentially increased in the energy range 1.6–12 MeV. Additional counts were related to protons recoiled from polyethylene. Useful for neutron flux monitoring detector sensitivity was obtained by the integration of the pulse height spectra in mentioned energy range. The improvement of detector sensitivity is shown in Figure 2. These studies show that total sensitivity of the detector block could be increased by ~1.8 times. Saturation at high radiator thickness takes place due to slowing down of the protons inside radiator.

The detector block consisting of 10 NDDs and radiators will have useful sensitivity $\sim 7\times10^{-3}$ counts/(n/cm^2). Being installed inside collimators of neutron cameras with flux $\sim 3\times10^{9}$ n/cm^2/s it will provide neutron flux and fusion power profiles measurements and control with temporal resolution 1 ms, accuracy 10% and dynamic range ~200.

Due to high energy required for atom displacement from diamond crystal lattice NDDs have higher radiation hardness then other semiconductor detectors. To determine NDD life-time in ITER-FEAT experiments we have studied the NDD radiation hardness during it in-situ neutron irradiation in nuclear reactor using ^{241}Am α-source[11]. Measured degradation of the detector charge collection efficiency (CCE) related to induced by neutrons and gamma rays crystal lattice defects is shown in Figure 3. NDD count rate that was estimated as a total number of counts in the α-source peak remained unchanged during the whole run of tests. NDDs show stable spectrometric performance during in-situ studies of detector properties in fission reactor up to accumulated neutron fluence $\sim 2\times10^{14}$n/cm^2. Even at fluence level as high as 10^{15}n/cm^2 NDD can be reliably used for rough spectrometric measurements, for example in counting mode. Taking into account measured restoration of the NDD spectrometric properties at elevated temperatures, applying external and internal excitation with red (λ=800 nm) light source on the base of the data obtained in ref.11 maximum acceptable DT neutron fluences could be estimated as $\sim 5\times10^{14}$n/cm^2 and $\sim 10^{16}$n/cm^2 for spectrometric detectors and NDD based neutron flux monitors. Design of ITER's NDDs should provide the possibilities of in-place thermal annealing at temperatures 250-300^{0}C and internal excitation with red light source.

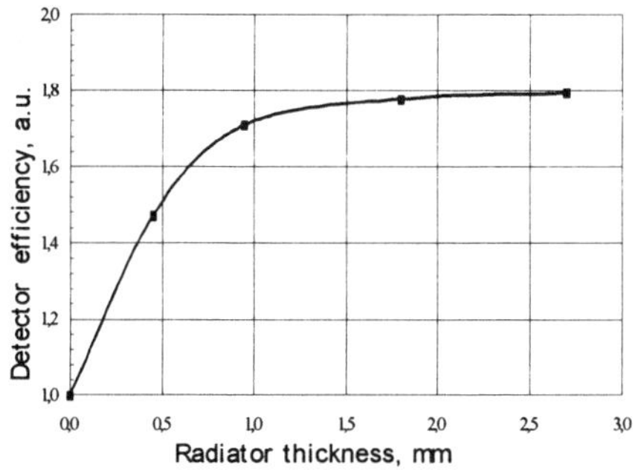

Figure 2. Dependence of the efficiency of NDD with polyethylene radiator upon radiator thickness.

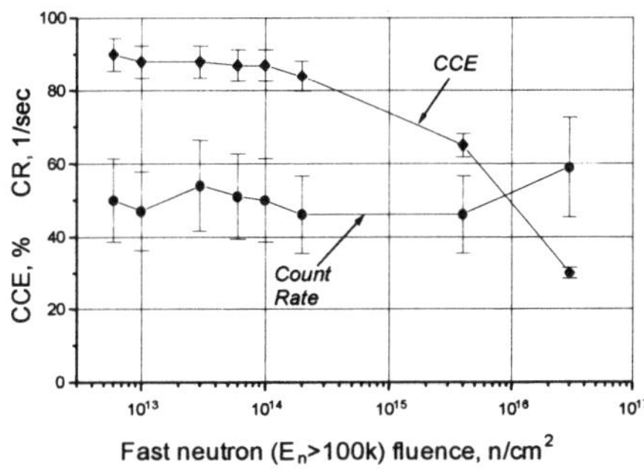

Figure 3. Charge collection efficiency and count rate variation upon the neutron fluence.

2. PERSPECTIVES OF THE NDD BASED CHARGE EXCHANGE ATOM SPECTROMETERS FOR FAST CONFINED ION STUDIES IN ITER-FEAT

NDDs provided spectrometry and flux monitoring of fast hydrogen charge exchange atoms during H^+ minority ICRH on TFTR[12] and LHD[13]. Transport and confinement of fusion α-particles in reactor could be essentially affected by MHD instabilities. The array of compact NDD based charge exchange atom spectrometers with lines of sight tangential to several magnetic surfaces could provide measurement of fast confined ion redistribution across the magnetic surfaces during ITER-FEAT toroidal Alfven eigen mode activity and other MHD instabilities. Installed behind stripping foil and analysing electric mirror NDD could provide fast charge exchange atom spectrometry due to such properties as high energy resolution and radiation resistance, low sensitivity to neutrons and gamma rays. Diagnostic arrangement with NDDs located behind analysing electric mirrors with varying electric field will provide measurements of fast atoms with energies in the ranges of interest and will reduce background induced by neutron and gamma rays.

3. POSSIBILITIES OF THE NDD APPLICATION IN ICF EXPERIMENTS

Large fusion product yield in major modern (OMEGA, GEKKO-XII, ISKRA-5) and especially future (National Ignition Facility and Laser Megajoule) laser driven inertial confinement fusion (ICF) experiments make it possible to use NDD as fusion product time-of-flight spectrometer for the characterisation of ICF plasma ion component, namely to provide information about fuel and pusher ρR and fuel temperature[14].

Several NDDs were developed and used for UV and X-ray flux measurements and fast electron time-of-flight spectrometry[15] carried out in TRINITI during liner compression experiments in ANGARA-5 installation and two experiments devoted to study the laser produced plasma of solid target in "TIR-1" and "Mishen-2" installations.

Large saturation velocity of charge carries drift in diamond (~10^7 cm/s) provides the possibility to create NDDs that have charge collection time ~0.5 ns and corresponding response rise/fall time. The examples of currents induced in 84 μm thick NDD by single 5.5 MeV α-particles are shown in Figure 4. Such unique NDD properties as high radiation resistance, blindness to visible light, relatively low sensitivity to gamma radiation and very small size make it possible to install and use it for fast charged particle time-of-flight spectrometry at short distance from the target in ICF experiments. Small size of NDD gives possibility to create the multi line of sight NDD based time-of-flight spectrometry systems enable to provide information about implosion symmetry.

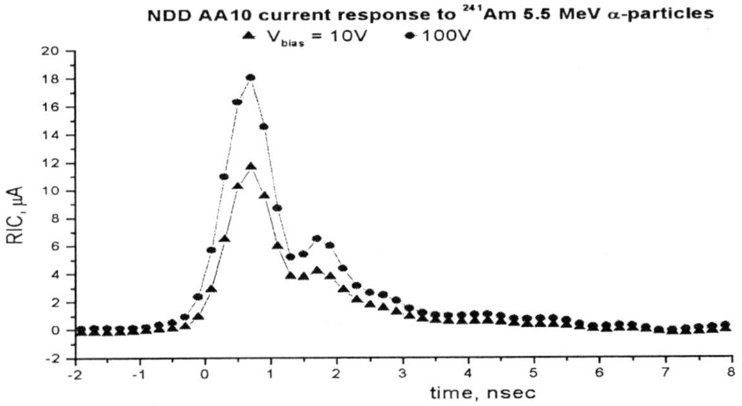

Figure 4. Single α-particle radiation induced current measured from NDD with thickness 84 μm.

4. REFERENCES

1. E.A.Konorova, S.F.Kozlov, Sov. Phys. Semiconductors **4**, 1600 (1971).
2. A.V.Krasilnikov, V.N.Amosov and Yu.A.Kaschuck, IEEE Trans. Nucl. Sci., **45**, 385 (1998)
3. A.V.Krasilnikov in "Diagnostics for ITER" P.E.Stott, et.al. Editors, p.435, Plenum Press, New York and London (1995).
4. A.V.Krasilnikov, in "Diagnostic for Experimental Thermonuclear Fusion Reactors 2", P.E.Stott, G.Gorini, P.Prandoni and E.Sindoni , eds, p.439, Plenum Press, New York and London, 1998.
5. A.V.Krasilnikov, J.Kaneko, M.Isobe, et. al., Rev. Sci.Instrum. **67**, n.11, (1996).
6. A.V.Krasilnikov, E.A.Azizov, V.S.Khrunov, A.L.Roquemore, K.M.Young, Rev.Sci.Insrtrum **68**, 553 (1997).
7. M.Isobe, T.Nishitani, A.V.Krasilnikov, et al., Fus.Eng.&Design **34-35** 573 (1997).
8. A.V.Krasilnikov, V.N.Amosov, P. van Belle, O.N.Jarvis and G.Sadler to be published in Nucl.Instrum&Methods in 2001.
9. A.Alekseyev and G.Martin, Nucl.Instr.&Meth.Phys.Research **A 417** 400 (1998)
10. V.N.Amosov, Yu.A.Kaschuck, A.V.Krasilnikov, D.V.Portnov and S.V.Siromukov, Rev Sci. Instrum., **70**, n.1 (1999).
11. A.G.Alekseyev, V.N.Amosov, Yu.A.Kaschuck, A.V.Krasilnikov, and D.V.Portnov to be published in Nucl. Instrum. & Methods in 2001.
12. A.V.Krasilnikov, S.S.Medley, N.N.Gorelenkov, R.V.Budny, D.S.Darrow, A.L.Roquemore, Nucl.Fus., **39** 1111 (1999).
13. M.Sasao, S.Murakami, M.Isobe, A.V.Krasilnikov, et.al. Presented at 18 IAEA Fusion Energy Conference, Sorento, October 2000.
14. A.V.Krasilnikov, A.G.Alekseyev, V.N.Amosov, D.V.Portnov, I.N.Rastyagaev, in Proceed. of Inertial Fusion Sciences and Applications 99, edited by C.Labaune, W.J.Hogan and K.A.Tanaka, p.937, (2000).
15. A.V.Krasilnikov, K.N.Makarov, I.N.Rastyagaev, Yu.A.Satov, Yu.B.Smakovskiy, A.G.Alekseyev, V.N.Amosov, "Time of Flight Electron Spectrometry on TIR Using Natural Diamond Detector" to be published in Nucl. Instrum. & Methods in 2001.

LASER FUSION FUEL ION TEMPERATURE DIAGNOSTICS

J. Chen, Z. Zheng, H. Peng, B. Zhang, Y. Ding, and C. Tang*

INTRODUCTION

The fusion fuel ion temperature is a very important implosion dynamics parameter in initial confinement fusion (ICF) research. It is often determined by the neutron time of flight (TOF) technique,[1, 2] which utilizes the neutron Doppler broadening due to the center-of-mass motion of the reacting ions in the fusion fuel. However, the relative yield method has an advantage when the neutron yield is low. On this condition the detector must be placed at a closer distance from the fuel-filled implosion capsule for enough statistical counts, which will cause a narrower neutron TOF width and thus a greater error as the detection system has a certain response time. But it is possible to employ a suitable fuel mixture in which different reactions occur simultaneously although the implosion neutron yield is sensitive not only to the ion temperature alone. Supposing that the final density ratio of the different fusion fuel components in the compressed microballoon core remains constant as compared with the initial one, the relative yield depends only on the ratio of averaged $<\sigma v>$ for the different reactions, where σ and v denote the reaction cross section and the relative velocity, respectively. Since the dependence of $<\sigma v>$ on temperature is different for different reactions, the relative yield ratio can reflect the ion temperature.[3, 4]

In the initial implosion experiment performed at the upgraded Shenguang laser facility (i.e., Shenguang II), exploding pusher targets and ablating targets were directly driven by incident laser. The former created higher yields (on the order of 10^9 neutrons per shot) than the latter (on the order of $10^7 \sim 10^8$ neutrons per shot). The DT-plasma ion temperatures of the exploding pusher targets have been obtained by the neutron TOF technique.[5] Not only neutron yields but also ion temperatures were lower for the ablating targets, which led to lower ratios of DT to DD neutron yields and does good to the determination of the ion temperatures by yield ratio. Now ion temperatures of the ablating targets are inferred by relative neutron yield measurements.

---

* J. Chen, National Key Laboratory of Laser Fusion, Research Center of Laser Fusion, China Academy of Engineering Physics. P.O. Box 919-986, Mianyang, Sichuan, 621900, China. E-mail: jbchen@263.net.

EXPERIMENTAL CONDITIONS

Shenguang II delivers eight beams of 1.06 μm light, each having about 800 joules of peak energy for 1 ns pulses. The target chamber is 160 cm in diameter and 5 cm in thickness. The wall where implosion neutrons passed through was a 0.5-cm-thick aluminum flange that had no appreciable effect on the energetic neutrons.

The detector was placed at 1 m from the center of the target chamber. It consists of a plastic scintillator optically coupled to a microchannel plate photomultiplier tube (MCP-PMT). It is 1 inch in thickness and 1.5 inches in diameter. The MCP-PMT has a 1.8 cm diam photocathode, whose range of spectral response is 300-650 nm and the peak wavelength is 375 nm. The typical gain is 10^5. The sensitivities of the detector to DD neutrons and DT neutrons were absolutely calibrated *in situ* with a DD pulse neutron tube and a DT pulse neutron tube. The distance between the detector and the neutron source *L* was 1 m, exactly the same as that in the experiment conducted on Shenguang II. This detector has a high sensitivity, fast response, low cost, and small volume. It is convenient to carry and to use. DD neutrons above several times 10^5 can be detected at a distance 1 m from the implosion capsule. The detector can not be closer to the capsule, otherwise the higher yield DT neutrons will make it saturated.

EXPERIMENTAL RESULTS

All the capsules employed as exploding pusher targets in the experiment were glass microshells containing an equimolar DT fuel mixture with 0.9 μm in thickness. They were driven directly by eight beams of incident laser with a pulse width of approximately 0.9 nanoseconds and a total energy of 4000 joules. Figure 1 was the oscilloscope trace of a measured typical neutron signal from such a directly-driven implosin. The detector was 1 m from the implosion capsule and the three peaks occurred at about 10 ns, 26 ns and 53 ns, respectively, which proved that they were the peaks of X-ray, DT neutrons and DD neutrons, respectively.

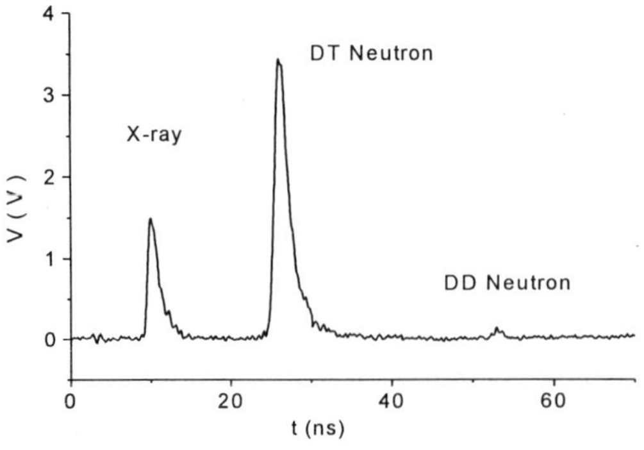

Figure 1. The oscilloscope trace of a measured typical neutron signal.

Table 1. The measurement results for ablating targets

shot No.	$D/\mu m$	P/Mpa	E_t /J	I_t /W	Y_{DT}	T_i/keV
14	495	0.63	3084	3.8×10^{12}	9.9×10^7	1.8
17	531	0.48	3683	4.0×10^{12}	3.5×10^8	3.2
21	508	0.49 + 4% Ne	3656	3.9×10^{12}	8.1×10^7	1.4
22	549	0.35	3694	3.8×10^{12}	1.3×10^8	2.9

Table 1 lists the measured results of fuel ion temperatures. D and P indicate diameter and pressure of DT-filled capsule, E_t and I_t indicate energy and power of the incident laser reaching the capsule surface, Y_{DT} and T_i indicate DT neutron yield and fuel ion temperature, respectively. On shot No. 21, about 4% inert gas Ne was mixed into the fuel. Large errors (about ±30% ~ ±50%) were introduced in the ion measurement as a result of the low DD neutron yields and poor signal-noise ratios in the DD neutron measurements.

ACKNOWLEDGEMENT

The authors would like to thank the staffs on Shenguang II for cooperative efforts. This work has been conducted under the auspices of National Hi-Tech ICF Committee of China under contract 416-3-1.12.

REFERENCES

1. B. A. Remington, R. A. Lerche, and M. D. Cable, Ion temperature analysis of implosionsof DT-filled capsules, *Rev. Sci. Instrum.* **61**(10), 3128-3130 (1990).
2. N. Izumi, K. Yamaguchi, T. Yamagajo, T. Nakano, T. Kasai, T. Urano, H. Azechi, . Nakai, and T. Iida, A highly efficient neutron time-of-flight detector for inertial confinement, *Rev. Sci. Instrum.* **70**(1), 1221-1224 (1999).
3. R. R. Goforth, F. J. Mayer, H. Brisk, and R. A. Cover, *J. Appl. Phys.* **47**(11), 4850-4850 (1976).
4. V. W. Slivinsky, H. G. Ahlstrom, J. H. Nuckolls, J. T. Larsen, B. W. Weinstein, K. G. Tirsell, E. K. Storm, and G. R. Leipelt, *J. Appl. Phys.* **49**(3), 1106-1109 (1978).
5. J. B. Chen, Z. J. Zheng, H. S. Peng, B. H. Zhang, Y. K. Ding, M. Chen, H. S. Chen, and T. S. Wen, Fusion fuel ion temperature diagnostic for directly driven implosions, *Rev. Sci. Instrum.*, **72**(9), 3534-3536 (2001).

X-RAY IMAGING SYSTEM WITH CONTINUOUS TIME RESOLUTION

Yefim Aglitskiy, Max Karasik, Victor Serlin, Stephen Obenschain and Carl Pawley*

1. INTRODUCTION.

A monochromatic x-ray imaging system based on the Bragg reflection from spherically curved crystals [1,2] has been successfully used on the NRL Nike KrF laser for the past five years [3,4,5]. Compared to other x-ray imaging techniques like pinholes, Kirpatrik-Baez microscopes, and Fresnel lenses, a curved crystal imager has several advantages. It is a high throughput, high spatial resolution and, by nature, monochromatic diagnostic technique all at the same time. At NRL, a curved crystal imager has been used in combination with a framing camera to provide sequences of high-spatial-resolution still images taken with a time resolution of 200 ps [3]. Up to four snap shots that correspond to four different times were available to study such fast process as Rayleigh-Taylor (RT) growth of mass perturbations in ablatively driven targets [6-8].

Undoubtedly, laser target distortion due to the Rayleigh-Taylor (RT) instability remains a critical issue of the inertial confinement fusion (ICF) program. However, our recent diagnostic development has been primarily motivated by the need to observe an evolution of the mass perturbations prior to the RT growth phase. The perturbations amplified by the RT instability grow from the so-called "seeds", which are linearly related to the non-uniformities of the target and the laser beam or the x-ray radiation field driving it. The seed amplitudes are known to form during the early-time period that includes a shock wave transit from the front to the rear surface of the target, and a rarefaction wave transit in the opposite direction. During this time interval, the areal mass perturbations caused by all sources of non-uniformity are expected to oscillate. The physical mechanisms driving the oscillations depend on whether the perturbations are

* Yefim Aglitskiy, Science Applications International Corporation, McLean VA, Max Karasik, Victor Serlin, Carl Pawley, Stephen Obenschain, Plasma Physics Division, Naval Research Laboratory, Washington DC 20375.

initially at the front surface of the target (laser imprint, front surface roughness) or at its rear surface ("feedout", which means the transfer of mass perturbations from the rear to the front surface of a driven target). Few experimental results are available to quantify the linear relationships between the measurable initial non-uniformities and the seed mode amplitudes. A good understanding of these relationships is necessary to ensure a reliable simulation of the overall perturbation growth, and hence, of the target performance.

However, measurements of mass perturbation evolution due to Richtmyer-Meshkov (RM) instability (ripples on the front surface) or feedout (ripples on the back side) related oscillations present new challenge to imaging diagnostics because of essentially non-monotonic evolution of the processes under study [9]. In these cases, diagnostics that take pictures incrementally over relatively long period of time are difficult to apply since it is easy to loose important information or even completely miss the effect. More snap shots require more crystal-backlighter pairs, which is technically difficult and ultimately unfeasible. To extend the capabilities of our diagnostic technique we modified the Nike imaging diagnostic setup by adding a streak camera to the system, which made it possible to analyze continuous time behavior of an x-ray image.

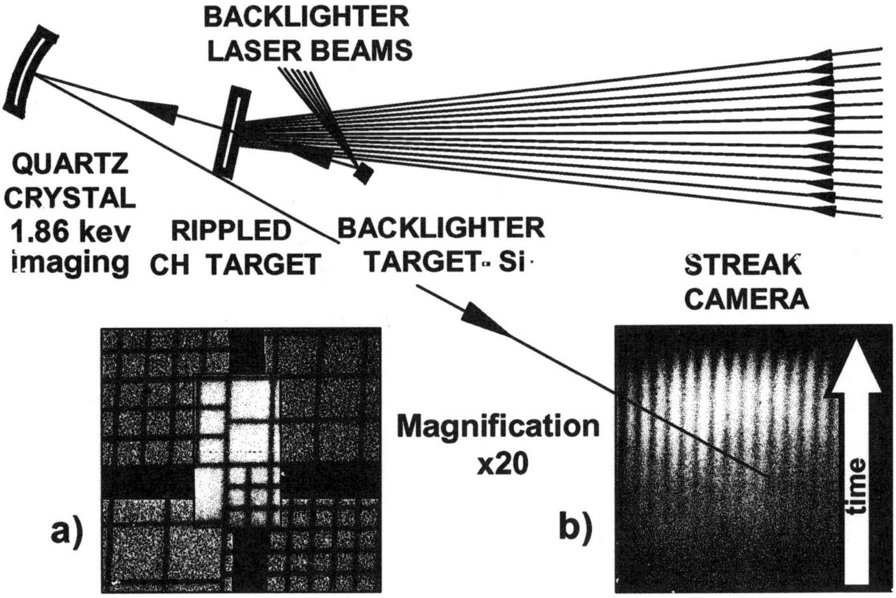

Fig. 1. The general scheme of the experiment. Also shown are: (a) test images of a 150-to-400 lpi composite mesh obtained on x-ray film (quadrants) and on an open photocathode of the streak camera (insert in the middle); the wires of 7 μm width are clearly seen on both images, though the streak camera image obviously has less spatial resolution; (b) an actual streak record showing the RT growth of areal mass variation – peaks and valleys are more pronounced at late time. Four-quadrant composite mesh was used for pointing of the streak camera slit onto the focal spot center with the accuracy of 20-30 μm.

Spherically bent crystal can provide: 1D spatially resolved x-ray spectra, monochromatic images of the laser plasma blowoff, etc.[3], but in our recent experiments we were particularly interested in the measurements of areal-mass-variation in rippled and flat plastic targets ablatively accelerated by the Nike laser. The setup provides spatial resolution in one relevant direction, i.e. along the wave vector of the ripple on the target surface, producing 1D streak records[10]. In addition to being capable to obtain qualitatively different experimental results the new modified system can do it using just one crystal and one backlighter. A similar set up with a thin slit as an imaging device has been used in NRL earlier for pioneering measurements of the Rayleigh-Taylor growth rates [11].

2. CHARACTERISTICS OF THE EXPERIMENTAL SETUP.

The energy of 12 Nike beams, ~500 J, was delivered to a silicon backlighter target, producing x-rays that backlight the main target for about 5 ns. The spherically curved quartz crystal selected the resonance line of the He-like Si (1.86 keV) and projected a monochromatic image of the target on the slit of an x-ray streak camera. We used a quartz crystal with the cut 1011 and a radius of curvature of 200 mm.

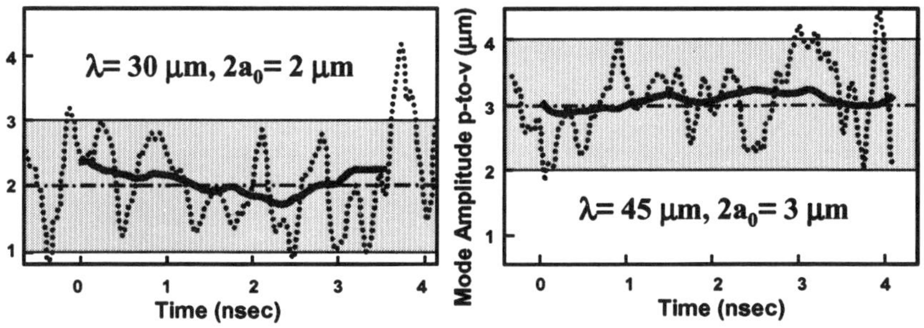

Fig. 2. MTF tests taken for two basic targets that were used in our experiments. A mode amplitude determination routine, which is based on the fast Fourier analysis of the experimental image, was applied to the streaked backlit images of undriven targets with known amplitudes. A variation of intensity predominantly due to noise is seen on the experimental lineouts (dashed lines). Being averaged over longer time (solid lines), the lineouts for both targets seem to reproduce expected amplitudes. Amplitude of the noise variation gives us an idea of an experimental error, with which we can measure the mass perturbations.

With 1.86 keV probing energy we were able to study 40 mm to 90 mm thick plastic (CH) targets either flat or rippled with perturbation wavelength λ = 30 µm or 45 µm. The streak records were taken with a time resolution of 170 ps. This is sufficient for the 0.5 ns characteristic times of interest. The large field of view (500 µm) combined with the large flat top (400 µm) of the laser focal spot gave us more ripples available for Fourier transform analysis, thus ensuring confidence in determining the dominant mode amplitude.

The spatial resolution of the x-ray optical system itself was tested with the help of x-ray film as a detector and was found to be of order 6-7 µm, (Fig. 1). High throughput of the x-ray optical system allowed us to magnify the images by a factor of 20 to compensate for the modest resolution of the streak camera.

Despite the high magnification of the system, the overall spatial resolution was still limited by the streak camera. In order to quantify the resolution of the x-ray crystal/streak camera system, a measurement of its modulation transfer function (MTF) was necessary. The MTF of the entire diagnostic system was obtained by imaging a knife-edge target. The derivative of the measured edge profile gave the line-spread function (LSF) of the system, from which the MTF could be obtained by Fourier transform[12]. In order to reduce the noise introduced by differentiation, a non-linear least-squares fit was applied to the LSF. It was found that the LSF is well fitted by the sum of a Gaussian and a Lorentzian profiles. The MTF was then obtained by applying Fourier transform to the resulting fit. For magnification of 20x, the MTF at λ=45 and 30 μm was found to be 0.6 and 0.4, respectively, and was used to correct the measured mode amplitude in the experiments. The MTF was also directly verified by imaging of undriven targets with predetermined amplitudes (Fig. 2). Since the spatial resolution of the system is limited by the spatial resolution of the streak camera it is possible to improve the MTF of the whole system by increasing of the magnification of the X-ray optical system. Fig. 3 shows significant improvement in the image quality when the magnification was changed from 20 to 30. Due to the decrease in the photon count we were able to use the

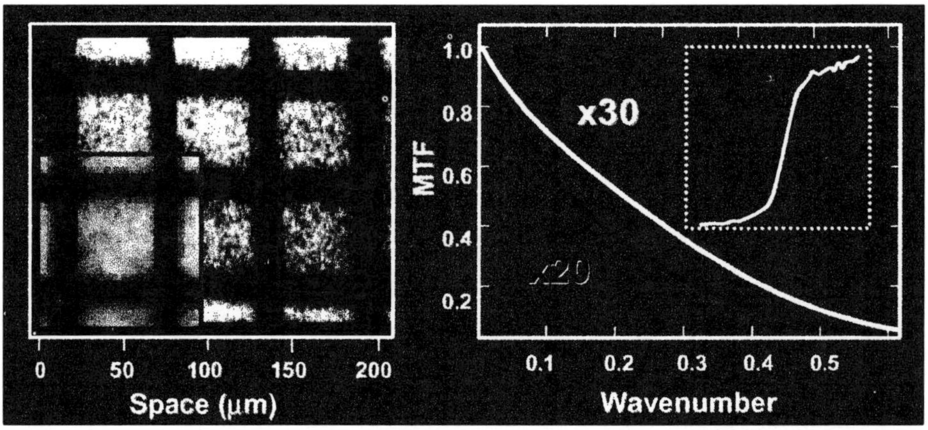

Fig. 3. Higher magnification of an x-ray imaging system leads to an improvement in the overall resolution of the diagnostics. Magnified (30x) image of a 400lpi test mesh is shown on the left side of the picture. Insert corresponds to the image taken with the lower magnification of 20x. Visibly it is more blurred than the image taken with higher magnification. The graph on a right side represents the MTFs measured for two magnifications. Insert shows the lineouts of the knife-edge images. The white curve corresponding to higher magnification is steeper than the black low-magnification one.

higher magnification only for targets with the thickness of 40 μm.

Under our experimental conditions it was very important to have an entrance slit of a streak camera aligned at the center of the target image as accurate as possible. We used an image of four-quadrant composite mesh as a test object in our pointing capability test (Fig.1. a). It was done with the streak camera operating in time integrated 2D mode. At the magnification of 20 and the crystal-to-detector distance of about 2 m were able to position the target image on the entrance slit of the streak camera with an accuracy that corresponds to 20-30 μm on the target. In 1D streak mode, with the help of the four-quadrant mesh we were able to adjust an image position along the slit with the same accuracy.

3. EXPERIMENTAL RESULTS.

An example of the measurements of RT exponential growth is shown on Fig.1b. The initial peak-to-valley amplitude of the sinusoidal ripples was in submicron region. For the diagnostics described it is the easiest case since initial mass perturbations grow very fast (up to two orders of magnitude) and become visible very soon after the beginning of the laser driving pulse. Relatively low thickness of the plastic target of about 40 μm allows for high signal-to-noise ratio, which makes subsequent Fourier analysis of dominant mode amplitude more reliable. A measurement of the areal mass evolution due to the RM instability has some similarity to the RT measurements (ripples on the front side of the target) but it is substantially different because of its oscillatory nature. The amplitude of the initial mass perturbation increases only by a factor of 5 and has to be high enough to be measured from the very beginning of the process under study.

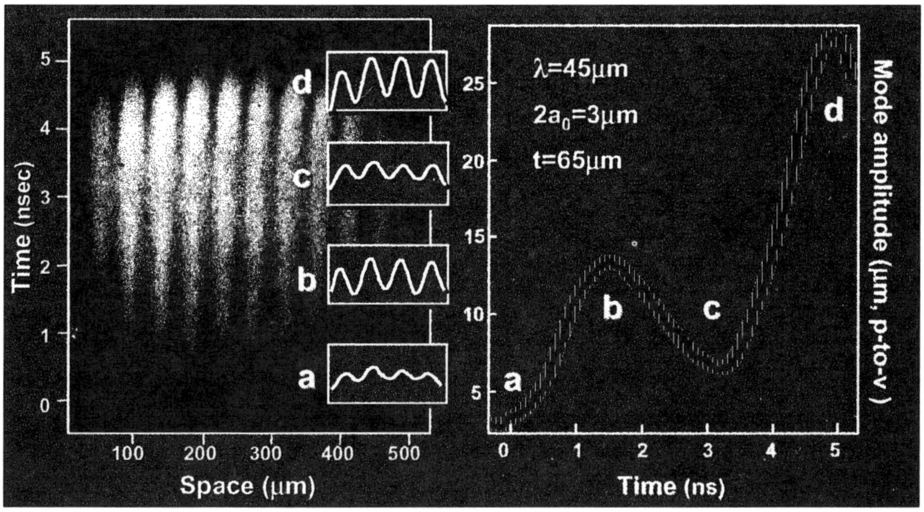

Fig. 4. Areal mass evolution due to the Richtmyer Meshkov instability. Original streak is shown at the left side of the picture along with the amplitude lineouts taken across the ripples at four benchmarking times. Time dependent Fourier amplitude (right) is also marked with four important times: beginning of the pulse, time of maximum growth, etc.

The figure 4 shows the RM instability for a 65 μm thick CH target rippled on the front side with perturbation wavelength $\lambda = 45$ μm. This was short enough to make the oscillation half-period fit into the Nike pulse duration, and yet long enough to be resolvable by our diagnostics. The initial peak-to-valley ripple amplitude was 3 μm. This combination of perturbation wavelength and initial amplitude was adequate for an experiment aimed at measuring the the early-time ablative RM phase, when the perturbations of all wavelengths oscillate rather than grow exponentially, so that the higher harmonics, being initially small, remain small until the onset of the exponential RT growth.

The figure shows an original streak record (left) and the time dependence of its MTF-corrected Fourier component (right, peak-to-valley amplitude) taken from a target, irradiated with peak intensity of 5×10^{13} W/cm^2. The thickness of the hatched area approximately corresponds to the experimental uncertainty. The peaks and valleys of mass variation are clearly visible on the streak record. Both light and dark stripes are less discernible between 2.5 and 3.5 ns, as if they were smeared in a vertical band ~ 0.5 ns wide, indicating a minimum in the perturbation amplitude. The fast rise in mass variation following the RM-related minimum is the observed RT perturbation growth. Note that the late-time peak at about 4 ns occurs when the laser pulse is over and the target decompresses.

Compared to Richtmyer-Meshkov and RT growth, feedout-related oscillations are much more difficult to observe since the perturbation amplitude is smaller than the initial one during much of the oscillation.

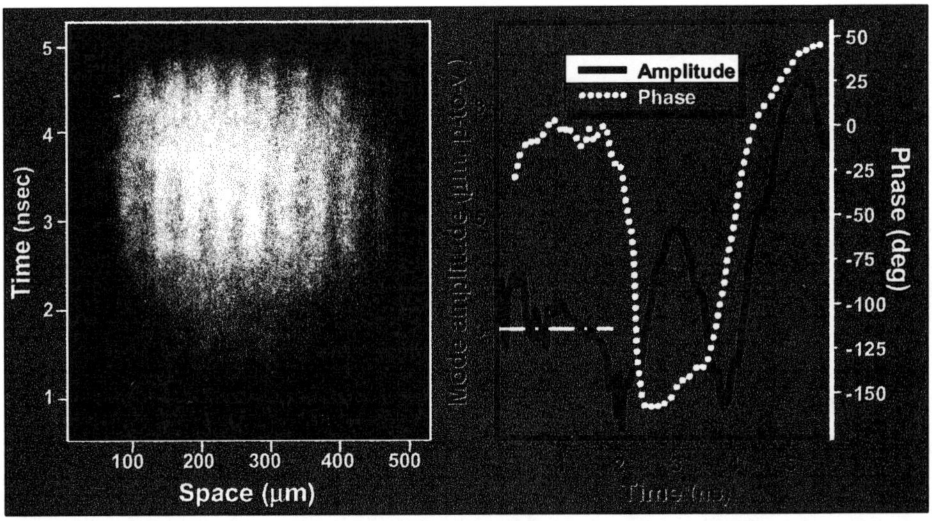

Fig. 5. Feedout-related oscillations detected with the help of our diagnostics. Two phase reversals (that appear as the half-period pattern shift in the image of a rippled target) are clearly visible on the picture (left). Also shown are time dependent Fourier amplitude (black solid curve) and phase (white dashed line) of backlit mass perturbations. Sharp changes of the phase are seen at the times that correspond to the zero amplitude crossings and to the disappearance of the ripples in the picture.

However, as Fig. 5 demonstrates, we were still able to observe this oscillation thanks to the high contrast capability of our diagnostics. Shown is a streak record (left) taken with a target thickness 60 μm, $\lambda = 46$ μm, and peak laser intensity 5×10^{13} W/cm^2. Also shown are time dependent Fourier amplitude and phase of the backlit mass perturbations (right). The two phase reversals illustrated by the lineouts are clearly seen on the original streak record. The light and dark stripes that correspond to low and high areal mass, respectively, are seen at early time, and then disappear, as if at about $t = 2$ ns they were smeared out in a horizontal band ~0.5 ns wide. Then the stripes reappear, but the light and dark ones change places, indicating a 180° phase reversal. Shortly after this, the RT growth in the positive direction (at the initial phase) begins, and the phase of areal mass modulation is reversed again, this time at about 4 ns. The RT growth is also clearly visible: dark and light stripes, representing peaks and valleys of areal mass distribution, respectively, become more pronounced at late time.

Applications of the curved crystal imager with a streak camera are not limited to objects with regular or periodic structure. It is also possible to image time development of random perturbations, as illustrated in Fig. 6. The right side of the figure shows a 2D snapshot of laser-imprinted perturbations on an initially flat ablatively accelerated plastic target taken with the original framing camera/curved crystal imager diagnostics.

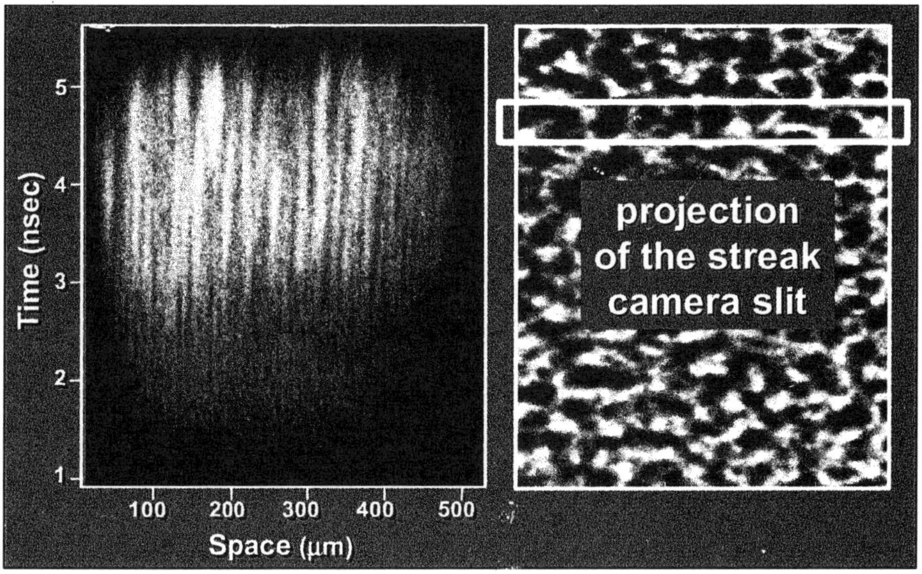

Fig.6. Streak image of the plain plastic target taken under conditions that enhanced mass perturbations seeded by laser imprint. Higher magnification of 30x and narrower streak camera slit of 250 μm will improve chances of detection of random perturbation as small as 10 μm. 2D snap shot of developing laser imprint taken with framing camera (right side) illustrates an idea.

Superimposed on it is a projection of the slit of the streak camera used to acquire the streak record on the left. As the right-hand image illustrates, if the projected width of the slit is on the order of the spatial scale of the perturbations, time development of individual fluctuations can be traced. This can be seen on the streak record, which shows perturbations appearing after about 1 ns, growing in amplitude, and finally coalescing into larger structures at late times. In this case, 20x magnification was used with a 500μm wide streak camera slit. Increasing the magnification to 30x and using a 250μm slit will allow us to detect perturbations of the order of 10μm.

4. CONCLUSIONS.

A novel x-ray imaging technique that uses spherically bent crystal imager coupled to a streak camera has been developed at NRL. It was applied to the studies of areal mass variations with small amplitudes (R-M, feedout) that otherwise could not be observed and measured with necessary quality. Future development of the diagnostics will include higher magnification of the x-ray imaging system and improvement of the spatial resolution of streak camera.

5. ACKNOWLEDGMENTS.

The authors wish to acknowledge an outstanding technical support of the Nike Laser Crew and important contributions of A. Faenov and T. Pikuz, VNIIFTRI, Moscow

This work was supported by the U. S. Department of Energy.

REFERENCES

1. A. V. Rode, A. M. Maksimchuk, G. V. Sklizkov, A. Ridgley, C. Danson, N. Rizvi, R. Bann, E. Forster and I. Uschmann, Intensity measurements of a quasi-monochromatic x-ray beam formed by a spherically bent crystal, *Optics Communications* **7**, 163-166 (1990).
2. S. A. Pikuz, T. A. Shelkovenko, V. M. Romanova, D. A. Hammer, A. Ya. Faenov, V. A. Dyakin and T. A. Pikuz, High-luminosity monochromatic x-ray backlighting using incoherent plasma source to study extremely dense plasmas, in: *Soft X-Ray Lasers and Applications*, edited by J. J Rocca and P. L. Hagelstein, Pros. SPIE, **2520**, 330-341 (1995); see also *Rev. Sci. Instr.* **68,** 740-744(1997).
3. Y. Aglitskiy, T. Lehecka, A.Deniz, J. Hardgrove, J. Seely, C. Brown, U Feldman, C. Pawley, K. Gerber, S. Bodner, S. Obenschain, R. Lehmberg, E. MacLean, M. Pronko, J. Sethian, A. Schmitt, C. Sullivan, G. Holland and M. Laming, X-ray emission from plasmas created by smoothed KrF laser irradiation, *Phys. Plasmas* **3**, 3438-3447 (1996).
4. Y. Aglitskiy, T. Lehecka, S. Obenschain, S. Bodner, C. Pawley, K. Gerber, J. Sethian, C. M. Brown, J. Seely, U. Feldman and G. Holland, High-resolution monochromatic x-ray imaging system based on spherically bent crystals, *Applied Optics* **37**, 5253-5261 (1998).
5. Y. Aglitskiy, T. Lehecka, S. Obenschain, C. Pawley, C. M. Brown and J.Seely, X-ray crystal imagers for inertial confinement fusion experiments, *Rev. Sci. Instr.* **70**,530-535 (1999).
6. C. J. Pawley, K. Gerber, R. H. Lehmberg, E. A. McLean, A. N. Mostovych, S. P. Obenschain, J. D. Sethian, V. Serlin, J .A. Stamper, C. A. Sullivan, S. E. Bodner, D. Colombant, J. P. Dahlburg, A. J. Schmitt, J. H. Gardner, C. Brown, J. F. Seely, T. Lehecka, Y. Aglitskiy, A. V. Deniz, Y. Chan, N. Metzler and M. Klapisch, Measurements of laser-imprinted perturbations and Rayleigh-Taylor growth with the Nike KrF laser, *Phys. Plasmas* **4**, 1969-1977 (1997).
7. C. J. Pawley, S. E. Bodner, J. P. Dahlburg, S. P. Obenschain, A J, Schmitt, J. D. Sethian, C. A. Sullivan, J. H. Gardner, Y. Aglitskiy, Y. Chan and T. Lehecka, Observation of Rayleigh-Taylor growth to short wavelengths on Nike, *Phys. Plasmas* **6**, 565-570 (1999).
8. A. J. D. Sethian, S. E. Bodner, D. G. Colombant, J. P. Dahlburg, S. P. Obenschain, C. J. Pawley, V. Serlin, J. H. Gardner, Y. Aglitskiy, Y. Chan, A. V. Deniz, T. Lehecka and M. Klapisch, Direct drive acceleration of planar liquid deuterium targets, *Phys. Plasmas* **6**, 2089-2094 (1999).
9. A. L. Velikovich, A. J. Schmitt, J. H. Gardner and N. Metzler, Feedout and Richtmyer-Meshkov instability at large density difference, *Phys. Plasmas* **8**, 592-605 (2001).
10. Y. Aglitskiy, A. L. Velikovich, M. Karasik, V. Serlin, C. J. Pawley, A. J. Schmitt, S. P. Obenschain, A. N. Mostovych, J. H. Gardner and N. Metzler, Direct Observation of Mass Oscillations Due to Ablative Richtmyer-Meshkov Instability in Plastic Targets, *Phys. Rev. Lett.* (2001), *in press*.
11. J. Grun, M. E. Emery, C. K. Manka, T. N. Lee, E. A. McLean, A. Mostovych, J. Stamper, S. Bodner, S. P. Obenschain and B. H. Ripin, Rayleigh-Taylor instability growth rates in targets accelerated with a laser beam smoothed by induced spatial incoherence, *Phys. Rev. Lett.* **58**, 2672-2675 (1987).
12. C. Pawley and A. V. Deniz, Improved measurements of noise and resolution of x-ray framing cameras at 1-2 keV, *Rev. Sci. Instr.* **71**,1286-1295 (2000).

Rayleigh Taylor and Laser Imprinting Diagnostics

H. Azechi[1], M. Nakai, K. Shigemori, N. Izumi*, H. Shiraga, M. Nishikino, T. Sakaiya, S. Fujioka, Y. Tamari, N. Miyanaga, and H. Nishimura

INTRODUCTION

The standard scenario of inertial fusion energy (IFE) is illustrated in Fig. 1. Laser beams symmetrically irradiate the surface of a fusion target that contains equimolar mixture of deuterium and tritium (DT) cryogenic layer; the target is imploded by rocket-like blow-off of the surface materials; when the target is stagnated, the kinetic energy of the target is converted to internal energy of the compressed core plasma which consists of a high-temperature, low-density hot spark and a low-temperature and, high-density main fuel. If the areal density of the hot spark is large enough to stop the fusion-generated alpha particles, the hot spark will ignite and the burn wave will propagates through the main fuel before hydrodynamic disassembly.

Primary obstacle of this success scenario is the Rayleigh-Taylor (RT) instability that may potentially quench the thermonuclear ignition and burn. (On the inwardly accelerated frame, the target feels outward gravity. Since the blow-off plasma is much less dense than the target, the interface becomes RT unstable.) We need to reduce the growth of perturbations based on very good understanding of the RT instability.

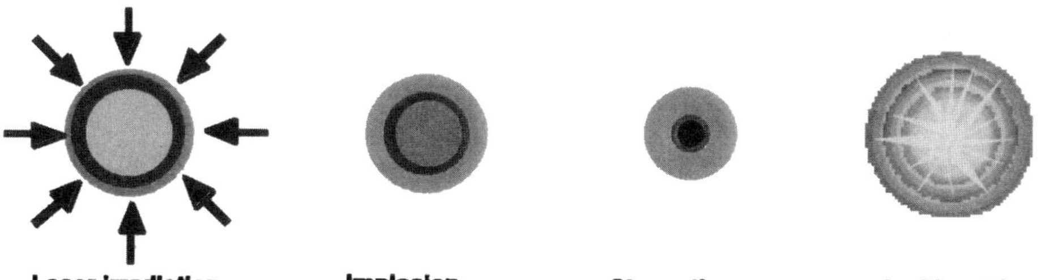

Figure 1. Schematic implosion process of IFE targets

[1] Institute of Laser Engineering, Osaka University, 2-6 Yamada-oka, Suita, Osaka 565-0879, Japan, *Present address: Lawrence Livermore National Laboratory, Livermore, CA 94551, USA.

Advanced Diagnostics for Magnetic and Inertial Fusion, Edited by Stott *et al.*
Kluwer Academic/Plenum Publishers, New York, 2002

It is generally accepted that the Takabe-like formula [1] with significant improvement by Betti [2] predict correct dispersion relation of the RT growth rates. However, there still exists theoretical uncertainty due to the different treatment of heat transport from the laser absorption region to the ablation region. (The ablation effect was first suggested by Bodner[3]. Kull, Sanz, and Piriz [4] also present the ablative RT theories.) Figure 2 shows the RT growth rates calculated using Spitzer-Harm (local) and Fokker-Plank (nonlocal) treatment for electron transport. In order to test these predictions, we need to diagnose short wavelength perturbation that is below spatial resolution of most x-ray imagers. Therefore advanced diagnostics with a few micron resolutions are required. In this report, we will present our RT data based on recently developed three techniques of high spatial resolution: moiré interferometry, penumbral imaging, and Fresnel-phase-zone-plateimaging.

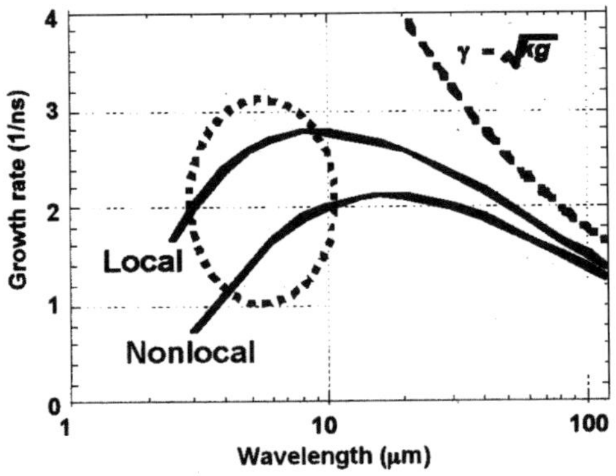

Figure 2. Rayleigh-Taylor dispersion relation for two different energy transports

MOIRÉ INTERFEROMETRY OF SHORT WAVELENGTH RAYLEIGH-TAYLOR INSTABILITY

We have innovated moiré interferometry [5] to overcome the problem of insufficient spatial resolution. The principle is schematically illustrated in Fig. 3. Suppose that one of the grid patterns is a corrugated RT target with certain wavenumber and another is a grid mask with slightly different wavenumber. Overlapping these patterns generates moiré interference with much longer wavelength than the original perturbation wavelength. The contrast of the moiré fringe is proportional to the contrast (and therefore the amplitude) of the original perturbation. The wavenumber of the moiré fringe is the difference of the two wavenumbers. Thus measuring the contrast and wavelength of the moiré fringe, one can deduce the original perturbation amplitude and wavelength.

Figure 3 also shows the experimental set up for the moiré interferometry, that is exactly the same as the standard x-ray backlighting except for the insertion of a grid mask behind the corrugated RT target. The distance between the RT target and the grid mask is short (0.3-1 mm) so that the x-ray diffraction at the grid becomes negligible.

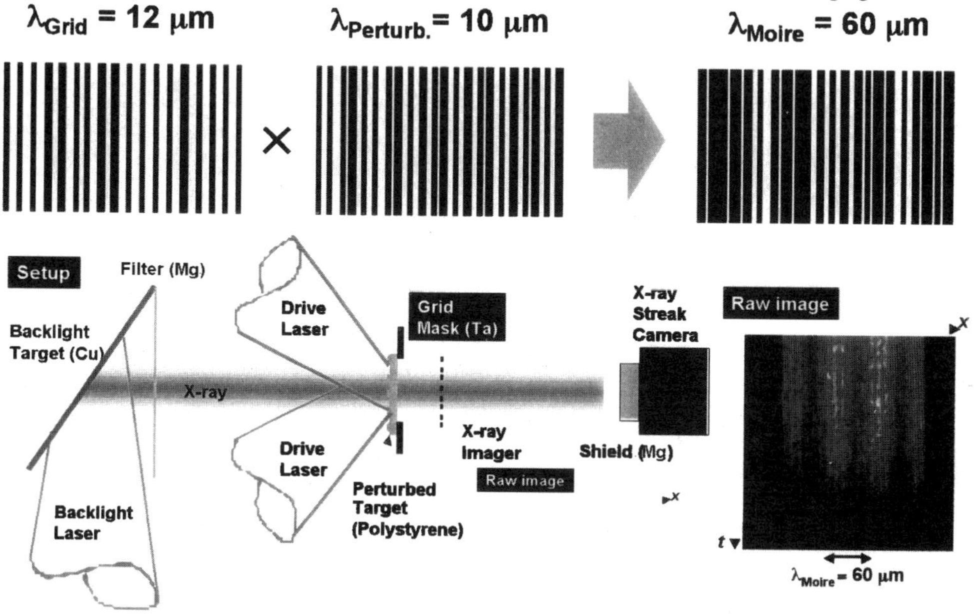

Figure 3. Principle and setup of moiré interferometry.

Before applying the moiré interferometry on the RT observation, we have tested the validity of this technique. A corrugated RT target with carefully inspected perturbation amplitude and wavelength has been diagnosed with the same moiré technique except that the target was un-driven by laser so that the perturbation was kept unchanged. In this particular test, the 30-micron perturbation wavelength together with the 20-micron grid mask generated the 60-micron moiré fringe. Good agreement between the deduced amplitude and the wavelength of the original RT target ensures the validity of the moiré interferometry.

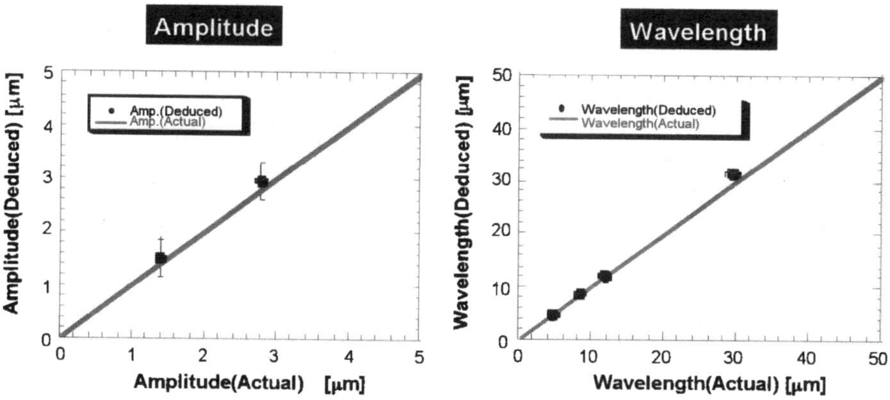

Figure 4. Comparison of the deduced amplitude and wavelength with actual values.

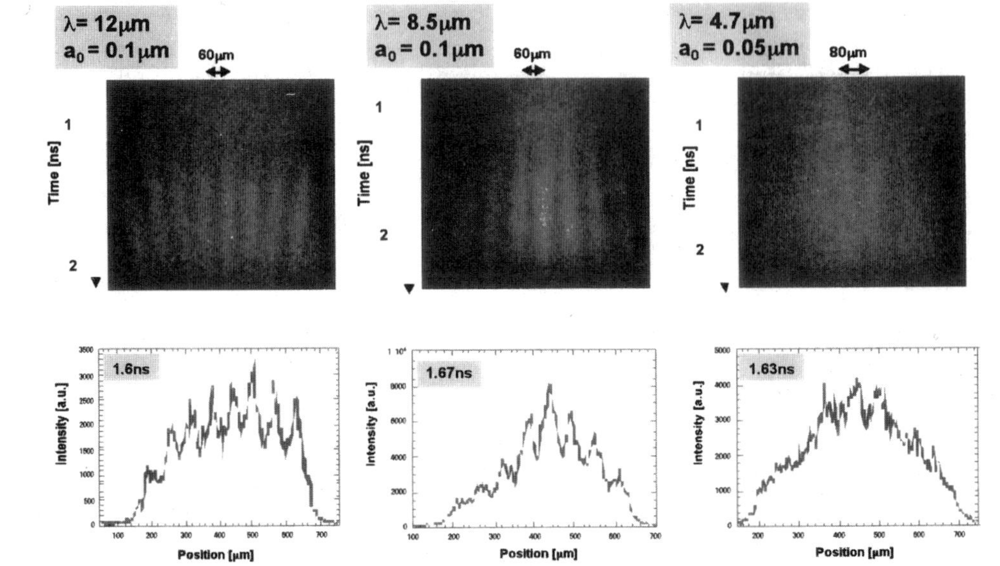

Figure 5. Results of short wavelength Rayleigh-Taylor observed with moiré interferometry.

Figure 5 shows the short wavelength RT observed as a contrast of the moiré fringe. Note that the observed modulation wavelength is much longer than the original wavelengths of 5-12 micron. (Note also that the moiré wavelength changes with time because the slight change of the RT wavelength is significantly amplified by the moiré conversion.)

Compilation of the RT growth data is shown in Fig. 6. These data suggest that nonlocal heat transport plays a major role in ablative stabilization. However, for unambiguous clarification, we need to ma k e independent observation.

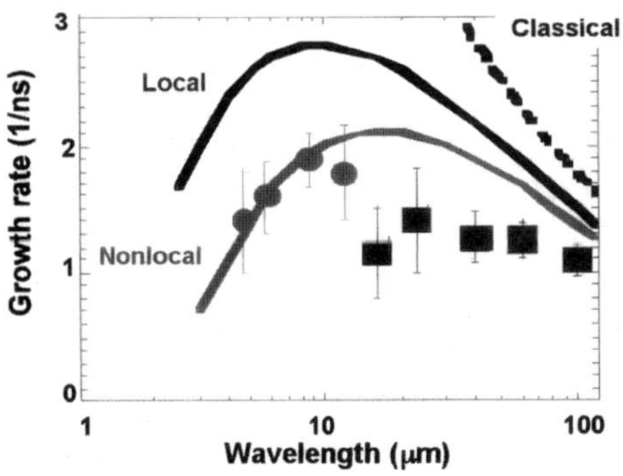

Figure. 6. Growth rate deduced from the areal-massperturbation.

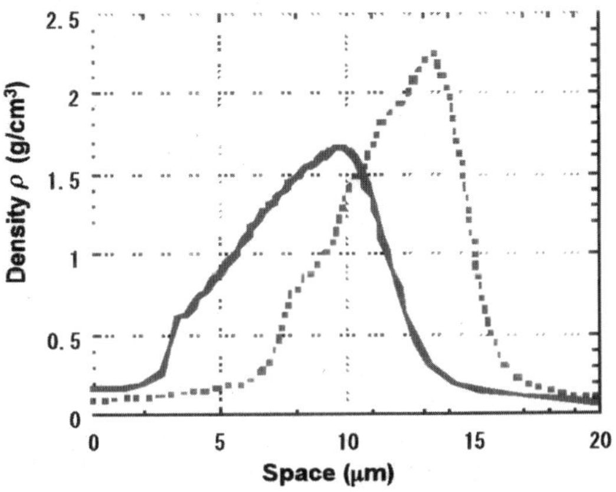

Figure 7. Plasma density profiles calculated based on two different transport treatment

Figure 7 shows that the distinct difference between the local and nonlocal transport calculations is the plasma density profile. The nonlocal transport calculation predicts much less density and gentle density gradient than the local transport calculation. This is because in the case of the nonlocal transport the tail electrons in Maxwell distribution penetrate deep into and preheat the target, thereby decompressing the target, increasing the ablation velocity and thus reducing the RT instability. We need to test which density prediction is correct.

Plasma density can be obtained with edge-on x-ray backlighting, where the x-ray irradiates the target from the side direction and the transmittance is measured. The plasma density is recorded as the transmittance of the x-ray. With known mass absorption coefficient and the target length along the x-ray pass, the density can be deduced. Since the target thickness is a few tens micron, we need about 3 micron resolution to measure the target density. To eliminate thc refraction of x-ray along the line of sight, about 5-keV photon energy is also needed.

PENUMBRAL IMAGING FOR DINSITY MEASUREMENT

The principle of penumbral imaging is that an x-ray spatial structure is recorded as a penumbral of a shadow of a knife-edge. The image is reconstructed from the penumbra by differentiation. Due to the large aperture size, half-infinity in this case, diffraction effect is minimized. The drawback is the necessity of differentiation process that generatesunwanted artifact.

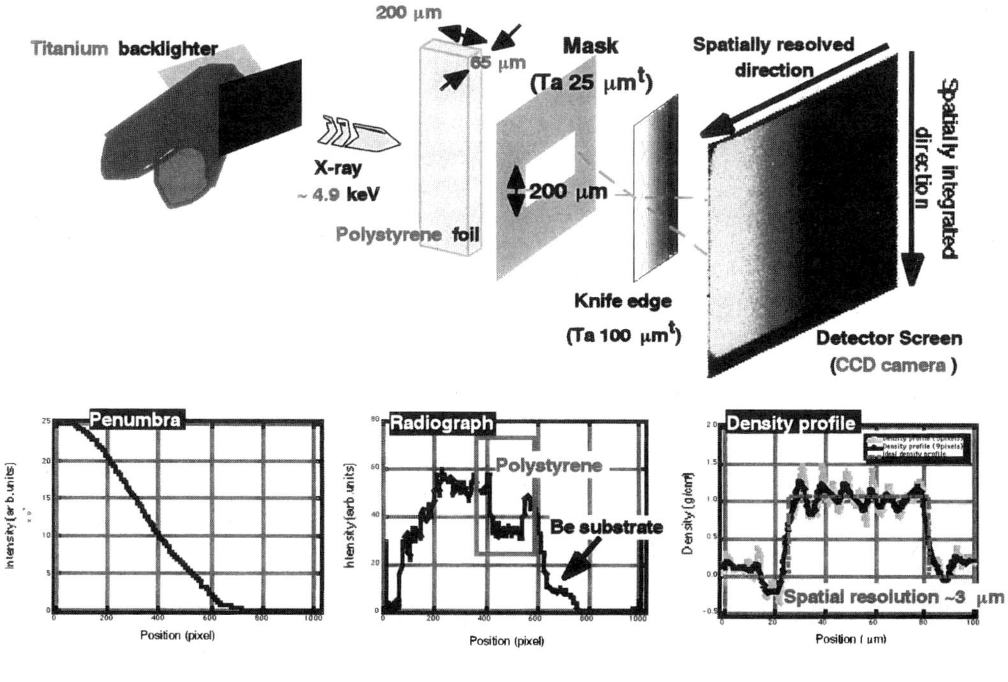

Figure 8. Proof of principle experiment of penumbral imaging

As a proof of principle experiment, we have observed polystyrene target without irradiation of the laser as shown in Fig. 8. The target density is unchanged by laser irradiation. The knife-edge and the target edge are both carefully aligned along the vertical direction. The differentiation of the penumbral image gives the x-ray distribution of the x-ray source and hence the density distribution of the target. Although there is some artifact generated in the reconstruction process, average density is in quite good agreement with the actual density of the target (1.06 g/cc). The spatial resolution can be estimated from the leading edge to be about 3-micron.

The penumbral imaging has been used to measure the in-flight target density. The deduced density is in good agreement with the simulation as shown in Fig. 9. In this particular case, due to short laser wavelength (0.35 micron) and low laser intensity ($7x10^{13}$ W/cm^2) nonlocal effect is NOT appearing. We will test the case where nonlocal effect is expected.

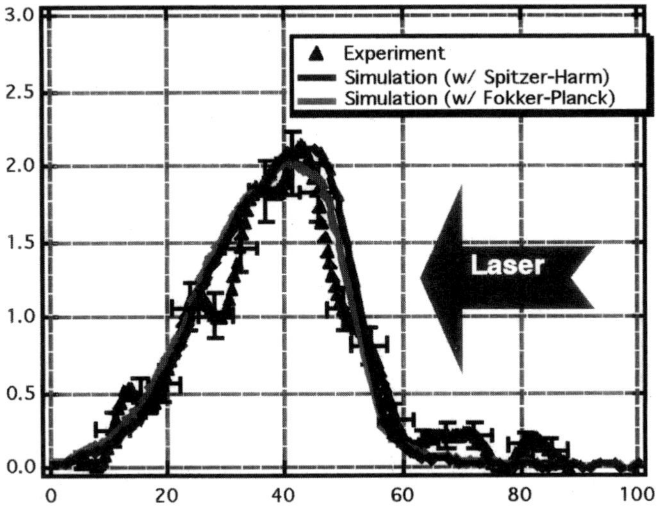

Figure 9. Density profile observed with the penumbral imaging

FRESNEL PHASE ZONE PLATE FOR DINSITY MEASUREMENT

For high resolution imaging, we have also developed Fresnel phase zone plate (FPZP). Historically, Fresnel zone plates (FZP) were introduced as a coded aperture [6]. The spatial resolution was therefore limited by x-ray diffraction at the outermost zone to be about 10 micron. We have made two advancements in the FPZP. One: This is DIRECT imaging just like imaging by lens. Thus the resolution is determined by x-ray diffraction of whole size of the zone plate. In principle sub-micron spatial resolution is readily obtained. The direct imaging was also demonstrated in elsewhere[7]. Two: FZPs have many zones so that the x-ray transmitted through one annular zone has a phase difference of pi compared to the x-ray transmitted through the neighboring zone. Generic

FZPs mask the pi-shifted zones. Instead, the FPZP adds pi-shift in these zones. This increases x-ray collection efficiency by a factor of 4. But more importantly, generic FZPs have a serious difficulty to get sufficient attenuation for high energy x rays.

The tested spatial resolution of the FZP is shown in Fig.10. A 10-micron pitch grid has been backlit with nearly monochromatic 1.25-keV Cu lines and imaged through the FZP on a CCD camera. The modulation transfer function (MTF) is obtained by Fourier transform of the image. The modulation wavelength at which the MTF becomes 5% is 2.2 micron. Figure 10 show that an almost identical resolution was obtained for the FPZP with the 4.8-keV Ti He-a line. This imaging technique will be used for the density measurements in the coming experimental champagne.

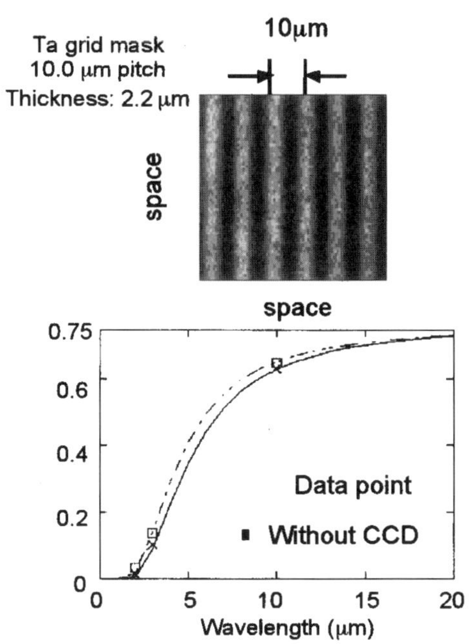

Figure 10. Resolution the Fresnel zone plate

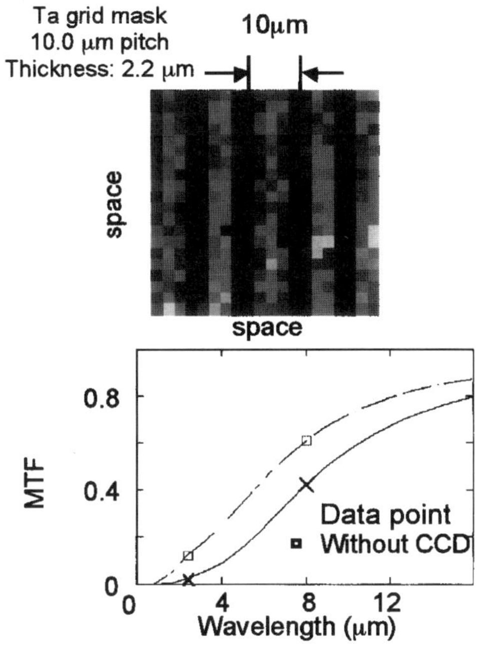

Figure 10.Resolution of the Fresnel phase zone plate

CONCLUSION

In summary, with advanced diagnostic techniques, we are approaching to better understanding of the Rayleigh-Taylor instability. The itemized summary is listed up:

- Rayleigh-Taylor (RT) is the critical physics for high-gain IFE
- Energy transport can modify the RT growth at short wavelengths.
- Moiré interferometry first observed the short wavelength RT growth.
- The observed RT growth suggests that nonlocal transport plays a role in ablative stabilization.

· For independent test of the transport effect, we are currently measuring the ablation density with high-resolution imaging techniques.
· Initial test result is favorable to the density measurement.with advanced diagnostic techniques, we are approachingto better understandingof the Rayleigh-Taylor instability.

Finally, we emphasize that our strategy is to measure all necessary quantities (growth rates, gravity, wavenumber of the perturbation, mass ablation rate, density profile) to test various theories of the Rayleigh-Taylor instability.

ACKNOWLEDGEMENT

We would like to acknowledge the dedicated technical support by the staff at the GEKKO XII facility for the laser operation headed by Mr. S. Urushihara, for the target fabrication by Dr. T. Norimatsu and Dr. K. Nagai, and for the plasma diagnostics by Mr. O. Maegawa, Ms. K. Shimada.

REFERENCES

[1] H. Takabe et al., Phys. Fluids 28, 3676 (1985).
[2] R. Betti et al., Phys. Plasmas 5, 1466 (1985).
[3] S. Bodner, Phys. Rev. Lett. 33, 761 (1974).
[4] H. J. Kull and S. I. Anisimov, Phys. Fluids 29, 2067 (1986); J. Sanz, Phys. Rev. Lett. 73, 2700 (1993); A. R. Piriz, Phys. Plasmas 8, 997 (2001).
[5] M. Matsuoka et al., Rev. Sci. Instrum. 70, 637 (1999).
[6] N. M. Ceglio et al., J. Appl. Phys. 48, 1566 (1977).
[7] C. Cauchon et al., Rev. Sci. Instrum. 69, 3186 (1998).

PLASMA DIAGNOSTICS IN INTERACTION OF POWERFUL LASER PULSES WITH INHOMOGENEOUS LOW-DENSITY MEDIA.

Alexey E. Bugrov, Igor N. Burdonskiy, Valeriy V. Gavrilov, Alexander Yu. Goltsov, Ildar K. Fasahkov, Vladimir N. Kondrashov, Sergey N. Koptyaev, Nikolay G. Kovalskiy, Mikhail I. Pergament, Viktor M. Petryakov, Georgiy M. Yankovskiy, and Evgeniy V. Zhuzhukalo*

1. INTRODUCTION

Utilization of low-density porous or foam-like materials in laser-irradiated target designs seems to be a very promising approach to the successful realization of such important research programs as ICF, elaboration of X-ray lasers, laboratory modeling of astrophysical phenomena, and EOS measurements of materials in Mbar pressure range[1-3]. To give some recommendations on application of low-density media in investigations mentioned above it is necessary to study laser light absorption, plasma formation, energy transport, and plasma homogenization in porous matter. These processes are very complicated and have some important features in comparison with the similar processes in the case of laser-irradiated solid-density targets. To have detailed and reliable experimental data we need to use in each experiment the combination of various diagnostics with high spectral, temporal, and spatial resolution. This paper presents the diagnostic complex developed on the "Mishen" facility as well as some obtained experimental results.

2. EXPERIMENTAL CONDITIONS

The experiments were carried out on the "Mishen" facility with the Nd-glass laser system[4]. The planar targets were irradiated in a vacuum chamber with one laser beam

* Alexey E. Bugrov, Igor N. Burdonskiy, Valeriy V. Gavrilov, Alexander Yu. Goltsov, Ildar K. Fasakhov, Vladimir N. Kondrashov, Sergey N. Koptyaev, Nikolay G. Kovalskiy, Mikhail I. Pergament, Viktor M. Petryakov, Georgiy M. Yankovskiy, Evgeniy V. Zhuzhukalo, Troitsk Institute for Innovation and Fusion Research, Troitsk, Moscow reg., Russia 142190.

Advanced Diagnostics for Magnetic and Inertial Fusion, Edited by Stott *et al.*
Kluwer Academic/Plenum Publishers, New York, 2002

(λ=1.054 μm, τ=2.5 ns, I≈10^{13}-10^{14} W/cm^2) or two laser beams focused into two 250 μm - diameter spots spaced by ~300 μm.

The planar targets with thickness from 100 μm to 1000 μm and average density in the range of 1-10 mg/cm^3 were fabricated of agar $(C_{12}H_{18}O_9)_n$ having a fibrous structure (randomly spaced filaments of 1-2 μm in diameter). It is very important to make adequate choice of the target design to provide suitable conditions for investigation of specific physical processes. Therefore the agar targets with and without $CuCl_2$ or Al_2O_3 admixtures (up to 50% by weight) of varied average density and thickness were used. In some experiments the 0.2 μm-thick Cu layer (X-ray converter) was deposited on the irradiated surface of pure agar samples. For diagnostic purposes in most experiments the 1.5 μm-thick Al foil was deposited at the target rear surface.

The following main diagnostics were used in our experiments: multiframe optical side-on shadowgraphy of accelerated matter at the rear side of irradiated targets, X-ray imaging of a hot plasma region with filtered pinhole cameras and X-ray streak camera providing 1D spatial and temporal resolution, optical streak camera recording of the luminosity at the rear surface of laser irradiated targets. Optical spectrographs in combination with streak cameras were also used to analyze time history of backscattered light spectra at ω_0, $2\omega_0$, and $3\omega_0/2$, as well as imaging X-ray spectrographs with bent mica crystals.

3. EXPERIMENTAL RESULTS AND DISCUSSION

Laser light absorption was investigated by X-ray pinholes, spectrographs, and streak cameras as well as by optical and X-ray calorimeters[5,6]. In experiments with pure and high-Z doped agar samples the volume character of laser light absorption and formation of a hot plasma region inside porous low-density targets were proven. As the average density of agar varied from 10 mg/cm^3 to 1 mg/cm^3 the absorbing layer thickness increased from ~150 μm to ~500 μm. When porous samples with high-Z admixtures were irradiated the dimensions of high-temperature region were about 10-15% in excess of above mentioned values. The plasma electron temperature was determined by processing the X-ray images of plasma and spectroscopic data recorded with spatial resolution. The measured electron temperature (T_e=0.6-0.8 keV) was by a factor of ~1.5 higher than that observed in experiments with solid-density targets. In experiments with the 2 mg/cm^3 porous samples the plasma ion temperature derived from Doppler broadening of multicharge ion spectral lines was found to be ~1.4 keV that was approximately by a factor of 2 higher than the electron temperature. This fact was earlier predicted by theorists[7] and explained by the complicated dynamics of numerous plasma streams produced under laser irradiation of small-scale target structure elements. Information on angular distribution of backscattered laser light was also obtained in these experiments. Laser-light absorption efficiency was as high as ~80%.

To get information on the important problem of plasma homogenization in porous media consisting of discrete small-scale structure elements the backscattered light at the fundamental laser frequency ω_0 as well as at the $2\omega_0$ and $3\omega_0/2$ harmonics was recorded with high spectral and temporal resolution. Under the complete ionization of 2 mg/cm^3 agar the electron density of homogeneous plasma should be essentially lower than the critical density, but the $2\omega_0$ harmonic emission is observed up to the end of laser pulse for relatively thick samples (~300 μm). It can be considered as an evidence of plasma inhomogeniety and presence of supercritical-density plasma regions in volume of agar sample even at the moment of laser pulse termination.

Multiframe side-on shadowgraphy of accelerated matter at the rear side of the irradiated targets of different thickness as well as streak camera recording of luminosity at the target rear surface were found to be very suitable diagnostic methods to study the longitudinal energy transport inward relatively cold regions of porous samples. The time histories of rear side luminosity for various types of porous samples show that energy transport inside these targets is of wave character with sharp front for all target types, but wave-front propagation velocities are strongly different. This indicates that the different mechanisms of energy transport in these types of agar samples are realized. The highest velocity of (4-5)$\times 10^7$ cm/s has been recorded in $CuCl_2$-doped agar samples (see Fig. 1).

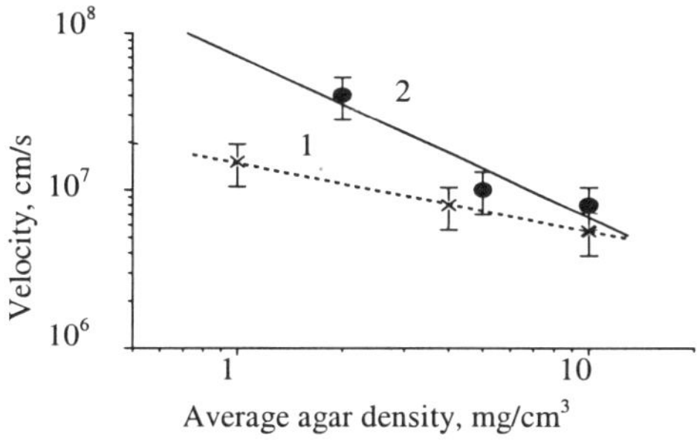

Figure 1. Heating wave velocity inside pure agar (1) and agar doped with $CuCl_2$ (50% by weight) (2) as a function of average agar density.

Valuable information on lateral energy transport in porous targets was also obtained in experiments with two laser beams focused into two separate spots ~250 μm-diam. spaced by ~300 μm on the irradiated surface of agar samples with and without high-Z admixtures. The experiments with two-beam irradiation of porous targets show that use of low-density media is efficient for smoothing of long-scale irradiation nonuniformities (see Fig. 2).

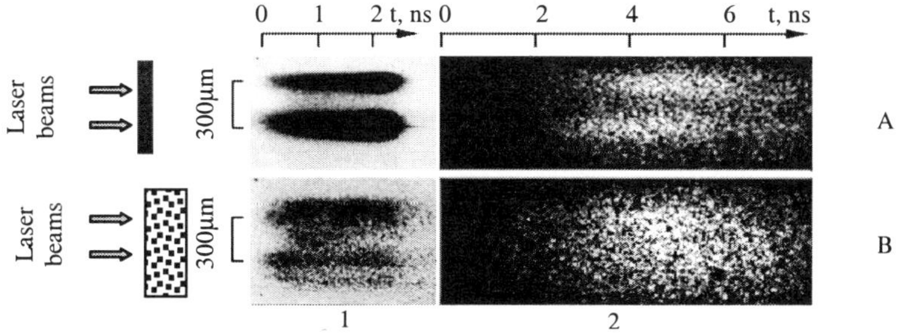

Figure 2. Streak records obtained by X-ray (1) and optical (2) streak cameras in experiments with two-beam laser irradiation of 6 μm-thick Al foil (A) and 500 μm-thick agar sample (ρ=2 mg/cm^3) (B). Cameras were placed behind the targets and recorded emission along the laser beam direction. In the case (2) the 1.5 μm-thick Al foil was deposited at the rear side of the porous target. Zero points on the t-axes correspond to the laser pulse beginning.

4. CONCLUSION

Complementary optical and X-ray diagnostic methods were developed and successfully used in experiments on detailed study of various physical processes in low-density (1-10 mg/cm^3) agar targets irradiated with powerful laser beams:

- X-ray imaging of hot plasma region in the agar target interior with pin-hole cameras, crystal spectrographs, and streak-camera allowed us to study features of laser light absorption and formation of plasma region in a porous medium, and to determine main parameters (including the ion temperature for the first time) of produced hot plasma;
- spectrally and temporally resolved measurements of fundamental laser frequency harmonics provided information on the problem of plasma homogenization in low-density porous media;
- multiframe optical side-on shadowgraphy of the accelerated matter in combination with streak-camera recording of the luminosity in visible at the rear surface of laser-irradiated agar samples were shown to be suitable and very informative diagnostics for investigation of energy transport inside porous targets.

It should be emphasized that the use of all above mentioned methods simultaneously in each laser shot is very desirable to provide reliable information for understanding of complicated physical processes and mechanisms being under investigation.

5. ACKNOWLEGMENT

This work was performed under the auspices of the Russian Foundation for Basic Research (Projects No. 01-02-17336 and 01-02-17361).

6. REFERENCES

1. M. Desselberger, M.W. Jones, J. Edwards, M. Dunne, and O. Willi, Use of X-ray Preheated foam layers to reduce beam structure imprint in laser-driven targets, *Phys. Rev. Letters* **74**(15), 2961-2964 (1995).
2. D. Batani, W. Nazarov, T. Hall, T. Lower, M. Koenig, B. Faral, A. Benuzzi-Mounaix, and N. Grandjouan, Foam-induced smoothing studied through laser-driven shock waves, *Phys. Rev. E* **62**(6), 8573-8582 (2000).
3. Y.-G. Kang, H. Nishimura, H. Takabe, H. Azechi, T. Norimatsu, M. Nakai, H. Nagamoto, A. Sunahara, K. Fujita, M. Nakatsuka, K. Mima, and H.G. Kim, Laboratory simulation of the collision of Supernova 1987A with its ring nebula, in *Inertial Fusion Sciences and Applications 99*, edited by Ch. Labaune, W. Hogan, and K.A. Tanaka (ELSEVIER, Paris, 2000), pp.1061-1066.
4. V.A. Bolotin, I.N. Burdonskii, V.V. Gavrilov, A.Yu. Gol'tsov, S.V. Zavyaletz, N.G. Koval'skii, V.N. Kondrashov, M.I. Pergament, and E.V. Zhuzhukalo, Diagnostic complex for investigations of laser fusion physical processes, *Rev. Sci. Instrum.* **61**(10), 3259-3261 (1991).
5. A.E. Bugrov, I.N. Burdonskii, V.V. Gavrilov, A.Yu. Gol'tsov, S.Yu. Gus'kov, N.G. Koval'skii, V.N. Kondrashov, S.F. Medovshchikov, M.I. Pergament, V.M. Petryakov, V.B. Rosanov, and E.V. Zhuzhukalo, Investigation of light absorption, energy transfer, and plasma dynamic processes in laser-irradiated targets of low average density, *Laser and Particle Beams* **17**(3), 415-426 (1999).
6. A.E. Bugrov, I.N. Burdonsky, V.V. Gavrilov, A.Yu. Goltsov, S.Yu. Guskov, V.N. Kondrashov, N.G. Koval'sky, S.N. Koptyaev, S.F. Medovshchikov, M.I. Pergament, V.M. Petryakov, V.B. Rozanov, and E.V. Zhuzhukalo, Interaction of powerful laser and X-ray pulses with porous materials for advanced ICF targets applications, in *Inertial Fusion Sciences and Applications 99*, edited by Ch. Labaune, W. Hogan, and K.A. Tanaka (ELSEVIER, Paris, 2000), pp.154-157.
7. S.Yu. Gus'kov and V.B. Rozanov, Interaction of laser radiation with a porous medium and formation of a nonequilibrium plasma, *Kvantovaya elektronika* (Russia) **24**(8), 715-720 (1997).

LASER AND X-RAY IRRADIATION DIAGNOSTICS THAT HAVE PAVED THE PATH TO SIGNIFICANTLY IMPROVED ICF TARGET PERFORMANCE

R. L. Keck, W. R. Donaldson, V. Yu. Glebov, P. A. Jaanimagi, F. J. Marshall, P. W. McKenty, D. D. Meyerhofer, S. P. Regan, W. Seka, C. Stoeckl, and R. Boni*

1. INTRODUCTION

Obtaining high performance from direct-drive inertial confinement fusion (ICF) targets requires careful control of the target irradiation. The temporal pulse shape and the on-target irradiation uniformity are of particular concern. A broad arsenal of techniques and diagnostics has been developed for the OMEGA laser[1] to ensure that the on-target irradiation meets the requirements. Most of these directly diagnose the laser itself. However, x-rays emitted from the target are also used to confirm that the required irradiation-uniformity levels have been achieved. The improvements to laser uniformity made possible with these diagnostics, combined with improvements to individual-beam uniformity provided by beam-smoothing techniques, have resulted in significantly improved target performance and reduced shot-to-shot variability.

2. IRRADIATION-UNIFORMITY REQUIREMENTS

To obtain high performance from direct-drive ICF capsules, numerical simulations show that the irradiation uniformity on the target surface must be at the 1%–2% rms level. Irradiation nonuniformity arises from several sources, each of which tends to show up at a different range of spatial wavelengths on the target. On the OMEGA system, beam-to-beam imbalance tends to produce longer wavelengths, showing up as spherical-harmonic ℓ modes 1 to 4. Since the beams overlap on the target, the beam-to-beam power

* All authors: Laboratory for Laser Energetics, University of Rochester, Rochester, NY 14623.

balance needs to be about 5% or better to meet the 1%–2% specification. The 60-beam geometry of the system tends to produce a weak mode at $\ell = 10$. The pattern produced by the distributed phase plates[2] (DPP's) used on each OMEGA beam to control the individual-beam uniformity produces a broad spectrum of modes up to about $\ell = 500$.

To obtain the required uniformity in the range of modes produced by the DPP speckle pattern, smoothing by spectral dispersion (SSD)[3] combined with polarization smoothing is employed. Polarization smoothing[4] uses birefringent wedges in each beam to produce an offset copy of each beam on the target, causing a $1/\sqrt{2}$ improvement in uniformity. SSD causes the speckle pattern to rapidly shift in time on the target, resulting in a time-dependent averaging. The 1-THz SSD system currently employed on OMEGA, combined with polarization smoothing, provides 2% rms nonuniformity after about 50 ps and 1% rms after about 200 ps.

3. CALORIMETERS

The ultimate reference for energy on OMEGA is a set of 60 swing-in, volume-absorbing, full-aperture calorimeters located at the stage-F alignment sensor package (F-ASP) after the frequency-conversion cells. The calorimeters can be absolutely calibrated because they are equipped with electrical heaters fed by precision electrical calibrators. Since knowing the relative energy in the 60 beams is more important than knowing the absolute energy, particular attention was paid to making sure all components are matched. The large size of the calorimeters makes them particularly sensitive to drafts, so care was taken to ensure that they seal to the surrounding structure when swung into place. A portable calibrator calibration fixture is used to check the performance of the electrical calibrators.

The calorimeters are read out by integrating their output voltage over time. This is done by numerically integrating the output of the initial waveform and then taking the integral of the tail to be the amplitude of the waveform at a cutoff point times the decay time constant. This has been found to minimize the sensitivity to long-term drift and to avoid the problem caused by differences in the electrical and laser-heated waveforms due to differences in energy deposition within the calorimeter.

The calorimeter performance is excellent. The repeatability on electrical calibrations is typically 0.5%. The relative accuracy of the calibrators is about 0.8%, due to a design problem that is being corrected. We expect to obtain about a 0.25% relative accuracy. The absolute accuracy is estimated at about 2% and is expected to improve with the new calibrator.

4. HARMONIC ENERGY DIAGNOSTIC

Since the calorimeters intercept the whole beam, they cannot be used to routinely measure the system energy; the harmonic energy diagnostic (HED) is used for on-the-shot energy measurements. This diagnostic has an additional advantage over the calorimeters: it can simultaneously measure the energy of each of the three wavelengths leaving the frequency conversion cells (FCC's). This diagnostic, shown in Fig. 1, picks off 4% of the whole beam with an uncoated surface, which, after additional pickoffs, is focused into an integrating sphere. The signals from the integrating spheres are sampled by optical fibers and brought to one of two spectrometers. The fibers are arrayed in a grid at the input to the spectrometer such that the dispersed image of the fibers at the output forms an array of nonoverlapping images, three for each beam. This image is captured on a scientific-grade, charge-coupled-device (CCD) camera, which is used as a diode array by summing the pixels in each fiber image. Thirty beams are diagnosed on each of two spectrometers to cover all 60 OMEGA beams.

The HED is calibrated by taking a series of shots into the calorimeters where the tuning of the FCC's and the energy of the laser are intentionally varied to produce a considerable variation in the output-wavelength mix from the FCC's. By fitting the HED data to the calorimeter data using a multiple linear regression, the calibration constant for each of the three wavelengths can be determined. The UV channel of the HED currently provides a relative accuracy of about 1.5%. This should improve to better than 1% with the improved calorimeter calibrator and some changes to calibration procedures.

5. TRANSPORT

A number of optical elements are located between the energy measurement point and the target: notably the blast window, the focus lens, and a DPP. The transmission of these elements must be accurately known to obtain the energy-on-target from the HED measurement. Since the transmission of the blast window is subject to change from deposition of target debris on its inner surface, the OMEGA in-tank integrating sphere (OTIS) is used to measure the transmission of the optics. OTIS consists of a diode sensor mounted in an integrating sphere that is inserted in the center of the target chamber. It is held by a motorized gimbal that allows it to be pointed at any of the 60 beams entering the chamber. A cw UV beam is injected into each beam using a robotic mirror assembly in front of each HED pickoff. An additional diode is used to factor out fluctuations of the cw laser. This system allows us to obtain the relative transmission of each beam in two groups of 30.

The existing system suffers from two limitations: It is cumbersome to insert the sphere into the chamber and the chamber must be vented to use it. Additionally, it is not

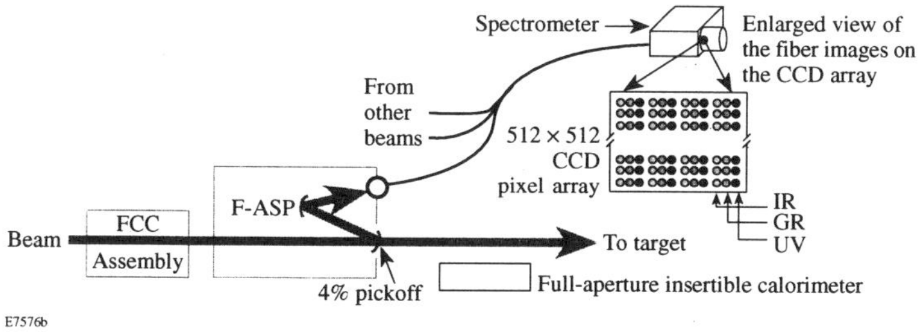

Figure 1. Two HED spectrometers are used to measure the energy in all 60 beams of OMEGA. Light from integrating spheres located on a pickoff near the end of the system is coupled to fibers arranged in array at the input of the spectrometer. Due to the dispersion in the spectrometer, each fiber produces three images at the spectrometer output CCD, one for each wavelength resulting from the frequency-conversion process. The signal in each spot is proportional to the energy in the corresponding beam at the corresponding wavelength. The HED is calibrated using the full-aperture flip-in calorimeter after the pickoff.

absolute, and since it measures two groups of 30 beams, some assumptions must be made to obtain the relative transmission of the 60 beams taken as a whole. To address these limitations, a replacement system is under construction that will utilize a 5-in.-diameter, uncoated absorbing-glass sphere in the center of the chamber as a 4% retroreflector. Using absorbing glass avoids problems from the rear surface reflection. The sensor will be located at the injection point, so that the measurement will be a double pass. By placing a flat uncoated surface immediately after the HED pickoff, it will be possible to calibrate the system to make an absolute transmission measurement. Both the input reference and back-reflected output measurement will be made using a single scientific-grade CCD camera as a diode array. Lab measurements show that by using this system we should be able to obtain results accurate to better than 0.5%. The reflective sphere, which can be placed in the chamber using a ten-inch manipulator (TIM), will be able to operate in vacuum without removing permanent diagnostics.

6. P510 STREAK CAMERAS

Although OMEGA uses a single injected pulse to feed all of its beamlines, the nonlinearities caused by gain saturation, the radial gain profile of the rod amplifiers, and the FCC's can result in mismatched pulse shapes at the output (power imbalance) even when the energy is balanced. In general, to obtain good power balance, the gains and losses of each stage of the laser must be closely matched and all the FCC's must be well tuned. While we have developed a scheme for setting up the system to obtain good power balance on the basis of matching the energy performance of the beamlines, it is important to directly measure the power of each of the 60 beams to confirm that good power balance is achieved. Since the pulse shape can influence the performance of the target, it is also of considerable interest in and of itself.

Power-balance measurements are made using six LLE-designed, ten-channel "P510" streak cameras,[5] where the P510 is the designation of the Phillips streak tube they employ. These cameras, which operate without an image intensifier, were designed with careful attention to the performance of the CCD used for readout and the coupling of the streak tube to the CCD. The streak cameras are fed from a pickoff located at the F-ASP via seven-fiber bundles. These bundles were constructed using matched fiber lengths to avoid significant pulse broadening beyond that caused by a single fiber. To sample the whole beam properly and avoid FM to AM conversion of the SSD modulation, the laser light is injected into the fibers using a dual engineered-diffuser assembly. At the streak cameras, an array of ten fiber bundles plus two fiducial fibers is imaged onto the photocathode using all reflective optics producing an image at the CCD similar to the one shown in Fig. 2. These cameras are unique because the input assemblies are on a motorized mount that allows different inputs to be selected via computer control. Special input assemblies allow the camera to be flat fielded and corrected for geometric distortion. This provision is essential for obtaining high-quality results. The cameras are calibrated against the HED energy measurement to provide power measurement.

At a standard speed of about 12 ps/pixel (512-pixel record length) the cameras provide a signal-to-noise ratio of about 0.5% at full scale and a bandwidth of about 10 GHz. Over most of their dynamic range, the signal is governed by Poisson statistics;

hence the cameras provide a signal-to-noise ratio of about 5% at 1% of full scale. By sacrificing bandwidth, which is acceptable for power imbalance due to gain saturation, the signal-to-noise ratio can be improved.

The presence of the fiducial channels allows the six P510's to be cross timed and other time-resolving target diagnostics to be time calibrated to the time of arrival of the laser pulse at the target. This can be accomplished to roughly 50-ps accuracy over spans of several months.

Power measurements of square pulses approximately confirm the balance expected based on the energy balance; however, residual time base errors and bandwidth differences among the channels tend to cause the streak cameras to overestimate the power imbalance wherever the time derivative of the pulse is not zero. To some extent, this can be mitigated by band limiting the measured pulse shapes before comparison, but the leading and trailing edges of the pulse remain a problem (Fig. 2). We are actively working on precision stabilization of critical streak-tube voltages so we can match the time bases with higher precision.

7. PREPULSE MONITORS

While the dynamic range of the P510's is high, targets are sensitive to prepulses at the 10^{-7} level, which are not only below the P510 detection capability, but may occur prior to the start of a P510 record. To measure these prepulses, two prepulse monitors are installed, one that monitors a UV beam going to the target, and one that monitors an IR beam before frequency conversion (the nonlinearity of the frequency-conversion process makes prepulses easier to detect in the IR). The prepulse monitors consist of a pickoff, a removable filter, an air spatial filter, and a vacuum photodiode detector readout that uses a 6-GHz oscilloscope. The detectors are calibrated by first taking a shot with the removable filter in place, allowing the entire pulse to be acquired without overdriving the detector. For subsequent shots, the removable filter is removed, allowing the detector to see prepulses at the 10^{-8} level in the UV (Fig. 3). In this mode, the air spatial filter breaks down during the main part of the pulse, protecting the photodiode from excessive energy.

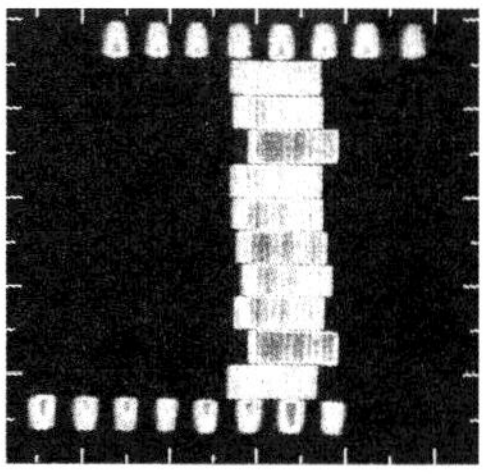

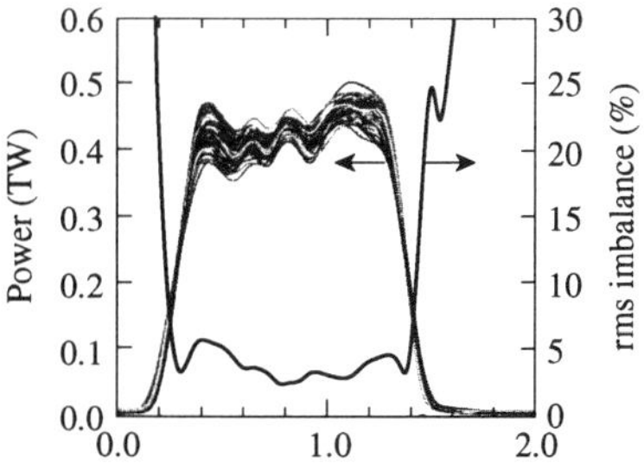

Figure 2. The photograph shows the raw output of a P510 streak camera. The top and bottom most streaks are fiducials used for timebase calibration and diagnostic cross timing. The middle ten streaks are of each of the beams in an OMEGA cluster. The plot shows the smoothed reduced data from all six P510 streak cameras (60 beams) and associated rms imbalance. Smoothing is required to avoid inflating the measured imbalance from time-based inaccuracies and bandwidth mismatches among the streak camera channels.

The detectors are set up to detect prepulses as far as 15 ns in advance of the main pulse. In practice, if prepulses occur, they are never more than about 2 ns in advance of the main pulse and are generally caused by improper setup of the pulse-shaping system.

8. UVETP MEASUREMENT

The profile of each beam on target is determined by the DPP placed immediately before each focus lens. While each DPP is characterized off-line using a cw laser before installation, it is important to characterize the performance of the DPP's on an actual beam to determine the effects of wavefront distortion of the beam, the nonlinear phase (B-integral), and the smoothing provided by SSD. To do this, an UV equivalent-target-plane (UVETP) monitor has been installed.[6] It allows a subset of beamlines to be sampled by inserting a moveable, uncoated pickoff. This beam is then transmitted through a DPP and focused by a standard focus lens into a vacuum vessel. The focal spot is then re-imaged with a magnification of about 40 onto a scientific-grade CCD.
To accurately assess the effect of SSD, it is important to ensure that the imaging system is capable of fully resolving the speckle pattern produced by the DPP. On a normal full-system shot, there is a sufficient phase change during the pulse to wash out the speckle contrast. For this reason, the reference measurement is made on low-power shots with the system carefully set up to avoid this effect. In this case, we observe the nearly 100% rms variation of the profile expected (Fig. 4).

By making the UVETP measurement of pulses of varying lengths, we can measure the time-dependent smoothing that SSD produces. We find that the measured and calculated smoothing are in very close agreement (Fig. 4).

9. X-RAY PINHOLE CHARGE-INJECTION-DEVICE IMAGER

For many years, x-ray pinhole images acquired on film have been used to monitor the pointing of the laser beams on the target. This is done by shooting large 4-mm targets so that all 60 beams, equipped with phase plates, show up as individual spots. The spot location is measured on multiple images showing the target from several views. The location data from all the images is combined to provide the location of each beam on the target.

More recently, the film-based cameras have been replaced by charge-injection-device (CID)-based cameras, which offer a number of advantages, notably rapid availability of the images. The generally higher quality of the CID images has made it apparent that some of the beam spots on pointing targets appear weaker than others. Since converting laser light to x rays is highly nonlinear (gamma is of the order of 4), the pointing images show only the brightest part of the focal spot, so the apparent mismatch could be due to either differences in spot shape or differences in the energy on target. We have performed experiments where the energy on target has been adjusted to match the intensities of the beam spots on the CID images. Results obtained to date show no significant change in performance with an x-ray balanced system, but this remains an area of active investigation.

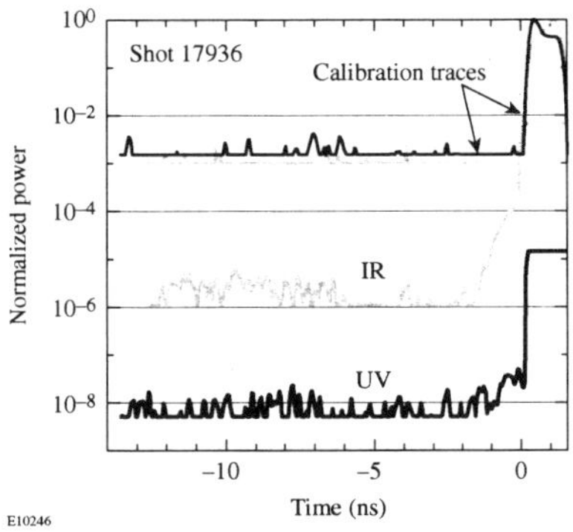

E10246

Figure 3. Prepulse diagnostics are used on both the UV and IR portions of the laser. In normal operation, the main portion of the pulse is driven off-scale. To calibrate the systems, filters are inserted that permit the diagnostics to measure the main pulse. Here, both the calibration and normal traces are plotted on a common scale, showing that the system can detect UV prepulses down to the 10^{-8} level and IR prepulses down to 10^{-6}.

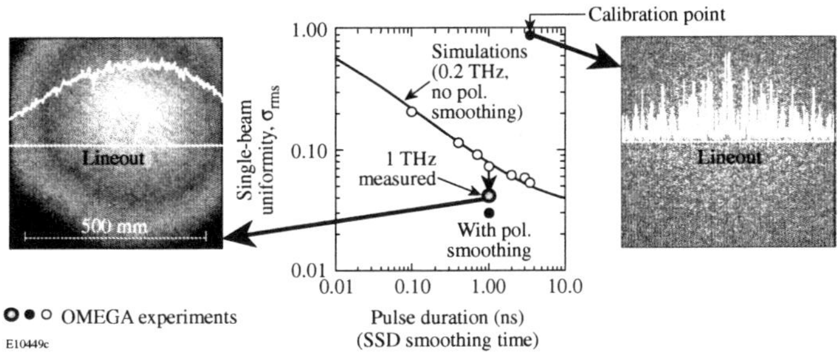

E10449c

Figure 4. The UVETP diagnostic provides a measurement of SSD smoothing. The upper photograph shows a DPP-only (SSD off) beam pattern. The lineout shows essentially 100% modulation, indicating the diagnostic is capable of fully resolving the speckle pattern. The lower photograph shows the effect of SSD on with a 1-ns smoothing time. The plot shows the results of a series of smoothing-time measurements made with the older 0.2-THz SSD system and the close agreement with the theory. Also shown are the results for a 1-ns smoothing time with the newer 1-THz system with and without polarization smoothing.

10. RESULTS

OMEGA is routinely operated with the IR energy into the FCC's balanced to about 1%. Small variations in frequency conversion and transport differences result in a UV energy balance on target of about 3%. For the square 1-ns pulses that have been used for most ICF experiments to date, the P510s show the UV power balance to be 5% or better throughout most of the pulse. The system repeatability is high enough that we should be able to improve the balance slightly, once the transport measurement is improved to a point that warrants doing so.

Over the past year, experiments on OMEGA have been performed using polarization smoothing, 1-THz SSD, and improved laser energy and power balance. As compared to earlier experiments, without polarization smoothing and using 0.35-THz SSD, we have observed about a factor-of-2 improvement in target performance in terms of neutron yield and fuel ρR. CH-shell targets (19-μm-thick) with 15-atm fills of DT or D_2 irradiated with 23-kJ, 1-ns square pulses achieve about 35% of calculated 1-D yields and better than 85% of 1-D ρR. Repeatability has improved to the few-percent level. These targets do not appear to be sensitive to the exact level of energy/power balance, with x-ray balanced shots, having a reported energy balance of about 5% performing essentially as well as the standard 3% balanced shots.

11. CONCLUSION

A substantial effort has been made to equip OMEGA with a full complement of precision diagnostics for evaluating on-target irradiation uniformity. These diagnostics have permitted us to accurately obtain beam-to-beam energy and power balance and confirm that we have obtained the expected levels of individual beam smoothing from the 1-THz SSD system. The improvements to on-target irradiation that have resulted in part from this effort have resulted in targets experiments that perform at a substantial fraction of calculated 1-D performance and are highly repeatable.

12. ACKNOWLEDGMENT

This work was supported by the U.S. Department of Energy Office of Inertial Confinement Fusion under Cooperative Agreement No. DE-FC03-92SF19460, the University of Rochester, and the New York State Energy Research and Development Authority. The support of DOE does not constitute an endorsement by DOE of the views expressed in this article.

13. REFERENCES

1. T. R. Boehly, D. L. Brown, R. S. Craxton, R. L. Keck, J. P. Knauer, J. H. Kelly, T. J. Kessler, S. A. Kumpan, S. J. Loucks, S. A. Letzring, F. J. Marshall, R. L. McCrory, S. F. B. Morse, W. Seka, J. M. Soures, and C. P. Verdon, Initial performance results of the OMEGA laser system, *Opt. Commun.* **133**, 495–506 (1997).
2. T. J. Kessler, Y. Lin, L. S. Iwan, W. P. Castle, C. Kellogg, J. Barone, E. Kowaluk, A. W. Schmid, K. L. Marshall, D. J. Smith, A. L. Rigatti, J. Warner, and A. R. Staley, Laser phase conversion using continuous distributed phase plates, in: *Second Annual International Conference on Solid State Lasers for Application to Inertial Confinement Fusion*, edited by M. L. André (SPIE, Bellingham, WA, 1997), Vol. 3047, pp. 272–281.
3. S. Skupsky, R. W. Short, T. Kessler, R. S. Craxton, S. Letzring, and J. M. Soures, Improved laser-beam uniformity using the angular dispersion of frequency-modulated light, *J. Appl. Phys.* **66**(8), 3456–3462 (1989).
4. T. R. Boehly, V. A. Smalyuk, D. D. Meyerhofer, J. P. Knauer, D. K. Bradley, R. S. Craxton, M. J. Guardalben, S. Skupsky, and T. J. Kessler, Reduction of laser imprinting using polarization smoothing on a solid-state fusion laser, *J. Appl. Phys.* **85**(7), 3444–3447 (1999).
5. W. R. Donaldson, R. Boni, R. L. Keck, and P. A. Jaanimagi, A self-calibrating, multichannel streak camera for inertial confinement fusion applications, to be submitted to *Rev. Sci. Instrum.*
6. S. P. Regan, J. A. Marozas, J. H. Kelly, T. R. Boehly, W. R. Donaldson, P. A. Jaanimagi, R. L. Keck, T. J. Kessler, D. D. Meyerhofer, W. Seka, S. Skupsky, and V. A. Smalyuk, Experimental investigation of smoothing by spectral dispersion, *J. Opt. Soc. Am. B* **17**(9), 1483–1489 (2000).

ABSORPTION AND FLUORESCENCE SPECTROSCOPY AS A DIAGNOSTIC OF LASER DRIVEN EXPERIMENTS.

D. J. Hoarty, C. C. Smith, O. Willi, G. Magelssen, R. Chrien, S. Caldwell, M. Wood, R. Watt.*

INTRODUCTION

Point projection absorption spectroscopy has been used for many years in the measurement of radiative opacity [1]. This powerful diagnostic technique allows both spectral and two-dimensional spatial information to be obtained by radiography of targets using a near point source of x-rays. Recently the authors have applied this technique to measure absorption spectra, not to study opacity, but as a diagnostic of a range of radiation-hydrodynamics phenomena relevant to inertial confinement fusion. Characterization of targets by absorption or emission spectroscopy at frequencies close to the peak of thermal emission is possible but can have problems with self-emission, high background and non-thermal emission and may reduce the size of targets which can be probed. Therefore K-shell absorption spectroscopy was chosen because the line spectra sampled are at much higher frequencies than the peak thermal emission of the targets thus avoiding the problems mentioned above. K-shell spectroscopy also allows the use of crystal spectrometers which tend to be simpler to implement than gratings of similar resolution. Examples of the use of the technique are given. For instance detailed characterization of conditions in shock and radiation waves have enabled models of transonic flows to be tested against experiment. Also the preheating effect of forming a plasma buffer to smooth laser beam imprint has been studied under conditions relevant to direct drive inertial confinement fusion. To infer the target conditions the measured spectra must be compared with synthetic spectra generated by modelling the opacity in the region of the diagnostic lines. The opacity model used in this work is briefly described.

---

* D J Hoarty, C C Smith, AWE, Plasma Physics Department, Reading UK, O. Willi Heinrich-Heine-Universitaet Dusseldorf, Germany, G. Magelssen, R. Chrien, S. Caldwell, M. Wood, R.Watt, Los Alamos National Laboratory, NM, USA.

Advanced Diagnostics for Magnetic and Inertial Fusion, Edited by Stott *et al.*
Kluwer Academic/Plenum Publishers, New York, 2002

In some situations absorption spectroscopy has disadvantages as a diagnostic. It may introduce an unacceptably large perturbation to the target temperature. A less perturbative technique, but with the advantages of K-shell spectroscopy, has been developed in which fluorescence emission is used to diagnose temperature. The technique and data modelling are described and examples shown of results obtained in hohlraum driven experiments.

EXPERIMENTS DIAGNOSED BY POINT PROJECTION ABSORPTION SPECTROSCOPY

To show the effectiveness of point projection absorption spectroscopy as a diagnostic of laser driven experiments examples of its use are given below. In the first example radiation or shock waves were driven through a low density target made from a cylinder of foam doped with chlorine. The target was radiographed perpendicular to the direction of wave propagation and analysis of the K-shell absorption spectra of the chlorine dopant in the material allowed the ionization distribution to be inferred [2]. A schematic of the experimental setup is shown in figure1. Density was inferred from the absolute transmission. Temperature was inferred by comparison of the data with a detailed model of the chlorine K shell absorption. A spatial resolution of 10μm was achieved by using a pin backlighter. The experimental temporal resolution was given by selecting the duration of the laser pulse onto the backlighter pin and was 90ps. By freezing the motion of a radiation wave or shock wave in this way a detailed measurement of the conditions was made. The technique has been used to study radiating shocks [3], transonic radiation waves [2], colliding shocks [4] and electron conduction, in the latter case in silicon aerogel. Figure 2 shows examples of radiographs

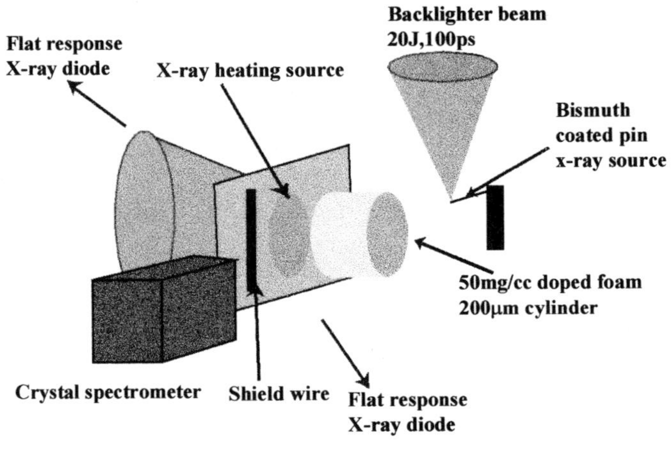

Figure 1. Schematic of the experimental setup for studying radiation waves diagnosed by point projection absorption spectroscopy.

obtained from studies of transonic radiation waves. Note the changes in the target absorption due to the presence of shocks and the changes in the absorption spectrum of the chlorine dopant. By changing the delay of the backlighter with respect to the heating beams the evolution of conditions over a number of times was measured. Radiation and

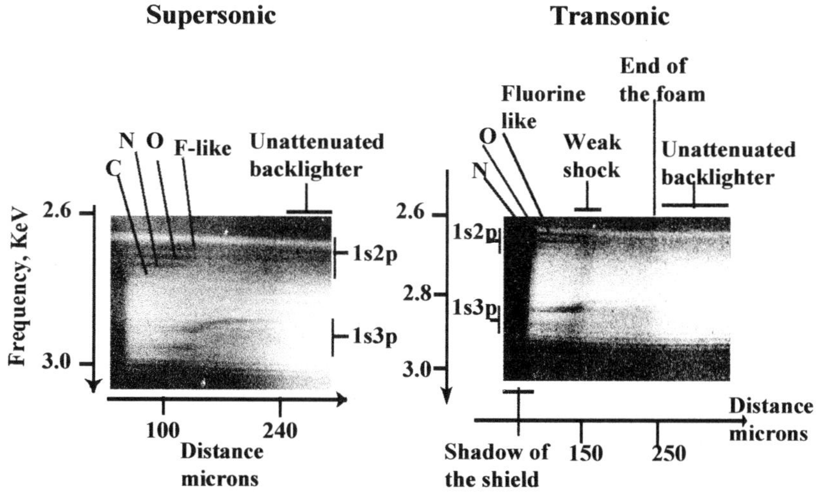

Figure 2. Examples of radiographs taken of radiation wave propagation in chlorine doped foam. The absorption features of the chlorine indicate the conditions in the target. By changing the delay of the backlighting pulse with respect to the main drive pulse the change in the conditions as the front slowed were observed.

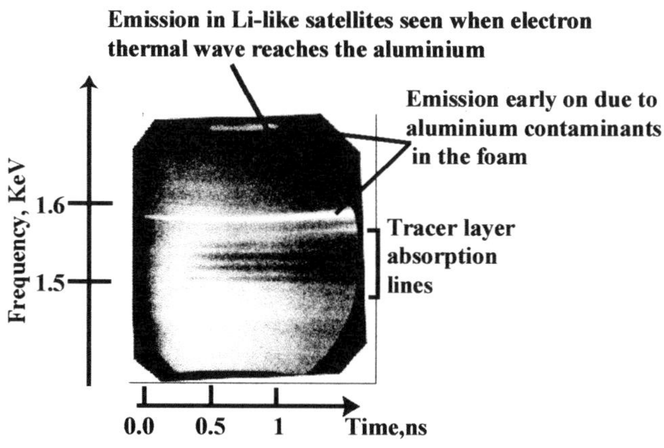

Figure 3. The effect of radiation, shock and thermal conduction heating was observed on the same shot using streaked point projection radiography of a buried layer of aluminium in the target.

shock wave studies have also been carried out in targets made from low density foams enclosed in high Z tubes. Here the conditions were characterized by changes in the absorption spectrum of an aluminium tracer layer placed at different points in the tube. In this case radiography was performed along the axis of the tube. The absorption spectrum was taken using a conical crystal spectrometer coupled to a x-ray streak camera. The absorption spectra were compared to simulated aluminium spectra to infer ionization.

Another example of diagnosis of target conditions using tracer layers is the measurement of preheat in foam buffered direct drive inertial confinement fusion [5]. In direct drive laser fusion laser imprint has been shown to seed instabilities which break up a capsule shell. Foam buffering is a scheme in which laser beam imprint is mitigated by creating a low density plasma around the capsule to thermally smooth the laser shock before it enters the capsule shell. A parameter study of planar foam buffered [6] foil targets measured the degree of target preheat when forming the plasma buffer. The conditions were measured by taking the absorption spectrum of an aluminium layer buried in the target. The experiment was able to discriminate between the effect of radiation, shock and thermal electron conduction heating of the target, on the same shot.

MODELLING OF THE ABSORPTION DATA

Models were constructed for aluminium and chlorinated foam for the tracer layer diagnosed experiments and the doped foam experiments respectively. The model predictions of synthetic absorption spectra of chlorinated foam are compared

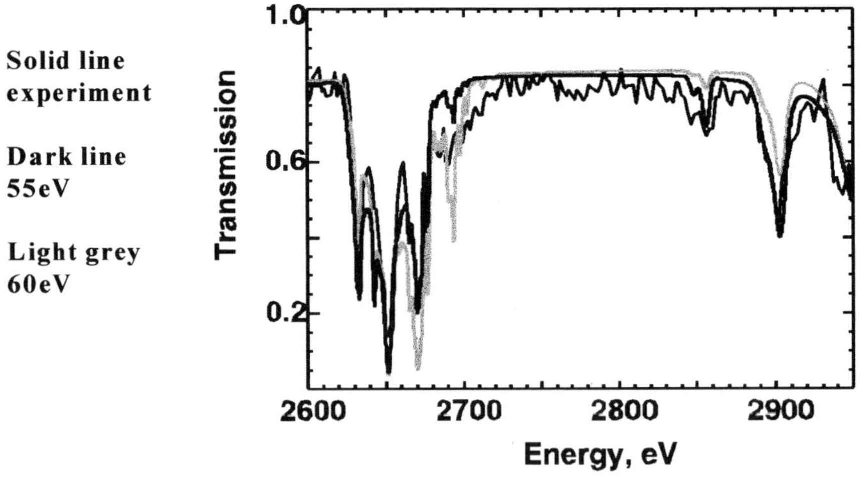

Figure 4. Comparison of measured transmission and synthetic spectra generated by an LTE model of the K-shell absorption of chlorinated foam. A similar model was constructed for aluminium. The sensitivity to temperature can be seen in marked difference in the two spectra 5eV apart

to experiment in figure 4. The sensitivity to temperature is shown by comparison of two calculations 5eV apart which show marked differences. Both the 1s-2p lines and the 1s-3p lines are fitted.
The effect of the other constituents of the foam, carbon, hydrogen and oxygen, are included in order to calculate the correct electron density, and continuum absorption. The synthetic spectra were generated using a local thermodynamic equilibrium (LTE) code, SEPSAHA. The SEPSAHA code reads a database of excitation energies, statistical weights, line energies and oscillator strengths. From this data the fractional populations in each ion stage and each energy level were calculated and a detailed spectrum constructed. The database was generated from ab initio calculations of line energies and oscillator strengths using a multi-configurational Dirac-Fock code, called GRASP[7] or from atomic physics data available in the literature. The atomic physics data included the permutations of L, M and N shell electrons. The myriad weak lines due to the many, low population, configurations with spectator electrons in M and N shells and their term structure were accounted for by including the lines as an unresolved transition array (UTA). Further approximations were used for configurations with spectator electrons in n=5-13. The weak lines were assumed to blend together into a continuous transition array with a Gaussian width determined by estimates of the energy shifts due to spectator electrons and their term structure. GRASP calculations of line shifts due to spectator electrons and width of the term structure in some of the weak lines were done to check that the widths chosen were reasonable. To check that the LTE populations were valid the ionization was also calculated using a collisional-radiative equilibrium (CRE) model. The CRE and LTE calculations were in agreement, LTE was maintained because of the high electron density, despite the radiation temperature exceeding the material temperature.

Lowering of the ionization potential due to the field of neighbouring ions in the plasma was included [2,3]. The effect of lowered ionization potential meant that high lying levels were no longer bound. The code calculated the smearing of the ionization edge using the nearest neighbour approximation and also the effect of broadening in the high lying lines on the position of the continuum.

EXPERIMENTS DIAGNOSED BY FLUORESCENCE SPECTROSCOPY

The fluorescence was induced by photopumping with a spatially separate broad band x-ray source. To prevent thermal radiation from the pump reaching the sample and changing it's temperature the pump radiation was filtered through 21μm thick plastic. The scheme described here, and shown below in figure 5, uses aluminium K-shell emission pumped by a quasi-continuum of gadolinium M-band radiation [8]. The M-band of the gadolinium is in the same frequency range as the 1s-2p transitions in aluminium ions. The gadolinium pump radiation field was measured during the experiments. Broad band pumping allows all of the ion distribution present to be sampled. Though the M-band radiation increases the population of states with a 1s vacancy by many orders of magnitude it has a negligible effect on the population of states with 1s full. The time resolved fluorescence was measured using a focussing crystal coupled to a x-ray streak camera. The heating of the aluminium by radiation waves in the target was evident by the changes in 1s-2p fluorescence emission from the aluminium. Figure 5b shows the aluminium fluorescence before, during and after the radiation front arrived at the position of the aluminium patch. Measurements were taken to establish that the pump radiation did not heat the aluminium. This was done by firing only the laser pulse onto the pump and not those producing the radiation wave inside the target. These experiments have been carried out on the NOVA and OMEGA lasers. The technique was developed in experiments using the AWE HELEN laser [8]. Spatial resolution can be achieved due to the size and placement of the fluorescent material. Temperatures measured with aluminium fluorescence have been between 15-70eV. It is intended to extend the technique to higher temperature.

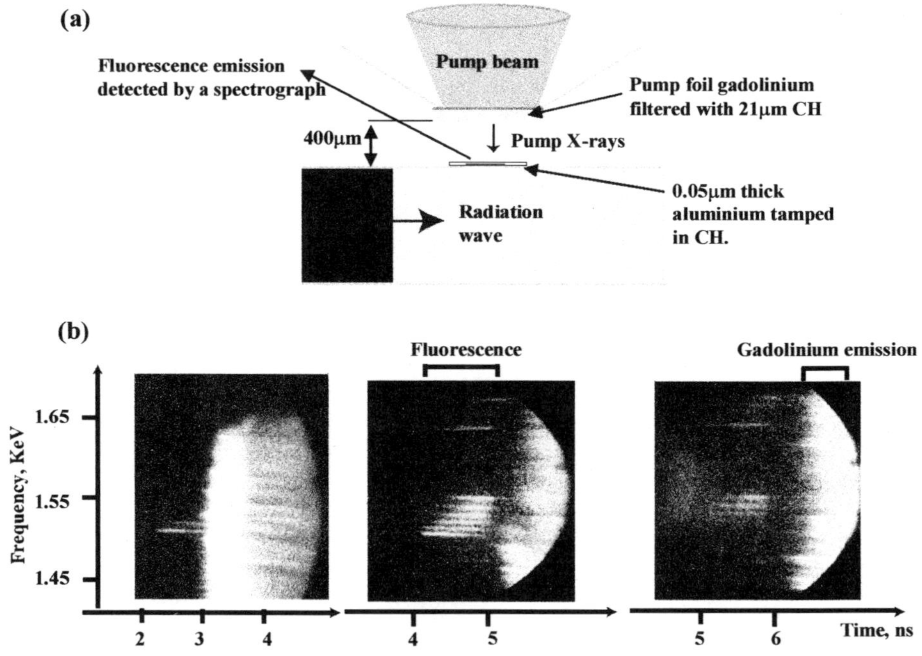

Figure 5. Fluorescence spectroscopy applied to radiation wave propagation in enclosed geometry. A 200μm aluminium patch is caused to fluoresce by radiation from a laser heated filtered gadolinium pump foil. Figure 5b shows how by changing the delay on the pump laser beam fluorescence before the radiation wave arrives at the patch (spectra on left), as the wave crosses the patch (middle) and immediately after the wave has passed the patch can be recorded. The turn on of each 1s-2p line shows the ionization of the aluminium sample as it is heated.

MODELLING OF THE FLUORESCENCE SPECTRA

Modelling of the fluorescent spectra in the HELEN experiments using a collisional radiative equilibrium model called NEBULA[9] showed that the important processes determining the intensity of the fluorescence were radiative excitation/deexcitation, photoionization, and autoionization. The NEBULA model alone could not reproduce the experimental results because it did not include radiation transport [8]. A code was written called RTRAN which includes the important processes in determining level populations for excited states and calculates the radiative transfer within a photopumped aluminum sample. It produces a synthetic spectrum which would be observed by photopumping of an aluminium plasma. It is this code which is used to infer temperatures from fluorescence spectra. The input to the RTRAN code was the detailed opacity over the frequency range of transitions from the K shells, calculated using the SEPSAHA code described above for modelling K-shell absorption, and the pump radiation field. The radiation field incorporated the pump spectrum measured in the experiment and extrapolated to 4keV. The code proceeded by calculating the number of photons from the pump source reaching each cell in the sample, at each frequency point in the radiation field on a 0.025eV mesh, by evaluating the angular integrals over the range defined by the pump source size and distance from the sample.

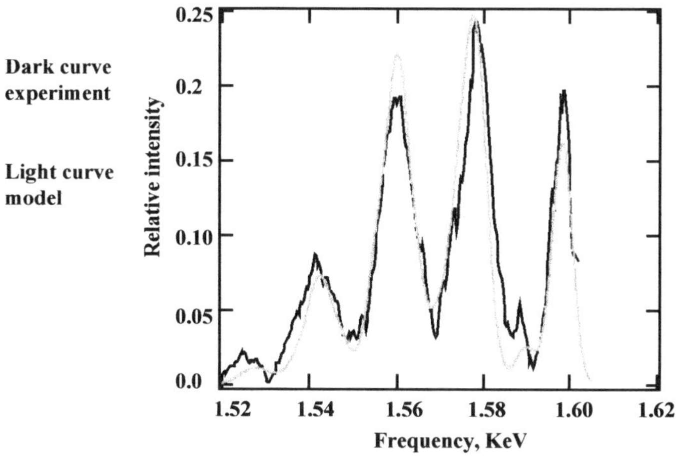

Figure 6. Comparison of measured and simulated spectra from the RTRAN code. The temperature input to the model is iterated until best agreement between prediction and experiment is obtained.

The population of states with 1s vacancies was calculated from the ground state populations defined as configurations with a full K shell, also supplied as input from SEPSAHA. All ions in the sample are available to be photopumped and the pump has a negligible effect on the population of states with a full K shell. The spontaneous emission was calculated from the excited state populations, assuming complete redistribution, (ie photon emission occurred with the thermal line profile and did not depend on where in the profile absorption took place) and the fluorescent efficiencies from detailed calculations by M. Chen. The number of absorbed photons were shared between excited states of all ion stages in proportion to the relative absorption strengths of the transitions which created the excited states, including n=1 to n=2 line transitions and photoionization. Angular integrals were then evaluated to calculate the extra number of photons available for photopumping in each zone because of reemission from all other zones. The calculation was repeated until the increase in photons produced by reemission was negligible. The effect of transport on the observed spectrum was then evaluated by calculating the emission and absorption at each photon energy through the different spatial zones in the direction of the spectrometer. The output spectra were inspected and the number of iterations and the number of zones chosen on a judgement that further increase of either did not change the output spectra. This method inherently calculates the time integrated fluorescence in that every photon which could photopump the sample plasma was allowed to do so and the probability of any emitted photon reaching the detector then calculated. The constant plasma conditions ensure that the solution is also equivalent to a steady state solution of the rate equations, equivalence being obtained simply by dividing by a pulse length which gives the required pump intensity. Comparion of the predicted and measured spectra are shown in figure 6. The measurement was taken from a OMEGA radiation transport experiment.

SUMMARY/CONCLUSION

Point projection absorption spectroscopy has been demonstrated to be a powerful diagnostic technique for use in laser driven targets, particularly in open geometries. It allows temperature and density distributions to be measured at spatial resolutions of around 10μm and temporal resolution of around 100ps. The use of a spectrograph coupled to a x-ray streak camera has been shown to increase temporal resolution to 20ps in experiments where the separate effects of radiation, shock and thermal conduction heating were observed on the same shot on a laser driven foil. In enclosed geometries, such as in hohlraum driven experiments absorption spectroscopy can introduce unacceptably large perturbations to the target conditions. In such cases a novel technique, that of fluorescence spectroscopy has been shown to give an accurate measure of temperature using aluminium as the fluorescent material. It is planned to extend the technique to higher temperatures by using higher Z fluor materials.

ACKNOWLEDGEMENTS

The authors would like to acknowledge the efforts of the laser and target area staff of the HELEN, VULCAN, NOVA and OMEGA lasers; the target fabrication groups at AWE and Los Alamos National Laboratories and Ted Perry and Steve Compton of Lawrence Livermore National Laboratory.

REFERENCES

1 T.S. Perry, P. T. Springer, D. F. Fields et al, Phys. Rev. E **54,** 5617 (1996) and references therein
2 D. Hoarty et al, Phys. Plasmas **6,** 2171 (1999)
3 D. Hoarty, A. Iwase, C. Meyer et al, Phys. Rev. Lett. **78**, 3322 (1997)
4 A. Schiavi et al, CLF report RAL-TR-1999-062
5 D. Hoarty et al in *Proceedings of Inertial Fusion Sciences and Applications 99* (ELSEVIER, Paris 1999)
6 R. J. Mason, R. A. Kopp, H. X. Vu, D. C. Wilson, S. R. Goldman, R. G. Watt, M. Dunne, O.Willi, Phys. Plasmas **5,** 211 (1998) and references therein.
7 K. G. Dyall et al, Comp. Phys. Comms., **55,** 425 (1989)
8 C.C. Smith, D. J. Hoarty and S. Mouncey, J. Quant. Spectrosc. Radiat. Transfer, **54,** 387 (1995)
9 S. J. Rose, J. Quant. Spectrosc. Radiat. Transfer, **54,** 333 (1995)

Measurements of the poloidal magnetic and radial electric field profiles in ASDEX Upgrade and JET

J. Hobirk, N. C. Hawkes, P. J. McCarthy, D. Merkl,
R. C. Wolf, ASDEX Upgrade Team
and contributors to the EFDA-JET Workprogramme*

1 Introduction

In magnetic confinement devices the plasma confinement and stability are determined by the magnetic field configuration consisting of toroidal and poloidal field components. In tokamaks the poloidal magnetic field is determined by the profile of the toroidal current. In addition to the Faraday rotation polarimetry [1, 2] the Motional Stark Effect (MSE) has become a routine diagnostic technique to determine the current density distribution [3, 4, 5, 6, 7, 8, 9, 10, 11]. The MSE diagnostic utilizes the Doppler shifted neutral beam emission spectrum which is dominated by the Stark effect due to the electrical field experienced by the fast atoms in their rest frame which is mainly the Lorentz electric field $\vec{v}_{\text{beam}} \times \vec{B}$. The orientation of this Lorentz electric field is measured, and incorporated in an equilibrium reconstruction, which yields profiles of poloidal magnetic field, current density, safety factor q or poloidal flux. In this paper the diagnostic principle and its implementation on ASDEX Upgrade and JET will be presented in sections two and three. The results of the diagnostic will be discussed in section 4 and the influence of another beam source on the measurement in section 5. Since the MSE diagnostic is sensitive to electric fields in general this opens the possibility to measure also internal electric fields such as the radial electric field [12, 13]. If E_r/v_{beam} becomes of the order of B_{pol} then E_r can no longer be neglected which is the case in particular for discharges with an internal transport barrier (ITB). The measurement of E_r may also be applicable to stellarators where the magnetic field configuration deviates only slightly from the known vacuum configuration. The stronger field line curvature could however increase the requirements regarding the spatial resolution of a MSE diagnostic. First measurements of the radial electric field will be presented in section 6 and a summary will follow in section 7.

*J. Hobirk, D. Merkl, R. C. Wolf, Max-Planck-Institut für Plasmaphysik, EURATOM Association, 85740 Garching, Germany, N. C. Hawkes, Euratom/UKAEA Association, Culham Science Centre, Abingdon,Oxfordshire,OX 14 3DB,UK, P. J. McCarthy, Physics Department, University College Cork, Association Euratom-DCU, Cork, Ireland

Advanced Diagnostics for Magnetic and Inertial Fusion, Edited by Stott *et al.*
Kluwer Academic/Plenum Publishers, New York, 2002

2 Principle of the measurement

The MSE diagnostic is a kind of beam emmission spectroscopy (BES)[14]. But in contrast the BES it is not the intensity of the light but rather the polarisation that is analysed. The injected fast neutrals experience in their rest frame a strong Lorentz ($\vec{E} = \vec{v}_b \times \vec{B}(+\vec{E}_{\text{internal}})$) electric field.

This gives rise to the Stark effect which dominates strongly over the Zeemann effect under usual experimental conditions in a tokamak ($E_b \approx 50 - 150\text{keV}$, ,$B_{\text{tor}} \approx 1 - 4\text{T}$). Due to the Stark effect the D_α or H_α beam emission lines split components perpendicular to the electric field linear polarized (σ) and parallel polarized (π).

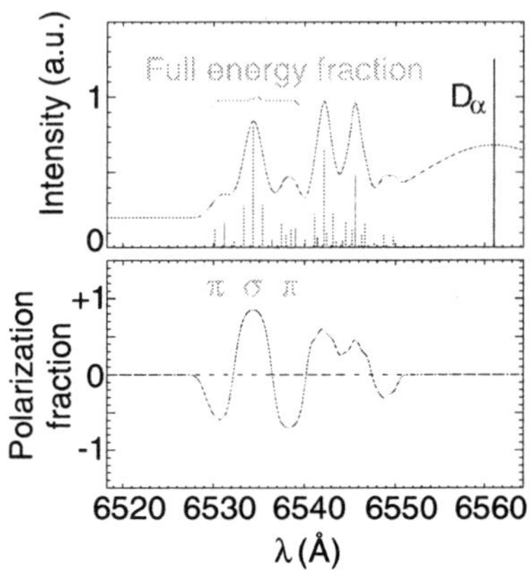

Figure 1: ASDEX Upgrade beam calculated emmission spectrum and polarisation fraction

An example of the spectrum for ASDEX Upgrade (calculated, maesured in figure 4) is shown in figure 1. Three lines each are very close together so that the spectrum looks more like a 3 line spectrum, two are π lines and one is a σ line. The spectrum presented in the upper part of figure 1 contains the 3 full energy component lines, the half, and the third energy component lines. The three beam components are doppler shifted corresponding to the energies of the accelerated D^+, D_2^+, D_3^+ particles in the neutral beam ion source. In the lower part of the figure the polarisation fraction is presented. +1 means only σ and the polarisation of the π lines is negative. The polarisation of the full energy σ line is the strongest and the line is not much overlapped by others. For the half energy the polarisation is still comparible to the full energy π lines.

To use the information carried by the spectral lines a diagnostic needs to measure the direction of the electric field by measuring the direction of the polarisation of one of the spectral lines. This can be done in a static way [15, 16] by splitting the light to two linear polarizers and a circular polarizer to get the complete stokes vector or it can be done in a dynamic fashion used in most tokamak diagnostics by means of two photo elastic modulators (PEMs).

3 Diagnostic setup

The diagnostic setup will be described for the JET motional Stark effect diagnostic in the follwing section. The ASDEX Upgrade diagnostic is very similar exept for the number of lines of sight, details of the monochromator and the data aquisition. The setup for JET is shown in figure 2. The optics has 25 lines of sight covering the whole minor radius on the low field side with a distance of 5cm between the channels and a spatial averaging of 3-7cm in one channel. The beam emitted light is reflected by a prism trough a vacuum window into a tube with an integrated transfer optic. The window and the optics effect the polarisation of the transmitted

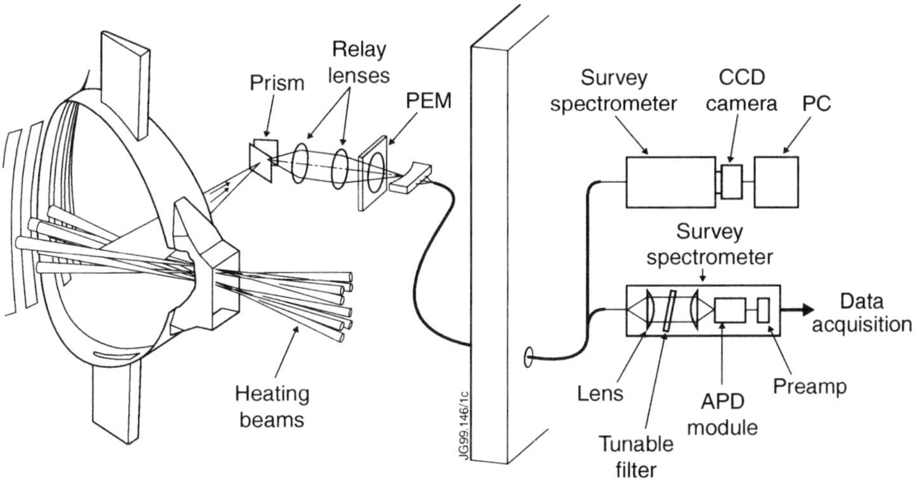

Figure 2: Diagnostic setup implemented at JET

light because in the optical components the Faraday rotation adds a magnetic field dependent offset to the measurement. This needs to be included in the calibration process. Behind the optics a set of two PEMs is placed followed by a linear polarizer which allows polarimetry by means of a lock-in technique. To collect the light only from the selected spectral line a small band width interference filter is placed before an Avalanche diode.

4 Experimental results

The discussed setup of the MSE diagnostic results in a measurement of a polarisation angle γ_m. For interpretation of this angle the complete observation geometry needs to be taken into account. For an arbitrary viewing geometry it results in the following formula:

$$\tan \gamma_m = \frac{A_1 B_R + A_2 B_T + A_3 B_Z + A_4 E_R / v_b}{A_5 B_R + A_6 B_T + A_7 B_Z + A_8 E_R / v_b + A_9 E_Z / v_b} \tag{1}$$

Equation 1 contains geometric coefficients A_n which are composed of the sin and cos of the observation angles and the inclination of the neutral beam against the midplane. Furthermore all magnetic field components (B_R, B_Z and B_T) and the components of a radial electric field (E_R and E_Z) are contained where R, Z, and T denote the radial, vertical and toroidal direction respectively. For a very simple viewing geometry in which beam and viewing lines lie in the magnetic midplane ($B_R = 0, E_Z = 0$) the equation 1 reduces to $\tan \gamma_m = \frac{A_3 B_Z + A_4 E_R}{A_6 B_T}$. This measurement is used as an additional constraint to the plasma equilibrium reconstructed by and equilibrium codes like EFIT on JET or CLISTE on ASDEX Upgrade. This code makes a fit to the measurement and minimizes the fit error. Additionally the magnetic data from the edge is included which gives mainly information about integrals (like plasma current) and a boundary condition (plasma shape). Without internal information from the MSE system the codes freely extrapolates the internal field structure. In figure 3 profiles for two different kinds of discharges from AS-

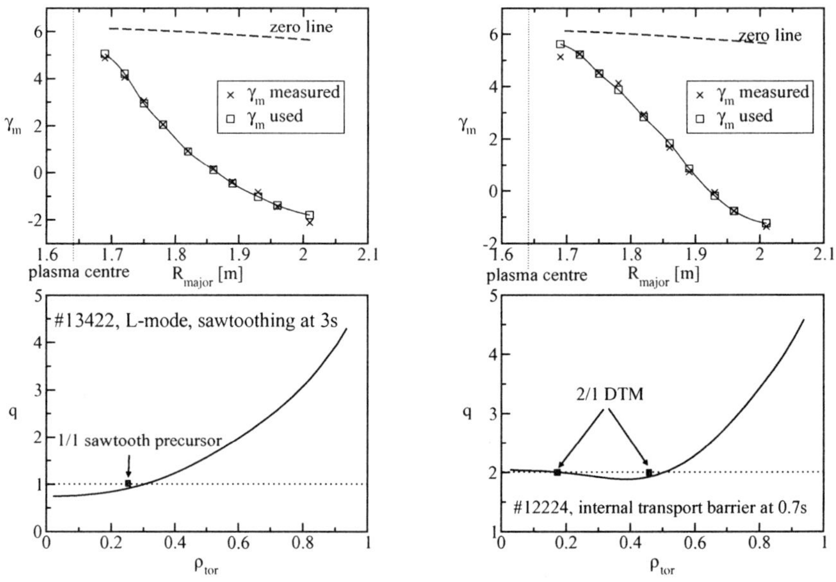

Figure 3: Measured and fitted polarisation angle profiles and corresponding q-profiles for ASDEX Upgrade discharges

DEX Upgrade are shown. On the left an L-mode profile during sawtooth oscillations taken in a stationary phase is given. In the upper part the measured MSE angles and the corresponding fit are plotted. For a help also the zero line (equivalent to poloidal field zero) is shown. In ASDEX Upgrade the MSE usually and unfortunately does not cover the plasma centre. In this example the fit is very good and the q-profile in the lower part of the figure represents the MHD (Magneto Hydro Dynamic) features of the discharge quite well. The discrepancy in the $q = 1$ radius is well inside the measurement errors. On the right an early time point ($t = 0.7$s) in a discharge with an internal transport barrier is presented. Already the measured angle profile shows that the poloidal field has different features. Already the smaller difference between γ_m and zero line. roughly the representing the poloidal magnetic field, shows that the poloidal field in the interior of the plasma is much smaller then in the other discharge. This results in reversed shear profile shown in the lower part. The measurement and the equilibrium reconstruction here also fit very well to the information from the MHD. In this reconstruction also the effect of the radial electric field has been included.

5 Influence of another beam source on measurement

One main difference between the JET and ASDEX Upgrade MSE diagnostic is the spectral line which is used for the measurement. ASDEX Upgrade uses the σ line because it has a much higher polarisation degree and a higher intensity, whereas the most Doppler shifted π line is utilized in the JET diagnostic. The reason for this is the influence of other beam lines (from the same beam box, overlaying in the plasma) to the measured polarisation angle. In figure 4 two ASDEX Upgrade spectra are shown. On the left the unperturbed spectrum with only the beam line

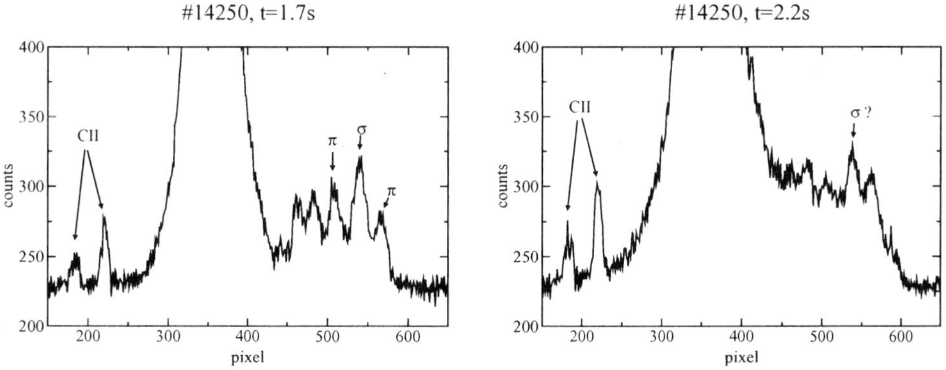

Figure 4: Beam emission spectra from ASDEX Upgrade. Left figure spectrum with only the measurement beam line and right spectrum with two beam lines.

which is used for the measurement is given. On the right the spectrum perturbed by additional beam is shown. The nicely separated lines in the unperturbed spectrum have smeared out by overlaying the spectrum of the other beam source which exhibits a slightly different Doppler shift. The σ line of the full energy component is still visible, but but the degree of polarisation is drastically reduced. The conclusion is that here π (and σ) lines from the second beam line overlay with the σ line from first source. The consequence is a jump in the γ_m which dependents on various plasma parameters. If the polarisation direction from one σ line changes, the polarisation from the other beam line changes also but differently because it corresponds to a different position in the plasma. The only conclusion which can be drawn is that the measurement is not useful if any other beam line is active in ASDEX Upgrade. For JET with 8 beam sources in one box this is a more serious problem. One possibility to prevent the disturbance is to use a spectral line which is not perturbed, ideally a different (higher) beam energy should be used for the measurement. A partly successful solution to the problem has been found at JET using the outermost π line from the beam source which has the largest Doppler shift. Ideally there should be no overlay between the spectra any more but the lines are very broad beam and viewing line divergence[17]. The goal achieved, is that the overlay now is mostly only between π and π components which can be included into the equilibrium reconstruction by a weighting function which resolves partly the problem.

6 First results of radial electric field measurements

The radial electric field plays an important role in tokamak transport theory as it determines the $\vec{E} \times \vec{B}$ shear flow which is one of the processes responsible for the suppression of micro turbulence [18]. To verify these theories various measurements of E_r have been developed in recent years: (A) From charge exchange recombination spectroscopy (CXRS) the different terms of the radial force balance are inferred and subsequently E_r is calculated [19]. (B) The change of the edge radial electric field in the H-mode transition has been derived by charge exchange measurements of high energy neutrals [20].(C) The measurement of the poloidal magnetic field by means of the motional Stark effect (MSE) is also sensitive to E_r [12], because the neutral beam atoms experience not only the Lorentz field, but also the radial electric field.

If two independent measurements of $\vec{E} = vecE_{\mathrm{Lorenz}} + \vec{E}_r$ are employed, this facilitates not only the deduction of B_{pol} but also of E_r. One possibility to measure E_r by means of MSE is to use two different lines of sight at the same radial position [12]. Another one is to use the same line of sight and to measure two different energy components of the neutral beam [13]. In ASDEX Upgrade the latter method is employed, which for machines with limited spatial access is easier to implement.

Neglecting the terms A_5B_R, A_7B_Z, A_8E_R/v_b, and A_9E_Z/v_b in the denominator of (1), which are small compared to A_6B_T, E_r can be approximated by

$$E_r = (\tan\gamma_m - \tan\gamma_{m/2})B_T v_b \frac{A_6}{A_4(1-\sqrt{2})} \tag{2}$$

In figure 5 this method has been used to derive E_r from the measured γ_m's at $\rho_{\mathrm{pol}} \approx 0.6$.

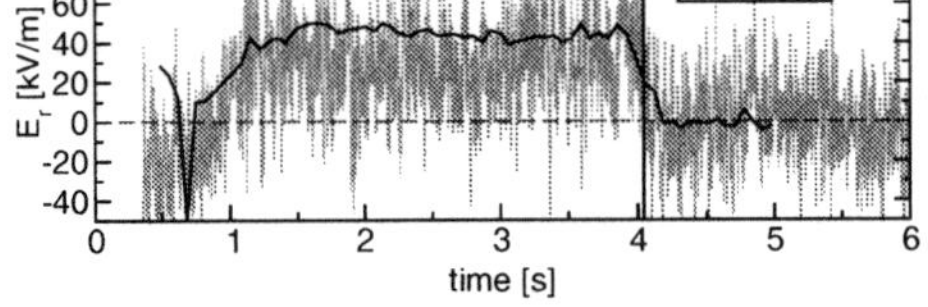

Figure 5: Time traces of γ_m,$\gamma_{m/2}$ (upper trace) and calculated E_r (not averaged,lower trace) together with CXRS E_r for comparison. After reaching the β-limit, a locked mode occurs approximately at 4s.

This is the only radial position at which the diagnostic has been upgraded so far. As the small difference $(\tan\gamma_m - \tan\gamma_{m/2})$ is multiplied by the large values of beam velocity and toroidal magnetic field, a large noise level is produced in the resulting E_r. Another problem is the absolute value of γ_m. The calibration procedure results in a minimum error of about 0.2 deg, corresponding to an uncertainty of 40kV/m in E_r. To improve the absolute accuracy, the calibration factors of the half energy channels are adjusted to produce $E_r = 0$ in the presence of locked modes. This assumption is justified, as the plasma toroidal rotation stops and the pressure gradient is balanced by the diamagnetic drift. In the example discharge of figure 5 a locked mode occurs at $t = 4$s.

ASDEX Upgrade is equipped with a CXRS diagnostic [19] which by means of poloidal and toroidal views measures the respective rotation velocities $v_{\rm pol}, v_{\rm tor}$ and the radial pressure gradient $\nabla_r p$ of the impurity ion species C^{5+}. Substituted into the radial force balance equation, E_r is calculated:

$$E_r = \frac{1}{Zen_Z}\nabla_r p_Z - v_{\rm pol}B_{\rm tor} + v_{\rm tor}B_{\rm pol} \tag{3}$$

where $B_{\rm pol}$ and $B_{\rm tor}$ are the poloidal and toroidal magnetic field components, n_Z the density, and Z the charge number of ion species used. In figure 5 the different E_r measurements are compared. The time evolution of E_r is qualitatively the same for both methods. While E_r from MSE measurement is calibrated to be zero when a locked mode is present, the CXRS value drops to zero independently. Considering that the E_r value calculated from the MSE polarisation angles directly is only an approximation, also the quantitative agreement is acceptable.

To study the influence of E_r on the q-profile evaluated by CLISTE using the MSE data, the radial electric field components in equation 1 are taken from the CXRS measurements. In figure 6 q-profiles without considering E_r and with the correction by CXRS are compared. In the first case, a discharge with reversed magnetic shear and an ITB is shown [21]. Including E_r the whole q-profile drops, but most prominently the central q-value (q_0). The minimum value of q ($q_{\rm min}$) decreases less, but nevertheless improves the agreement between the appearance of a ($m = 2, n = 1$) double tearing mode and the time when $q_{\rm min}$ reaches 2. The same trend in the q-profile can be observed in the second case, which shows the monotonic q-profile of an improved confinement H-mode and is characterized by an extended low shear region in the plasma core. Here the central magnetic shear is not affected, but rather the absolute values of q in the core region of the plasma.

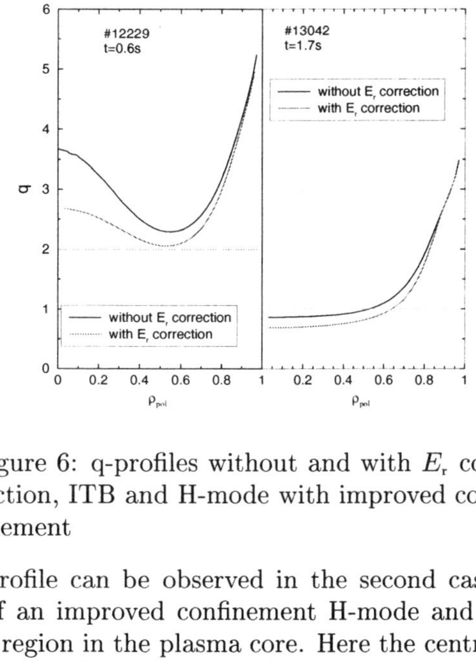

Figure 6: q-profiles without and with E_r correction, ITB and H-mode with improved confinement

7 Summary

The MSE diagnostic is after some years of development a routinely used diagnostic on many tokamaks. It gives reliable results on the plasma current profile and is an important ingredient in the equilibrium reconstruction. For discharges with high radial electric fields it is important to include also the radial electric field into the equilibrium reconstruction. On the other hand the MSE diagnostic can also be used to measure the radial electric field directly giving crucial informations on the physics of internal transport barriers.

References

[1] H. Soltwisch, Rev. Sci. Intrum. **57**, 1939 (1986).

[2] H. R. Koslowski and H. Soltwisch, Fusion Engineering and Design **34-35**, 143 (1997).

[3] F. M. Levinton *et al.*, Phys. Rev. Lett. **63**, 2060 (1989).

[4] D. Wróblewski *et al.*, Rev. Sci. Instrum. **61**, 3552 (1990).

[5] D. Wróblewski and L. L. Lao, Rev. Sci. Instrum. **63**, 5140 (1992).

[6] F. M. Levinton, S. H. Batha, M. Yamada, and M. C. Zarnstorff, Phys. Fluids B **5**, 2554 (1993).

[7] B. W. Rice, Fusion Engineering and Design **34-35**, 135 (1997).

[8] R. C. Wolf, P. J. Mc Carthy, F. Mast, and H. P. Zehrfeld, Europhysics Conference Abstracts (Proc. 24th Eur. Conf. Berchtesgaden 1997) **21A**, 1509 (1997).

[9] B. C. Stratton, D. Long, R. Palladino, and N. C. Hawkes, Rev. Sci. Instrum. **70**, 898 (1999).

[10] N. C. Hawkes *et al.*, Rev. Sci. Instrum. **70**, 894 (1999).

[11] F. M. Levinton, Rev. Sci. Instrum. **70**, 810 (1999).

[12] B. W. Rice, K. H. Burell, L. L. Lao, and Y. R. Lin-Liu, Phys. Rev. Let. **79**, 2694 (1997).

[13] M. C. Zarnstorff, F. M. Levinton, S. H. Batha, and E. J. Synakowski, Phys. Plasmas **4**, 1094 (1997).

[14] W. Mandl, R. C. Wolf, M. G. von Hellermann, and H. P. Summers, Plasma Phys. Contr. Fusion **35**, 1373 (1993).

[15] D. Voslamber, Rev. Sci. Instrum. **66**, 2892 (1995).

[16] T. Soetens, R. Jaspers, and E. Desoppere, Rev. Sci. Instrum. **70**, 890 (1999).

[17] R. C. Wolf, D. Ciric, and R. W. T. König, JET-P **68**, (1994).

[18] K. H. Burrel, Phys. Plasmas **4**, 1499 (1997).

[19] H. Meister *et al.*, this conference (P3.035) (2000).

[20] W. Herrmann, Phys. Rev. Lett. **75**, 4401 (1995).

[21] S. Günter *et al.*, Phys. Rev. Lett. **84**, 3097 (2000).

RECENT PROGRESS IN BEAM EMISSION AND CX SPECTROSCOPY

M.G. von Hellermann, R.Jaspers[1], H.P.Summers[2], K-D.Zastrow[3]

1. INTRODUCTION

Beam Emission spectroscopy (BES) [1,2,3,4] has been considered for quite a while as an important collateral tool to active charge exchange spectroscopy (CXRS). In addition to its application in magnetic field measurements via the Motional Stark effect [5,6] it is of equal interest for the deduction of local neutral beam densities. For higher levels of accuracy in CXRS data evaluation, that is, the deduction of local impurity ion densities from measured CX spectral intensities and (usually) calculated neutral beam densities, contributions of excited state populations need to be taken into account. The Charge Exchange Analysis Package (CHEAP), which has been developed at JET, calculates the beam n=2 population and in a second step a hybrid emission rate, which is the sum of ground-state and excited state contributions.

Beam Emission Spectroscopy

$$I_{bes}(E) = \frac{1}{4\pi} n_e \cdot Q_{BES}(E, n_e, T_i, Z_{eff}) \cdot \int n_b(E) ds$$

Charge Exchange Spectroscopy

$$I_{cx} = \frac{1}{4\pi} n_z \cdot \sum_k Q^z_{cx}(E_k, n_e, T_i, Z_{eff}) \cdot \int n_b(E_k) ds$$

Neutral beam attenuation process

$$n_b = n_b(0) \exp\left[-\int ds \cdot n_e \sum_{k>1} \frac{n_z}{n_e} \sigma_{stop,k} \right]$$

---

[1] FOM Institute for Plasma Physics 'Rijnhuizen', EURATOM/Fusion Association, 3430 BE Nieuwegein, NL
[2] Department for Appl.Physics, University of Strathclyde, Glasgow, G4 0NG,UK
[3] UKAEA Culham, EURATOM/Fusion Association, Abingdon, OX14 3EA, UK

Since the beam attenuation calculation requires the input of all impurity ions contributing to the beam-stopping process, CX ion densities have to be derived iteratively. The loop converges usually in a few steps. Note also that the fractional beam densities, for beam species E, E/2 and E/3, need to be established either from BES data or calorimetric neutral beam test-bed measurements. In the paper presented here the BES deduced values are used.

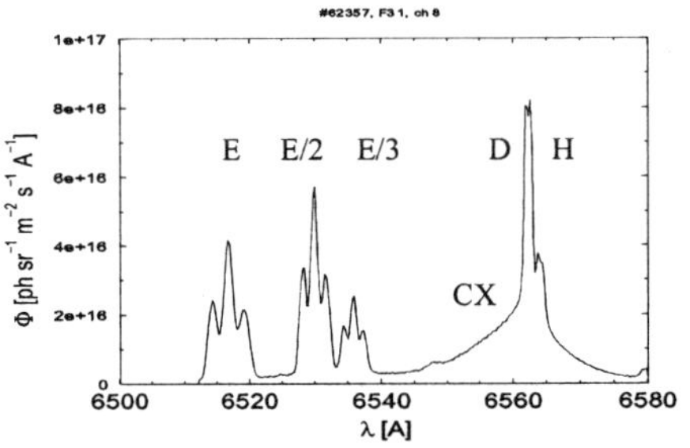

Figure 1. Balmer-Alpha spectrum measured at TEXTOR. BES analysis provides local beam line densities, radial positions, as well as the energy species fractions.

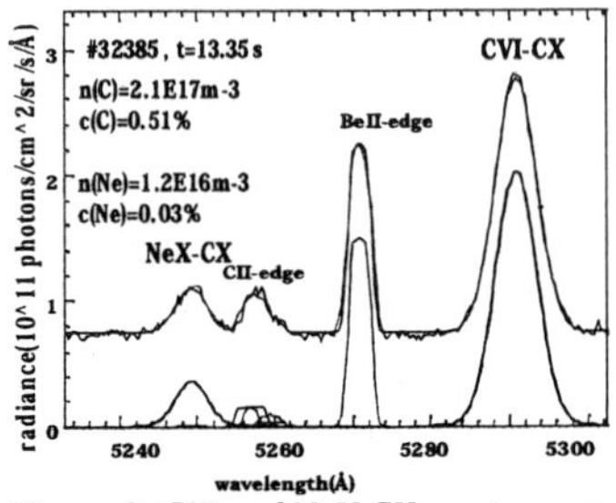

Figure 2. CVI and NeX CX spectrum at JET. The simultaneous extraction of T_i, v_{tor} and n_z for the 2 impurities allows the same BES-CXRS analysis in an energy scan.

2. COMBINATION OF CXRS AND BES:

There are several reasons which recommend the combination of CXRS and BES results. On the one hand, BES offers an attractive alternative to beam attenuation calculations which imply exponential error propagation, in particular, in high-density plasmas and large dimensions (such as ITER-FEAT). On the other hand, offers the combination of a fully consistent treatment comprising the beam-target interaction

$$n_z = \frac{4\pi \cdot I_{CX}}{\sum_k Q_{cx}(E_k) \int n_b(E_k) ds} = \frac{I_{CX} \cdot n_e}{\sum_k Q_{cx}(E_k) \cdot I_{BES}(E_k) / Q_{BES}(E_k)}$$

processes. Moreover, the fact that both spectroscopic methods share the same viewing geometry means that geometry errors cancel out, and also unknown window transmissions.

A further aspect, which may play an increasing role in burning fusion plasmas, is the need for a direct and local measurement of the main fuel ions, that is deuterons and tritons. However, a full blown least-square minimisation which is needed to extract CX D and CX T as well as all BES components (cf. Fig. 7) is a challenging task. It is obvious from the present understanding that only a much advanced modelling will enable good enough initial values on the search towards a global minimum.

3. HYBRID EFFECTIVE EMISSION RATES FOR LOW-Z CX SPECTRA

The atomic data base ADAS [7] provides the effective CX emission rates for the main low-Z transitions including that of the hydrogenic Balmer-Alpha series. The charge capture process is a near resonant process and reaches a maximum when the collision energy coincides with the orbit velocity of the quantum shell of the receiving ion. For most of the low-Z impurity ions with transitions in the visible range this maximum is typically near 50 keV/amu. In the case of an excited neutral beam population (n=2) the maximum occurs at a much lower collision energy. For this very reason in CX diagnostic systems with neutral beam energies below 30keV/amu excited populations need to be considered. In the case of the TEXTOR tokamak, the neutral beam energy can be scanned in a shot-to-shot basis from 20keV/amu to 50 keV/amu. The actual n=2 population is derived from collisional emission rates (cf. [2]).

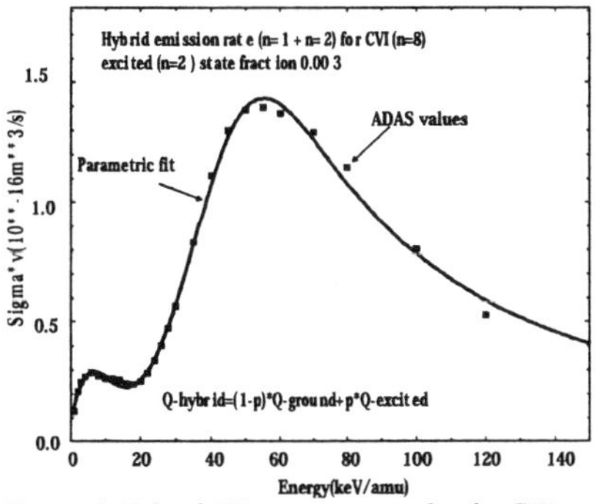

Figure.3 Hybrid CX emission rate for the CVI (n=8 to n=7)transition. The sum of n=2 and n=1 contributions is calculated for an assumed n=2 fraction of 0.003. The parametric fit is used for an analytical treatment of the CX-cross-section-effect (cf. [8]).

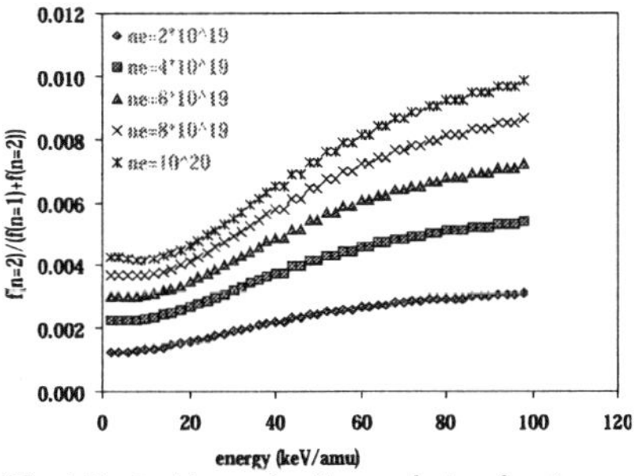

Fig.4 Excited beam (n=2) population fraction versus beam energy (cf.[2]). The electron density as approximately linear parameter changes between $2\cdot 10^{19} m^{-3}$ and $10^{20} m^{-3}$. Characteristic excitation levels for n=2 are about 0.5% of the ground state, n=1, population.

4. CROSS-SECTION EFFECTS IN THE PRESENCE OF A N=2 POPULATION

In the region below the resonant charge capture maximum significant distortions of the observed CX spectrum may arise leading to 'apparent' Doppler shifts and Doppler widths. Hybrid emission rates need to be introduced for a calculation of 'true' velocities and temperatures. We use an empirical parametric fit for the characterisation of the energy dependence:

$$Q_{hyb} = (1-p)\cdot Q^0_{grd} \frac{(E/E_{m,grd})^{\alpha,grd}}{1+(E/E_{m,grd})^{\beta,grd}} + p\cdot Q^0_{exc} \frac{(E/E_{m,exc})^{\alpha,exc}}{1+(E/E_{m,exc})^{\beta,exc}}$$

Observed 'apparent' velocities and temperatures can be represented as first and second order derivatives of the function Q_{hyb} (cf. [8]). The actual excited beam population fraction p is calculated for each energy and plasma background density and Z_{eff}.

5. ENERGY SCANS AT TEXTOR-94

Recent experiments at TEXTOR-94 have demonstrated the role of n=2 populations in a controlled energy scan from 20 keV/amu to 50 keV/amu of a diagnostic neutral beam. An adjustable shutter in the beam path allowed the power deposited to the plasma to be kept on a constant level, thus ensuring the same heating level. An additional heating neutral beam was kept at a power level of 1 MW providing the momentum for plasma rotation and heating. The enhancement factor Q-hybrid/Q-ground reaches quite appreciable values as can be seen in Fig. 5 and Fig.6, where C^{6+} and Ne^{10+} ion densities are deduced once in a ground-state only case, which gives clearly erroneous results and once for the fully modelled case respectively. BES data for beam-densities and species mix were used in each case.

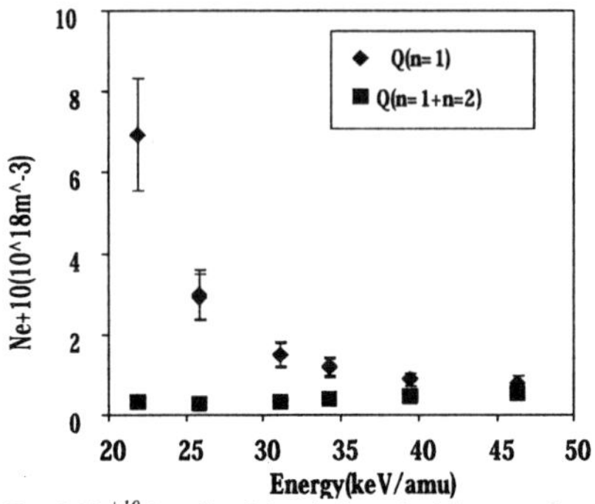

Fig.5 Ne^{+10} ion densities deduced with ground state-only emission rates and hybrid emission rates. The enhancement factor for the effective emission rate ranges between 1.8 at 46 keV/am and 33 at 22 keV/amu respectively.

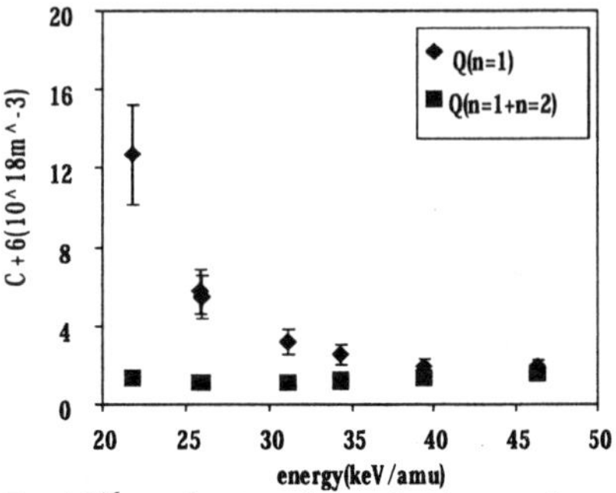

Fig.6 C^{+6} ion densities deduced with ground state-only emission rates and hybrid emission rates. The enhancement factor for the effective emission rate ranges between 1.4 at 46 keV/am and 14 at 22 keV/amu respectively.

6.FUTURE WORK AND APPLICATIONS

Clearly the routine use of BES in combination with CXRS needs to be further encouraged and exploited. There is one further example, where the two are inseparably interwoven, that is the direct deduction of deuteron densities from an active Balmer Alpha spectrum, which compromises both the characteristic Motional Stark beam emission features and also the local CX Balmer Alpha spectrum representing the bulk plasma ions. One of the remaining and most challenging tasks for the diagnosis of future DT experiments is the spatially resolved measurement of the local fuelling mixture (D:T).

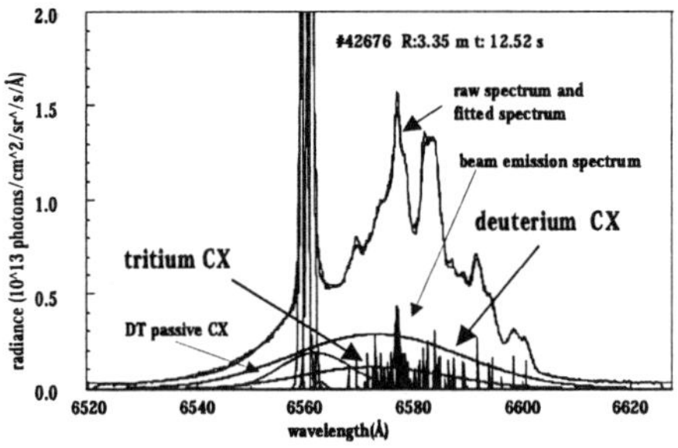

Fig.7 Combined CXRS and BES spectrum in a JET DT plasma. The local D:T density ratio can be extracted in an iterative procedure where alternatively the BES part or the CX part is solved in a non-linear least square minimisation. This requires comprehensive modelling of initial values. Note, the local D:T in the confined plasma region is not equal to spectroscopic plasma boundary values.

[1] W.Mandl *et al.*, Plasma Phys. Control. Fusion,**35**(1993)1373

[2] R.J Fonck et al.,Rev.Sci.Instr.61(1990)3487

[3] H.Anderson *et al.*, Plasma Phys. Control. Fusion,**42**(2000) 781-806

[4] I.Hutchinson, subm. to Plasma Phys. Control. Fusion 2001

[5] A.Boileau et al. J.Phys.B:At.Mol.Opt.Phys.,**22**(1989) L145

[6] B.W.Rice *et al.*, Phys. Rev. Lett. **79** (1997)2694

[7] Atomic Data and Analysis Structure, http://adas.phys.strath.ac.uk

[8] M.von Hellermann *et al.*, Plasma Phys. Control. Fusion, **37**(1995)71

LINE RATIO METHOD FOR POLOIDAL MAGNETIC FIELD MEASUREMENT USING Li-MULTIPLET ($2\,^2S - 2\,^2P$) EMISSION

Anatoli Korotkov, Kent McCormick, Philip Morgan, Josef Schweinzer, John Vince* and contributors to the EFDA-JET workprogramme

1. INTRODUCTION

Measurement of the current profile at the plasma edge in modern tokamaks remains a key issue for investigating the physics of ELMs and achieving advanced performance. Existing methods based on polarisation measurements[1,2] require that the initial direction of polarisation is maintained. This is of great concern, as the image is transmitted through a significant thickness of glass and thus subject to plasma-induced Faraday rotation.[3]

We propose an alternative method based on the measurement of the intensity ratio of the π and σ components of the Li-multiplet ($2\,^2S - 2\,^2P$) emitted by a neutral Li-beam. The technique utilises a difference in dependence of these intensities on the inclination angle between the total magnetic field vector **B** and the line of sight. The method does not require a polarisation measurement but can provide similar accuracy to the Motional Stark Effect diagnostic (MSE) for the inclination angle, $\sim 0.2^o$. An important advantage of the proposed method is that there is no interference from the plasma radial electric field.

In the paper, the physical background of the method and its implementation for measurement on JET are discussed. The relationship between the accuracy of the deduced poloidal magnetic field B_P and the statistical accuracy of the line ratio measurement is presented, and the required Li-beam density is established. The influence of the non- statistical population of the 2pm-states on the intensity ratio of the π and σ components is considered. Analysed results from preliminary measurements on JET are presented.

* Anatoli Korotkov, Philip Morgan, John Vince, Euratom /UKAEA Fusion Association, Culham Science, Centre , Abingdon, Oxon., OX14 3DB, UK. Kent McCormick, Josef Schweinzer, Euratom / IPP Fusion Association, Max Planck Institute für Plasmaphysik, D-85148 Garching, Germany.

2. PRINCIPLE OF THE MEASUREMENT

In plasmas of modern tokamaks the 2p jm_j – states of the lithium atom are subject to very strong mixing (Paschen-Back effect) in the magnetic field B, due to $\eta = B/B_0 >> 1$, where $B_0 = A / 2\mu_B = 0.24$ tesla, A is the fine splitting parameter for the 2p ^2P- states, μ_B is Bohr magneton. As a result the relatively simple Zeeman pattern of the allowed 2 ^2S – 2 ^2P-transitions forms a Lorentz triplet (Fig.1). The central peak is due to π- transitions, while the two flank peaks are nearly pure σ- transitions. The relative intensities in the multiplet are accurately simulated for arbitrary values of the parameter η.[4]

The intensities of the π and σ components have essentially different angle distribution: $J_\pi(\theta) \sim \sin^2\theta$, $J_\sigma(\theta) \sim 1+\cos^2\theta$, where θ is the inclination angle of the line of sight to the total magnetic field vector **B**. Measurement of $\xi(\theta) = J_\pi(\theta) / J_\sigma(\theta)$ provides a determination of the direction of **B**.

In the general case two components of B_P (B_Z, B_R) must be determined simultaneously. This implies the measurements of $\xi(\theta)$ at two inclination angles. Sight lines and beam trajectory applicable for JET are presented in Fig.2. The Li-beam is injected from the top port in oct.7 along a vertical line (R=3.25 m). Radiation from the Li-beam is detected from 24 points (z_j) along the beam direction using two periscopes with turning mirrors M_1 and M_2. B_Z and B_R are determined from scalar products of **B** and line of sight unit vector in which $\cos\theta_i$ is input from the measurements:

$$B\cos\theta_i = B_T \cos\beta_i \cos\varphi_i + B_R \cos\beta_i \sin\varphi_i - B_Z \sin\beta_i \tag{1}$$

$$\xi(\theta_i) = \frac{Ck\sin^2\theta_i}{1 + k\cos^2\theta_i + f(B,\theta_i)} \quad , \quad i = 1,2 \tag{2}$$

All parameters in Eqs. (1)-(2) are known. β, φ are checked using Doppler shifts of the radiation. k, the ratio of mirror reflectivity for P and S polarisation, is measured during calibration in gas. $f(B,\theta)$ is a reliably-calculated correction parameter (<<1) taking into account a small contribution of the π components into the flank peaks of the triplet. The polarisation Π ($C = (1+\Pi)/(1-\Pi)$) can be measured using the ratio of total intensities at two inclination angles. The polarisation, which is small in magnitude, is due

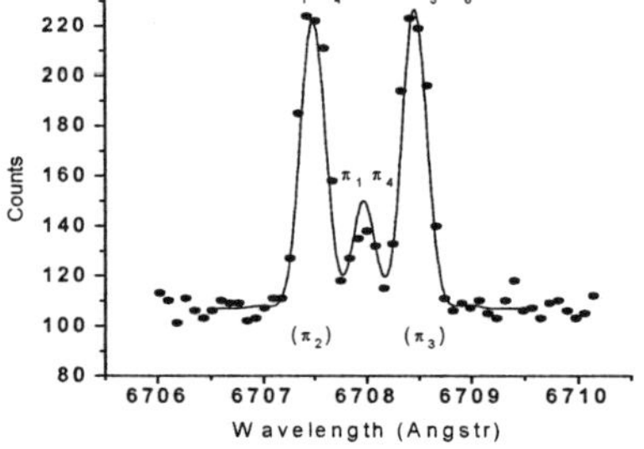

Figure 1. Lorentz triplet of thermal lithium .

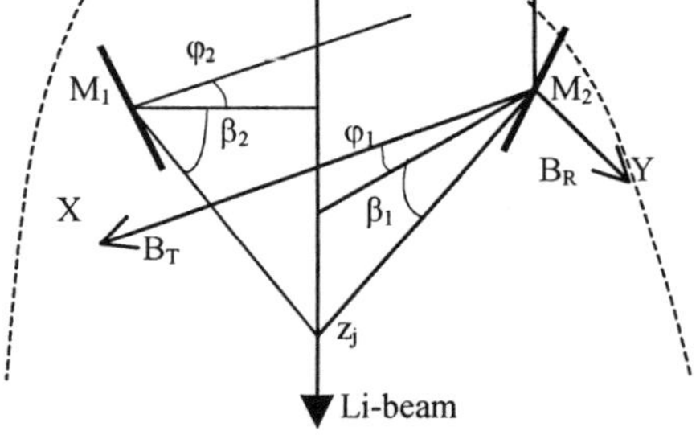

Figure2. Geometry diagram of the measurement.

to excitation by ions. The effect decreases with increasing beam energy.

We have found that polarisation of Li(2 ^{2}P-2 ^{2}S)-emission is $|\Pi| \leq 1.4\%$ at Li-beam energy $E_{beam} \geq 50$ keV and perpendicular injection to **B**. Therefore a small systematic error for the inclination angle $\sim 0.1^0 - 0.3^0$ is expected if Π is not measured.

Notice that radiation from the fast Li-beam atoms as well as from thermal Li-(or Na-) atoms coming from the neutraliser of the injector can be used for the measurement.

The accuracy of the B_P- measurement is determined by statistics of counts collected and depends on geometry of the measurement:

$$\frac{\Delta B_P}{B_P} = F(\eta,\beta,\varphi)\frac{\sqrt{\xi}}{1+\xi}\frac{\Delta\xi}{\xi} = F(\eta,\beta,\varphi)\frac{1}{\sqrt{N_\Sigma}} \quad , \qquad (3)$$

where N_Σ is the total number of counts collected for π and σ components, $\eta = B_P / B_T$. $F(\eta,\beta,\varphi)$ must be minimised to obtain the smallest statistical error in the measurement. We have found that for B_Z- measurement the line of sight should have parameters: $\varphi \leq 10^0$, $\beta \leq 40^0$ ($\varphi \geq 80^0$ for B_R- measurement). The condition $\beta \leq 40^0$ is also beneficial for reduction of systematic error due to polarisation of the radiation. Fig.3 presents ln$F$ for the existing periscope on JET. The geometry is close to optimal for B_Z- measurements.

3. PRELIMINARY EXPERIMENTS ON JET

The line-ratio method has been applied to the measurement of B_Z / B_T in a number of JET shots using radiation from thermal Li-atoms coming from the neutraliser of the injector. The measurements were performed in the beginning of the pulses for reliable comparison with EFIT data and because of enhanced spread of the thermal lithium atoms inside the plasma. Only one periscope was used to observe the plasma in the toroidal direction ($\varphi_1 = 1.57^0$). It was adequate for measurement of B_Z since the contribution of B_R to $\xi(\theta)$ is $\sim \sin\varphi / \mathrm{tg}\beta \approx 0.05$ of that from B_Z. A grating spectrometer SPEX1.25m (F=9) with linear dispersion ~ 4 Å/mm equipped with a CCD camera was used. Fig.1 is an example of the spectrum detected. The exposure time τ and the spatial resolution ΔR were 0.5 s and ± 15 mm. Agreement of the measurement with EFIT data within the existing error bars is observed (Fig .4). Error bars correspond to an accuracy for the inclination angle of ~ 1^0. The main errors are due to the small Li-beam

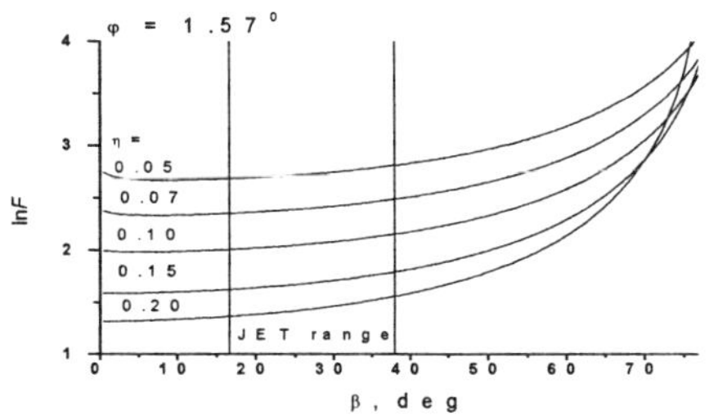

Figure 3. ln$F$ as a function of β for the geometry of Li-beam system on JET.

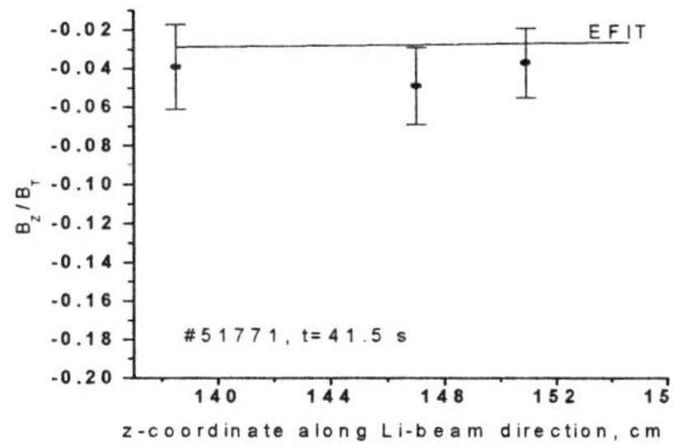

Figure 4. Result of the measurement at $\tau = 0.5$ s, $\Delta R = \pm 15$ mm (mapped onto midplane).

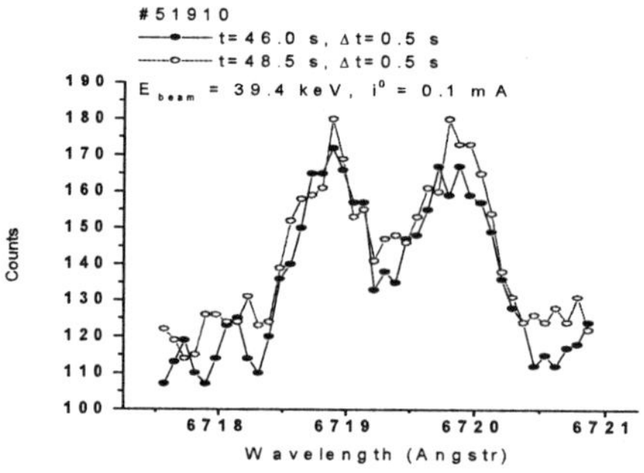

Figure 5. Lorentz triplet of Li-beam atoms.

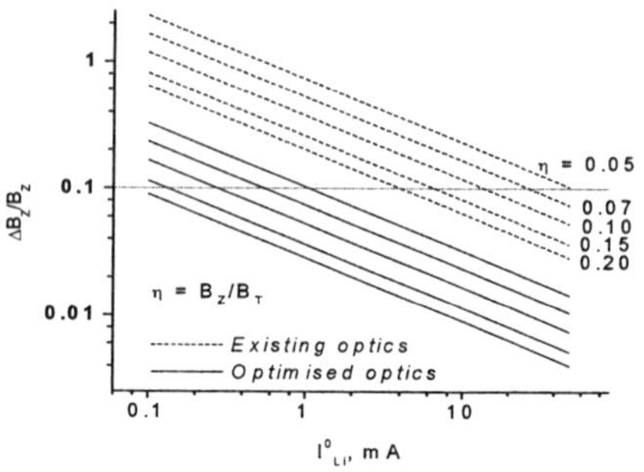

Figure 6. Accuracy of B_Z-measurement as a function of Li-beam current. $\tau = 0.1$ s, $\Delta R=\pm 5$ mm.

current during the measurement (0.1 mA) and the small étendue of the spectrometer. The instrument makes use of only 0.5% of the radiation relayed by the optical fibres.

The Lorentz triplet has also been detected for the emission of 40 keV Li-beam atoms (Fig.5). Additional widening of the peaks due to Doppler effect at finite aperture $\Delta\beta$ of the periscope was observed: $\Delta\lambda = (v_{beam}\, \lambda / c) \cos\beta\, \Delta\beta$. The aperture should be reduced by factor of 2 at least for an accurate measurement of B_P.

We have analysed the accuracy of the measurement for the optimised optical system using Fabry-Perot interferometers with étendue equal to that of the optical fibres (2×10^{-3} cm^2 sr). The result is presented in Fig.6 as a function of Li-beam current I^0_{Li}. Optimisation of the optics reduces required I^0_{Li} by one order of magnitude.

4. CONCLUSIONS

A line - ratio method has been proposed for B_P- measurement utilising the intensities of π and σ components of the Li-multiplet (2 ^{2}P- 2 ^{2}S). The method does not suffer from plasma – induced Faraday rotation in the optical elements and no interference from the plasma electric field exists. Preliminary experiments have demonstrated agreement within 40% with EFIT data for B_Z. A poloidal angle resolution of 1^0 is obtained. A statistical accuracy of 5-10% for B_P and poloidal angle resolution of $\sim 0.2^0$ can be reached for the optimised optical system and Li-beam current in the range 0.5-2 mA. A two-periscope scheme has been proposed for a simultaneous measurement of B_Z and B_R. It allows also the measurement of Π, hence avoiding systematic errors.

5. REFERENCES

1. K. McCormick, Measurement of poloidal field distribution in a tokamak plasma via neutral lithium beam spectroscopy, *Proceedings of the course and workshop*, vol II, 635-656 (Varenna, Italy, 1986).
2. F..M. Levinton, The multichannel motional Stark effect diagnostic on TFTR, *Rev.Sci.Instrum.* **63** (10), 5157-5160 (1992).
3. B. C. Stratton, D. Long, R. Palladino and N. C. Hawkes, Instrumentation for the joint European torus motional Stark effect diagnostic, *Rev. Sci. Instrum.* **70**(1), 898-901 (1999).
4. I.I. Sobelmann, *Atomic Spectra and Radiative Transitions* (Springer Verlag, Berlin , 1979).

Acknowledgements: This work was performed under the European Fusion Development Agreement and was partly funded by EURATOM and the UK Department of Trade and Industry.

X-RAY EMISSION FROM LASER IRRADITED CLUSTERS

Tara Desai

Dipartimento di Fisica "G.Occhialini" and INFM Universita degli Studi di Milano - Bicocca, Piazza della Scienza, 3, Milano-20126, Italy

Clusters are an aggregate of large number of identical atoms/molecules and occupy an intermediate state between atoms/molecules and condense phase[1]. They can be produced by various technique like, as a tiny fragment from solid material, gas jets, laser ablation etc.. Although clusters are classified as an intermediate state of matter, many of their properties differ from their quantum objects(atoms/mol) and the bulk[2]. One of the property we have studied is the X-ray emission spectra from laser irradiated clusters.

In this report, we present our experimental study on the X-ray emission from aluminum and gold clusters at $I \sim 10^{12}$ W/cm^2. These spectra are compared with the respective bulk materials. Clusters are commercially available as fine particles and their density is similar to that of solid material. Results are analyzed on the basis of laser pulse duration and cluster size. To study the behavior of clusters of different size at high intensity (10^{17}W/cm^2) of different λ_L, PIC simulations are performed. Analysis of both these studies reveal the correlation between cluster size and laser wavelength.

Two types of targets were used viz.1. Slab targets of Al and gold. 2. Cluster targets. A known quantity (50 mg) of al clusters with an average diameter $C_{da} \sim 0.4$ µm or gold clusters of average diameter $C_{dg} \sim 1.25$ µm were homogeneously mixed in one cc of a polymer base solution (low X-ray emitter) and thin films of~10 µm were fabricated for the present study. Here we consider two cases 1. $C_d < \lambda_L$ like Al cluster and 2. $C_d \geq \lambda_L$ like gold cluster where C_d refers to cluster diameter. Experiments were performed using $\lambda = 1.06$µm with $E_L \sim 1$ J in 10 ns (FWHM) duration. X-ray emission spectra in the range of 5-22 nm range were recorded at 45° to target normal using grazing incidence spectrometer. Details of the set up are also described earlier (Desai et al 2000). Since, we have used 50 mg of Al clusters by weight per cc of polymer, the no. of clusters $\sim 5.5 \times 10^{11}$ /cc and number of atoms per Al clusters $\sim 8 \times 10^9$. Inter cluster spacing (center to center)

Advanced Diagnostics for Magnetic and Inertial Fusion, Edited by Stott *et al.*
Kluwer Academic/Plenum Publishers, New York, 2002

A= $(3/4\pi n)^{1/3} \approx 0.755$ µm. Similarly density of gold cluster (dia~1.25µm) is ~$2.6x10^9$/cc, and number of atoms in each Au cluster ~10^{11} with A ~ 4.0 µm. Since A > dia of theclusters in both the cases, we assume clusters behaves independently. With the onsof laser beam, each clusters within the focal spot area experiences uniform radiation field.

Aluminum clusters (ave.dia.~0.4 µm) embedded in polymer show negligible X-ray intensity in 5-22nm spectral range (fig.1 a) compared to pure Al slab target (fig.1b). Shape and intensity of the spectra of Al cluster is not identical to bulk material indicates, the plasma ionization level in cluster is different than that of the bulk material under identical experimental conditions. Contrary to this gold cluster target (fig.2a) with average diameter ~1.25 µm, emit continuum which is similar nearly in shape to bulk gold target (fig.2b) but has lower intensity. This means the process of X-ray emission from gold cluster of diameter~1.25 µm >λ_L, is approximately similar to that of bulk material.

Cluster is a small ball of solid material and the mechanism of plasma production is similar to that of bulk material[3]. Cluster is heated due to electron-ion collisions and the hydrodynamic expansion of the plasma from cluster surface results in the decrease of plasma density. The inward energy transport in to the cluster heats the cluster as a whole. For planar targets plasma density falls to 1/e times the critical density at a distance[4] L~2R where R is laser focal spot radius but in the case of clusters, this distance depends on 2 factors viz. 1).Plasma density decreases exponentially from the cluster surface due to hydrodynamic expansion of the surface plasma, similar to slab target with a velocity V ~$(KT_e)^{1/2}$ where KT_e is temperature of the plasma on cluster surface. 2) Simultaneously, cluster as a whole expands in all directions due to its internal pressure P_i+P_e= $(nkT)_i+(nkT)_e$ and increases in size. This leads to cluster disassembly. Cluster disassembly time $t_d = C_d/2C_S$ where C_d is cluster diameter, $C_S= (ZKT_e/m_i)^{1/2}$, T_e depends on the effective laser energy deposition on the cluster surface on a time scale ~ t_d, cluster size, its mass density, laser intensity, pulse duration etc.

These experimental results raise a question; why a cluster, a fragment from a bulk, having a dimension smaller than the incident laser wavelength behave differently than the bulk in the present investigation reported here? While with increased cluster diameter (>λ_L), cluster behavior evolves towards the bulk material. Therefore, we analyze our results on the basis of cluster dimension, laser wavelength and pulse duration. We believe there can be several reasons to explain these results.

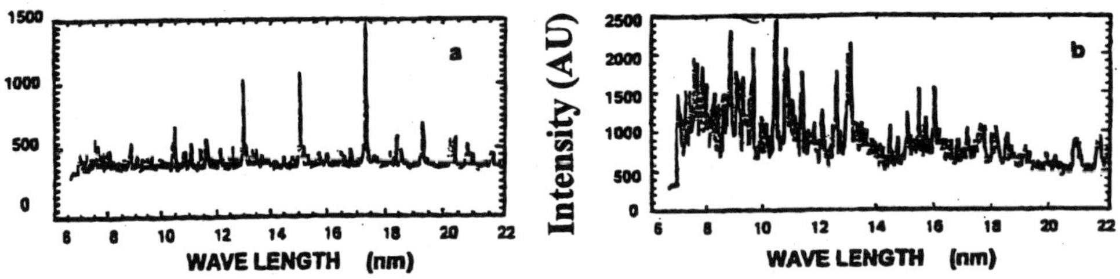

Fig.1a. X-ray emission from aluminum cluster target.

Fig.1b. X-ray emission from aluminum slab target.

$C_{da} < \lambda_L$, Aluminum cluster: In our present case Al cluster dia, $C_{da} < \lambda_L/2$. At the beginning of laser pulse, electrons of the entire cluster as a whole will respond to the electric field. Since clusters are independent entities, free electrons will oscillate in the direction of EF in half the cycle and in the opposite direction in the remaining half of the cycle of the wave. This will create a giant dipole[5]. As the cluster is similar to solid surface, various interaction processes that occur on the time scale of the laser pulse are laser ionization, laser absorption, X-ray generation, cluster expansion, etc. Laser radiation is predominantly absorbed by inverse bremsstrahlung. The incident laser radiation becomes evanescent at a distance δ $(=C/\omega_p) > C_{da}$. Ablation depth[6] for Al slab target at $I \sim 10^{12}$ W/cm^2 is ~ 0.4 μm which is comparable to cluster diameter. As discussed above, due to cluster disassembly, plasma density falls below the critical value n_c and cluster disassembly time ~ 1 ns < laser duration (10 ns). Since laser is still on, the remaining laser radiation penetrates the cluster plasma without being absorbed. Therefore, no significant laser energy is deposited on the cluster surface and a large fraction of radiation transmits through the diluting cluster plasma.

In the case of Al cluster of dia~0.4 μm, 1) plasma density and temperature are not very high hence a low X-ray yield. 2) Radiation mean free path[7] for a slab target at $I \sim 10^{12}$ W/cm^2 is $\lambda_R \sim 3$ μm which is larger than the dimension of the cluster; implies initial X-rays will transmit through the plasma. Thus, further X-rays can not be generated due to down conversion mechanism similar to solid targets. For the above mentioned reasons, we believe that X-ray yielding process in cluster plasma is not same as that in the solid aluminum target. Therefore, the shape of the X-ray emission spectra is different and X-ray intensity is less than that of solid target.

$C_{da} \geq \lambda_L$, Gold cluster: Here the gold cluster diameter is $C_{dg} > \lambda_L$. i.e. Laser wave experiences evanescence inside the cluster. Ablation depth for planar gold target at $I \sim 10^{12}$ W/cm^2 is ~1.2 μm, which was estimated from ablation pressure[8] and is comparable to cluster dimension. Therefore, the laser interaction and the process of X-ray emission from gold cluster could be similar to bulk material. Since each cluster is a source of X-rays and the number of gold clusters within the laser irradiated area being ~2.6×10^9/cc, X-ray intensity from gold clustered target is lower than the bulk gold target.

PIC simulations: We have performed preliminary particle-in-cell (PIC) simulations of clusters irradiated by intense laser pulses. For simplicity we used a two-dimensional (Cartesian) geometry. The most serious limitation comes from the different asymptotic behavior of the electrostatic force in 2D and 3D clusters [9], so that for instance only

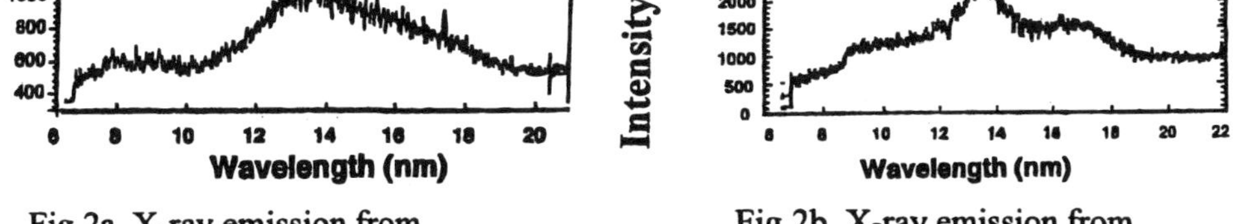

Fig.2a. X-ray emission from gold cluster target.

Fig.2b. X-ray emission from solid gold target.

qualitative features of Coulomb explosion may be described in 2D. Particle simulations of clusters in different regimes and geometries are reported[9,10]. The important parameter which discriminates cluster behavior with respect to laser irradiation is $\xi = k(n_e/n_c)(R/\lambda_L)$ where k is the dimensionality of the cluster (2 in the presnt simulation), n_e = electron density, n_c = critical density, R = cluster radius and λ_L = laser wavelength. For $\xi >> 1$ the cluster is opaque to the laser radiation and behaves as a small piece of bulk plasma, so that cluster heating may occur only via surface interaction, i.e. via collisionless mechanisms which are well known in the irradiation of solid targets (anomalous skin effect or vacuum heating). This is the case for the simulation of Fig.3a (n_e/n_c=10, I=10^{17} W/cm^2, R=λ_L=1.0 μm) where we clearly see that the clusters screens the laser light. No significant energy transfer to the ions is observed. In contrast, for a case with R=0.01 μm, λ_L=0.4 μm and I=4x10^{17} W/cm^2, the laser field penetrates the clusters and rip off the electrons. Within few laser cycles an electron cloud is formed outside the ion ball. Due to the unscreened electrostatic field of ions, the Coulomb explosion begins leading to a radially symmetric acceleration, expansion of the ions and finally disintegration (fig.3b).

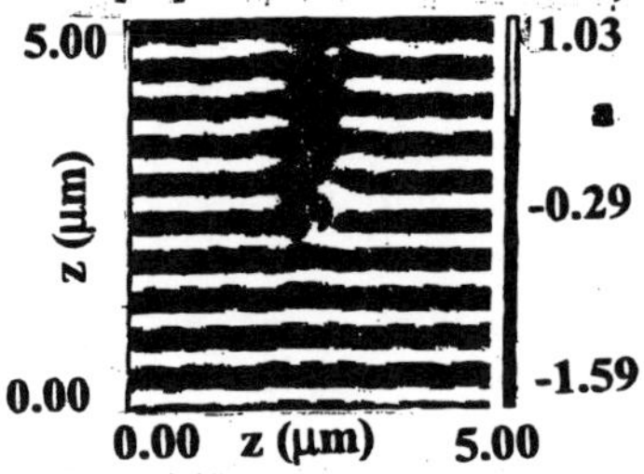

Fig. 3a. Laser -Cluster interaction

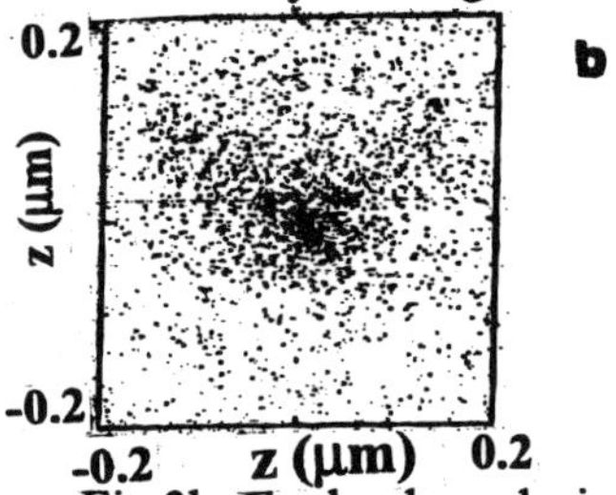

Fig.3b. Coulomb explosion

Clusters offer an interesting behavior with respect to laser wavelength. At moderate laser intensities I~10^{12} W/cm^2, experimental study shows that the clusters evolve towards bulk if cluster size ≥ laser wavelength and small clusters with dimension less than the laser wavelength show cluster disintegration due to thermal pressure. At high laser intensity I~10^{17} W/cm^2, PIC simulations predict the surface interaction for large clusters and small clusters disintegrate due to Coulomb pressure. We believe there is a direct correlation between the size of the cluster and the incident laser wavelength. Further study using various size of the cluster and laser parameters is necessary to under stand the cluster properties. Details of the work will be published elsewhere.

We sincerely acknowledge the support received from Drs. A.Machhi (Univ. of Pisa, Italy) and H. Ruhl (MBI for Nonlinear Optics, Berlin, Germany) in PIC simulation.

References

[1] O.F.Hagena, 1992, Rev. Sc.Instru.63, 2374
[2] T.D.Mark, 1987, Int.J of mass spectrometry and ion processes. 79,1
[3] T.Ditmire et al, 1996, Phy. Rev. A. 53, 3379
[4] C.E.Max, 1982, Phys. of Laser Fusion, Vol.1, UCRL-53107
[5] H.Hutchinson 1998, Science, 280, 693
[6] T.Desai and H.C.Pant, 2000, Laser and Part.beams, 20(1)
[7] Ya.Zeldovich and Yu. Raiser, 1966, Phys. of Shock waves and high temp. hydrodynamics
[8] B.K.Godwal et al, 1989, J. Appl.Phys., 65(12), 4608
[9] T. Esirkepov et al., 2000, Laser and Part. Beams 18, 503
[10] M. Eloy et al., 2001, Phys. Plasmas 8, 1084.

DETERMINATION OF THE ELECTRON DENSITY AND ION COMPOSITION HISTORIES IN A NONEQUILIBRIUM PLASMA USING SELF-CONSISTENT COLLISIONAL-RADIATIVE CALCULATIONS

Vladimir Bernshtam, Lev Gregorian, and Yitzhak Maron*

Knowledge of the electron density, electron temperature, and ionic charge-state composition is essential for the understanding of the atomic-physics and hydrodynamic processes in laboratory and astrophysical plasmas. Determination of these parameters with sufficient accuracy in the case of nonequilibrium, highly-dynamic plasmas interacting with strong electric and magnetic fields, is in many cases not possible. For example, determination of the electron density using direct measurements of Stark line-broadening in dense, high-current-carrying plasmas is extremely difficult due to significant contributions of additional mechanisms, such as thermal and nonthermal electric fields, large Doppler broadening, line opacities and Zeeman line-splitting.

We have developed a highly accurate calculation method for a self-consistent determination of the electron and ion densities and the composition in multi-species dynamic plasma, based on collisional-radiative analysis of experimental spectral-line emission data. This method, requiring only the knowledge of the electron temperature T_e and the use of an absolutely calibrated spectroscopic system, can be used as a means to examine the uncertainties of other methods such as Stark broadening, particularly in cases when other line-broadening mechanisms contribute significantly to the width[1].

Here the method is demonstrated for the imploding, optically thin CO_2 plasma produced in a gas-puff Z-pinch experiment[2]. We used the radial distribution of charge-state in the plasma for measurements of the absolute line intensities from the various oxygen charge states up to the OVI giving a complete mapping in (r,t) of the absolute upper-level populations n_u. The lines used here were optically thin, therefore the upper-level populations are given by $n_{u(exp)}= I_{ul}/A_{ul}g_u VK(\lambda)$, where I_{ul} is the line intensity, A_{ul} is

* Vladimir Bernshtam, Lev Gregorian, and Yitzhak Maron, Faculty of Physics, Weizmann Institute of Science, Rehovot 76100, Israel

Advanced Diagnostics for Magnetic and Inertial Fusion, Edited by Stott *et al.*
Kluwer Academic/Plenum Publishers, New York, 2002

the Einstein coefficient, g_u is the degeneracy of the upper level, V is the light-emitting plasma volume, and $K(\lambda)$ is the system collection efficiency and absolute calibration factor. For the analysis, we used time-dependent collisional-radiative (CR) calculations[3], in which the input values include the electron density, the electron temperature, and the source function. The electron density is used as a variable parameter, and T_e is obtained independently from measurements of line intensity ratios, found to be insensitive to n_e. The source function simulates the time-dependent supply of new particles to the plasma. In our case, we use an instantaneous (at $t = 0$) population of the ground state of the ion in question, referred to as the δ-source, chosen based on the weak dependence of the calculation results on the ionization history in the plasma.

The CR calculations yield a complete set of time-dependent values of $n^{\alpha}_{u(theory)}$, and ratios of the form $\chi\,(T_e, n_e, t) = (n^{\alpha}_{u} / n^{\alpha})_{theory}$ for all charge states α, and the total charge $Z(t)$. (Here, "theory" stands for a parameter obtained from CR calculations, and "calc" stands for a parameter calculated using the CR and experimental results.) The dependence of χ on n_e is given in Fig. 1.

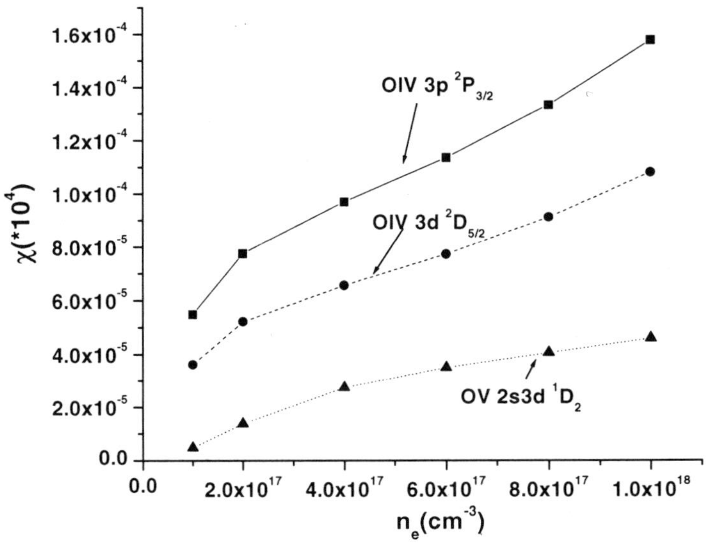

Figure 1. Theoretical ratios χ of an upper-level population to the total ionic density in charge states α as a function of n_e, for several upper levels used in the analysis. Here, T_e =10 eV, and t=20 ns.

CR calculations using varying δ-source specification resulted in weak dependence on time of $\chi(t)$ (uncertainties of up to 20% throughout the calculation interval) for the various upper levels used here, thus justifying the use of these calculation procedures for the time varying conditions of the Z-pinch plasma.

We then use the experimental data for calculating the time dependent ion density for the charge-state α using $N^{\alpha}_{calc}(t) = n^{\alpha}_{u} / \chi^{\alpha}(t)$. The time t for which the calculations can be interpreted correctly should be longer than t_0, the typical upper-level equilibration time. For plasma regions occupied by several ion species the total density is $N_{calc}(t) = 3/2\sum_{\alpha} N^{\alpha}_{calc}(t)$, where the 3/2 factor is due to the contribution of carbon. Using the mean charge Z_O and Z_C obtained from CR calculations for oxygen and carbon, respectively, we then calculate the electron density: $N_{e(calc)}(t) = \{Z_O(t) + 0.5Z_C(t)\}\sum_{\alpha} N^{\alpha}_{calc}(t)$.

This calculation procedure is performed for several values of n_e. A self-consistency is achieved for the value of n_e which simultaneously (at some $t = t_{min}$) gives minimum values for $\delta(n_e) = \left|N_{e(calc)}(t) - n_e\right| / n_e$ and $\delta(\gamma) = \left|\gamma^{\exp} - \gamma^{theor}\right| / \gamma^{theor}$ where $\gamma^{\exp} = n^{\alpha}_{u} / n^{\beta}_{u}$ and $\gamma^{theor} = n^{\alpha}_{theor} / n^{\beta}_{theor}$ (see example in Fig. 2). This procedure is then carried out by calculations in which variable T_e, n_e are used.

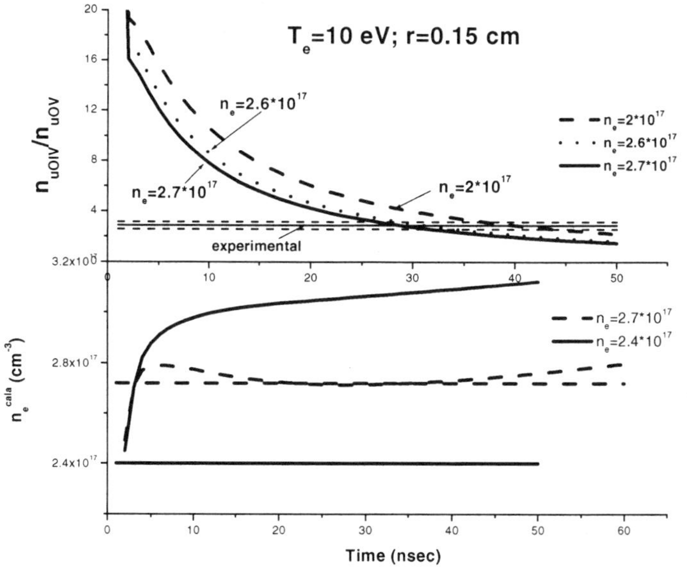

Figure 2. Calculated ratio of the 3d ^{2}D level of **OIV** to the 3d ^{1}D of **OV**, together with the calculated electron density. The source function used is a δ-source with 80% to 20% distribution of population of the **OIV** and **OV** ground states, respectively.

The simultaneous minimum of $\delta(n_e)$ and $\delta(\gamma)$ is shown to be obtained in Fig. 2 for the value of n_e=2.7*10^{17} cm^{-3}. Using these results, the ratio of the total charge-state densities is calculated giving us a self-consistent ion composition. This procedure is repeated for all charge states in our plasma, allowing us to obtain the radial-distribution history of both n_e and the ion composition across the plasma shell. The calculation uncertainties, including the absolute-intensity calibration of the spectroscopic system, the experimental shot-to-shot irreproducibility, and the uncertainty in T_e are all accounted for. The complete results are given in papers[2,4], where it is shown that the value of n_e and the calculated total ion mass well agree with the results of other methods and with the energy and momentum balance calculations. We believe that this method can be applied to investigations of various laboratory and astrophysical plasmas, requiring high-accuracy measurements of the plasma density and composition

REFERENCES

1 E. J. Iglesias and H. R. Griem, Ion-motion broadening of the Delta n=0,n=2 transitions of C V in a carbon laser-produced plasma *Phys. Rev. A* **38** (1), 301–307 (1988)

2 L. Gregorian, G. Davara, E. Kroupp, V. Bernshtam, S. Alexiou and Y. Maron, Plasma density and thermal pressure distribution in an imploding Z-pinch plasma, (to be published) Presented as an invited talk at the ICOPS 2000 (June 2000, USA)

3 Yu. V. Ralchenko and Y. Maron, Accelerated recombination due to resonant deexcitation of metastable states, JOSRT 71, 609 (2001)

4 V. Bernshtam, L. Gregorian, and Y. Maron, Determination of the time-dependent particle density and composition in a nonequilibrium plasma using collisional-radiative modeling of spectral-line intensities, (to be published)

USE OF 1D D_α CAMERA TO MEASURE EDGE NEUTRAL AND ELECTRON DENSITY EVOLUTIONS

M. R. Tournianski, P. G. Carolan, G. F. Counsell, M. G. O'Mullane †

INTRODUCTION

One of the clearest signatures of H–mode transitions in START and MAST plasmas is the sharpening of the edge visible radiation, indicating decrease in the edge neutral density and an increase in the edge electron density gradients[1]. These densities are also very important factors in understanding the fuelling efficiency, H–mode access and dynamics, NBI edge loses, as well as interpretation and cross–calibration of a number of diagnostics. The neutral densities are obtained from the absolute D_α emissivity profiles unfolded from a linear camera[1] and employing a collisional–radiative modelling of emissivity rates from the ADAS[2] rate coefficients database and using T_e and n_e from the multipoint Thomson scattering[3]. The near insensitivity of the D_α emissivity profiles to electron temperatures can be turned to good advantage in extracting electron densities. Here we describe that simulations and observations show an almost reciprocal dependence of the D_α spatial FWHM on the electron density at the position of the D_α peak. These values and the absolute D_α emissivity allow the neutral density to be determined at the D_α peak. This is being used in H–mode threshold studies as well as to optimise the fuelling and heating efficiencies. We also outline some anticipated developments and applications.

INSTRUMENTATION

An absolutely calibrated D_α linear CCD camera (256 elements) images the midplane of the START and MAST plasmas with a time resolution, Δt, as low as $125\mu s$ and with $\sim$10mm spatial resolution. The walls of both START and MAST were blackened using colloidal graphite to suppress wall reflections which compromise the inversion. In unfolding the line–of–sight integrated intensities we assume toroidal symmetry and divide the diameter into the same number of toroidal annuli as viewing chords. The contributions from the

†M. R. Tournianski, P. G. Carolan, G. F. Counsell, EURATOM/UKAEA Fusion Association, Abingdon, UK.
M. G. O'Mullane, Department of Physics and Applied Physics, University of Strathclyde, Glasgow, UK

Advanced Diagnostics for Magnetic and Inertial Fusion, Edited by Stott *et al.*
Kluwer Academic/Plenum Publishers, New York, 2002

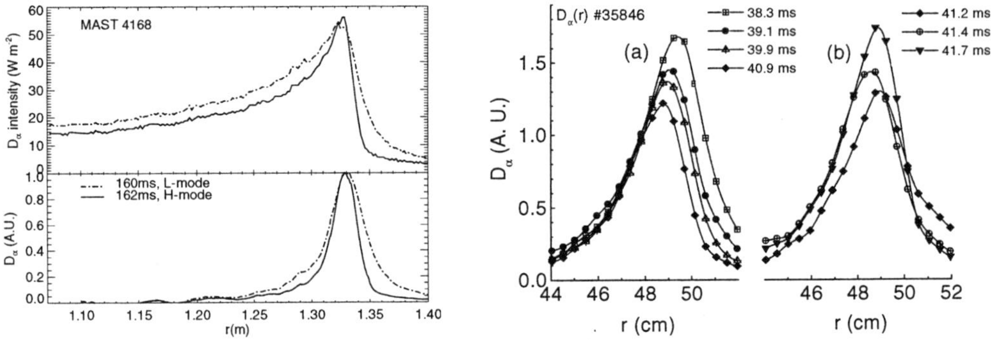

Figure 1: The line–integrated D_α intensity(Δt=1ms)(top) and radially unfolded D_α emissivity profiles during H and L mode of operation in MAST device(bottom).
Evolution of D_α emissivity profiles in START during the inter–ELM period(a) and when there was an ELM(b).

annuli to each of the viewing chords is calculated from the camera optics producing a set of linear equations. Matrix inversion algorithms are used to produce the solution matrix. We generally use "Singular Value Decomposition" to avoid numerical round-off problems. The unfolding automatically incorporates the radial smearing from the optics including the finite CCD element size.

METHOD AND EXPERIMENTAL RESULTS

A typical example of results from the D_α linear camera is shown in Fig. 1(left) during an L to H mode transition in the MAST plasma (MAST #4168, t = 160ms and t = 162ms). Inverting the line–of–sight integrated intensities gives the local D_α emissivity distribution. In the example presented at the bottom of Fig. 1(left) the inverted D_α emissivity profile is seen to narrow to ~2cm FWHM in 2ms during the L to H transition strongly suggesting an increased ionisation rate from a steepening of the edge electron density.

Another important example obtained on the START tokamak (#35846) is shown in Fig. 2(right) showing the time evolution of the D_α emissivity profiles during an inter–ELM period. The D_α FWHM gradually reaching a minimum of ~2.4cm (t=40.5ms), *cf.* Fig. 2 (a) but broadening again starting at t=41ms when there was an ELM (*cf.* Fig. 2 (b)).

The neutral density may be obtained from the inverted D_α emissivity, using the T_e and n_e edge data from Thomson scattering, and collisional–radiative modelling of the hydrogen n=3 fractional population from the ADAS database. The results are compared with those obtained from neutral particle simulations using a 1D Monte–Carlo neutral transport code[1]. The code simulations incorporate the measured profiles of T_e and n_e from Thomson Scattering (TS) and T_i from multichord charge-exchange Doppler spectroscopy[4]. An estimate of the neutral influx velocity was derived from the D_α Doppler measurements of edge neutrals ($T_0 \approx$1–2eV)[1]. The steepening of the electron pressure, particularly from the density component, accounts quantitatively for the narrowing of the D_α emissivity profile due to increased ionisation rates. The simulations show that the narrowing is primarily due to the increased electron densities at the edge. Narrowing due to changes in the $T_e(r)$ profile is less important because of the slow variation of rate coefficients in the temperature range of interest (*i.e.* the effective $< \sigma v >$ for populating the $n = 3$ principal quantum level, varies by only $\sim$ 15% for T_e = 30eV to 100eV, - a typical T_e range for most of the D_α emissivity in START and MAST plasmas).

Because of the D_α emissivity profile is predominantly dependent on the edge $n_e(r)$ and neutral influx velocity for MAST and START plasma edge conditions, the electron densities

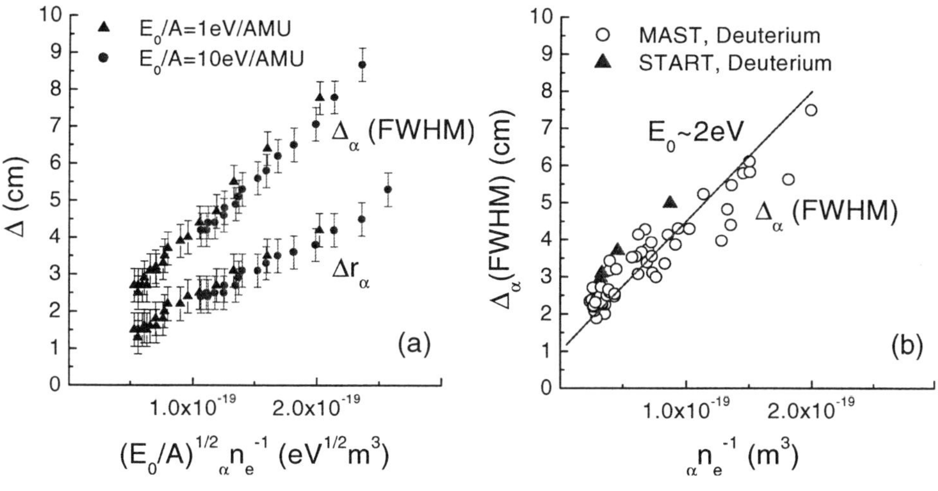

Figure 2: (a) Simulations of D_α FWHM , Δ_α, and Δr_α, for two neutral influx energies.
(b) Experimental observation of the D_α FWHM , Δ_α for a number of MAST and START Deuterium discharges.

at the position of D_α emissivity profile can be estimated. Experimental observations and computer simulations based on ADAS database show an almost reciprocal dependence of both the D_α FWHM, Δ_α and Δr_α (the distance from the D_α peak to the separatrix), on the electron density, ${}_\alpha n_e$, at the D_α peak position. Results of the modelling with linear approximation for the the edge electron density profile and two different neutral particle influx energies are presented in Fig. 2(left). They show a linear relationship between both Δ_α(FWHM) and Δr_α with electron density and neutral influx velocity. The experimental data for the conditions here (*e.g.* no gas puff) give $\Delta_\alpha = a_\Delta + b_\Delta\sqrt{E_0/A} \cdot {}_\alpha n_e^{-1}$ where $a_\Delta = \sim 0.07$m, $b_\Delta = \sim 3.5 \cdot 10^{17}$ m^{-2}eV$^{-1/2}$ and E_0 is a neutral particle influx energy. The right graph in Fig. 2 shows the experimentally obtained Δ_α(FWHM) for a number of MAST and START deuterium discharges taken at the TS time as a function of ${}_\alpha n_e$ obtained at D_α peak position using TS measurements. It demonstrates a reciprocal dependence of the D_α FWHM on ${}_\alpha n_e$, with energy of influx neutrals of E_0 ~2eV, which is in agreement with experimental observations from Doppler Balmer measurements. Armed with values of the ${}_\alpha n_e$ and the absolute D_α emissivity, we can estimate the neutral density, ${}_\alpha n_0$ at the D_α peak position. A good agreement between experiment and modelling is also supported by analytical estimates of Δ_α and Δr_α based on a simple 1D approximation for neutral density, uniform T_e profile and no pedestal linear edge n_e profile.

The merit of fast time response in investigating H–mode dynamics was demonstrated in the START device, particularly by the evolution of the edge electron and neutral densities and the poloidal rotation associated with the radial electric field. The D_α peak is typically about 3cm inboard of the separatrix. The C^{2+} impurity emissivity shows a peak within ~1cm of the D_α one. Thus, we can estimate the behaviour of the electron and neutral densities as the poloidal velocity, v_θ, of C^{2+} changes at much the same location. The results of these considerations are shown in Fig. 3. Between the ELMs ($t \sim 38.4$ms to 41ms) the increases in the v_θ are accompanied by a $\sim 40\%$ increase in electron density, ${}_\alpha n_e$, from $\sim 2.3 \cdot 10^{19}$m^{-3} to $\sim 3.2 \cdot 10^{19}$m^{-3} and about a halving in neutral density, ${}_\alpha n_0$, from $\sim 1.8 \cdot 10^{16}$m^{-3} to $\sim 0.9 \cdot 10^{16}$m^{-3}. Fig.3 shows that, the formation of an edge pedestal, the increase in radial electric field associated with changes in v_θ and the reduction of neutral density are occurring together, indicating a symbiotic dependence and none can be identified as a precursor into ELM–free H–mode transition.

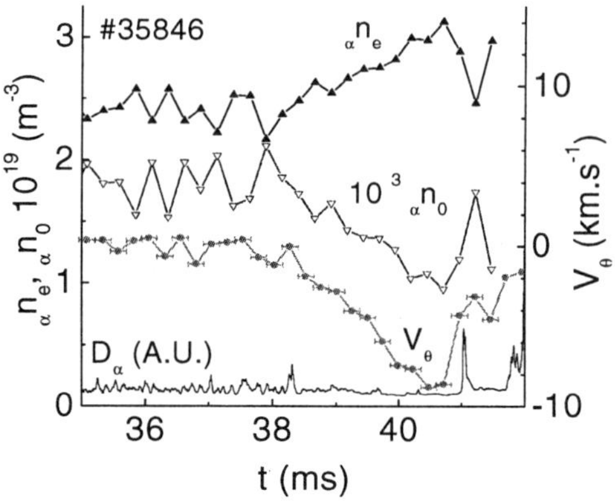

Figure 3: Comparisons, during H-mode evolution, between the poloidal velocity, v_θ from C^{2+}, and the estimated electron and neutral densities, $_\alpha n_0$ and $_\alpha n_e$ at the D_α emissivity peak.

DISCUSSION

The versatility of using linear D_α camera extends beyond extracting n_e and n_0 at discrete edge positions. For example, the neutral influx velocity may be obtained given the knowledge of the edge electron density, and may be an important factor in understanding of H–mode dynamics in contrasting wall recycling and gas feed sources of neutrals. Another area of ongoing work is to use the measured D_α radial emissivity profile together with simulated edge neutral density profile in a fitting routine using the edge electron density profile and the neutral influx velocity as free parameters. This approach may facilitate the extraction of the electron density not only at the peak of D_α but also the full n_e profile at the edge. The improvement in accuracy of atomic rate coefficients combined with advances in CCD technology is allowing us to design the next step diagnostics. For example, simultaneously monitoring of D_α, D_β and D_γ emissions can also yield edge electron densities. Finally, the simple relationship between D_α emissivity and the position of the plasma boundary allows for online analysis in the feedback systems to control the plasma positions and shapes. This monitoring of the edge shape and position avoids problems associated with pick–up coils being remote from the plasma edge (as often the case on MAST) with long time integration required from the edge pick–up coils in quasi continuous plasmas, such as ITER.

ACKNOWLEDGEMENT

This work is funded by the UK Department of Trade and Industry and EURATOM.

REFERENCES

1. M. R. Tournianski, P. G. Carolan, N. J. Conway, G. F. Counsell, A. R. Field and M. J. Walsh, Poloidal rotation and associated edge behaviour in START plasmas, *Nuclear Fusion* **41**(1), 77–89 (2001).
2. H. P. Summers, Atomic Data and Analysis Structure, *JET Report*, JET–IR(94) **06** (1994).
3. M. J. Walsh, N. J. Conway and M. Dunstan *et al.*, Interactive optical design and realization of an optimized charge coupled device Thomson scattering system for the spherical tokamak START, *Rev. Sci. Instrum.* **70**(1), 742–746 (1999).
4. N. J. Conway, P. G. Carolan and M. R. Tournianski, The unfolding of multichord Doppler spectroscopy, *Rev. Sci. Instrum.* **70**(1), 934–938 (1999).

X-RAY FIELD ASYMMETRY EFFECT ON THE TIME OF NEUTRON GENERATION IN EXPERIMENTS WITH INDIRECT DRIVE TARGETS

S.A.Bel'kov, A.V.Bessarab, G.G.Kochemasov, S.P.Martynenko*

1. INTRODUCTION

The investigations of the effect of controlled X-ray field asymmetry on the compression degree and the neutron yield in the targets of indirect (X-ray) drive were performed in 1998 at ISKRA-5 facility[1]. The field asymmetry was induced by shift the central capsule (microballoon) with respect to the center of spherical box-converter. The experiments demonstrated neutron yield reduction with increasing of asymmetry degree. In parallel with this fact, shifting of microballoon about central position caused the rise of compression time, which was determined by indirect method using the compressed region X-ray luminescence[2].

Our simulations used 2D code MIMOSA ND[3] showed that theoretically calculated asymmetric X-ray field allows give satisfactory explanation for the experimentally observed neutron yield reduction in relation to the microballoon shift value. The simulations showed also that there is the time delay of neutron generation relative to the self generated X-ray pulse. It was interestingly to perform the direct measurements of the time moment of neutron yield in experiment. For these purposes authors designed the special diagnostic.

2. THE RESULTS OF MEASUREMENTS OF TIME MOMENT OF NEUTRON YIELD

Experiments were performed with indirect drive (X-ray) targets. The target consisted of 2 mm in diameter spherical cupper box and glass sphere (microballoon) with DT gas

* Russian Federal Nuclear Center – VNII of Experimental Physics, 607190, Sarov, Prospect Mira 37, Nizjny Novgorod Region, Russia

Advanced Diagnostics for Magnetic and Inertial Fusion, Edited by Stott *et al.*
Kluwer Academic/Plenum Publishers, New York, 2002

that was placed inside the box. The laser radiation input was on the inner surface of the box through the six holes of 600 micron in diameter. To increase the laser to X-ray conversion the inner surface of the box was covered by thin gold layer (about 1 micron thickness). X-ray generated on the wall of the box irradiated the glass microballoon. To induce the controlled asymmetry of the X-ray field the center of the microballoon was shifted relative to the center of box in forward special seventh hole on different distances. So the closer microballoon was to the seventh hole the larger asymmetry of X-ray.

In presented diagnostic to measure the value $\tau_{\gamma n}$ of the time delay between neutron yield and self generated X-ray pulse the silicon semiconductor p-i-n diode SPPD11-12 and super-high-speed oscilloscope SRG-7 were used. To increase the registration efficiency of neutrons the polyethylene converter of 2 mm thickness was placed in front of surface of diode. The ratings impulse response of diode SPPD11-12 has duration about 2 ns. The detector has more sensitivity to hard X-ray generated by plasma of box walls than to neutron. Therefore to reduce the influence of hard X-rays on the measurements the corresponded X-ray filters were placed before detector. In presented results we used the following filters: tungsten and cooper with thickness 21 mm and 7 mm correspondently. To have reference point we used the X-ray signal detected by vacuum X-ray diode (VXD). This detector can measure the self-generated soft X-ray of plasma of box walls with the energy of photons close to $h\nu\sim1.5$ keV. The distance between detector and the target was $R_l=115.3\pm0.7$ mm. The both signals of neutron detector and VXD were sent to the same input line of super-high-speed oscilloscope SRG-7. This oscilloscope has the time resolution not worse as 0.1 ns. The time delay of generation of neutron yield $\tau_{\gamma n}$ was called the time interval between of moments of appearances of X-ray and neutron impulses on oscilloscope registration. The moments of appearances of impulses detected as the intersection point of the tangent line to front of each pulses and zero line of oscillogram. Figure 1 shows the typical oscillogram of the signal of neutron detector obtained in one of the experiment when the shift of the center of microballoon from the box center was one half of radius of the box.

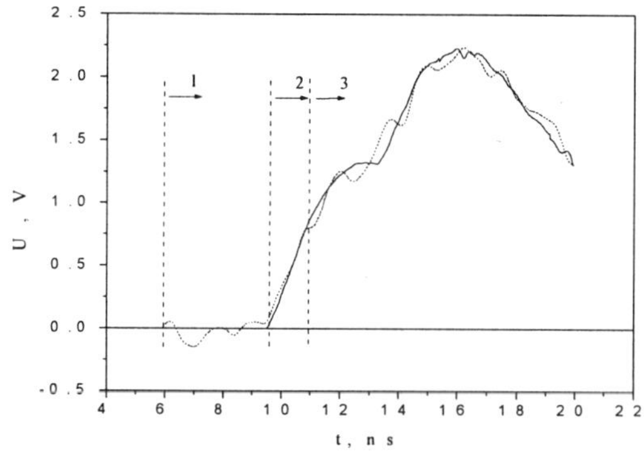

Fugure 1. The typical oscilloscope view of the neutron detector's signal.

It is obvious that the duration of neutron pulse in experiment was negligible less in comparison with the impulse response of the detector mentioned above. According to our hydrodynamic simulation the typical value of the neutron pulse duration during compression of microballoon is 0.1 ns or less. Therefore the treatments of the detector signals were performed using its impulse response. To measure impulse response of the detector $G(t)$ experimentally the special series of calibration experiments were fulfilled.

In these experiments the only hard X-ray with photon energy close to hν~100 keV and pulse duration about 0.3 ns was generated. These hard X-ray impacted to the detector and its response was studied.

Then the detected in indirect drive target asymmetrical experiments with shifted microballoon signals of neutron detector (see for example figure 1) were fitted by superposition of two impulse response as follows:

$$u(t) = p_1 \cdot G(t - p_2) + p_3 \cdot G(t - p_4) . \qquad (1)$$

Parameters p_i, i=1,2,3,4 calculated using least-squares method. The parameter p_2 is the moment of appearance of the first pulse at oscillogram and p_4 is the moment of second pulse. According to experimental results the first pulse corresponds to the moment when neutron flux with neutron energy 14.1 MeV arrives at the distance where detector is situated. The possible nature of the second pulse will be discussed below.

The procedure mentioned above allows us to increase the accuracy of determination of the time moment of the generation of the neutron yield. The results of treatment of the experimental oscillograms are shown in the table 1.

Table 1. The results of measurements

#	ΔR/R	τ_{yn}, ns	Δt, ns	p_1/p_3	N_{exp}	N_{est}	N^*
1	0	0.09±0.082	3.60±0.04	1.5	$1.1\cdot10^9$	$7\cdot10^8$	$5.6\cdot10^8$
2	1/3	0.10±0.073	1.3± 0.04	4-5	$1.6\cdot10^9$	$1.4\cdot10^9$	$1.2\cdot10^9$
3	1/2	0.25±0.082	3.75±0.04	0.9	$8.8\cdot10^8$	$1.3\cdot10^9$	$7.5\cdot10^8$
4	2/3	0.18±0.10	3.81±0.04	0.3	$4.3\cdot10^8$	$9\cdot10^8$	$2.5\cdot10^8$
5	1	-	-	-	$<3\cdot10^4$	$<10^8$	-

Here: ΔR/R is a shift of DT capsule from the center of spherical box-converter normalized to its radius; τ_{yn} is the value of time delay of neutron pulse generation; Δt is the delay between first and second pulses if the experimental signal is fitted by the dependency Eq. (1); p_1/p_3 is the ratio of amplitudes of first and second components; N_{exp} is the average over different diagnostics experimental neutron yield according to paper[1] ; N_{est} is the estimated neutron yield obtained by using signals amplitudes of the detector and its estimated sensitivity ; N^* is estimated neutron yield obtained by using the only amplitude p_1 of first components of the signals. For experiment #5 neutron yield corresponds to the threshold of the diagnostics.

As we told above the first pulse had neutron nature. The additional proof of this fact is the result obtained in experiment #5. In this experiment the shift of the microballoon was the largest value so the asymmetry of the X-ray filed was the biggest in all experiments. The neutron yield was less then the threshold of the integral neutron diagnostics and there was no first pulse at oscillogram. Unfortunately in this specific experiment the oscilloscope had very high time resolution (the shortest time trace) so we can detect only first pulse, therefore we could not answer was the second pulse in this experiment or not.

Now we want to say a few words about possible origin of the second pulse at the oscillogram of the neutron detector. It draws attention that time delay between first and second pulses is nearly constant value. So one can suppose that the origin of the second pulse causes by charge particle burning simultaneously with neutrons but having less energy (and velocity) then it. If so then we can estimate a velocity of these particles as 10^9 cm/s using the measured time delay between first and second components of the detector signal. One of the more probable candidates to this role is the proton, which always exists in laser plasma[4].

Now discuss the results of measurements of neutron yield generation time. One can see that in experiments with zeros or small values of shifts of the central microballoon the average value of $\tau_{\gamma n}$ is smaller in comparison with the experiments with large shifts. Analogous dependency was obtained in behavior of neutron generation time in simulations of these experiments using 2D hydrodynamic code MIMOZA-ND[3]. In simulations the compression of the central microballoon was studied in conditions when asymmetry of the X-ray field corresponded to the experimental conditions[1].

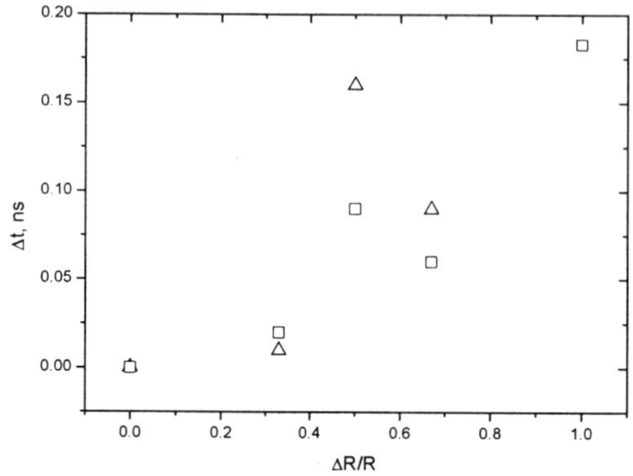

Figure 2. The experimental (triangles) and calculated (squares) dependences of relative time delay of neutron generation against the shift of center target.

Figure 2 shows the comparison experimentally measured value of neutron generation time $\tau_{\gamma n}$ and simulations results obtained from 2D MIMOZA-ND calculations for different values of shift of the center of microballoon. It is necessary to pay attention that experimental value $\tau_{\gamma n}$ on Figure 2 are calculated relatively the value of $\tau_{\gamma n}$ obtained in experiment with zeros shift (so the neutron generation time for symmetrical case corresponds to zeros time and we talk about time delay of neutron generation due to asymmetry of the X-ray field). One can see that simulation results qualitatively are rather close to the experimental dependency of time delay upon the shift normalized to the radius of the box-converter. From our point of view some quantitative discrepancy is unessential. For very large asymmetry corresponded to the shift $\Delta R/R \sim 1$ we obtained in simulation very sharp increase of time delay. Unfortunately in experiment the whole neutron yield less then the threshold of sensitivity of neutron detector and we could not see this effect directly. Besides experimental data shows more strong dependence of time delay upon the shift for shifts value in diapasons $\Delta R/R = 0.5 \div 0.7$. Nevertheless in spite of these different we can conclude that there are good qualitative and quantitative agreements between our MIMOZA-ND simulations and experimental data. We believe that this different can be understood if more detailed information about experimental conditions will be included in simulation.

3. REFERENCES

1. L.S. Mkhitarian, S.A. Bel'kov, A.V. Bessarab et al. in Proceedings of Inertial Fusion Sciences and Applications 99, Editors: C. Labanne, W.J. Hogen, K.A. Tanaka, p. 142, Elsevier, 2000.
2. F.M. Abzaev, S.A. Bel'kov, A.V. Bessarab et al. Zh. Eksp. Teor. Fiz. **114,** 155-170 (1998).
3. I.D. Sofronov, S.A. Bel'kov, O.A. Vinokurov, L.S. Mkhitarian, N.A. Ryabikina, VANT, series: Mathematical modeling of physical processes, part.1, p.8, (2000).
4. Skupsky and S. Kasenjiar, J.Appl. Phys. Vol.52, No.4, 2608 (1981).

X-Ray Multi Camera Tomography System of Wendelstein 7-X

S. KLOSE, A. WELLER, P. SCHÖTZ, A. WERNER, H. GREVE[1]

1 Introduction

A X-ray Multi Camera Tomography System (XMCTS) will be constructed for the measurement of the spatio-temporal structure of the spectral-integrated X-ray emissivity (0.1 ... 10 keV) of the W7-X plasma. The position (SHAFRANOV-shift) and shape of the equilibrium magnetic flux surfaces as well as of MHD mode structures can be determined with this diagnostics. A particular objective is the verification of the improved equilibrium and stability properties resulting from the stellarator optimization. As a reference system, the well working tomography system of the W7-AS with 8 pinhole cameras and 32 diodes each is used [Goe98a, Goe98b, Wel99, Wel01].

2 Physical Background of the X-Ray Emission

The emission in the X-ray spectral range is caused by collisions of energetic electrons with plasma ions. The radiation intensity increases strongly with the atomic number, and therefore, the X-radiation due to impurities is typically significantly higher than the hydrogen bremsstrahlung. The distribution of impurity ionization states and the emissivity depends on the electron temperature and the radial impurity transport. If the impurities are only partially ionized the electron impact excitation of resonance lines (e.g. from He- and H-like ionization states) exceeds the power due to the continuum radiation (free-free and free-bound radiation). The X-ray emission can be considered as homogeneous on magnetic flux surfaces, since the thermal speed of electrons and ions involved in the generation of X-rays is relatively large. Therefore X-ray measurements can provide information on the magnetic flux geometry, in particular by tomographic reconstructions of plasma equilibrium surfaces and the spatial structure of MHD instabilities.

3 Geometrical layout of XMCTS

The following quantities are to be determined with the XMCTS

- Spectral integrated distribution of the X-ray emissivity $\epsilon(r, \theta)$ (0.1 ... 10 keV)
- Plasma position, e.g. SHAFRANOV-shift (radial resolution $\leq 2\,\text{cm}$)
- Poloidal mode number $m \leq 10$, radial position of modes, mode amplitude and displacement

[1]Max-Planck-Institut für Plasmaphysik, Boltzmannstr. 2, D-85748 Garching b. München, Germany

Advanced Diagnostics for Magnetic and Inertial Fusion, Edited by Stott *et al.*
Kluwer Academic/Plenum Publishers, New York, 2002

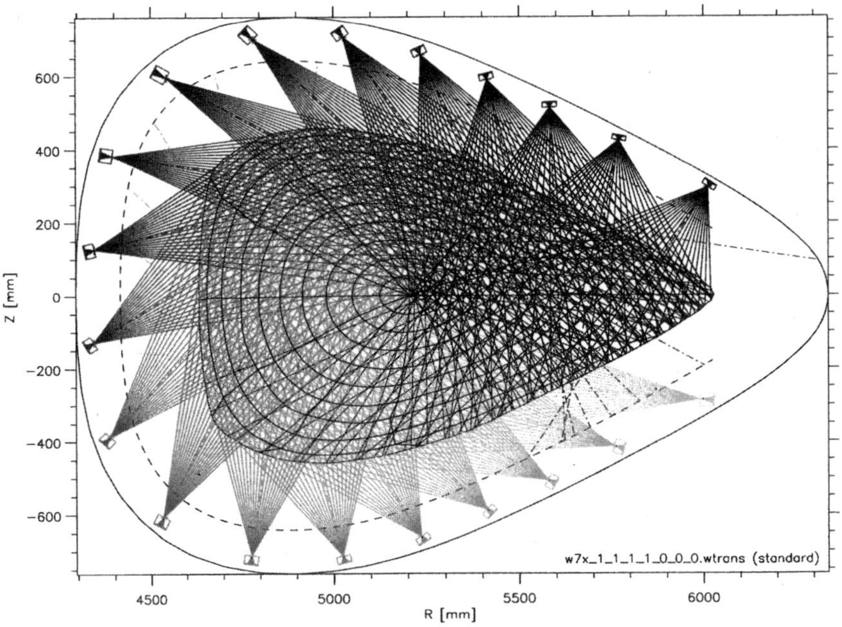

Fig. 1: Geometrical layout of XMCTS.

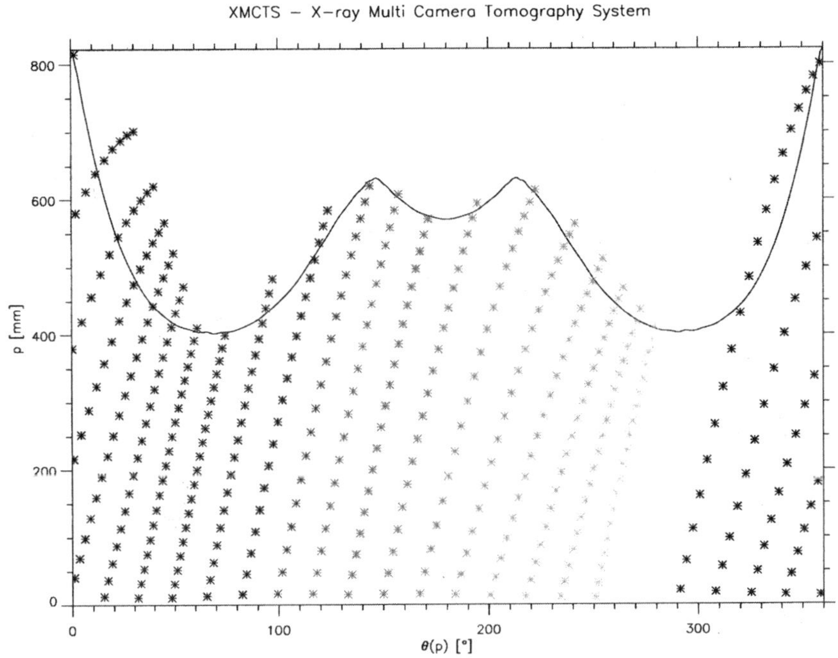

Fig. 2: Ray-coverage of XMCTS.

- Rotation frequency of modes (NYQUIST-frequency $\geq 500\,\mathrm{kHz}$)
- Phase velocity of modes (incl. direction)
- Toroidal mode number n by the help of additional cameras and mirnov-coils,

therefore the system will be consist of

- 20 cameras around the triangular plane ($\varphi = 36°$)
- each camera covers only the half plasma with 20 rays (400 rays total)
- IRD AXUV-20 Si-diode array (p-bulk)
- $6\,\mu m$ Be filter $\Rightarrow E = 0.1 \ldots 10\,\mathrm{keV}$
- $f_{samp} = 1 \ldots 2\,\mathrm{MHz}$.

Figure 1 shows a tentative draft of the geometrical layout of XMCTS. The cameras are equidistantly arranged around the triangular plane. Each camera covers only the half plasma. The rays are nearly homogenious distributed, and there are no redundant rays (Fig. 2). Due to constructive circumstances there is a void zone around $\theta = 270°$. They will be closed with an additional camera in the future.

4 Camera Design

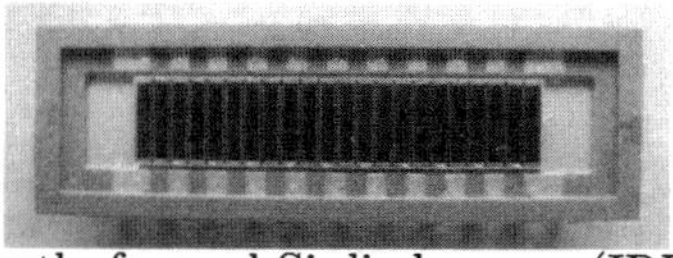

Fig. 3: Currently favored Si-diode array (IRD AXUV-20).

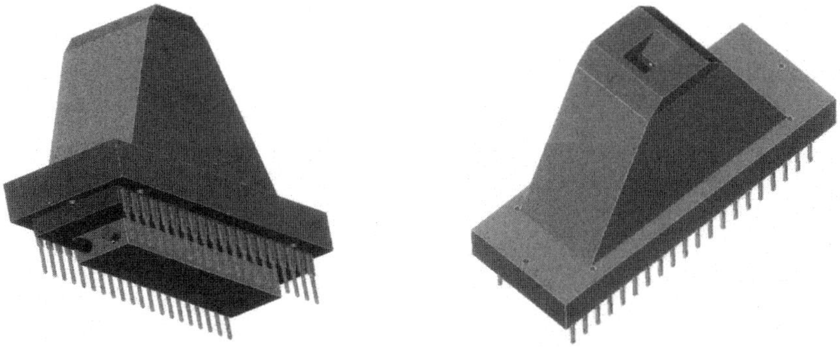

Fig. 4: Provisional camera design.

The favored diode array (Fig. 3) consist of 20 (+ 2 blind elements) p-bulk Si-diodes with an active area of 0.75 x 4 mm^2 each. The distance between the diodes is 0.2 mm, the risetime is $0.2\,\mu\mathrm{s}$, and the spectral range is $0.2 \ldots 20\,\mathrm{keV}$ with a nearly constant spectral response of 0.2 A/W. For enough sensitivity in the edge channels the incident angles of the rays on the diode surfaces should be smaller than 45°. If the poloidal slit size equal to the distance between the diodes (0.2 mm), the ray-cones are even

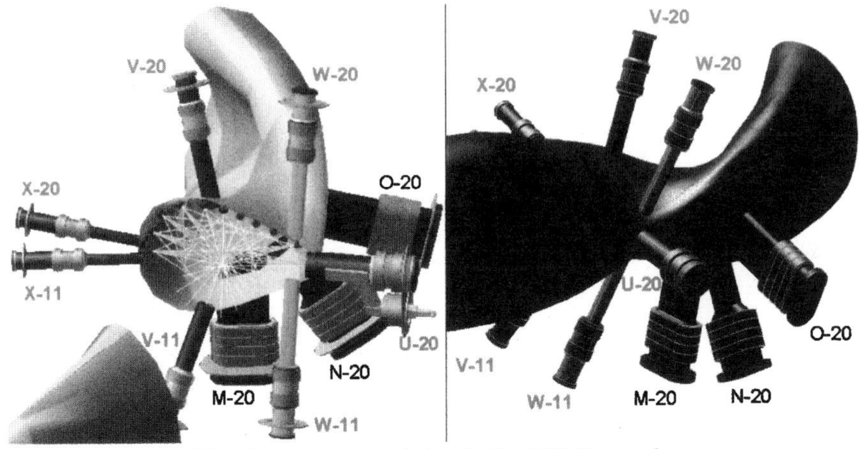

Fig. 5: Arrangement inside the W7-X vessel.

separated and their coverage of the plasma crossection has an optimum. The etendue (solid angle) can be adjusted with the toroidal slit size.

Due to very limited space inside the vessel, the camera housing should be as small as possible (Fig. 4). Their height will be only $\approx 6\,\mathrm{cm}$. For a constant dark current, the diode array (and some electronics) will be peltier cooled. A curved Be-filter of few microns thickness selects the lower energy cut-off. With a shutter (pressure or electric driven) the dark current can be measured during long plasma pulses.

5 Vessel Integration

XMCTS must be located inside the vacuum vessel in the triangular plane ($\varphi = 36°$, Fig. 5). Only in this divertor-free plane it is possible to view the plasma from all directions. Because of changes of the vessel temperature and stress, a special support-structure is needed to maintain the camera positions. Preamplifiers and cables must be in a secondary vaccum inside the plasma vessel.

References

[Goe98a] C. Goerner. *Tomographische Untersuchung von globalen Alfven-Eigenmoden am Stellarator Wendelstein VII-AS.* PhD thesis, IPP III/230, 1998.

[Goe98b] Goerner, C., Anton, M., Geiger, J., Linden, von der, W., Weller, A., Zoletnik, S. Soft X-ray tomography of plasma equilibria and MHD modes at Wendelstein 7-AS. *Stellarator News*, 55:2–5, January 1998.

[Wel99] Weller, A., Görner, C., Gonda, D. X-Ray Diagnostics on Wendelstein 7-AS. *Rev. Sci. Instr.*, 70:484, 1999.

[Wel01] Weller, A., Anton, M., Geiger, J., Hirsch, M., Jaenicke, R., Werner, A. Survey of magnetohydrodynamic instabilities in the advanced stellarator wendelstein 7-as. *Phys. of Plasmas*, 8(3):931–956, March 2001.

The Atomic K- and L-shell Ionization Cross-sections by Electron Impact

Changhuan Tang Zhengming Luo Zhu An Fuqing He
Xiufeng Peng Xianguan Long *

INTRODUCTION

The experimental and theoretical study of atomic inner-shell ionization cross-sections by electron impact, a subject of scientific study for many years, is important both for understanding the interaction between electrons and atoms and for technology application.[1] The cross-section for the ionization of atoms and molecules by electron impact is one of the essential sets of data needed in a wide range of applications. For example, this can be used to the impurity diagnosis of magnetic fusion devices. It is well known that radiative cooling by low- and high-Z impurities accounts for a significant fraction of the power loss from tokamak plasmas. Several investigations have shown that the impurity radiation can seriously impair tokamak performance. The study of soft x-rays emitted between 1-20keV has, therefore, increased in importance, furnishing electron temperature data and a wealth of impurity information. In general, the power radiated by an impurity ion, i in a charged state, j can be expressed as [2]

$$P(T_e)_{ij} = n_e n_i \left(\frac{n_{ij}}{n_i}\right) \langle \sigma v(T_e) \rangle_{ij} E \qquad keV \ /(cm^3 \cdot s) \tag{1}$$

where n_e and n_i denote the electron density and the total density of the ith impurity element, respectively. The density ratio n_{ij}/n_i is the relative abundance of the jth ionic charge state of the impurity i. E is the x-ray energy emitted by impurity, and $<\sigma v(T_e)>$ refers to the total excitation cross-section averaged over a Maxwellian velocity distribution. From Eq. (1), we can see that the atomic inner-shell ionization cross-sections are essential parameters in calculation of the power loss in the tokamak.

Unfortunately, there are still inadequate experimental ionization cross-sections data

---

* Changhuan Tang, National Key Laboratory of Laser Fusion, Research Center of Laser Fusion, CAEP, P. O. Box 919-986, Mianyang, Sichuan 621900, P. R. China; Zhengming Luo, Zhu An, Fuqing He, Xiufeng Peng, Xianguan Long, Key Laboratory for Radiation Physics and Technology of Education Ministry of China, Institute of Nuclear Science and Technology, Sichuan University, Chengdu 610064, P. R. China

Advanced Diagnostics for Magnetic and Inertial Fusion, Edited by Stott *et al.*
Kluwer Academic/Plenum Publishers, New York, 2002

at present, especially in lower incident energy region.[3] We think that preparing the thin and uniform self-supporting target is a main difficulty existing in the experimental measurement. In order to avoid the difficulty, the thin targets with thick substrate, which could be easily prepared by vacuum coating techniques, are used in the experiment. The electrons reflected from the substrate, whose energy exceed the ionization energy of the target atoms, would induce ionization of the target atoms once more, which increase the counting number of characteristic X-ray, requiring a correction on measured data. We developed a method to correct the measured data. The method need to know the energy spectra of the electrons reflected from the substrate. Using the so-called bipartition model of electron transport, the energy spectra of the reflected electrons can be precisely calculated, we can eliminate the influence of the substrate.[4]

In this paper, we reported the experimental data of cross-sections of V, Mn, Ni, Ti, Cr and Zn elements, which usually appear in tokamak as impurities. For elements of V, Cr and Zn, none of the experimental data of K-shell ionization cross-sections by electron impact were reported in near threshold until the method presented.[5,6] Comparison of the experimental data with the results measured by others (if existing) and semi-empirical or empirical formula as well as theoretical results is presented.

EXPERIMENT

The experimental set-up consists of an electron beam production system, target chamber and an X-ray analysis system. 0-60 kV dc power supply is used to accelerate the electrons. The characteristic k_α X-ray emitted from the targets is detected by a Si(Li) detector with an energy resolution of 170 eV FWHM (full width at half maximum) for the k_α X-ray of ^{55}Mn, and charges of the electron beam are collected by a deep Farady cup and recorded with a digital current integrator. The electron beam is incident upon the targets placed at 45° with respect to the electron beam direction and the Si(Li) detector is positioned at 90° with respect to the electron beam direction. The detection efficiency calibration of this system is performed with standard radioactive sources, i.e. ^{241}Am, ^{137}Cs, ^{55}Fe and ^{54}Mn, provided by China Institute of Atomic Energy. The uncertainty of calibrated efficiency is believed to be less than 5%.

Targets enriched to 99.9% with the thickness about 10 μ g/cm^2 are used in our experiments. The thickness of targets is monitored and measured by a quartz-oscillator. And the thickness uncertainty dose not exceed 10%. The homogeneity is tested by recording the intensity of the X-ray from the different spots on the targets. The reproducibility of measured data is also checked, and the repeatedly measured data are found consistent within an uncertainty of 1.5%.

RESULTS

As mentioned in Section I ,the data of the measured cross-sections should be corrected due to the re-ionization of the electrons reflected from the substrate. For the correction method used here, a detailed description has been given in Ref. 4. Here, we only present the final formula:

$$\sigma_k = \left[\frac{4\pi}{\eta\Omega}\right]\left[\frac{N_x A\cos\theta}{N_A N_e \omega_k \rho d}\right]\left[1+\frac{I_\beta}{I_\alpha}\right] - \cos\theta \int_{E_k}^{E} \Phi_{ref}(E')\sigma_k(E')dE' \quad (2)$$

Where ω_k is the K-shell fluorescence yield, $N_x(E)$ denotes the K_α X-ray counts. ρ and d are target density and thickness, respectively, and N_e means the number of incident electrons. N_A and A are Avogadro constant and atomic weight. K_α to K_β X-ray intensity ratios are referred to as I_α/I_β. E_k represents the threshold energy of target element. Φ_{ref}, the energy spectrum of the reflected electron at the surface of the substrate, is precisely calculated by using bipartition model of electron transport and is given in Ref. 4. η, the efficiency of the detector, Ω, the solid angle subtended by the detector's effective area to the beam spot. The angle between the incident beam direction and the normal direction of target plane is denoted as θ. After performing an iteration calculation in Eq. (2), the corrected K-shell ionization cross sections, $\sigma_k(E)$, can be given. If Φ_{ref} are assumed to vanish, then, uncorrected cross sections are obtained. In general, the corrected cross sections compared with uncorrected ones will decrease 10% -20%. The relevant constants, used here for the calculation of the K-shell ionization cross-sections, are given in Table 1.The present results are demonstrated in Figs. 1-6

Table 1. Relevant constants for the calculation of the K-shell ionization cross-sections[10]

Elements	Z	A	ω_k	E_k(keV)	I_α/I_β	ρd ($\mu g/cm^2$)
Ti	22	47.90	0.219	4.966	0.134	5.5
V	23	50.94	0.250	5.466	0.134	26.1
Cr	24	52.00	0.282	5.991	0.135	6.0
Mn	25	54.94	0.314	6.538	0.135	7.0
Ni	28	58.69	0.414	8.332	0.135	4.5
Zn	30	65.37	0.479	9.661	0.138	9.3

In Figs. 1-6, our measured data as well as the others experimental data are drawn and compared with the predictions given by classical theory of Gryzinski[7] and results of the empirical formula of Casnati *et al* [8]and semi-empirical formula of Green and Coslett.[9] We can see that the present data are in good agreement with the existing data measured by others and in acceptable agreement with the results of the empirical formula of Casnati *et al.* Powell has made a comparison of several widely used empirical formulae and of some theoretical results with experimental data of K-shell ionization cross sections for C, N, Ne, Al, Ar, Fe, Ni, Cu, Mo and Ag.[11] It was found that the empirical formula of Casnati *et al* was superior to that of the others. From Figs.1-6, we can see that the empirical formula of Casnati *et al* somewhat overestimates the measured data and on the contrary the theoretical results of Gryzinski underestimate the experimental data. Based on the present data, we find that the results, given by Green and Cosslett, are more closer to the experimental data for medium Z atoms (i.e.$19\leq Z\leq 40$) in lower energy regions (i.e.$1<U<4$) in which the equation of Bethe[12] is thought to be invalid. In the low Z ($6\leq Z\leq 18$)or high Z ($Z>40$) region, Casnati *et al' s* empirical formula provides better agreement. Through comparisons in Figs. 1-6, we also can see that the experimental data measured with our method are reliable. Recently, using the same method, we carry out the study of atomic L-shell ionization, for which the experimental data are quite scarce and poorer accuracy.

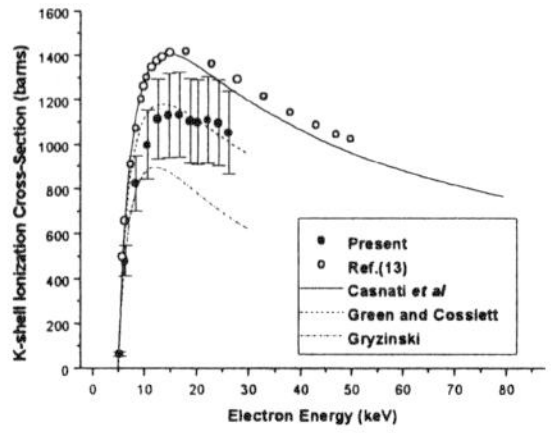

Figure 1. Ti

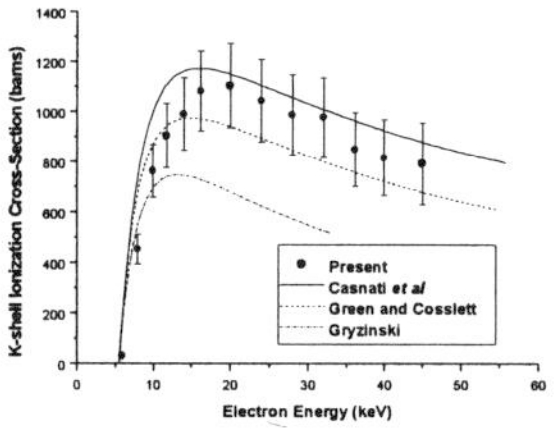

Figure 2. V

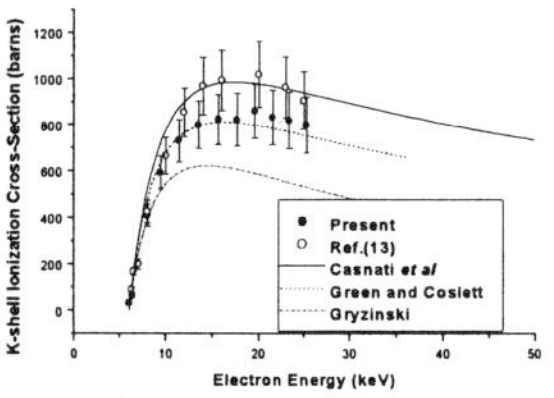

Figure 3. Cr

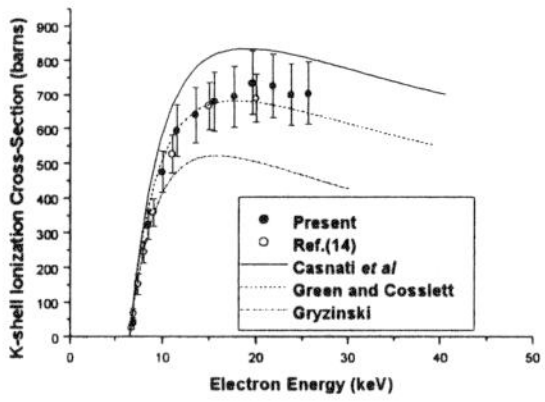

Figure 4. Mn

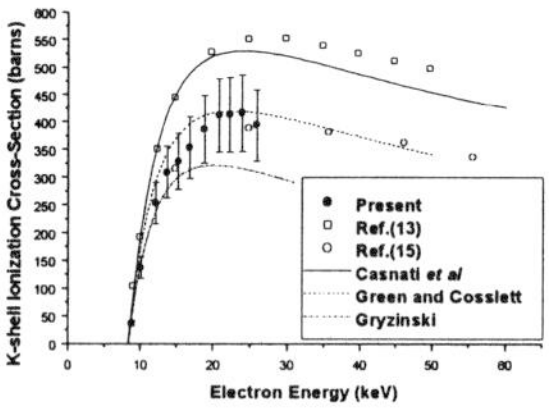

Figure 5. Ni

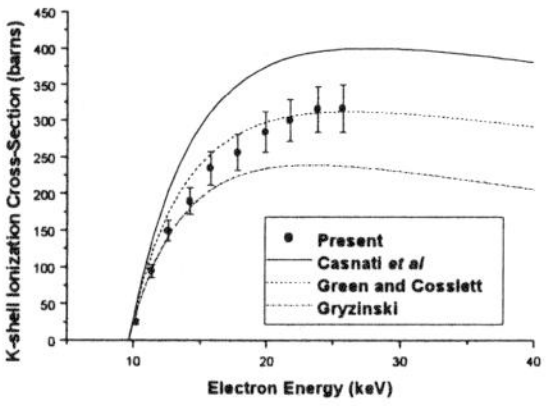

Figure 6. Zn

ACKNOWLEDGMENT

This work was supported by the National Science Foundation of China under grant No. 19874045

REFERENCES

1. C. J. Powell, "Electron Impact Ionization", Chapter 6, Springer-Verlag, New York, (1985).
2. E.H. Silver, M. Bitter, K. Brau *et al*, Rev. Sci. Instrum. 53(8), 1198-1213 (1982)
3. Long Xianguan, Liu Mantian, He Fuqing *et al*, At. Data Nucl. Data Tables, **45**, 353-366, (1990)
4. Zhengming Luo, Zhu An, Fuqing He *et al*, J. Phys. **B**: At. Mol. Opt. Phys. **29**, 4001, (1996)
5. Changhuan Tang, Zhengming Luo, Zhu An *et al*, Chin. Phys. Lett. **16**, 505-507, (1999)
6. Z. An, C.H. Tang, C.G. Zhou *et al*, J. Phys. **B**: At. Mol. Opt. Phys. **33**, 3677-3684, (2000)
7. Gryzinski, M., Phys. Rev. **138**, A336, (1965)
8. E. Casanati, A. Tartari, and C. Baraldi, J. Phys. **B**: At. Mol. Phys. **15**, 155, (1982)
9. M. Green and V. E. Cosslett, Proc Phys Soc London, **78**, 1206, (1961)
10. Y. Z. Liu, Decay Scheme of the Usual Radioisotopes, Atomic Energy Press, Beijing, (1982) (in Chinese)
11. C. J. Powell, Cross sections for inner-shell ionization by electron impact, Private communication (1998)
12. H. Bethe, Ann. Phys., **5**, 325, (1930)
13. J. Jessenberger and W. Hink, Z. Phys. A **275**, 331, (1975)
14. K. Shima, T. Nakagawa, K. Umetani *et al*, Phys. Rev. **A 24**,72, (1981)
15. L.P. Pockman *et al*, Phys. Rev. **71**, 330, (1947)

INNOVATIVE NEUTRAL BEAM BASED DIAGNOSTICS ON THE MST REVERSED-FIELD PINCH

D. J. Den Hartog, D. Craig, G. Fiksel, J. C. Reardon, V. I. Davydenko, and A. A. Ivanov*

1. INTRODUCTION AND SUMMARY

Charge-exchange recombination spectroscopy (CHERS), motional Stark effect (MSE), and Rutherford scattering diagnostics are now in operation on the Madison Symmetric Torus (MST) reversed-field pinch (RFP). This is the first time MSE has been applied to measure |**B**| in the core of a low-field (≤ 0.5 T) magnetic confinement device, an accomplishment made possible by a high quality diagnostic neutral beam and carefully designed collection optics. The Rutherford scattering diagnostic has demonstrated the robustness of this technique for reliable measurement of majority (H) ion temperature.

MST is a large RFP (R = 1.5 m, a = 0.52 m) operated at moderate current ($Ip \leq 500$ kA), with n_e typically 1–2 × 10^{19} m^{-3} and T_e, $T_i \leq 1$ keV.[1] Two compact and reliable diagnostic neutral beams are installed on MST.[2] Each beam contains an ion beam source that is electrostatically produced and focused, and neutralized by passing through a gas cell. These beams are short pulse (3 ms), intense, mono-energetic, and low-divergence. The first, a 30 keV, 4 A, 0.4 A/cm^2 neutral H beam is used in combination with visible and ultraviolet spectroscopy to make the CHERS and MSE measurements. MSE measurements are made by recording the Doppler shifted Hα Stark spectrum emitted by the beam with an imaging spectrometer and CCD camera. We record the separation of the π components of the Stark manifold for accurate measurements of |**B**| in the range 0.2 – 0.5 T. A second beam, 20 keV, 4 A, 0.4 A/cm^2, neutral He, is used for Rutherford scattering measurements of majority ion temperature. Two electrostatic neutral particle energy analyzers are used to observe the Rutherford scattered He atoms from the beam. This diagnostic has a time resolution of 100 μs and a radial range of 0 – 0.5 r/a.

* D. J. Den Hartog, D. Craig, G. Fiksel, and J. C. Reardon, Department of Physics, University of Wisconsin-Madison, Madison, Wisconsin 53706, U.S.A.; V. I. Davydenko and A. A. Ivanov, Budker Institute of Nuclear Physics, Novosibirsk, Russia.

2. CHARGE-EXCHANGE RECOMBINATION SPECTROSCOPY

To date, the most successful tool for localized minority ion measurements in fusion research plasmas is CHERS.[3,4] We conducted an extensive survey of ultraviolet and visible transitions in MST to find an appropriate candidate for CHERS measurements.[5] The C VI transition at 3433.69 Å is relatively strong, gives the largest charge-exchange signal/background ratio of any line surveyed, and is well-isolated in MST. Some possibility of "contamination" due to O VI emission at 3434 Å exists but measurement of nearby O VI lines at 3426 and 3438 Å indicate that this should not be a severe problem.

Although the C VI emission from charge exchange with the beam is strong, the background signal from the plasma is still of the same order or larger in some circumstances. High background neutral density in MST results in substantial charge exchange of C^{+6} with thermal neutrals. In order to distinguish between beam-induced and background emission, we have implemented a dynamic background subtraction technique with two optical views through each viewing port: one intersecting the beam and one to the side (Fig. 1).

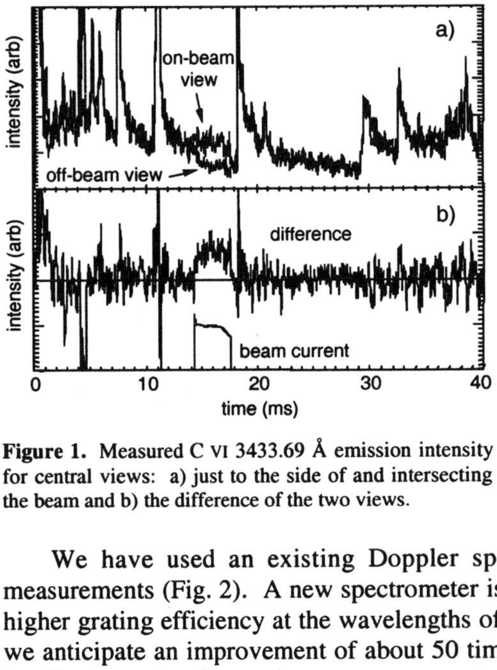

Figure 1. Measured C VI 3433.69 Å emission intensity for central views: a) just to the side of and intersecting the beam and b) the difference of the two views.

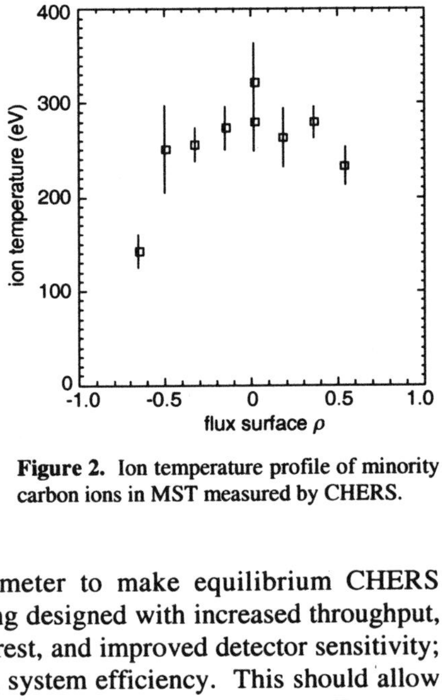

Figure 2. Ion temperature profile of minority carbon ions in MST measured by CHERS.

We have used an existing Doppler spectrometer to make equilibrium CHERS measurements (Fig. 2). A new spectrometer is being designed with increased throughput, higher grating efficiency at the wavelengths of interest, and improved detector sensitivity; we anticipate an improvement of about 50 times in system efficiency. This should allow localized core ion fluctuation measurements to be made for the first time in the RFP.

3. MOTIONAL STARK EFFECT

In many low-field (< 0.5 T) magnetic fusion energy research devices such as reversed-field pinches, magnetic mirrors, and spherical tori, the magnitude of the magnetic field in the core of the plasma is an important constraint for equilibrium modeling. In these devices the vacuum magnetic field profile can be greatly altered by dynamo, diamagnetic, or other plasma driven mechanisms. In order to measure core |**B**| in MST, we have implemented a spectral MSE diagnostic.[6] This is a direct measurement of the Balmer-α Stark spectrum emitted by the neutral hydrogen atoms injected as a 30 keV beam into the plasma. The beam atoms are excited by collisions with the plasma and emit statistically mixed line radiation. We measure the separation of the π manifolds (polarized parallel to the motional $\mathbf{E} = \mathbf{v} \times \mathbf{B}$) of the Stark split Balmer-α line at 6563 Å. Since the beam velocity is accurately known, the measured separation is directly analyzed for local |**B**|. As a side note, measuring field line pitch via MSE filter polarimetry[7] is not suitable for low-field (<0.5 T) devices. The π and σ manifolds are not sufficiently spaced in wavelength to allow for separation.

Figure 3 shows a typical Stark spectrum from MST, measured with a very low noise CCD spectrometer. Each of the individual components is fit as a Gaussian-broadened line with the relative amplitudes of the components fixed. The progression of central |**B**| through a sawtooth crash is shown is Fig. 4. Time resolution is provided by a ferroelectric liquid crystal (FLC) shutter capable of an exposure time of < 100 μs.

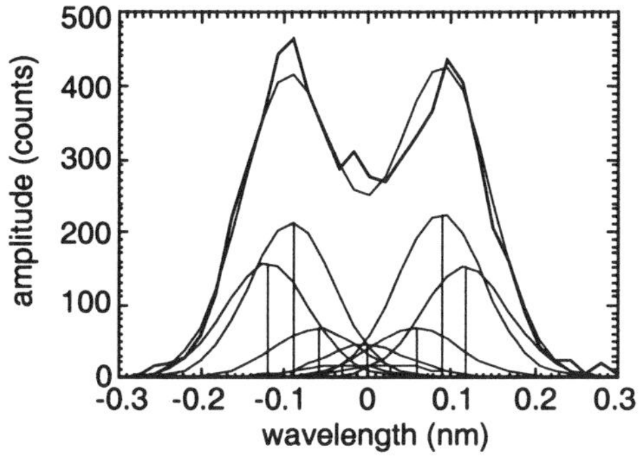

Figure 3. Hα motional Stark spectrum recorded in MST. Also shown is the composite fit and the nine individual components of the fit.

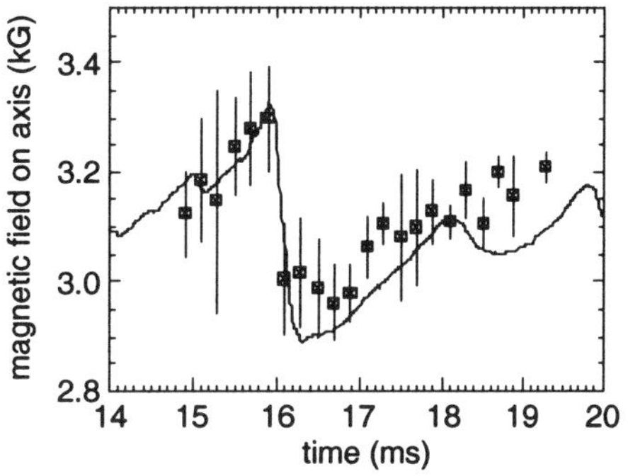

Figure 4. Time variation of central |**B**| in MST through a sawtooth crash. Data are from MSE diagnostic, solid line from equilibrium modeling.

4. RUTHERFORD SCATTERING

In a plasma diagnostics context, Rutherford scattering refers to the scattering of injected beam atoms by plasma ions. The energy distribution of the scattered atoms is determined by the velocity distribution of the ions. Thus, the ion temperature is determined from the measured width of the scattered atom energy distribution, the ion flow velocity from the shift of the energy spectrum, and the ion density from the intensity of the scattered beam (Fig. 5). The technique has been used to measure the equilibrium ion temperature[8] and equilibrium flow velocity[9] in mirrors and tokamaks. In MST, local measurement of equilibrium majority ion temperature has been done for the first time in an RFP (Fig. 6).[10] Our goal is to extend this technique and make the first measurement of majority ion fluctuations. The key to reaching this goal is our high intensity neutral beam, which is about two orders of magnitude more intense than that used for Rutherford scattering experiments on the TEXTOR tokamak.

Rutherford scattering and CHERS are complementary diagnostics in several respects. CHERS has very good spatial resolution (~1 cm) while Rutherford scattering is somewhat coarse (~10 cm). Rutherford scattering, however, measures majority ion properties while CHERS typically measures minority ions. Both can have excellent time response if signal levels are sufficient. On MST, the ability to make both measurements simultaneously will enable us to directly compare the flow, temperature, anomalous heating, and transport of majority and minority ions—a valuable tool to discriminate among various theoretical models.

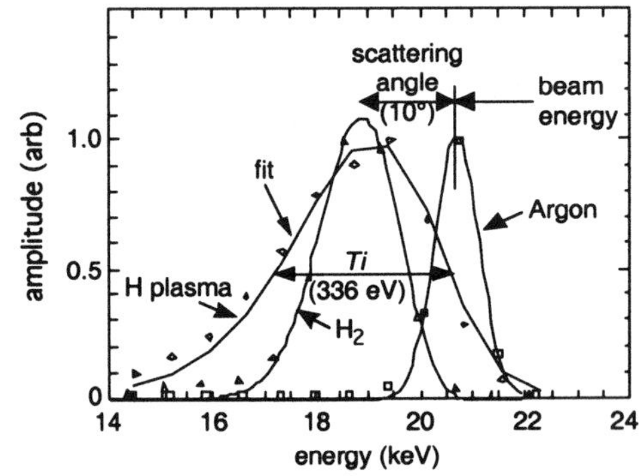

Figure 5. Rutherford scattering spectrum of beam He neutrals off H plasma, recorded in MST. Also shown are the calibration peaks on argon and neutral hydrogen.

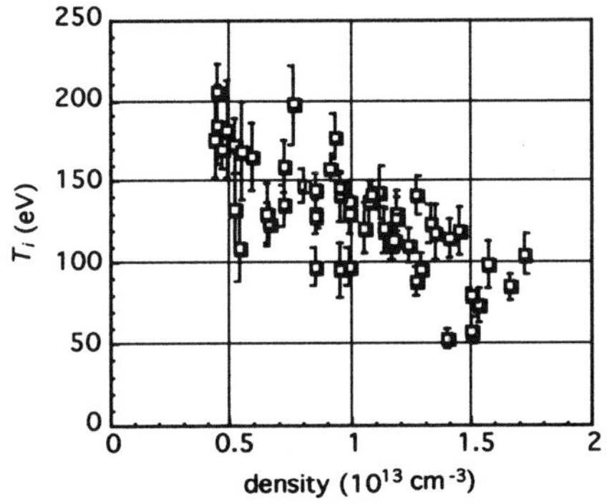

Figure 6. Variation of majority T_i with n_e, measured with the Rutherford scattering diagnostic.

5. REFERENCES

1. R. N. Dexter, D. W. Kerst, T. W. Lovell, S. C. Prager, and J. C. Sprott, The Madison Symmetric Torus, *Fusion Tech.* **19**, 131 (1991).
2. G. F. Abdrashitov, V. I. Davydenko, P. P. Deichuli, D. J. Den Hartog, G. Fiksel, A. A. Ivanov, S. A. Korepanov, S. V. Murakhtin, and G. I. Shulzhenko, A diagnostic neutral beam system for the MST reversed-field pinch, *Rev. Sci. Instrum.* **72**, 594 (2001).
3. R. J. Fonck, D. S. Darrow, and K. P. Jaehnig, Determination of plasma-ion velocity distribution via charge-exchange recombination spectroscopy, *Phys. Rev. A* **29**(6), 29 (1984).
4. R. C. Isler, An overview of charge-exchange spectroscopy as a plasma diagnostic, *Plasma Phys. Control. Fusion* **36**, 171 (1994).
5. D. Craig, D.J. Den Hartog, G. Fiksel, V.I. Davydenko, and A.A. Ivanov, First charge exchange recombination spectroscopy and motional Stark effect results from the Madison Symmetric Torus reversed field pinch, *Rev. Sci. Instrum.* **72**, 1008 (2001).
6. A. Boileau, M. von Hellermann, W. Mandl, H. Weisen, and A. Zinoviev, Observations of motional Stark features in the Balmer spectrum of deuterium in the JET plasma, *J. Phys. B: At. Mol. Opt. Phys.* **22**, L145 (1989).
7. F. M. Levinton, R. J. Fonck, G. M. Gammel, R. Kaita, H. W. Kugel, E. T. Powell, and D. W. Roberts, Magnetic field pitch-angle measurements in the PBX-M tokamak using the motional Stark effect, *Phys. Rev. Lett.* **63**, 2060 (1989).
8. E. L. Berezovskii, A. I. Kislyakov, S. Ya. Petrov, and G. V. Roslyakov, Measurement of ion temperature in a hot plasma by scattering of fast atoms, *Sov. J. Plasma Phys.* **6**, 760 (1980).
9. H. F. Tammen, A. J. Donné, T. Oyevaar, and F. C. Schüller, Rutherford scattering at TEXTOR: measurements on the ion velocity distribution *Plasma Phys. Reports* **20**, 158 (1994).
10. J. C. Reardon, G. Fiksel, C. B. Forest, A. F. Abdrashitov, V. I. Davydenko, A. A. Ivanov, S. A. Korepanov, S. V. Murachtin, and G. I. Shulzhenko, Rutherford scattering diagnostic for the Madison symmetric torus reversed-field pinch, *Rev. Sci. Instrum.* **72**, 598 (2001).

A VACUUM X-RAY CRYSTAL SPECTROMETER FOR THE HANBIT MAGNETIC MIRROR DEVICE

S. G. Lee, J. G. Bak, and M. Bitter*

INTRODUCTION

A vacuum X-ray crystal spectrometer was fabricated for the Hanbit magnetic mirror device[1, 2] utilizing a spherically-bent mica crystal and an X-ray sensitive CCD camera. The main diagnostic application for this spectrometer is to observe the spectra of helium-like neon, NeIX, in the wavelength range from 13.4474 Å to 13.6980 Å. Spherically bent crystals and two-dimensional detectors like a CCD camera are the essential components of a new type of imaging crystal spectrometer[3].

For tokamak plasmas, X-ray crystal spectrometers[4] have been used to measure the central ion temperature and plasma rotation velocity from the Doppler width and Doppler shift of X-ray lines, and measurements of line intensity ratios, which provide information on the central electron temperature, the ionization equilibrium and the impurity ion transport. Hence, the development of X-ray crystal spectrometers for tokamaks is certainly feasible based on the present experiences.

However, the design of a crystal spectrometer for a mirror device like Hanbit appears to be challenging due to the fact that the X-ray emissivity of Hanbit plasmas is rather low as a result of the low values of the electron temperature and density. Also, the emitted low energy X-rays will be strongly attenuated even by very thin foils or windows on the detector and the vacuum vessel. The low X-ray emissivity can be overcome by arranging a spectrometer sightline parallel to the long axial length of Hanbit. The attenuation problem for the low energy X-rays also can be overcome by using an X-ray sensitive vacuum CCD camera as a detector.

In this paper, we present the fabricated X-ray crystal spectrometer for the Hanbit device and the results of calibration measurements include an integrated reflectivity of the spherically bent mica crystal and measurements of the spectral resolution of the spectrometer.

---

* S. G. Lee, Korea Basic Science Institute, Yusung, Taejeon 305-333, Korea. J. G. Bak, Korea Basic Science Institute, Yusung, Taejeon 305-333, Korea. M. Bitter, Princeton Plasma Physics Laboratiry, Princeton, NJ 08543, U. S. A.

HANBIT X-RAY CRYSTAL SPECTROMETER

As shown in Fig. 1 (a), the fabricated Hanbit X-ray crystal spectrometer consists of a spherically-bent mica crystal and an X-ray sensitive CCD camera to observe the spectra of helium-like neon, NeIX. The mica crystal had a radius of curvature of 1524 mm and a reflecting surface of 70 mm x 17 mm. The CCD detector has no beryllium window so that its detection efficiency for the X-ray energy of 900 eV, which corresponds to the energies of the NeIX spectral lines, is as high as 40 %. The spherically bent mica crystal is installed inside of the crystal holding vacuum chamber as shown in Fig. 1 (b).

(a) (b)

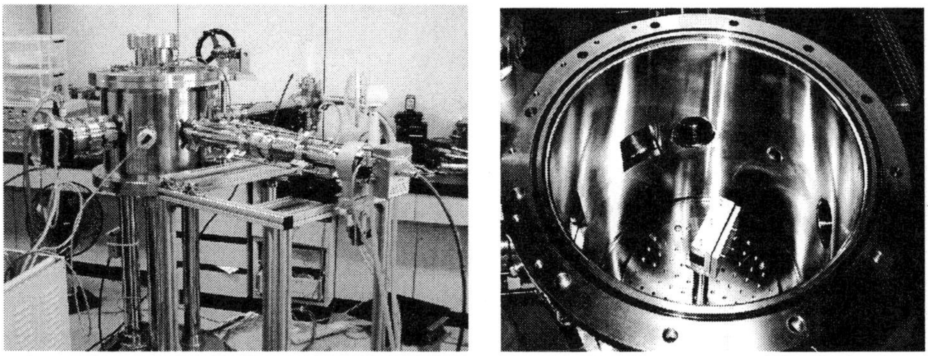

Figure 1. (a) A photo of the fabricated Hanbit X-ray crystal spectrometer. (b) An inside view of the chamber.

CALIBRATION MEASUREMENTS

The calibration measurements include measurements of the integrated reflectivity of the spherically bent mica crystal and measurements of the spectral resolution. These measurements were performed with an X-ray tube using the bremsstrahlung continuum and characteristic line radiation from the copper anode of the X-ray tube. The Cu L-radiation is close to the energy of the spectral lines of NeIX.

The experimental arrangement for the integrated reflectivity measurement is shown in Fig. 2 (a). The spherically bent mica crystal, X-ray tube, and an energy-resolving Si(Li)-detector were placed on the Rowland circle in the Johann configuration. Both the spherically bent mica crystal and the X-ray tube were mounted in a common vacuum chamber, while Si(Li)-detector was located outside of the vacuum chamber.

A 20 μm thick polypropylene foil of 10 mm diameter served as a vacuum window in front of the detector. The Si(Li)-detector had a sensitive area with a diameter of 5 mm and was covered by a protective 8 μm thick beryllium foil. Between the detector and the polypropylene foil was an air gap of a few millimeters, in which a vertical slit of a 1 mm width was placed on the Rowland circle. The width of the slit defined the angular resolution of the instrument.

Figure 2 (b) shows the energy-resolved spectrum of the Bragg reflected radiation from the mica crystal for Bragg angles of 43 , 45 , 47 and 50 . The observed spectral features represent the Bragg reflections from the 1st to the 12th order. The width of the peaks is determined by the energy resolution of the Si(Li)-detector of 155 eV at FWHM. The energy resolution of the crystal spectrometer is much higher than that of the Si(Li)-detector.

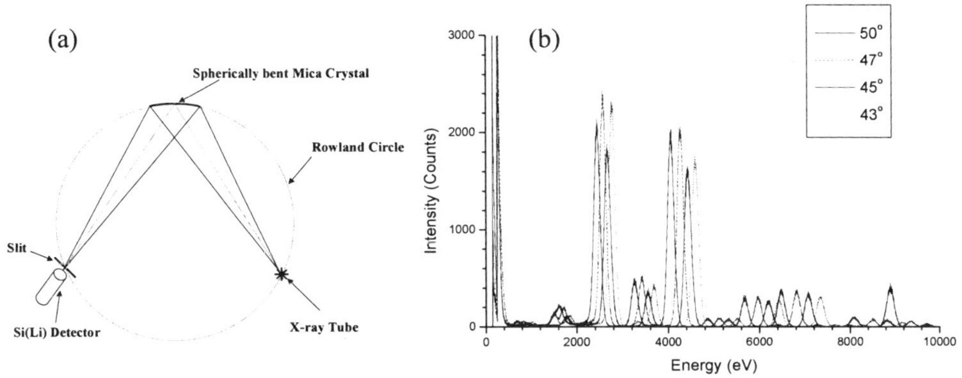

Figure 2. (a) Experimental arrangement. (b) Bragg reflected spectrum.

Measurement of the spectral resolution requires a position-sensitive proportional counter (PSPC) or a CCD camera. For this measurement, the CCD camera was installed instead of the Si(Li)-detector and 200 μm wide slit was put on the Rowland circle in front of the X-ray tube. The width of a Bragg reflected peak is determined as a function of the distance between crystal and X-ray source and the distance (R) between the crystal and CCD camera.

Figure 3 shows a Bragg reflected images of 200 μm wide slit from the CCD camera for various positions when the Bragg angle was set up 45°. Figure 4 shows the measured FWHM of the observed Bragg reflected peak as a function of the distance between crystal and CCD camera. The minimum width was observed for a distance of 1099 mm, which agrees well with the calculated value for the position on the Rowland circle. The measured FWHM of the Bragg peak is 15.6 pixels, which corresponds to a spectral resolution of 3070, which agrees well with the theoretically predicted value[5].

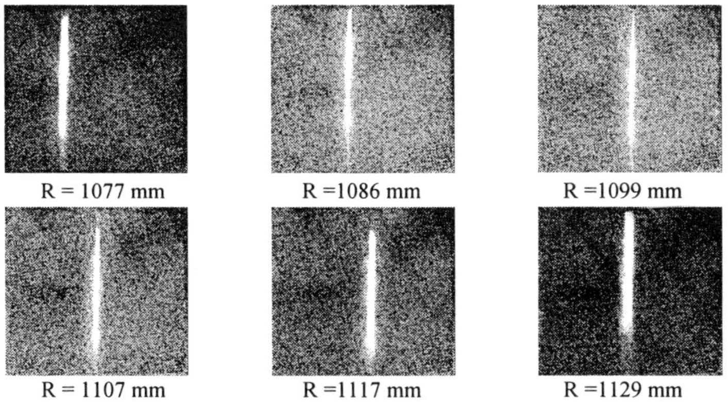

Figure 3. Bragg reflected images for the various CCD camera positions.

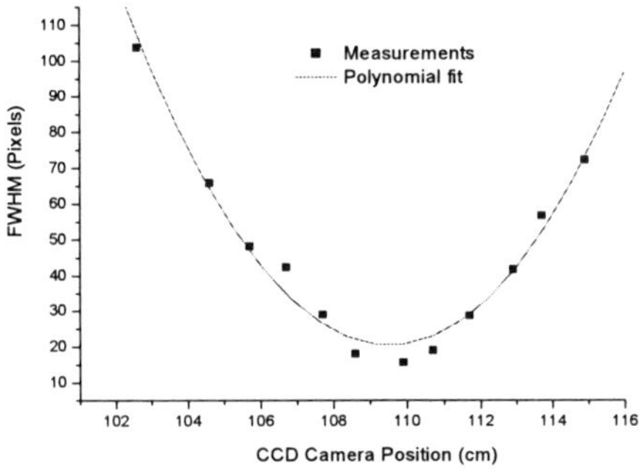

Figure 4. The measured FWHM of the Bragg reflected peak vs. CCD camera position.

CONCLUSION

X-ray crystal spectrometer for the Hanbit mirror device was fabricated by using a spherically bent mica crystal and CCD camera. Calibration measurements include an integrated crystal reflectivity and spectral resolutions were carried out. Four different Bragg angles for the integrated reflectivity measurements from the spherically bent mica crystal have performed. The results were compared with theoretical predictions[5].

The Bragg reflected image from a CCD camera showed that the width of the Bragg reflected image is a function of the distance between crystal and X-ray source. From this imaging characteristic measurement, the spectral resolution demonstrated that the Hanbit X-ray crystal spectrometer has a high spectral resolution. The resolution may also be sufficient for ion temperature measurements if the ion temperatures are above 2 keV.

ACKNOWLEDGEMENT

This work was supported by the Korean Ministry of Science and Technology under the Hanbit and KSTAR project contracts.

REFERENCES

1. S. G. Lee, S. M. Hwang, and M. Bitter, X-ray crystal spectrometer for the Hanbit mirror machine, Rev. Sci. Instrum. **70**(1), 299-301 (1999).
2. S. G. Lee, and J. G. Bak, A vacuum chamber with a radial rotating port, Rev. Sci. Instrum. **70**(11), 4437-4438 (1999).
3. M. Bitter et al., Imaging x-ray crystal spectrometer for the NSTX, Rev. Sci. Instrum. **70**(1), 292-295 (1999).
4. M. Bitter, S. von Goeler, R. Horton, M. Goldman, K. W. Hill, N. R. Sauthoff and W. Stodiek, Doppler-broadening of x-ray lines for determination of the ion temperature in tokamak plasmas, Phy. Rev. Lett. **42**(5), 304-307 (1979).
5. G. Holzer *et al.*, Flat and spherically bent muscovite (mica) crystals for x-ray spectroscopy, Physica Scripta **57**, 301-309 (1998).

X-RAY SPECTROSCOPY AND CALIBRATIONS IN THE 50 EV TO 60 KEV RANGE

John Seely, Uri Feldman, James Weaver, and Lawrence Hudson*

INTRODUCTION

Spectrometers have been developed to record x-ray spectra in the energy range 50 eV to 60 keV. The dispersion elements are transmission crystals for energies higher than approximately 11 keV, reflection crystals for 1 keV to 20 keV, and transmission gratings for energies less than 1 keV. The spectra are recorded on CCD or CMOS detectors with phosphor conversion screens. Silicon photodiodes are positioned in front of the detectors to provide absolute x-ray flux calibrations. The diodes have 1 mm^2 area and subnanosecond time response. The diodes, transmission gratings, and attenuation filters were absolutely calibrated using synchrotron radiation. In addition, the diodes were calibrated in pulsed mode using the soft x-ray (70 eV to 250 eV) pulses from individual electron bunches circulating in the storage ring, and a self-calibration model extends the calibration to higher energy. X-ray spectra in the energy range >12 keV were recorded at the OMEGA laser using a transmission crystal. Spectra in the energy range <1 keV were recorded at the NIKE laser using a transmission grating and fast silicon photodiodes. A NIF core diagnostic is being built that covers the 1-20 keV range with one transmission crystal and four reflection crystals.

HARD X-RAY SPECTROMETER (HXS) FOR LLE

A Hard X-Ray Spectrometer (HXS) was developed for the purpose of recording the time-integrated x-ray spectra from targets irradiated by the OMEGA laser at the Laboratory for Laser Energetics (LLE).[1] The 12 keV to 60 keV energy range was covered with moderate spectral resolving power (typically in the 100 to 200 range). As shown in Fig. 1, the spectrometer consists of a quartz (10-1-1) crystal operating in transmission and bent to a radius of curvature of 112 mm. The target to crystal distance is

---

* John Seely, Uri Feldman, and James Weaver, Naval Research Laboratory, Washington DC 20375 USA. Lawrence Hudson, National Institute of Standards and Technology, Gaithersburg MD 20899 USA

500 mm. The dispersed x-rays pass through a slit and are imaged on a flat dental x-ray CCD sensor with a phosphor conversion screen. Filters are inserted at the cross-over slit and provide absorption edges for *in situ* energy scale calibration. The filter elements and the absorption edges that were used on various shots were Se (12.65 keV), Y (17.037 keV), Zr (17.996 keV), and Cd (26.713 keV). These absorption edges, as well as other edges used during developmental testing at NIST, were fitted to an analytical formula that was derived from the geometry of the bent crystal spectrometer.[2]

A pinhole on the front face of the spectrometer provides an image of the source, and if desired the image intensity can be controlled by an attenuation filter that covers the pinhole. The location of the pinhole image provides information on the instrument pointing and target position. An additional Be filter is mounted in the spectrometer's nosecone and protects the crystal from target debris. The CCD sensor is covered by a thin plastic filter that is opaque (black) to visible light.

The spectrometer, CCD drive electronics, and rechargeable battery pack are housed in a Ten Inch Manipulator (TIM). Communication with the external control computer and the diagnostic interface unit (DIU) is by two fiber optic cables. The DIU receives the timing fiducials from the LLE data acquisition system and triggers the CCD integration time to span the laser shot. The x-ray image is automatically downloaded to the external control computer after the laser shot and is saved and displayed. The instrument is capable of completely autonomous operation throughout the shot day. The instrument operation and the recorded x-ray images may be viewed using a remote computer on the local LLE computer network.

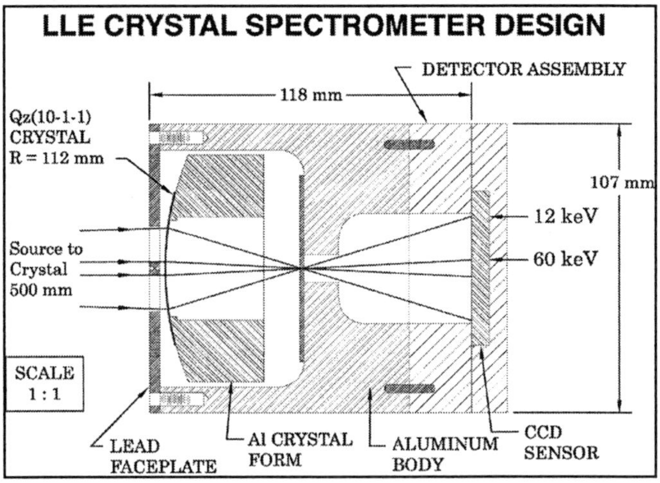

Figure 1. Optical design of the Hard X-Ray Spectrometer.

A typical spectral image and a lineout of the spectrum, from a krypton filled CH (plastic) shell target, are shown in Fig. 2. The spectral image consists of two mirror spectra on either side of the pinhole image (faint in Fig. 2 owing to attenuation by the pinhole filter). The He-like krypton resonance lines appear in the spectra. Spectra from krypton filled hohlraum targets are extremely bright compared to the CH shell spectrum shown in Fig. 2. Additional spectra were recorded from uranium, lead, gold, and silver foil targets.

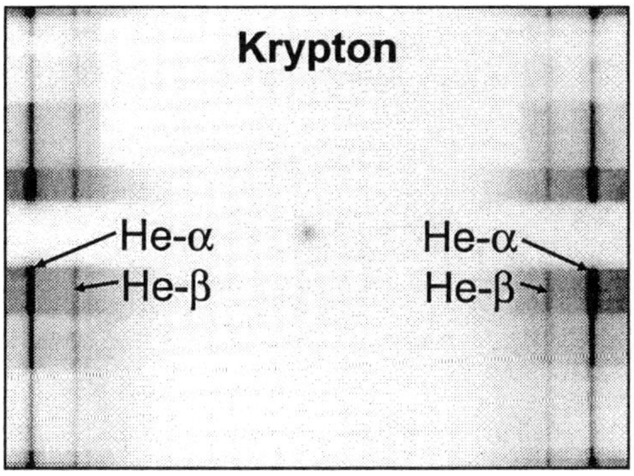

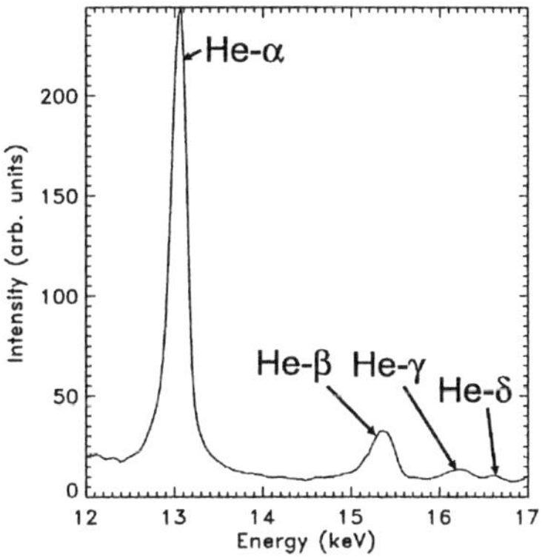

Figure 2. Left: The spectral image from a krypton filled CH shell target. Right: The lineout from the spectral image through the unfiltered region.

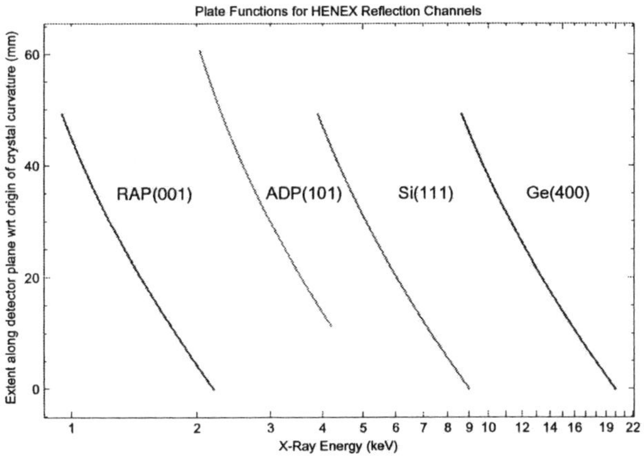

Figure 3. The four HENEX reflection crystals and their energy coverage.

HIGH ENERGY ELECTRONIC X-RAY SPECTROMETER (HENEX) FOR NIF

A National Ignition Facility (NIF) core diagnostic instrument is being fabricated to record x-ray spectra primarily in the 1.0 keV to 20 keV energy range.[3] The High-Energy Electronic X-Ray (HENEX) instrument has four reflection crystals with overlapping coverage of 1.0 keV to 20 keV with spectral resolving power from approximately 2000 at low energies to 300 at 20 keV. One transmission crystal covers 11 keV to 40 keV with lower resolving power. The spectrum produced by each crystal is recorded by a modified commercial CMOS detector with a phosphor screen and a dynamic range of at least

2500. The HENEX transmission crystal channel is similar to the one shown in Fig. 1. The four reflection crystals and their energy coverages are shown in Fig. 3.

SOFT X-RAY POWER DIAGNOSTIC (SXPD)

The soft x-ray power diagnostic fielded at the NRL NIKE laser is a two-channel spectrometer covering 50 eV to approximately 1 keV.[4] Each channel consists of a slit aperture, transmission grating, filter, and detector. One channel sends the spectral data to a streak camera, and the other channel to a CCD and overlying linear array of photodiodes as shown in Fig. 4. Two mirror spectra on either side of the zero diffraction order are produced. The diodes are absolutely calibrated at the Brookhaven synchrotron and have subnanosecond time response.

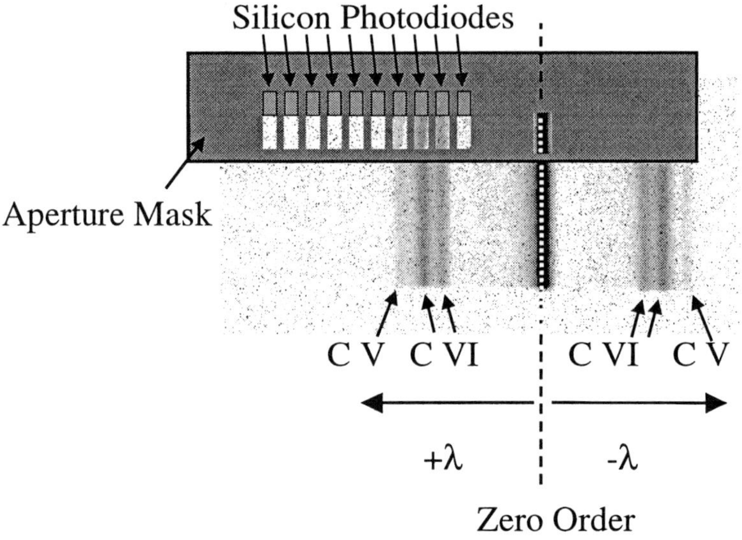

Figure 4. SXPD CCD spectral image and the overlying linear diode array.

REFERENCES

1. J. Seely, R. Deslattes, L. Hudson, G. Holland, R. Atkin, D. Meyerhofer, and C. Stoeckl, X-ray spectra in the 12-60 keV energy range from plasmas produced by the OMEGA laser, in *Applications of X Rays Generated from Lasers and Other Bright Sources II*, ed. by G. Kyrala and J.-C. Gauthier, SPIE vol. **4504** (in press).
2. R. D. Deslattes, J. C. Levin, M. D. Walker, and A. Henins, Noninvasive high-voltage measurement in mammography by crystal diffraction spectrometry," *Med. Phys.* **21**, 123 (1994).
3. J. F. Seely, C. Back, R. Deslattes, L. Hudson, G. Holland, P. Bell, and M. Miller, Hard x-ray spectrometers for the National Ignition Facility, *Rev. Sci. Instr.* **72**, 2562 (2001).
4. J. Weaver, G. Holland, U. Feldman, J. Seely, C. Brown, V. Serlin, A. Deniz, and M. Klapisch, The determination of absolutely calibrated spectra from laser produced plasmas using a transmission grating spectrometer at the NIKE laser facility, *Rev. Sci. Instr.* **72**, 108 (2001).

INTEGRATED SPECTROSCOPIC EDGE DIAGNOSTICS ON COMPASS-D

N. J. Conway, P. G. Carolan, A. R. Field, H. Meyer and M. J. Walsh*

INTRODUCTION

The investigation of edge physics in tokamaks often requires high spatial resolution, for example: transport barrier scale lengths in COMPASS-D H-modes are ~1cm. With this comes a requirement for accurate inter-diagnostic spatial mapping, since alignment uncertainties can make it very difficult to combine data from different sources. This is especially so when dealing with poloidally separated regions, but when working at the highest resolution, even toroidal symmetry may not be adequately preserved.

On COMPASS-D we have addressed this problem by combining the visible light collection for two separate spectroscopic diagnostics into a single assembly[1]. "HELIOS" (measuring HeI line ratios and Balmer D_{α} light) and "CELESTE" (HeII Doppler measurements) share a collection lens with a poloidal view through a thermal helium jet, and have optical fibres located in the same ferrule (see Figure 1). The extreme proximity of the collection fibres and their respective collection volumes in the plasma ensures a near-perfect mapping, permitting the combination of the data to yield the radial electric field E(r). This greatly enhances studies of H-mode dynamics. A further benefit of the combined collection system is that only one viewing port is required on the tokamak.

The diagnostics are used for H-mode transition studies, for which good time resolution is essential. The use of a localised source ensures excellent signal strength, allowing high-speed operation, and also means that no spatial inversion is required.

DIAGNOSTIC DESCRIPTION

A schematic of the combined diagnostics is shown in Figure 1.

A nozzle provides a collimated jet of thermal helium atoms (influx ~$10^{19}s^{-1}$), and the emission plume is imaged onto the combined fibre array. HELIOS has 10 lines of sight each with a single fibre. CELESTE has 19 lines of sight each with a fibre pair and a 20th

* N. J. Conway, P. G. Carolan, A. R. Field, H. Meyer, EURATOM/UKAEA Fusion Association, Culham Science Centre, Abingdon, OXON, OX14 3DB, UK. M. J. Walsh, Walsh Scientific, Culham Science Centre, Abingdon, OXON, OX14 3EB, UK.

Advanced Diagnostics for Magnetic and Inertial Fusion, Edited by Stott *et al.*
Kluwer Academic/Plenum Publishers, New York, 2002

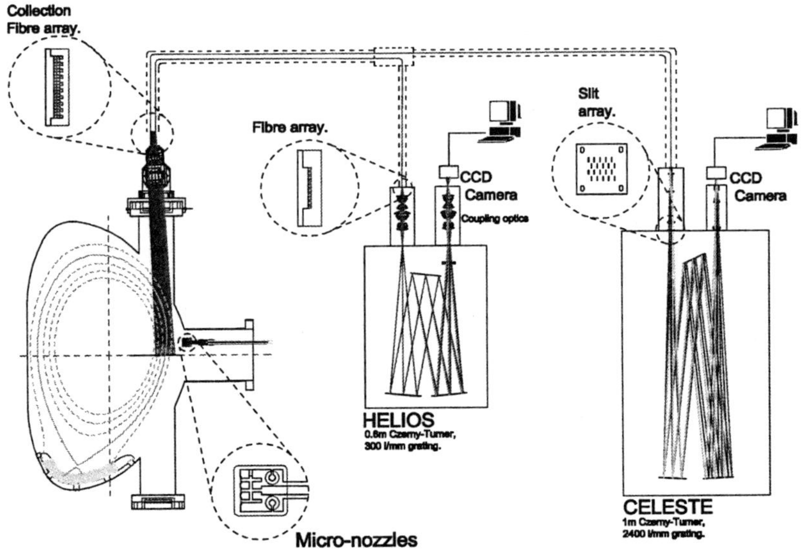

Figure 1. Schematic of the combined diagnostics

fibre pair coupled to a spectral lamp providing a wavelength reference (permitting absolute velocity measurement). All fibres are 400μm in diameter. The spatial resolution is ~2.5mm for CELESTE and ~5mm for HELIOS with total coverage ~50mm.

HELIOS records the intensities of three HeI emission lines (667.8, 706.5, 728.1nm) as well as D_α (656.1nm), with time resolution 1-5ms.

CELESTE measures the shape of the HeII resonance line at 468.6nm (other lines are occasionally used, e.g.: CIII 464.8nm) with time resolution 1-2ms, though the system can also be configured to run at extremely high speeds ~100μs with 10 lines of sight.

ANALYSIS DESCRIPTION

During the first stage of the analysis the diagnostics are treated separately. A transformation to normalised flux coordinates (ψ_N) is performed using EFIT[2], to remove the effect of changes in plasma size, shape and position.

HELIOS: two ratios are used – R1 (singlet-singlet 667.8/728.1) and R2 (singlet-triplet 728.1/706.5). The combination of singlet and triplet transitions is the key to the operation of the diagnostic (Fig. 2). This is because the triplet states are primarily excited from the singlet states' ground state by resonant spin exchange with background electrons while excitations within the singlet states are provided by the electron energy continuum above the excitation energy. As a result, R1 is sensitive mostly to n_e and R2 to T_e. Using effective emission rates from the ADAS[3] database, profiles are obtained of n_e and T_e, and thus p_e. The D_α measurement is non-local, and so a spatial inversion is performed using a maximum-entropy algorithm to yield $D_\alpha(\psi)$, which is combined with the n_e and T_e profiles and a collisional-radiative (CR) model[4] to provide the neutral density $n_0(\psi)$.

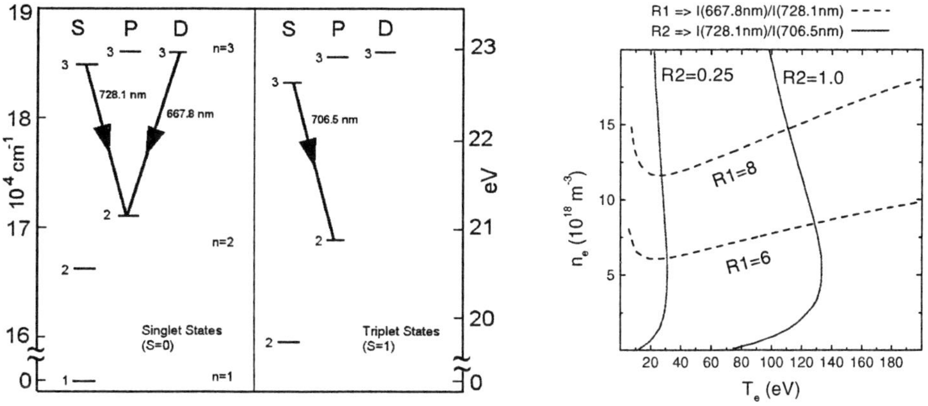

Figure 2. Term structure of the HeI transitions, and the variation of the line ratios with T_e and n_e.

CELESTE: the HeII line-shapes are analysed for intensity and Doppler shift and width, providing direct profiles of impurity ion poloidal velocity $v_{z\theta}$, temperature T_{zi} and intensity I_z. This completes the first stage.

In the second stage of the analysis the data-sets are combined. A simplified radial force balance equation (see ref. 5 for more detailed description) relates the radial electric field E_ψ to $v_{z\theta}$ and a diamagnetic term $v_{z,dia} \propto \nabla p_z/n_z$. By taking n_e, T_e, T_{zi} and I_z, in conjunction with a CR model, it is possible to calculate the $v_{z,dia}$ term and hence obtain the profile of E_ψ.

PHYSICS STUDIES

The diagnostics have principally been designed to study the dynamics of H-mode transitions, by measuring most of the important parameters connected with the edge transport barrier. The commonly understood reason for confinement improvement in H-mode is suppression of turbulent transport due to sheared ExB flows. Although there are many theories exhibiting H-mode-like solutions, none of them describes the transition satisfactorily.

A virtue of COMPASS-D is that many types of H-mode transition can be observed, such as slow transitions via a dithering phase, fast transitions and ECRH-assisted transitions. All of these have been studied, as well as the effect of ELMs on the transport barrier and the effect of different fuelling positions (inboard vs. outboard).

It was found that in high density (n_e ~60% of Greenwald density n_g) Ohmic SND discharges the radial electric field shear ∇E_ψ changes marginally or not at all prior to the L-H mode transition whereas large changes are observed after the transition. In these discharges the electric field shear and the electron pressure gradient evolve simultaneously and peak in the same spatial region, just inside the separatrix (ψ_N ~ 0.96). In contrast, lower density (n_e ~30% of n_g) ECRH-assisted H-modes show somewhat different electric field structure. Furthermore, the electron pressure gradient peaks further inward compared to the Ohmic case.

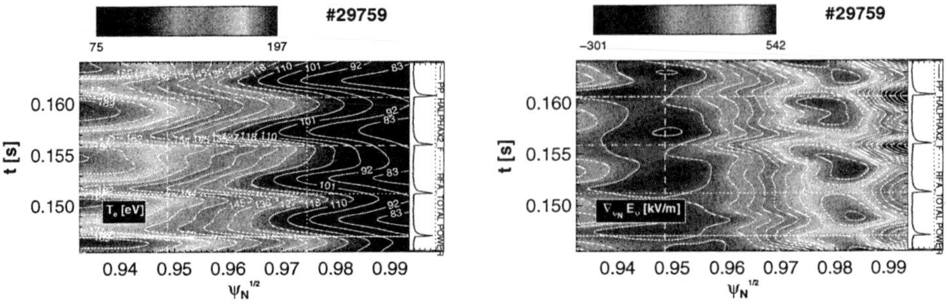

Figure 3. Evolution of T_e and ∇E_ψ profiles during an ELMing ECRH-heated discharge

Examples of T_e and ∇E_ψ profiles from an ELMing ECRH-heated discharge are shown in Figure 3. It can be seen that each ELM destroys the transport barrier, reducing both electric field shear and electron temperature over the whole of the observed region.

CONCLUSIONS AND FUTURE WORK

The integration of the HELIOS and CELESTE systems has been very valuable on COMPASS-D, and it is planned to adopt the same approach on the MAST tokamak for these and other diagnostics. Possible future enhancements to the system include: toroidal views; the operation of photomultiplier detectors in parallel with HELIOS to improve speed (CCD system to permit background subtraction); the exploitation of Zeeman splitting to measure magnetic field direction; the use of a lithium beam.

ACKNOWLEDGEMENTS

This work was funded jointly by the UK Department of Trade and Industry and EURATOM.

REFERENCES

1. P. G. Carolan, N. J. Conway, A. R. Field, P. B. Jones and H. F. Meyer, Synergy of multiviewing spectroscopic edge diagnostics on COMPASS-D, Rev. Sci. Instrum **72**, 881 (2001).
2. L. L. Lao *et al*, Reconstruction of current profile parameters and plasma shapes in tokamaks, Nuclear Fusion **25**(11), 1611 (1985).
3. H. P. Summers and M. von Hellermann, in *Atomic and Plasma-material Interaction Processes in Controlled Thermonuclear Fusion*, edited by R. K. Janev and H. W. Drawin (Elsevier Science Publishers B. V., Amsterdam, 1993).
4. K. Sawada, K. Eriguchi and T. Fujimoto, Hydrogen-atom spectroscopy of the ionizing plasma containing molecular hydrogen: Line intensities and ionization rate, J. Appl. Phys. **73**(12) 8122 (1993).
5. H. Meyer, P. G. Carolan, N. J. Conway. A. R. Field, S. J. Fielding and P. Helander, Profile evolution of edge parameters during L/H and H/L mode transitions on COMPASS-D, Czech J. Phys **50**(12), 1451 (2000).

CONCEPTUAL DESIGN OF THE CXRS SYSTEM FOR ITER-FEAT.

S.Tugarinov, M.vonHellermann, I.Beigman, V.Dokouka, R.Khayrutdinov, A.Krasilnikov, A.Malaquias, I.Tolstikhina, L.Vainshtein*.

1. INTRODUCTION.

The following report is the result of several iterations and discussions between the authors in the course of the last year. The originally apparent differences for the projected performance of a CXRS diagnostic for ITER could be successfully resolved by sorting out differences assumptions, choosing eventually optimised parameters for the future ITER geometry and instrumentation. At the same time, have extrapolations of existing CXRS diagnostic systems, e.g. JET, provided a firm guide-line for a realistic frame-work of expected performance parameters.

Active plasma spectroscopy, or spectroscopy based upon local signal enhancement via a penetrating atomic beam, is employed as an essential diagnostic on almost all present-day tokamaks. This technique is utilized for a wide variety of measurements in the plasma edge and core, including ion temperature (via Doppler broadening of intrinsic impurity lines, which are efficiently populated by charge exchange from beam atoms), plasma rotation (via Doppler shift of the same impurity lines), and impurity density profile information (via quantitative spectroscopy of the impurity line intensities). The feasibility of CXRS for ITER-FEAT relies on the extrapolation of data from present-day devices to the conditions expected in ITER-FEAT. Two fundamental limitations are considered in the following assessment. First is the detectability of a weak CX signal against a strong background of plasma continuum radiation, and the second, equally important, requirement is the accuracy by which local neutral beam densities can be established in order to derive absolute ion densities from the extracted active CX signals. The second problem can be approached either by experimental data from Beam Emission Spectroscopy (BES) or beam attenuation codes making use of electron and ion density profiles, includes both bulk and impurity ions.

* S.Tugarinov, V.Dokouka, R.Khayrutdinov, A.Krasilnikov, Troitsk Institute for Innovating and Fusion Research (TRINITI), Troitsk, Russia; I.Beigman, I.Tolstikhina, L.Vainshtein, FIAN, Moscow, Russia; M.vonHellermann, FOM Institute for Plasma Physics, Rijnhuizen; A.Malaquias, Garching JWS, ITER.

Advanced Diagnostics for Magnetic and Inertial Fusion, Edited by Stott *et al.*
Kluwer Academic/Plenum Publishers, New York, 2002

2. EMISSION MODELING AND BEAM ATTENUATION.

At the presence of neutral beam injection, intense charge exchange recombination reaction appear in plasma volume intersecting by the beam.

The equation, that describe this process is:

$$H^0 + A^{z+} \Rightarrow H^+ + A^{*(z-1)+} \qquad (1),$$

where is A^{z+} is a fully stripped ion and $A^{*(z-1)+}$ is a hydrogen-like ions in an excited state, that subsequently decays by prompt emission.

The effective emission rate is determined by the actual charge capture cross-section and the subsequent redistribution processes following cascading from higher n-shells and collisional redistribution. The equation for emission rate simulation is:

$$v\sigma_{n-n'} = \frac{A_{n-n'}}{A_n} v \sum_{m=0}^{N} c_{mn}\sigma_{n+m} \qquad (2),$$

where is : $A_{n-n'}$ – probability for (n-n') transitions; A_n – probability for radiation decay from level (n); σ_{n+m} – charge-exchange cross-section to the (n+m) level; C_{mn} – Sitons matrix elements.

The results of the emission rate simulation was obtained at FIAN Institute of Moscow[1] is in a very good agreement with calculation by the ADAS code[2,3] routines taking into account the plasma environment and also contributions from excited neutrals. The comparison of obtained emission rate values for 100 keV neutral beam shown in a Table.1

At the presence of neutral beam injection extra bulk neutrals appear inside the plasma beam intersected volume, due to charge exchange with bulk ions. This "extra neutrals" called – "halo". Halo atoms contribute to charge exchange with impurity nuclei. Results of the "total emission rates coefficient" calculation including "halo" radiation due to cascade charge exchange, versus ion temperature presented in a Table 1:

Table 1. Total emission rates [10^{-8} cm^3/sec]

Ions	*Trans.*	$v\sigma_d$, (FIAN)	$v\sigma_d$, (ADAS)	$v\sigma_t$, 10 [keV/amu]	20	25
He^+	4-3	0.094	0.088	0.11	0.13	0.15
Be^{3+}	6-5	0.338	0.354	0.34	0.34	0.37
C^{5+}	8-7	0.609	0.688	0.61	0.61	0.62

Column 3 and 4 in a Table 1 is an emission rates without "halo" effect. Column 5-7 is a total emission rates for various plasma ion temperature values. Conclusion from Table 1 data is following: "halo" effect important for HeII ion only and influence from "halo" atoms increase simultaneously with plasma ion temperature.

Simulation of CXRS signal from the DNB has been made including the beam attenuation processes by ionization due to electron and ion impact and charge exchange with the plasma ions and impurities. Also the multistep contribution, taking into account the excited level of the beam atoms, was included.

The beam attenuation along the elementary path dr is determined by the formula:

$$dP_0(r) = -\sigma_s{}^{(N)} n_e P_0(r) dr \qquad (3),$$

where is, σ_s^N is determined by the formula (4) from reference[4] and $P_0(r)$ is power density at the beam axis.

$$\sigma_s{}^{(N)} = \frac{\exp S_1(E_0, n_e, T_e)}{E_0}\left[1 + \sum_{q=1}^{N} c_q z_q (z_q - 1) S_{z_q}(E_0, n_e, T_e)\right] \qquad (4),$$

If DNB modulation and "lock-in" detection are used, the fundamental limitation for experimental measurement is the value of continuum emission fluctuation. Thus, the signal to noise ratio can be calculated by formula from reference[5].

$$SNR = \frac{I'_{cx}\sqrt{\tau}}{\sqrt{2(2I'_{ff} + I'_{cx})}} \qquad (5),$$

where τ - integration time, I'_{cx} and I'_{ff} is a number of photoelectrons for CX and continuum signal, respectively.

Taking into account CX spectral line and continuum intensity integrating along viewchord, equation (7) can be rewritten as :

$$SNR = \left[\frac{\Omega_\alpha S^0_{cx}(r_\alpha)\delta r \Delta z \delta t \eta \tau}{2\left(2\frac{S_{ff}(r_\alpha)}{S^0_{cx}(r_\alpha)} \cdot \frac{\Delta p_\alpha \cdot \sin\alpha}{G(\sigma)\Delta z} + 1\right)}\right]^{1/2} \qquad (6),$$

Coefficient η takes into account: losses connected with transparency and reflection coefficients of the optical elements, light flux limitation by the entrance slit size and efficiency of detection system.

We plan to use vertical viewchord observation and horizontal DNB injection, along major radius. For DNB attenuation and SNR simulation, we assume flat electron density profile with average values 7.5, 10 and 13 x 10^{13} cm^{-3}, cases 1,2,3 at Fig.1-4, respectively. We assume also, constant plasma impurity composition: He=4%, Be=2%, C=1.2%, which is correspond to Z_{eff}=1.7 and integration time τ =0.1 s.

DNB has a cross section of 30x30 cm^2, current density at the axis 60 mA/cm^2 and 100 keV atoms energy.

4. **RESULTS AND CONCLUSIONS.**

The profiles of DNB attenuation and SNR value for CVI, BeIV and HeII ions shown at Fig.1-4. The most important conclusion, which we can make from the results of our modeling simulation, is following. A workable CXRS diagnostic based on the present DNB parameters appears to be feasible with acceptable signal levels. Recent optimisations of viewing geometry and spectroscopic equipment have led to more favourable numbers, which are compatible with extrapolations from existing CXRS diagnostics of today's fusion devices. Fundamental limitations at high plasma densities introduced by non-linear error propagation in neutral beam calculations can be counter acted by further progress and application of BES (beam emission spectroscopy). More importantly, the helium ash measurement at ITER-FEAT by CXRS must be seen as a comprehensive diagnostic tool encompassing all of the main low-Z impurity ions.

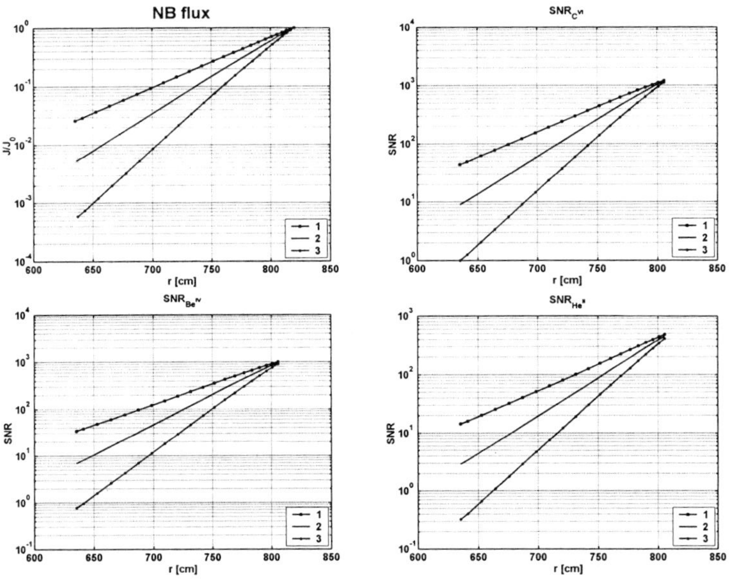

Fig. 1-4. DNB attenuation and SNR profiles for CVI, BeIV and HeII ions.

REFERENCES :

1. S.Tugarinov et al., 14th Meeting of the ITER Physics Expert Group on Diagnostics Jülich, Germany, March 2001.

2. M.G. von Hellermann et al. Plasma Physics and Controlled Fusion, 33, 1805 (1991).

3. M.G. von Hellermann et al., 1st International Workshop on ITER Diagnostics, Varenna, Sept. 1995 Plenum Press 1995, JET-P (95) 63.

4. R.K.Janev, C.D.Boley, D.E.Post, Nuclear Fusion, Vol. 29, No. 12 (1989)

5. M.G. von Hellermann and H.P. Summers, 1993, in "Atomic and Plasma-Material Interaction Processes in Controlled Thermonuclear Fusion".

X-RAY BACKLIGHTING OF CURRENT - DRIVEN MULTI WIRE ARRAY ON ANGARA-5-1 FACILITY

Alexandrov V.V., Grabovsky E.V., Mitrofanov K.N., Oleinik G.M., Porofeev I.Ju., Smirnov V.P., Volkov G.S.*

INTRODUCTION

Z pinches and tungsten wire arrays, compressing under action of current, are widely studied in ICF nowadays[1,2,3]. The wires have typical diameter 5-10 microns. Duration of output soft x-ray emission about 5-10ns. Spatial distribution of mass inside current - driven array during implosion has crucial importance for process of array implosion.

One of capabilities to carry out such research is to backlight[4,5] the contracting array by a separate point - like soft x-ray source and to record obtained shadow on a film. For adequate temporal resolution of such technique the backlighting x-ray pulse should have a short duration. Energy region of x-ray quanta for backlighting of a wire array, compressing under action of current, on Angara-5-1 facility should be 2-10keV. The backlighting source should be sufficiently bright in this spectral range, more bright, than investigated compressed array. Spatial resolution of technique is determined by size of backlighting source.

EXPERIMENTAL ARRANGEMENT

The experiment arrangement is presented in Fig.1.

X-pinch was used as x-ray backlighting source. In our experiments X-pinch was produced by two or more crossed molybdenum wires with diameter of 20μm. These wires have electric contact in crossing point in the middle of wires. The crossing point of wires under action of current pulse of ~300kA flashes up brightly in x-ray region.

* Alexandrov V.V., Grabovsky E.V., Mitrofanov K.N., Oleinik G.M., Porofeev I.Ju., Smirnov V.P., Volkov G.S., GSC RF TRINITI, 142190, Troitsk, Moscow reg., Russia.

Advanced Diagnostics for Magnetic and Inertial Fusion, Edited by Stott *et al.*
Kluwer Academic/Plenum Publishers, New York, 2002

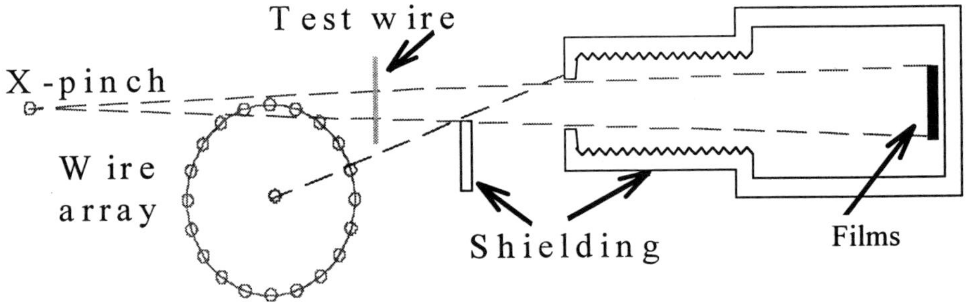

Figure 1. Experimental arrangement

X-pinch was located near the array and was synchronized with driving current pulse through investigated array. In order to satisfy these conditions one of 8 current - return posts (at distance of 45mm from the liner axis) was replaced by X-pinch. Current through the X-pinch was 300-400kA (about 1/8 of total driving current).

Shadow image was recorded on two types of films (УФ-ШС and РФ-3) placed closely in series on each other. To protect films from direct and scattered radiation of Z-pinch shielding system was installed between Z-pinch and films. To protect films from x-rays, which appears in the moment of final pinching, and from a flux of microparticles, which appears after the moment of final pinching, the films were protected by titanium foil of 16μm thickness. Step attenuator from this of a material, as liner (tungsten in our case) was situated in front of film for opportunity of transition from photometric density to density of substance.

A silicon p-i-n detector behind a titanium 16μm filter was used to measure a moment and duration of X-pinch radiation in spectrum range of 3-5keV. Energy response of the detector with a filter is presented in Fig.2.

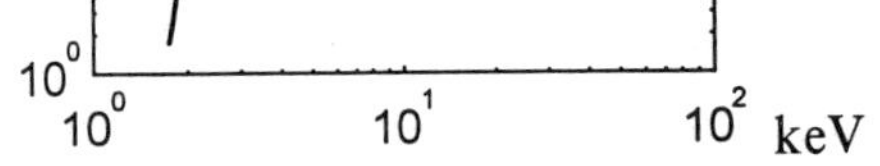

Figure 2. Response of a silicon p-i-n detector behind a titanium 16μm filter.

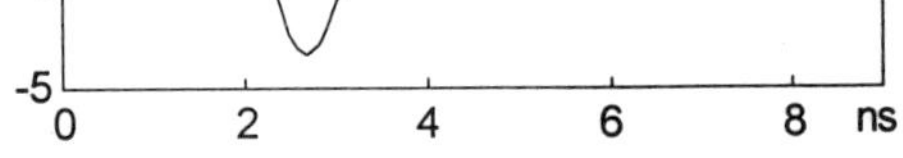

Figure 3. Temporal profile of X-pinch radiation, recorded by a silicon p-i-n detector behind a titanium 16μm filter.

Temporal profile of X-pinch radiation is presented in Fig.3. The FWHM of a probing x-ray pulse of X-pinch recorded in quanta energy region 3-5 keV does not exceed 2 ns.

To check spatial resolution of technique and to determine the size of x-ray source the test wires (diameter of 5μm) were set near the array. Magnification of array wires image was varied from 10 to 25; for test wires - from 5 to 10. The shadows of test wires and shadows of current driven array were obtained on same film simultaneously.

EXPERIMENTAL RESULTS

Presented technique was used to investigate the implosion of multiwire arrays. X-ray probing pulse was produced in 50-90 ns after driving current start. Duration of array implosion was 100-110ns. Typical profile of driving current, soft x- ray radiation of array, and x- ray radiation of X-pinch are presented in Fig.4.

The part of wire array shadow images (vertical dark strips) and shadows of test wires (inclined relatively to vertical dark strips) obtained by x-ray probing and recorded in 60 ns after current start are shown in Fig.5. By means of this measurements it was verified that X-ray source size of X-pinch for selected quanta energy region does not exceed 2 microns. The spatial resolution of technique evaluated at wire position is ~ 1.7 microns.

The analysis of obtained experimental data of x-ray backlighting shows that in 50-60 ns after current start the wires are recorded as the object with diameter up to 20 μm. The evaluation of expansion velocity of dense cores gives the value $\sim 10^4$cm/s. In 50-60ns after a beginning of a current through a load, the diameter of wires has increased in 3 times. Such effect was observed both for wires with diameter of 5μm at level of current ~30kA/wire and for wires with diameter of 8mm at level of current ~50kA/wire.

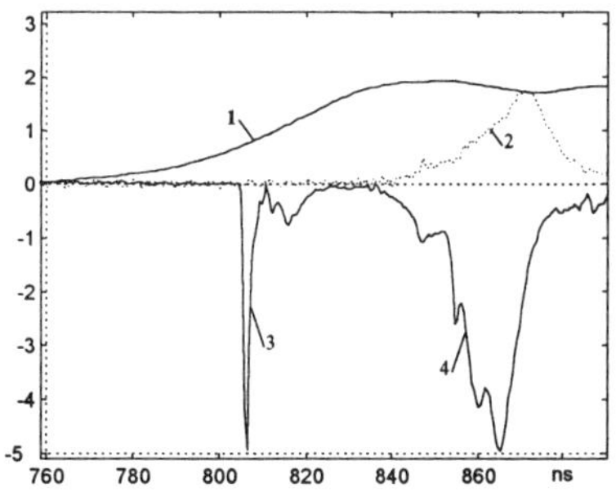

Figure 4. Temporal profiles of: 1 - current [2MA/div], 2 - soft x-ray radiation power [2TW/div] of wire array (20mm diameter, consisted of 60 tungsten wires of ∅5μm), 3,4 - radiation power [10MW/div], registered by silicon p-i-n detector behind a titanium 16μm filter in range of 3-5keV (3 - of X-pinch radiation, 4 - of Z-pinch radiation).

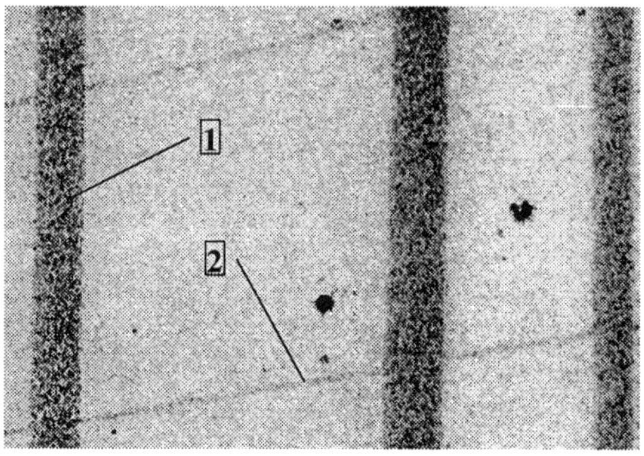

Figure 5. The shadow images: 1 - wires of the current driven array in 60ns after current start, 2 - test wires (∅6.5μm). The initial diameter of array wires 5μm. The diameter of wires of the array at the moment of a of the X-pinch flash 20μm.

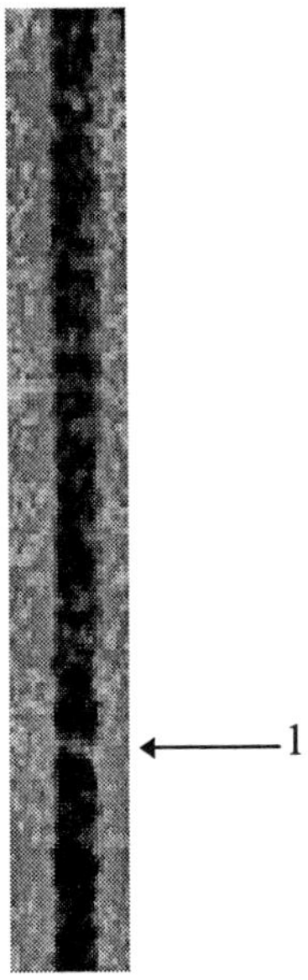

Figure 6. An image of a wire with axial inhomogeneity. 1 - one of points, in which wire seems breakup.

At some images the axial spatial non-uniformity of dense substance of wires with spatial scale of 5-15μm is remarkable (see Fig.6).

The breakup of wire density were recorded at some wire array images. In breakup points substance density is smaller and possibly «hot spots» may be generated in these places. From the other hand, the scale of a non-uniformity of a dense cores does not coincide with the scale of axial non-uniformity of wire surrounding low density plasma with size 250μm presented, for instance, in [6].

CONCLUSION

As the result of investigation of current - compressed multiwire array was found that in 50-60ns after current start at current level of ~30kA/wire the initial diameter of current - driven wires was increased from 6μm to 20μm. The velocity of dense substance cores expansion is ~ 10^4 cm/s. Internal wires structure was observed. The spatial non-uniformity of dense substance of wires with scale of 5-15μm was recorded.

This work was supported in part by grand of RFBR # 01-02-17319.

REFERENCES

1 Grabovskii E.V., Smirnov V.P., Zakharov S.V., Vorob'ev O.Yu., Dyabilin K.S., Lebedev M.E., Fortov V.E., Frolov A.A. JETP. 82(3), 445 (1996).

2 R. B. Spielman, C. Deeney, G.A. Chandler et.al. Dense Z pinches: 4th International conference, Vancouver, Canada May 1997 edited by N. R. Pereira, J. Davis, and P.E. Pulsifer, p. 101.

3 M. G. Haines, S. V. Lebedev, J.P. Chittenden et.al Phys. Plasmas 7, 1672 (2000).

4 D. N. Kalantar, D. A. Hammer, Phys. Rev. Lett. 71, 3806, 1993.

5 I. Yu. Skobelev, S. A. Pikuz, A. Yu. Faenov et.al., JETP 81, 692, 1995.

6 V.V.Aleksandrov, A.V.Branitskii, E.V.Grabovskii, et al. Plasma Physics Reports, Vol. 27, No. 2, 2001, pp. 89–109. Translated from Fizika Plazmy, Vol. 27, No. 2, 2001, pp. 99–120.

PRELIMINARY RESULTS OF CHARGE EXCHANGE q MEASUREMENTS IN T-10 TOKAMAK

Alexander A.Medvedev, Vyacheslav S.Strelkov
Nuclear Fusion Institute, RRC KI, Moscow, RF

1. INTRODUCTION

The great attention in tokamak research is paid to study of the magnetic field structure now. The MSE method is used for q measurements on most large tokamaks. Numerical analysis and a series of preliminary experiments have demonstrated that for T-10 conditions - vertical beam and short (1 ms) duration of the diagnostic beam pulse - use of the MSE is inefficient. We consider a charge exchange technique described below as an alternative method.

A theoretical possibility of the charge exchange q measurements was mentioned rather long ago[4]. Preliminary experiments using some other approaches, were performed at the TUMAN[5] and at the TEXT[6] tokamaks.

2. PHYSICAL BASIS OF THE TECHNIQUE

Technique is based upon a beam of energetic (tens keV) hydrogen atoms injected into plasma vertically through the center of the plasma column (Fig.1). The beam axis is nearly perpendicular to all magnetic flux surfaces at this geometry.

Advanced Diagnostics for Magnetic and Inertial Fusion, Edited by Stott *et al.*
Kluwer Academic/Plenum Publishers, New York, 2002

Atoms of the beam are ionized due to interactions with plasma particles. The originated ions safe the energy and momentum of initial beam atoms since the energy transfer in relevant atomic processes is small.

The fast ions will be captured by the magnetic field and move along spiral trajectories, pitch of that depends on the ion toroidal velocity component. If the ripple depth is large enough, the ions will be local trapped. The nondegenerate ripple exist in many of T-10's performance regimes[7].

A part of the trapped ion's trajectory lies inside the beam volume and the ion can be charge exchanged with a beam atom. The generated secondary charge exchange atom will escape plasma along a straight trajectory almost at right angle to the magnetic field direction. Therefore, one can get information about the magnetic field direction in a plasma region, measuring the prevailing angle of the secondary charge exchange atoms' escape from there (Fig.1).

3. RESULTS OF PREIMINARY EXPERIMENTS ON T-10

We use the DINA-6 diagnostic injector for the experiments at T-10. Injector has the following main parameters: energy of atoms (main component) – 30 keV, total equivalent beam current – 2.2 A, effective diameter of the beam – 5.5 cm, duration of the injection – 1 ms, and effective divergence – 0.9-1.1°.

The layout of the preliminary experiments that were carried out at T-10 is shown in Fig.2. A silicon detector was used for the measurements. The surface of the detector was covered by 200 nm aluminum film to reduce the background due to plasma light and low energy atomic flux. The detector was arranged by a collimator (vertical aperture - 1.2°, horizontal - 0.5°) and installed at a gimbal that enabled to scan plasma in vertical and horizontal directions.

A first series of the experiments were carried out with the hydrogen filled vessel ($P_H=2\cdot10^{-4}$ torr) at a toroidal magnetic field of 1.5 T. The flux of secondary charge exchange atoms with density of about $1\text{-}2\cdot10^{8}$ $cm^{-2}\cdot s^{-1}$ was detected. The signal vanished if the magnetic field was turned off.

In plasma experiments specific fluxes of $3\text{-}9\cdot10^{7}$ $cm^{-2}\cdot s^{-1}$ were measured. The dependence of the signal against detector's horizontal angle is exhibited in Fig.3.

Results of measurements performed with detector rotating round a remote axis do not enable reliable deducing of the magnetic pitch angle, unfortunately. Firstly, as simulations show, a position of the maximum of the measured angle distribution depends upon the ripple depth and the angle distribution of initial fast trapped ions, even at homogeneous charge exchange target. And secondly, the neutral target density in the observation region changes due to moving the region across the beam at the horizontal scanning.

However, the performed numerical modeling shows that direct pitch angle measurements can be done by scanning round the beam axis.

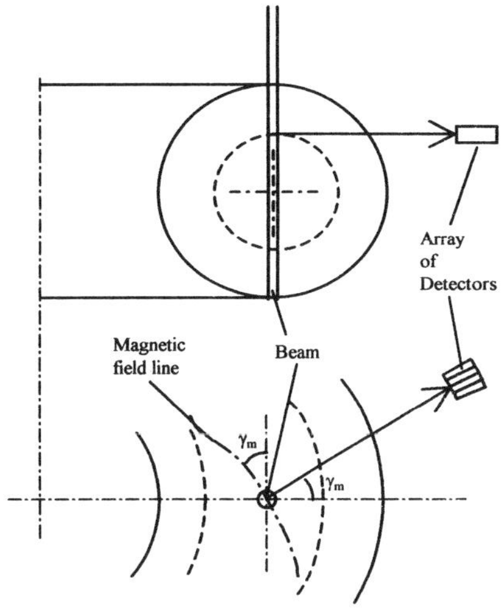

Fig.1 Algorithm of magnetic pitch angle γ_m determination from angle distribution of secondary charge exchange atoms' flux.

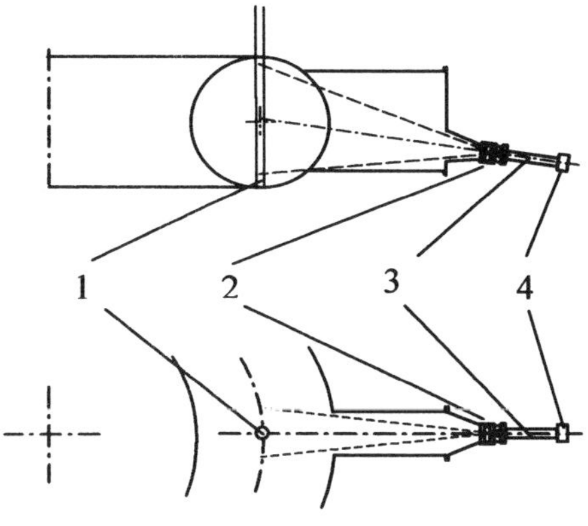

Fig.2 Scheme of preliminary measurements.
1 – beam;
2 – gimbal;
3 – collimator:
4 – detector.

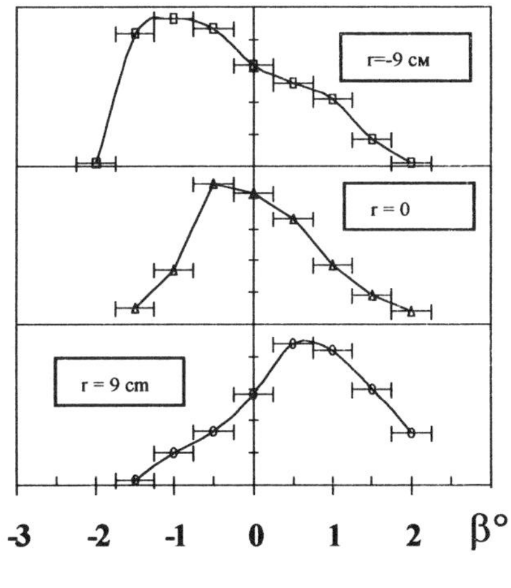

Fig.3 Measured flux of secondary charge exchange atoms versus horizontal angle of the detector.

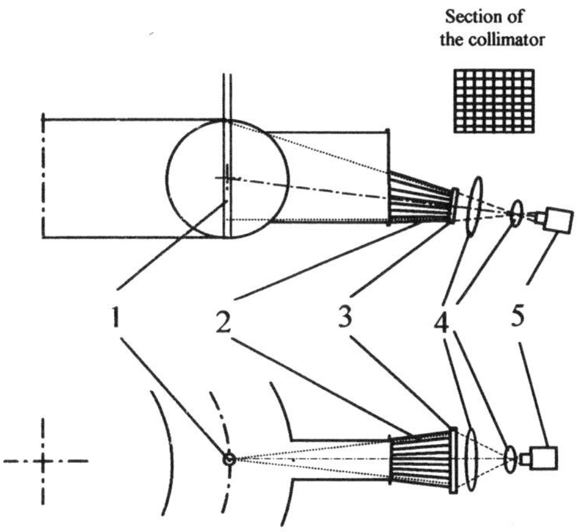

Fig.4 Layout of multi-chord q measurements for T-10.
1 – beam;
2 – collimator;
3 – scintillation converter;
4 – optical system;
5 – ICCD detector.

4. APPARATUS FOR MULTI-CHORD q MEASUREMENTS IN T-10

The conception of apparatus for multi-chord q measurements in T-10 (Fig.4) was developed on the basis of the preliminary measurements.

A multi-channel collimator is installed at a vertical diagnostic flange of the machine. A crosscut section of the collimator looks like a matrix, horizontal 'lines' of which serve for angle separation of fast atoms, and vertical 'rows' - for space resolution along the beam.

The scintillation converter is installed beyond the collimator. It is a glass plate covered by a thin layer of scintillator. To reduce background related to plasma light and low energy atomic flux the scintillator is shielded by a thin aluminum film. The image of the scintillation screen is projected by a simple optical system onto the photocathode of an ICCD detector.

The numerical modeling shows that the following main parameters of the measurements could be achieved: number of radial points - 20, space resolution 1.4 cm, statistical accuracy - $5 \cdot 10^{-2}$, absolute error of the magnetic pitch angle value – 0.3-0.8°.

5. CONCLUSION

A diagnostic technique based upon detection of secondary charge exchange atoms was proposed for q measurements at T-10.

A numerical modeling of underlying processes and a series of preliminary experiments have been performed. Experimental and simulation data are in a good agreement. Large specific flux of the secondary atoms enables to perform measurements with high space (~1 cm) and temporal (~10^{-4} s) resolution. Rather detailed measurements of q(r) profile can be done by means of relatively simple apparatus.

A conception and a composition of apparatus for multi-chord measurements of q(r) profile in T-10 have been developed.

At present we analyze a possibility to use the technique in ITER relevant conditions.

The work was supported by the Russian Fund of Fundamental Investigations (Grant No.96-02-18547).

References

1. F.Levinton et al, Rev. Sci. Instr., 61 (1990) p.2914.
2. R.C.Wolf, JET Report JET-IR (93) 08.
3. V.A.Krupin, S.N.Ivanov, A.A.Medvedev, T.B.Myalton, V.S.Strelkov, E.A.Chaklin, Preprint of RRC KI IAE-59 40/7, 1995 (in Russian).
4. F.C.Jobes. 2-nd Topical Conf. on High-Temperature Plasma Diagnostics, Santa Fe, 1978, LA-7160-C, p.101.
5. V.I.Afanasjev, S.L.Lebedev, F.N.Chernyshev, S.Ya.Petrov, A.I.Kislyakov, K.G.Shakhovets, Preprint of Ioffe Physical-Technical Institute 1406, 1989.
6. P.M.Valanju, L.Duraiappah, Roger D.Bengtson, Y.Karzhavin, and A.Nikitin, Rev. of Sci. Instrum, **66** (1) 1995, p.369.
7. Yu.V.Gribov, S.V.Tsaun, P.N.Yushmanov, Preprint of IAE, IAE-3681/71982, 1981 (in Russian).

THERMAL SOFT X-RAYS INTERACTION WITH FOAM LAYERED GOLD TARGETS

D.Batani[1], T.Desai[1], Th.Löwer[2], T.A.Hall[3], W.Nazarov[4], M.Koenig[5], A.Benuzzi-Mounaix[5].

[1]Dipartimento di Fisica "G.Occhialini" and INFM, Università di Milano-Bicocca, P.za della Scienza 3, 20126 Milano, Italy

[2]Max Planck Institut fur Quantenoptik, D-85740 Garching, Germany

[3]University of Essex, Wivenhoe Park, Colchester, Essex, CO4 3SQ, UK

[4]Department of Chemistry, University of Dundee, Scotland, UK.

[5]LULI, UMR n. 7605 CEA-X-Paris VI, Ecole Polytechnique, 91128 Palaiseau, France

The purpose of the experiment presented in this paper was to study the effect of low density foams on the generation of shocks driven by thermal soft X-rays because of their applications in Inertial Confinement Fusion (ICF)[1-4]. Experiments were performed at MPQ, Garching using Asterix laser energy up to 400 J at 0.44 μm (3ω of Iodine laser) in ≈ 450 ps (FWHM). A conical shaped shield was constructed inside the Labyrinth cavity[5] so that the laser irradiated area and the shocked material are not in direct view of each other (Fig. 1). Cavity surface area was equivalent to a sphere surface area of radius 0.5 mm. Target specifications are, 1) Aluminium base of thickness 16.95 μm with Al step of thickness 6.23 μm. 2) gold base of thickness 4.96 μm with a 1.95 μm gold step. The front surface of the gold foil was layered with low density foam (TMPTA-trimethylol propane triacrylate, $C_{15}H_{20}O_6$) of 50 μm thickness, ρ =20 mg/cc, and also 50, 100 and 150 μm thick with ρ = 50 mg/cc). The areal densities were d = 0.1, 0.25, 0.5 and 0.75 mg/cm^2 respectively. A visible streak camera recorded the shock breakout signal.

The shock transit time provides the measure of shock velocity through the target rear surface. Shock pressures for Al and gold targets were evaluated according to Sesame report[6] and empirical scaling law[7] respectively. Although recent EOS measurements for

Advanced Diagnostics for Magnetic and Inertial Fusion, Edited by Stott *et al.*
Kluwer Academic/Plenum Publishers, New York, 2002

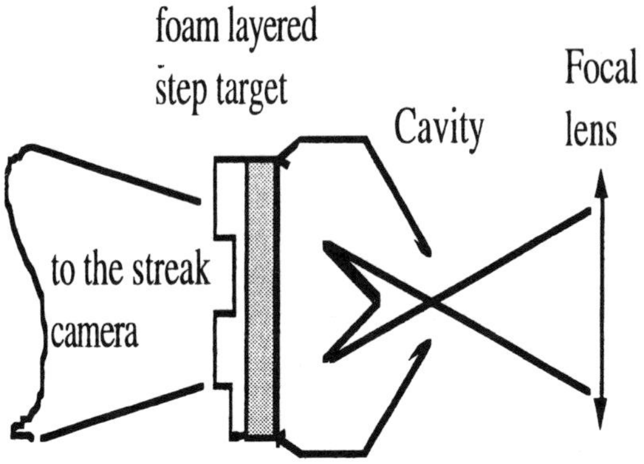

Fig. 1: Experimental set-up showing the labyrinth cavity and the schematic of the target design.

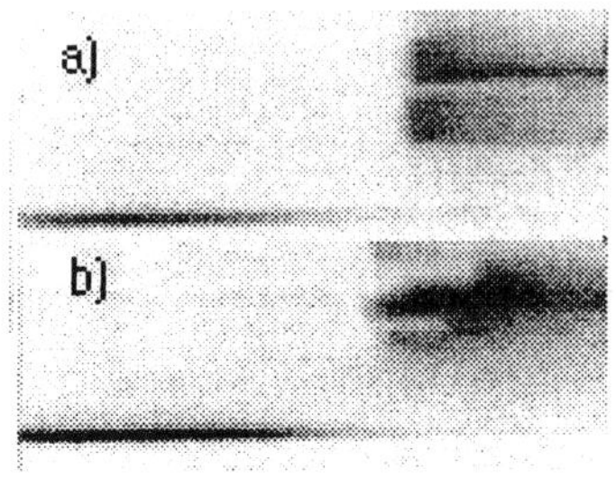

Fig. 2: Streak camera images of a) Al target b) foam-Au target

gold[8] suggest that Sesame is not correct at high pressure, the change in the pressure values (10 %) do not affect our conclusions.

Fig. 2a shows the streak camera image obtained with an Al target. In this case, the shock velocity in the Al step was D= 6.23 μm / 259 ps = 24.1 km/s which corresponds to a shock pressure 9.6 Mbars according to the sesame table. Pressure induced by the X-rays in a target do not depend on the target material. Therefore, shock pressure induced in a gold target is P_o = 9.6 Mbar. The signal on the left is a time fiducial. The time interval between the arrival of laser peak and the shock breakout on target rear side is measured from the streak image. Fig. 2b shows the shock breakout from foam/gold target (ρ = 50 mg/cc and d = 50 μm). In this case the shock velocity is 14.1 km/s which corresponding to an average pressure of 20.2 Mbars.

Fig. 3 shows the pressure enhancement (P/P_o) in foam/Au targets which increases with increasing ρd. Fig. 3 also shows, the time delay Δt, is larger for larger values of d. The time delay plotted here corresponds to the foam d only, i.e. we have subtracted the shock transit time (d/D) in the metal thickness from the original values of the delay time. Using a simple error propagation evaluation, and including reading errors, a maximum error of ± 7 % on D, implies an error of ± 15 % on shock pressure.

In indirect drive, the radiation temperature is given by Stefan Boltzman's law, $T = (\eta E_L / \sigma 4R_c^2 t_L)^{1/4}$ where is $\sigma = 10^5$ when T is in eV, E_L in J, R_c in cm, and t_L in seconds. For E_L = 400J, we get T = 120 eV in good agreement with direct experimental measurements5. The conversion efficiency η = 80 % for gold. The cavity temperature corresponds to a peak X-ray wavelength = 0.25/T(eV) = 20 Å according to Wien's displacement law. When such radiation impinges on targets (without foam) we can estimate the pressure[9,10]:

$$P\,(\text{Mbar}) = 44\,(\eta E_L / 4R_c^2\, t_x)^{10/13}\, t^{-3/26} \qquad (1)$$

where t is in nsec and the quantity in bracket is in the units of 10^{14} W/cm^2. Since there is no explicit dependence of pressure on the material we compare the pressure obtained with Al and foam/Au targets. In fig. 3, d=0 corresponds to the values for the aluminium target without foam layer while shock pressure and shock transit time increases with increasing d for foam/Au targets. According to eq. (1), P = 11 Mbars for E_L =300J which is in fair agreement with our results on Al targets (P_o = 9.6 Mbars).

What is the origin of the pressure increase with foam layer in foam/Au target? Let's first notice that the foams are clearly undercritical to the thermal X-ray radiation. However the foam is initially made of cold atoms which strongly absorb X-rays by bound-bound transitions. Transparency of cold TMPTA foam was estimated using the X-ray data from ref. 11 and shown in fig. 4. Part of X-rays are absorbed in the foam, ionising it. As ionisation goes on, the absorption coefficient reduces (for completely ionised foams it is only due to bremsstrahlung and recombination and is much smaller). We evaluate the total ionisation energy spent in the foam material as ΔE = 0.3 J using an average atom model[12] for 50 mg/cc and d =100 μm. This is compared with the fraction of the total X-ray energy (ηE_L) impinging on the cavity hole (= 12.8 J) which shows that the ionisation losses are negligible.

Two scenarios are then possible depending on foam density. If this is very low, an ionisation wave propagates supersonically in the foam. At the same time, hard X-rays (as shown in fig. 4) penetrate to the Au-foam interface and produce shock pressure. A similar behaviour of X-rays has been reported[8]. However, now the ablating gold plasma is confined due to the presence of foam, the plasma energy which would have been spent in the motion of the plasma, is retained as thermal energy of the confined plasma. Thus the total pressure is due to ablation of gold surface by (harder) X-rays and also due to gold plasma confinement (because of the presence of foam). An enhanced pressure P due to confinement of the plasma has been reported[13]. The pressure increment should be given by the material pressure of the foam plasma $P = (Z+1)\, n_i\, kT$, where k is Boltzman's constant and the average ion density is $n_i = \rho\ N_A / A$. In the case of 20 mg/cc and T = 120 eV we find ΔP = 1.5 Mbars, against an experimental value ΔP =1.4 Mbar (or P/P_o = 1.1, where P = 11 and P_o= 9,6 Mbar.).

At higher foam densities, the scenario is quite different. Here a real radiation-driven shock is formed and propagates in the foam, reaches the interface with gold and undergoes a strong amplification due to impedance mismatch, i.e. the density difference between foam and gold. This has been previously studied in direct drive[14,15]. The pressure increment ΔP = 27 Mbar in the case of a foam of ρ =50 mg/cc, d = 100μm, is much higher than the material pressure while, within our error bars, it is in fair agreement with the maximum amplification expected from impedance mismatch (P/P_o = 37/9.6 = 4).

In our experimental conditions, we think that the transition between the two scenarios takes place between 20 and 50 mg/cc. Indeed the fraction of the X-ray energy directly transmitted to the foam/metal interface varies from 36%, for 20 mg/cc and 50 μm, to 1.5% for 50 mg/cc and 150 μm. Although only qualitative, being calculated using the absorption coefficients of the cold foam, these value seem to suggest a change of regime. But the decisive point is the experimental observation of the shock delay Δt. In Fig. 3, the delay in the shock breakout scales linearly with foam thickness. For the case of 50 mg/cc density, the slope of the linear interpolation gives a velocity 100 μm/ns. This corresponds to the order expected for a shock wave and does not match with the hypothesis of a supersonic ionisation wave. [Also, in the foam-confinement scenario, a shock is generated at the interface by X-rays penetrating practically instantaneously, so that a large shock delay seems difficult to be justified]. This conclusion seems to be in disagreement with ref. 16 who refer to the first scenario even in the case of a 100 mg/cc foam (with practically the same cavity temperature as in our experiment).

We notice a negative value for shock breakout delay in the case of foam/Au target which can be explained as follows. The foam layer of ρd = 0.25 mg/cm2 corresponds to a non stationary shock, as can be seen by looking at the delay in shock breakout Δt (which implies that the shock reaches the interface before the full laser pulse energy has been deposited on target). Here the pressure amplification ($P/P_O = 2.3$) agrees with the result of analytical formulas[8].

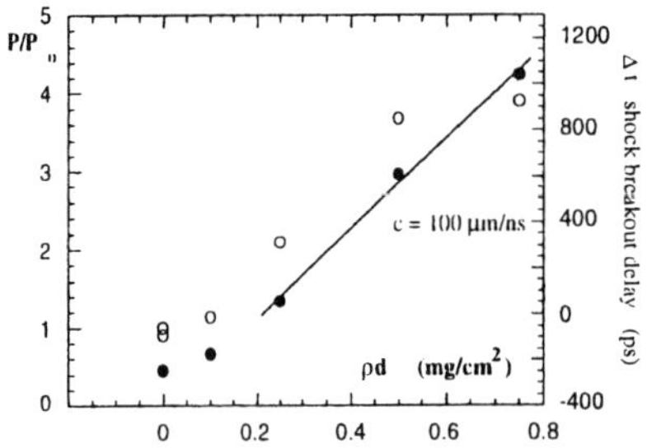

Fig. 3: Pressure Increment P/P_O (empty circles) and delay of the shock breakout in ps, vs. foam layer areal mass density (mg/cm^2).

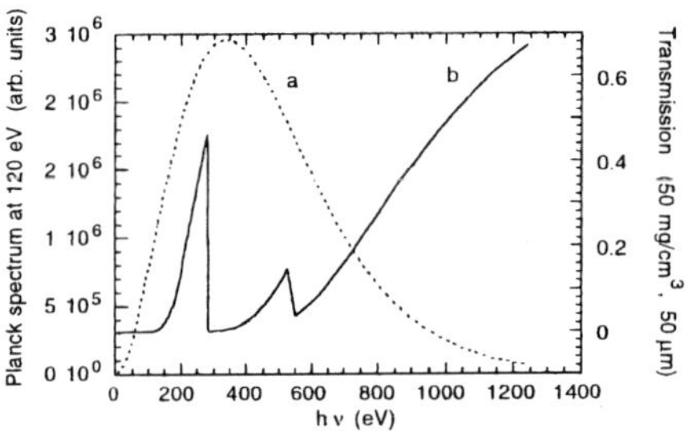

Fig. 4. Planckian spectrum for T = 120 eV (a), and transparency of a 50 μm, 50 mg/cc cold foam.

We have shown how shock pressure can be amplified in X-ray driven foam-layered targets. At low foam densities, a supersonic ionisation wave is produced and the pressure amplification is due to the foam plasma applying a material pressure on the ablating target plasma and effectively confining it. For denser foams, a real shock propagates in the foam and the pressure increase is due to impedance mismatch at the foam-metal interface, similar to the directly-driven foam-metal targets[10]. In both cases, such amplification is dependent on the foam density and also on the foam thickness. In the case of thin foams, a rapid decompression may dump the confinement effect. Details of the work will be published elsewhere.

This experiment was performed at MPQ and fully supported by the E.U. in the framework of the Programme "Access to Large Scale Facilities

References

1. Dunne M., *et al.*, Phys. Rev. Lett., **75**, (1995) 3858.
2. Batani D., *et al.*, Plasma Phys. Cont. Fusion, **40**, (1998) 1567.
3. Lindl J., Phys. Plasmas, **2**, (1995) 3933.
4. M. Desselberger et al., Phys. Rev. Lett., 74, (1995), 2961
5. I.B. Foldes et al., Phys. Rev. E., 64, (2001),
6. SESAME Report, (1983), Los Alamos Eq.-of-State library, No. LALP-83-4 (T4 Group, LANL, Los Alamos)
7. Al'tshuler L.V., *et al.*, (1981) J. Appl. Mech. Techn. Phys., **22**, 145 (1981) [Zh. Prikl. Mekh. Tekhn. Fiz., **2**, 3 (in Russian].
8. Batani D., et *al.*, (2000) Phys. Rev. B, **61**, 9287.

HIGH SPECTRAL RESOLUTION X-RAY IMAGING CRYSTAL SPECTROMETER FOR TOKAMAKS AND STELLARATORS

G. Bertschinger, M. Bitter*, D. Rusbüldt

*Institut für Plasmaphysik, Forschungszentrum Jülich, EURATOM Association, Trilateral Euregio Cluster, 52425 Jülich, FRG; *Princeton Plasma Physics Laboratory, Princeton University, Princeton, New Jersey 08543, USA*

INTRODUCTION

High resolution x-ray spectroscopy is frequently being used for the measurement of basic plasma parameters in high temperature plasmas. These include measurements of the ion and electron temperature, the plasma motion and the abundance of different charge states. Early experiments have been done on PLT and TFR, and more recently on TFTR, JET, JT60, LHD, W7-AS, TORE SUPRA, ALCATOR-C-MOD, TEXTOR and NSTX.

He-like ions are the most abundant species over a large range of the electron temperatures, and therefore most of the experiments have been done with the Kα lines of He-like ions. The Kα spectra are dominated by transitions in the He-like ion and the associated well separated satellites, which originate from doubly excited Li-like and Be-like ions[1,2]. These spectral lines provide information on the ion and electron temperature and the relative abundance of Be-like and Li-like to He-like ions. The accuracy of the spectroscopic method is comparable to other diagnostic methods used on thermonuclear plasma physics experiments [1]. At TEXTOR it has been shown [2] that with sufficiently high time resolution, x-ray spectroscopy can also support transport studies in the plasma center, by studying the time evolution for the different ionization stages of injected impurities.

REQUIREMENTS FOR HIGH RESOLUTION X-RAY SPECTROSCOPY

The spectral range of a x-ray spectrometer includes all medium and high-Z impurities, which are expected in an experiment for magnetic fusion. Besides the main components of the plasma, the hydrogen isotopes and helium that results from thermonuclear burn, both low- and medium-Z elements are expected in the plasma. These elements are either released from the plasma facing components, or they are purposely added to the plasma in order to increase the radiation losses and hence reduce the power load to the plasma limiters and divertors. Additionally, short pulses of impurities are applied to the plasma either by fast valves or by laser blow off, to study the transport properties of the background plasma. The impurities released from the walls of the plasma are mainly low-Z elements as carbon, beryllium and boron, but also

medium-Z metals of the iron group are released from antennae and structure materials. In addition, some high–Z impurities are expected, since molybdenum and tungsten are used for selected wall components with a high thermal load.
For plasma edge cooling, neon has been successfully used on TEXTOR; for larger devices, silicon, argon or krypton are the preferred elements for operation with a "radiative mantle".

The x-ray spectrometers for magnetic fusion devices must therefore cover the spectral range of the He- and H-like ions from silicon to about krypton, which includes all elements used for the vacuum vessel and the mechanical structures exposed to the plasma. This is the spectral range from 1.8 keV to above 15 keV. The spectral resolution must be sufficient to resolve the Doppler broadening of the lines and the sensitivity of the instrument has to be as large as possible, in order to detect very small concentrations of impurities. High resolution spectrometers in this range of low energy x-rays are built by using the Bragg reflection from perfect crystals. The resolution $\lambda / \Delta\lambda > 10000$, which is required for temperature measurements, is easily obtained by a proper choice of Bragg diffraction crystals. High quality crystals of quartz, silicon and germanium are available and have been successfully used.

At most of the present experiments, Johann spectrometers are used as they provide high spectral resolution and high sensitivity, and no entrance slits are necessary. The key element of a Johann spectrometer is a cylindrically bent crystal, which focuses x-rays from a point on the Rowland circle, to an image which is also positioned on the Rowland circle (see Fig. 1a). Only radiation, which fulfills the Bragg condition is reflected,

$$n\lambda = 2d * \sin(\alpha) \qquad \text{[eq. 1]}$$

with **n**, λ, **d**, and α being the order of reflection, the x-ray wavelength, the spacing between the crystal planes, and the Bragg angle, respectively. (Note the Bragg angle is the grazing angle to the surface, contrary to the convention in geometrical optics). The bending radius **R** of the crystal is twice the radius of the Rowland circle.

BRAGG REFLECTION ON SPHERICALLY BENT CRYSTALS

Spherically bent Bragg crystals combine the focusing properties of a cylindrically bent crystal in the Johann mount and the light gathering properties of a concave mirror. The strong astigmatism of a concave mirror can be utilized to produce an image of the plasma on the detector. The imaging properties of a spherical mirror are described by the lens equation:

$$1/a + 1/b = 1/f \qquad \text{[eq. 2]}$$

where **a** is the distance between the source and the crystal, and **b** the distance between the crystal and the detector; **f** is the focal length of the crystal. Due to the astigmatism, the focal lengths for the meridional (dispersion) plane and for the vertical (sagittal) plane are different. The meridional and the sagittal focal lengths are

$$f_m = R/2*\sin(\alpha), \qquad f_s = R/(2*\sin(\alpha)),$$

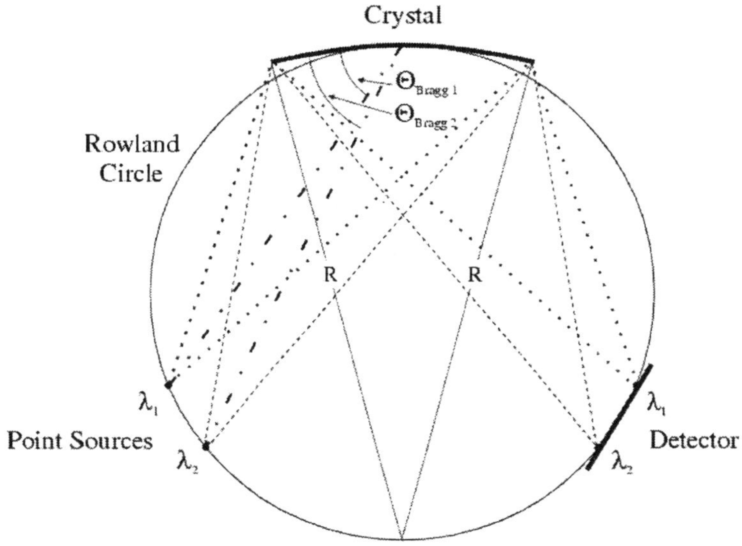

Fig. 1a) Principle of Johann spectrometer, the crystal in cylindrically bent, x-rays of certain wavelengths λ_1, λ_2, are focused onto the Rowland circle, the source can be set arbitrarily (Fig. 1b).

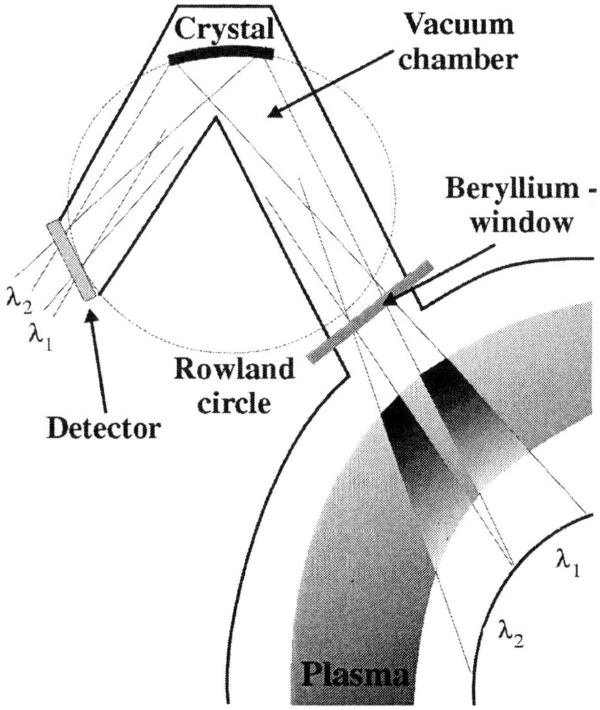

Fig. 1b) Typical setup of a Johann spectrometer at a tokamak experiment

respectively. For a x-ray spectrometer, the Bragg condition [eq. 1] has to be fulfilled for the whole crystal and the detector must therefore be placed on the Rowland circle, i. e

$$b_m = R*\sin(\alpha).$$

The distance to the meridional focus is then $a_m = R*\sin(\alpha)$, i. e. it is on the Rowland circle as well. The distance of the sagittal focus is

$$a_s = R*\sin(\alpha)/\cos(2\alpha)$$

(see Fig 2). If the plasma is placed at the sagittal focus such that the symmetry plane of the plasma is parallel to the dispersion plane of the spectrometer, an image of the plasma on the detector is obtained. The plasma parameters must not vary along the astigmatic sagittal image line of a point on the detector. For a tokamak, the dispersion plane is therefore chosen parallel to the equatorial plane, for a stellarator, the dispersion plane must be parallel to the flux surfaces. Using a 2-D detector, spectral resolution is obtained in the dispersion plane and spatial resolution in a direction perpendicular to the main dispersion plane.

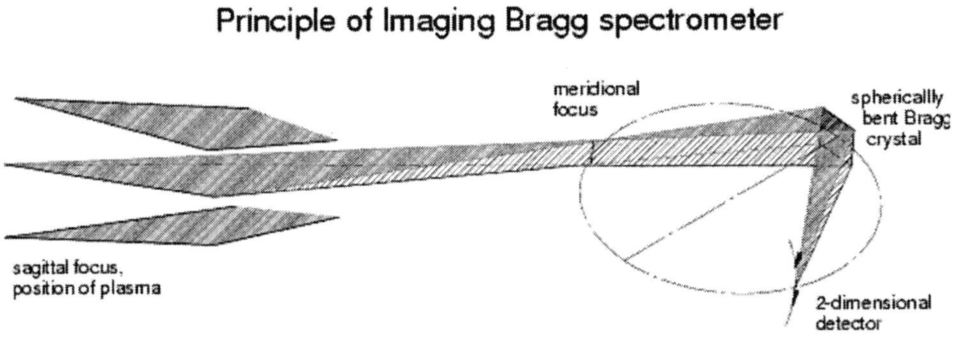

Fig. 2) Imaging by a spherically bent crystal. For a spherically bent crystal, the properties are rotation symmetric to the normal of the crystal

In order to calculate the geometrical errors of imaging and to maximize wavelength resolution, a ray-tracing program has been developed. The program allows for different shaping of the crystal layers, as well as the crystal surfaces.
The distance of the sagittal focus depends on the Bragg angle. In principle, the sagittal focal length can be adjusted by using toroidal instead of spherical crystals. The calculations show however, that the spectral resolution decreases for toroidal surfaces.

A crucial point is the availability of suitable Bragg crystals and x-ray detectors. In order to obtain the required resolution of $\lambda/\Delta\lambda > 10\,000$, about 10 000 crystal layers must contribute to the Bragg reflection. Therefore the absorption depth must be between 10 μm and about 100 μm. For larger absorption depths, the resolution is higher than needed and hence the width of the rocking curve is too low for efficient reflection. For low x-ray energies, silicon and quartz crystals provide high reflectivities, whereas for higher energies, germanium crystals are more efficient (see Fig. 3). For x-ray detectors, the absorption length must be adequate to deposit the full energy of the x-ray radiation in the detector. In solid state (CCD) detectors, low energy x-ray are absorbed close to the surface, where the electrons are not collected efficiently, high energy x-rays are lost if they are not absorbed in the depletion layer. With depletion layers about 50 μm deep, Si detectors loose efficiency above 6 keV. For x-ray energies below 10 keV, x-ray CCD detectors have the highest sensitivity, for higher energies, high-Z scintillators and CCD detectors in the visible spectral range provide higher performance.

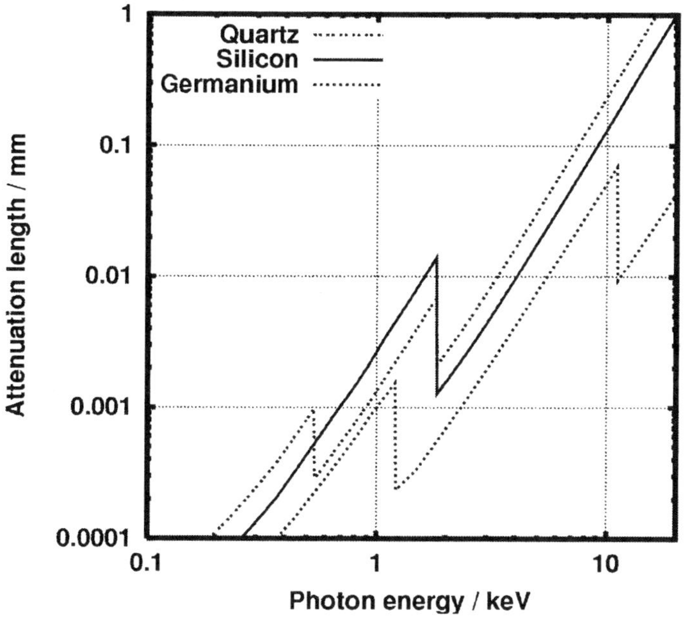

Fig. 3) Absorption length of x-rays in quartz, silicon, and germanium [3].

For the x-ray radiation from different elements, crystals with adequate spacing between the crystal layers must be available. For Kα lines of He like ions, crystal cuts in quartz, silicon and germanium can be found, which provide Bragg angles around 53°. Quartz crystals have many useful crystal cuts, but silicon and germanium have higher reflectivity if the reflections are allowed. A selection of suitable crystal cuts is shown in table 1. Except for phosphorus and chlorine, the Bragg angles for all elements are in a narrow range $51.5° < \alpha < 54.5°$. Some of the crystals are useful to measure the Lyman series of H-like ions, too. Convenient tools for the selection of suitable crystal cuts and the calculation of the rocking curves of a large number of crystals can be found in [4].

LAYOUT OF AN IMAGING SPECTROMETER FOR TEXTOR AND W7-X

For a tokamak or a stellarator plasma, it is sufficient to measure from the center up to 0.7*a, with a being the plasma radius. At the plasma edge the temperature is too low as to excite x-ray lines. For TEXTOR with a plasma radius of 46 cm, the measurement extends from the center up to about 35 cm. A design with a Rowland radius of 78 cm, Bragg angles around 54° and x-ray CCD detectors was chosen. This ensures that commercially available CCD detectors are long enough to measure the full spectra of

Table 1.

	Wavelength / m	Element		Crystal	Cut	Bragg Angle
14	6.68805E-10	Si	Silicon	Quartz	100	51.884
15	5.79326E-10	P	Phosphorus	Quartz	100	42.961
				Quartz	101	60.151
16	5.06640E-10	S	Sulphur	Si	111	53.892
17	4.46793E-10	Cl	Chlorine	Si	111	45.436
				Quartz	110	65.554
18	3.96930E-10	Ar	Argon	Quartz	110	53.975
19	3.54948E-10	K	Potassium	Quartz	111	52.598
20	3.19270E-10	Ca	Calcium	Si	220	56.242
				Ge	220	52.943
21	2.88693E-10	Sc	Scandium	Quartz	112	52.649
22	2.62290E-10	Ti	Titanium	Si	311	53.216
23	2.39334E-10	V	Vanadium	Quartz	211	51.001
24	2.19251E-10	Cr	Chromium	Quartz	212	52.569
				Si	400	53.845
				Ge	313	57.626
25	2.01512E-10	Mn	Manganese	Si	313	53.968
26	1.85949E-10	Fe	Iron	Quartz	204	54.687
				Ge	224	53.615
27	1.72000E-10	Co	Cobalt	Si	333	55.369
				Ge	333	52.169
28	1.59655E-10	Ni	Nickel	Quartz	223	51.945
				Ge	404	52.953
				Ge	513	56.586
29	1.48526E-10	Cu	Copper	Si	513	53.996
30	1.38531E-10	Zn	Zinc	Ge	533	53.397

the Kα lines, including forbidden lines and satellite lines [2]. By operating the CCD detectors in the vertical binning mode, the detectors are reasonably fast to follow impurity injection experiments. Readout times of 2 ms for a spectrum have been obtained. In the binning mode, the detectors are operated as linear arrays, spatial resolution is obtained by combining a stack of 6 to 8 detectors. Alternatively, the CCDs can be operated as 2-D detectors with increased spatial resolution but slower readout frequency of about 10 Hz. The magnification in the sagittal (imaging) direction is about 0.3, therefore the detector height which is required for TEXTOR is 12 to 15 cm, the spatial resolution in the plasma will be about 6 cm.

For a certain choice of the Bragg crystal, the sensitivity of the spectrometer is proportional to the crystal area. It is desirable to use crystals as large as available, on the other hand, imaging errors increase with increasing crystal size. In order to optimize the crystal shape and to analyze the geometrical errors, detailed ray-tracing calculations have been performed. The dominant error is a second order term, which depends on the square of the crystal length, similar to the Johann error in the case of a cylindrical crystal. For a spherical surface, the second order term in the height of the crystal vanishes, therefore the imaging properties decrease only slightly with the crystal height (see Fig. 4). Contrary to standard Johann mounts, where the longer side of the crystal is parallel to the diffraction plane, the imaging errors are minimized if the longer side of the crystal is directed perpendicular to the diffraction plane. If larger crystals are required for high sensitivity, the surface of the crystal must be shaped toroidally. In sagittal direction the curvature has to be R, whereas in meridional direction the curvature must be R/2 (see Fig. 5). If the surface of the crystal is shaped spherically with radius R/2, the errors due to the finite height of the crystal are of the same order as the Johann errors. However, for spherically shaped crystals, the imaging errors are reduced for the standard mount with a rectangular crystal and the longer side in meridional direction [5].

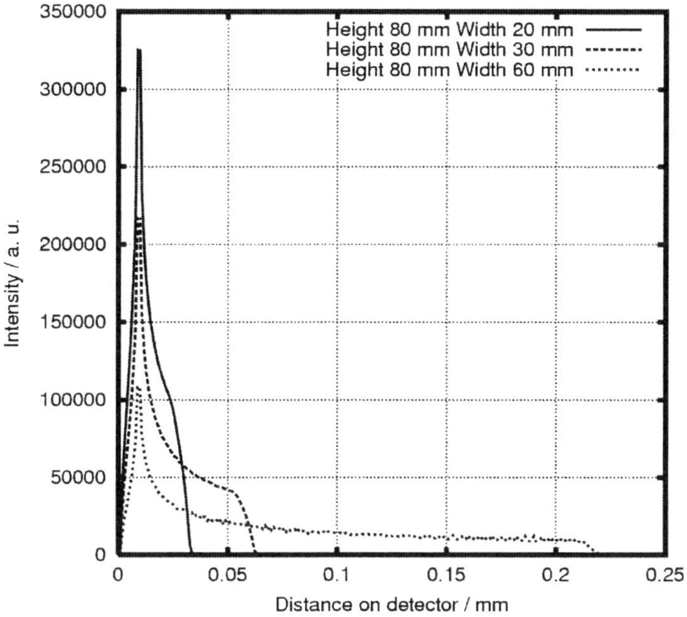

Fig. 4) Intensity distribution of monochromatic x-ray radiation due to geometrical errors by finite crystal size. The illumination of the crystal is uniform, the detector is tangential to the Rowland circle. Broadening due to the rocking curve of the crystal is not taken into account. Note the large defocusing due to crystal width.

For a spherical crystal, the height can be increased significantly, without compromising the spectral and the spatial resolution. The sensitivity of the instrument is then maximized and the area of the crystal is mainly limited by the size of available high quality crystals.

Additionally, the resolution is limited by the rocking curve of the crystals. This can be taken into account by applying more sophisticated ray-tracing codes such as SHADOW[6] or XRT[1]. These codes take into account the changes in resolution of the crystal due to bending and finite thickness. For crystals of high quality and bending radii larger than 1 m, a convolution of the geometrical errors with the rocking curve, as obtained e. g. from [4], is a save approximation.

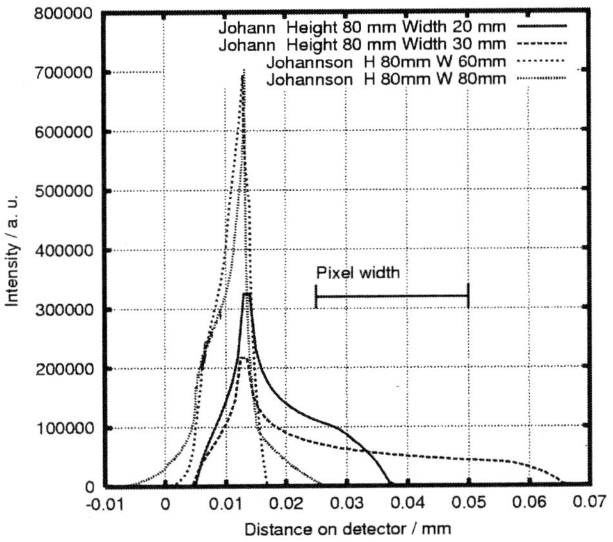

Fig. 5) Comparison of geometrical errors with CCD pixel width, for spherical surface (Johann) and narrow crystals. For a toroidally shaped surface (Johannson), the crystal size can be much larger.

CONCLUSIONS

It is shown that spherical Bragg crystals have several advantages compared to the traditional cylindrically bent crystals. They can provide higher sensitivity, as the errors due to the size of the crystal are reduced significantly. Using the astigmatism of off axis reflection on a spherical surface, it is possible to obtain simultaneously both spatial and spectral resolution. Spatial resolution can be obtained either with available 2-dimensional solid state detectors, which have however only slow readout capability, or by combining several linear arrays to a 2-D detector without compromising the time response. If very large crystals are required, the resolution can be retained by toroidally shaping the surface of the crystals. Imaging spectrometers will be installed on NSTX, KSTAR, TEXTOR and W7-X. For ITER the optimization of x-ray spectroscopy can be done following the same guidelines.

References:

1) Weinheimer, J., I. Ahmad, O. Herzog, and H.-J. Kunze, G. Bertschinger and W. Biel, G. Borchert, M. Bitter, Rev. Sci. Instrum. 72, 2566-2574 (2001)
2) Bertschinger, G., W. Bicl, the TEXTOR-94 Team, O. Herzog, J. Weinheimer, H.-J. Kunze and M. Bitter, Physica Scripta. T83, 132-141 (1999)
3) B.L. Henke, E.M. Gullikson, and J.C. Davis. Atomic Data and Nuclear Data Tables Vol. 54 (no.2), 181-342 (July 1993). http://www-cxro.lbl.gov/optical_constants
4) S. Stepanov, http://sergey.bio.aps.anl.gov/x0h.html
5) B. Fraenkel, M. Bitter, S. von Goeler, and K. W. Hill, J. X-Ray Sci. Technol. 7, 171 (1997)
6) F. Cerrina, B. Lai, K. Chapman, C. Welnak, P. Runkle, http://www.nanotech.wisc.edu/shadow/

TRITIUM ISSUES IN NEXT STEP DEVICES

C. H. Skinner and G. Federici*

1. INTRODUCTION

Tritium issues will play a central role in the performance and operation of next-step deuterium-tritium (DT) burning tokamaks and the safety aspects associated with tritium will attract intense public scrutiny. Tritium fuel has been successfully used in the Tokamak Fusion Test Reactor (TFTR) and the Joint European Torus (JET) producing 10 and 16 MW of fusion power respectively[1,2]. This experience together with focussed laboratory studies, has illuminated the challenges[3]. The orders-of-magnitude increase in duty cycle and stored energy will be a much larger change than the increase in plasma performance necessary to achieve high fusion gain and ignition[4,5]. Erosion of plasma facing components will scale up with the pulse length from being barely measurable on existing machines to cm-scale. Magnetic Fusion Energy (MFE) devices with carbon plasma facing components will accumulate tritium by co-deposition with the eroded carbon and this will strongly constrain plasma operations. We report on a novel laser-based method to remove codeposited tritium from carbon plasma facing components in tokamaks. The operational lifetime of alternative materials such as tungsten has significant uncertainties due to melt layer loss during disruptions. Production of dust and flakes will need careful monitoring and minimization, and control and accountancy of the tritium inventory will be critical issues. The relevant plasma material interactions are comprehensively reviewed in [4]. Operation of next step experiments will help resolve key tritium issues in the design of a magnetic fusion reactor. Many of the tritium issues in Inertial Fusion Energy (IFE) are similar to MFE, but some, for example those associated with the target factory, are unique to IFE.

The plasma-edge region in a tokamak has greater complexity than the core due to lack of poloidal symmetry and nonlinear feedback between the plasma and wall. Sparse diagnostic coverage and low dedicated experimental run time has hampered the development of predictive models. Diagnostic advances are urgently needed to better characterize the plasma edge and wall and improve our predictive capability.

---

* C.H. Skinner, Princeton Plasma Physics Lab., Princeton N.J. 08543. G. Federici, ITER Garching Joint Work Site, Boltzmannstrasse 2, D-85748 Garching, Germany.

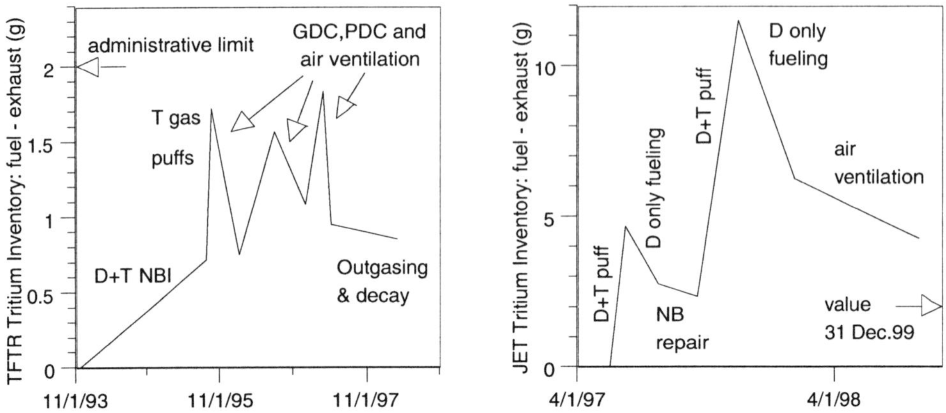

Fig. 1 Tritium Retention on TFTR and JET

2. TRITIUM EXPERIENCE IN CONTEMPORARY TOKAMAKS

A large fraction of tritium was retained during DT plasma operations in TFTR and JET [6,7] and a comparison shows striking similarities and contrasts[8]. Tritium was retained by co-deposition with eroded carbon and by isotope exchange with previously retained deuterium (fig. 1 & table 1). When the tritium in-vessel inventory approached the administrative safety limit, it was removed by extensive campaigns involving several weeks of glow discharge cleaning and deuterium operation. An unexpectedly large amount of tritium was transported to the JET sub divertor region (sect. 6). In summary, while tritium retention was not a significant limit to plasma operations in these machines, the high rate of tritium retention and slow rate of tritium removal would be unacceptable in a next step device and in a fusion reactor.

Table 1. Tritium Retention in TFTR and JET.

	TFTR	JET (DTE1)
Total tritium injected by NBI	3.1 g	0.6 g
Total tritium injected by gas puff	2.1 g	34.4 g
Total tritium retained during DT operations	2.6 g	11.5 g
Initial % retention during T puff fueling (wall saturation + isotope exchange)	≈ 90%	≈ 40%
Longer term % retention including D only fueling (mostly co-deposition)	51%	17%
Tritium remaining in torus	0.85 g (4/98)	4.2 g (7/98)
Long term retention	16% (4/98)	12% (7/98) 6% (12/99)
Average deuterium retention (for comparison)	≈ 40%	≈ 10-15%

The retention experience of Alcator C-mod is particularly interesting since it is lined with Mo tiles and there are no carbon plasma facing components. Although no external tritium was supplied, nuclear reactions in the deuterium plasma produce tritium and neutrons and the amount of tritium generated was deduced from the measured neutron fluence. The amount of deuterium and tritium retained in the vessel was measured by analysis of tiles removed from the vessel[9]. The analysis showed that most of D inventory was implanted (not codeposited) on main chamber wall. Very sensitive measurements of tritium gave a lower limit for the fractional tritium retention of less than 0.2%. This is encouraging confirmation of the drastically lower in-vessel tritium inventory possible in the absence of carbon plasma facing components.

3. TRITIUM IN ITER AND FIRE

Safety issues and fuel economy will be primary drivers in the operation of next step DT machines. Fusion is considered to be 100 to 1000 times less hazardous than fission, non the less public sensitivities are acute[10]. Tritium permeation and leakage during normal operation must be strictly limited. The total tritium inventory on site is restricted to avoid the need for public evacuation under the worst postulated accident. Tritium (as hydrogen) can pose a deflagration or detonation risk. Present guidance from the US Department of Energy for a site with a 1 km radius, would limit the accidental release of HTO to 15 g at ground level or 150 g if elevated[11]. Tritium inventories larger than this will need highly reliable robust confinement barriers. Vacuum penetrations either for diagnostic purposes, or for IFE beams can be problematic.

The availability of fuel is the most fundamental requirement of any power source and independent of safety considerations, the available tritium supply is limited. In the growth of a future DT powered economy, tritium for future reactors will be bred in a lithium blanket but the margin for tritium breeding is relatively low: the tritium breeding ratio for MFE is 1.05-1.18, and for IFE is 1.05 — 1.75 [12]. This factor also limits tritium inventories in a reactor design to approximately 1 kg.

ITER employs a variety of plasma facing materials selected for their suitability to regions of the vessel with different power and particle flux characteristics[4]. The first wall is Be, the baffle regions of the target, the dome and the liner in the private flux region below the dome are W and the divertor strike-point regions are carbon fiber composite (CFC). CFC is chosen because of its high thermal shock resistance and tolerance to off-normal events, however carbon erosion will facilitate retention of tritium via co-deposition of tritium with carbon. The in-vessel tritium limit is ~350 g limit and migration of hydrocarbon radicals may result in accumulation of tritium in inaccessible areas (as in JET). The number of plasma pulses before the inventory limit is reached is estimated to be 70 — 170 pulses (Fig. 2)[13]. After this, the subsequent availability of the machine for plasma operations will depend on efficient and fast tritium removal techniques that have yet to be fully demonstrated in a tokamak (see sect. 6). For comparison, in TFTR several weeks were needed for tritium removal after

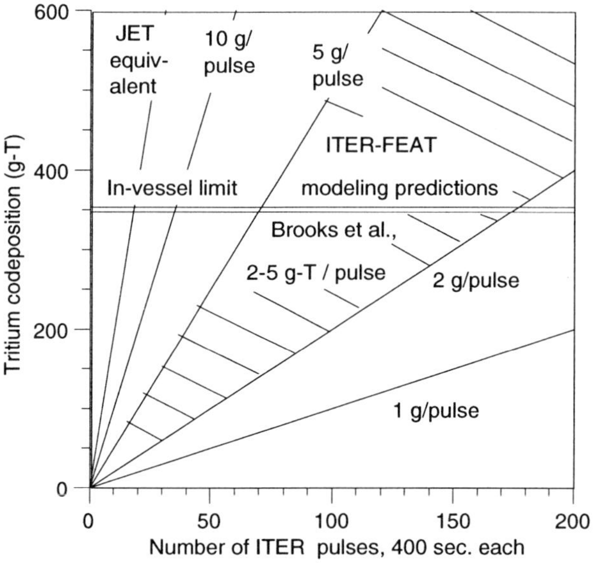

Fig.2 Tritium retention for ITER, showing modeling predictions[5] and JET DTE equivalent rate. The inventory limit (shown by double line) is predicted to be reached in approximately 100 pulses.

only 10-15 min of cumulative DT plasma operation. To address this concern, ITER will maintain the option to switch from CFC to W armour on the divertor targets prior to DT operation. This change will depend on the frequency and severity of ELMs and disruptions in the initial hydrogenic plasmas and the effectiveness of techniques to mitigate tritium codeposition.

FIRE is a next step option being designed in US, and is aimed at understanding and optimizing alpha dominated burning plasmas. While a smaller machine, the heat loads

Table 2 Comparison of FIRE and ITER parameters.

	FIRE	ITER
major radius, R	2.14 m	6.2 m
magnets	Liquid N_2 cooled, 10 T	Superconducting, 5.3 T
fusion power	150 MW	500+ MW
pulse duration	~ 20 s	400 s
duty cycle	0.2%	10%
disruption energy on divertor	4 — 96 MJ/m^2	10-100 MJ/m^2
divertor material	W	CFC*
divertor lifetime	70 — 300 disruptions	~ 4000 pulses
in-vessel T inventory limit	15 g	350 g
pulses before T removal	not an issue	70 — 170 *
cost	≈ $1 billion	≈ $5 billion

* For a ITER W divertor, the lifetime would be similar to FIRE and T removal not an issue.

on the FIRE divertor are similar to ITER because of the much higher fusion power density. FIRE plans to have a Be coated Cu wall and W divertor and no carbon plasma facing components[14]. W brush materials have been successfully tested at heat fluxes up to 25 MW/m^2 however erosion and melt-layer loss during off normal events such as disruptions and ELMs are key issues for the lifetime of W clad divertor targets (Table 2). Techniques for mitigating or avoiding disruptions are under development. Since tungsten is a very efficient radiator and its core concentration must be less than ($n_i/n_e < 10^{-5}$) impurity influx must be severely limited. ASDEX Upgrade has shown encouraging results in this regard[15]. The major benefit of the absence of carbon is the very low tritium retention, which allows a major reduction in tritium inventory. The in-vessel tritium inventory limit for FIRE is only 15 g and allows FIRE to be classified in the U.S. as a Category III Low Hazard Nuclear Facility.

5. TRITIUM IN IFE DESIGNS

Many tritium issues in IFE are similar to those found in MFE[16]. For dry wall chamber designs such as Sombrero, codeposition of carbon and tritium is a concern along with deposition of target ablator materials. The extended boundaries associated with IFE beam tubes may make this deposition difficult to control. The target factory in IFE has no MFE equivalent. At present permeation is used to fill IFE targets and straightforward extrapolation of this technique could result in very large tritium inventories, however some innovative techniques such as non diffusive capsular fill and cryogenic assembly are being explored to reduce this. Fast ignitor target designs also offer prospects of reduced inventory. Robust confinement barriers for accident conditions may be 50 m up the beam lines[11] and control of such a large confinement boundary may be difficult. On the positive side the simpler geometry of IFE offers potential for higher tritium breeding ratios.

6. DIAGNOSTIC NEEDS FOR MFE NEXT-STEP DEVICES

There is significant uncertainty in predictions of tritium behaviour in next step devices, uncertainty that is much larger than predictions of core plasma performance. In view of the potential for loss of plasma operational time due to time spent on tritium removal there is an urgent need to address this. While sophisticated models are available to predict edge plasma behavior in tokamaks, there are typically many more adjustable parameters in the models than constraints due to experimental data. Agreement between observed deposition in JET and DIVIMP modeling was only achieved after including drift in the scrape off layer, additional interaction with the main chamber wall and enhanced erosion of the re-deposited carbon films at the inner target. Together, these altered deposition patterns in JET by more than an order of magnitude[17]. Carbon transport and tritium codeposition in the inner divertor/louver is under detailed analysis using the CARJET/REDEP, MOLDYN and ADAS codes with the goal of validating on JET the model used to predict tritium retention on ITER[18].

Diagnosis of the edge / wall region is difficult because of the lack of symmetry due to open flux surfaces intersecting plasma facing components. Understanding of the complex non-linear feedback between the plasma boundary and materials is limited by the available diagnostic information and low priority given to dedicated run time for plasma wall interaction studies[19]. Higher priority for divertor and wall diagnostics, is needed e.g. sensitive Langmuir probes, CXN measurements, IR camera, erosion / deposition collector probes, microbalances, laser desorption to measure D/C ratio of hydrocarbon films, sticking probes, molecular spectroscopy together with dedicated run time for experiments on e.g. SOL drift with reversed grad B and thick Be wall coatings.

Two other diagnostic areas are highly relevant to next step needs. In-vessel dust can be highly reactive, for example, a loss of cooling accident may result a steam reaction with Be dust and the hydrogen generated can pose a detonation risk. There are strict regulatory limits on vessel dust for these reasons, however diagnostics to demonstrate compliance with these limits remain a major challenge[20]. As can be seen above, disruption erosion is a major driver in the choice of plasma facing materials and diagnostic advances to confidently identify disruption precursors would facilitate disruption mitigation and avoidance. More information on MFE diagnostic needs is given in ref. [19]. For diagnostic considerations for a large scale IFE facility we refer the reader to ref. [21].

7. TRITIUM REMOVAL BY A SCANNING LASER

Efficient and rapid tritium removal will be essential to future devices with carbon plasma facing components. Present candidate techniques involve oxidation of the codeposit at temperatures above 250 C but raise several issues. The introduction of oxygen into a tokamak deconditions the walls and requires lengthy reconditioning to restore high performance plasma operations. It also raises issues of collateral damage on in-vessel components. The product of the oxidation is tritiated water that requires expensive reprocessing[3]. A further complication is that the temperature required is beyond the range of pressurized water cooling. However, to thermally desorb tritium it is not necessary to heat the whole vacuum vessel since most of the tritium is codeposited on the surface. The present method is based on surface heating with a scanning laser beam. Modeling showed that a laser power density of 3 kw/cm^2 and heating duration of 20 ms could heat a 50 micron codeposit to 1000 C — 2000 C appropriate for tritium release[22].

The laser scanning method for tritium release has been successfully tested on graphite and carbon fiber composite (CFC) tiles with tritiated codeposits from the TFTR DT campaign. The focussed beam from a continuous wave 325 watt Nd:yag laser was scanned in a raster pattern over the codeposited surface transiently heating it up to 1500 - 2000 C for 10 — 200 ms releasing a major fraction of the tritium [23]. The surface temperature was measured by a pyrometer operating in the 1.58 — 1.8 micron band, with a 0.7 mm measuring spot and response time of 0.3 ms. Interestingly the temperature excursion of the codeposited layer was significantly higher than a bare graphite surface (1770 C compared to 1080 C), implying a very different thermal con-

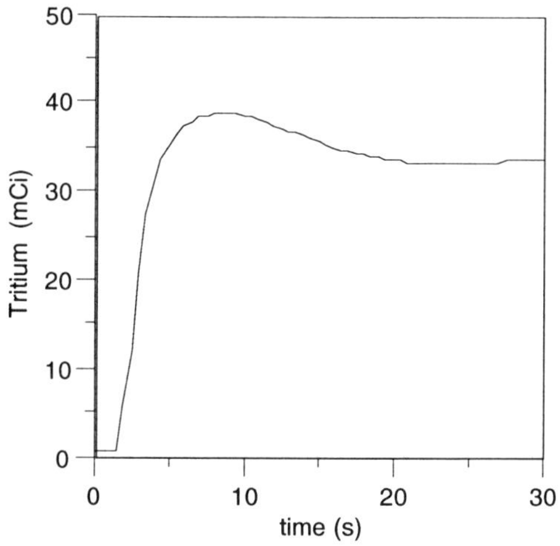

Fig.3 Tritium recorded by an ion chamber after laser irradiation. The delay is due to the pumping time.

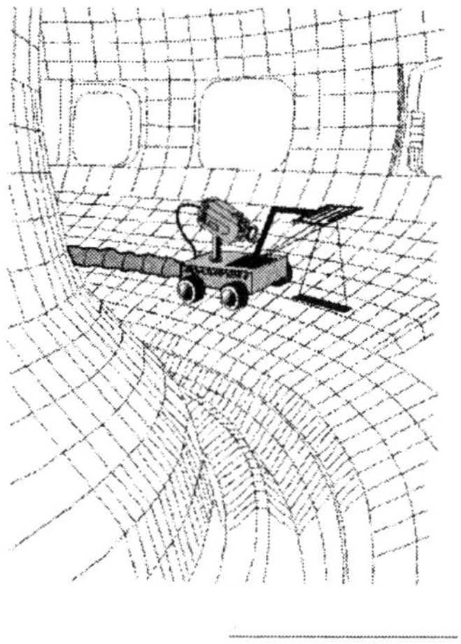

Fig.4 Potential implementation in a next step device.

ductivity. This underlines the importance of testing heat flux loads on tokamak generated codeposits.

The released tritium was measured by circulation in a closed loop system to an ion chamber (Fig.3). A movie of the laser in action may be seen on the web at ref. [24]. Microscopic images of the incandescence from the laser spot showed micro "hot spots" presumably due to the inhomogeneous granular structure of the codeposits. Any change in appearance of the codeposited layer was recorded with a microscope. While slow scans (> 50 ms heat pulse) resulted in visible ablation of the surface, fast scans (10-20 ms) showed almost no change to the surface appearance, indicating that the release of tritium was not accompanied by ablation of the codeposit. Optimal conditions were laser intensity of 8 kw/cm^2, scan speed 2 m/s and local heating duration 10-20 ms, resulting in surface temperatures 1300 — 2000 C.

Tritium remaining after the laser scan was released by heating with an unfocussed stationary laser spot in an air atmosphere for 30 minutes, which oxidised the remaining codeposit. Two experiments at different focal intensity showed a large fraction (46% and 84%) of the tritium released by the laser scan. This is an encouragingly high number and further optimization may be possible.

High power Nd lasers with fiberoptically coupled scanning heads are a proven technology in manufacturing. For a next step device we envisage a scaled up system with a multi-kilowatt laser beam delivered by fiberoptics on a robotic arm or mobile robot inside the vessel (fig.4). Energetically 30 MJ are needed to heat the top 100 microns of a 50 m^2 codeposited area from 400 to 2000 C and this corresponds to the output of a 3 kw laser for 3 hours. Clearly much work remains to be done to develop this technique at an engineering scale, however the initial results show clearly that tritium can be released by a scanning laser without the deleterious effects of oxidation.

ACKNOWLEDGEMENTS

We acknowledge the invaluable contributions from our colleagues to this work, in particular C. Gentile, K. Young and the PPPL tritium group for the laser detritiation study and J. Brooks, J.P.Coad, D. Meade and M. Ulrickson. Support for the Nd laser work was provided by the Annex IV to the JAERI/DOE Implementing Arrangement on Cooperation in Fusion Research and Development, U.S. DOE Contract Nos. DE-AC02-76CH0307. One of us (GF) acknowledges support from the ITER EDA Agreement.

REFERENCES

1. A. Gibson et. al., Phys. Plasmas, 5 (1998) 1839.
2. R. J. Hawryluk, Reviews of Modern Physics, 70, (1988) 537-587.
3. C. H. Skinner, et al., Nuclear Fusion 39 (1999) 271.
4. G. Federici, C. H. Skinner et al., "Plasma-material Interactions in Current Tokamaks and their Implications for Next-step Fusion Reactors", PPPL-3531/IPP-9/128 January 2001, http://www.pppl.gov/cgi-bin/pubrepcpyreq_2001.pl Nuclear Fusion in press.
5. G. Federici et al., J. Nucl. Mater., 290-293 (2001) 260.
6. C. H. Skinner, et al., J. Nucl. Mater. 290-293 (2001) 486.
7. P. Andrew et al., Fusion Eng.& Des. 47 (1999) 233.
8. C. H. Skinner et al., "Long Term Tritium Trapping in TFTR and JET" Proceedings of the 28th EPS Con ference on Controlled Fusion and Plasma Physics, Madeira, Portugal, 18-22nd June, 2001.
9. W.R. Wampler, et al, "Long Term Retention of Deuterium and Tritium in Alcator C-mod" Proceedings of the 18th IEEE/NPSS Symposium on Fusion EngineeringAlbuquerque, October 6-10, 1999, p.267 IEEE, Piscataway, NJ, USA (1999).
10. R P Crease, "True Tritium Tales" Physics World, August 2001 p.18.
11. D. Petti, "Tritium Safety Issues" ARIES Town Meeting on Tritium and the DT Fuel Cycle March 6,7, Livermore CA. http://michelle.ucsd.edu/LIB/MEETINGS/0103-ARIES-TTM/
12. L. El-Guebaly, "Tritium Breeding Issues for MFE and IFE" ARIES Town Meeting on Tritium and the DT Fuel Cycle March 6,7, Livermore CA. http://michelle.ucsd.edu/LIB/MEETINGS/0103-ARIES-TTM/
13. G. Federici et al., Physica Scripta T91 (2001) 76.
14. M. A. Ulricksen et al., Fusion Technology 39 (2): 378-382, Part 2 MAR 2001
15. R. Neu et al., J. Nucl. Mater., 290-293 (2001) 206.
16. K. Schultz et al., "Meeting Summary Session 2" ARIES Town Meeting on Tritium and the DT Fuel Cycle March 6,7, Livermore CA. http://michelle.ucsd.edu/LIB/MEETINGS/0103-ARIES-TTM/
17. J. P. Coad et al., J. Nucl. Mater., 290-293 (2001) 225.
18. J. N. Brooks et al., personal communication.
19. G.F. Counsell et al., J. Nucl. Mater., 290-293, (2001) 255.
20 G.T. Razdobarin, et al., J. Fusion Science and Technology (in press).
21. P. Bell et al., Rev. Sci. Instrum., 72 (2001) 492.
22. C. H. Skinner et al., Proceedings of the 17th IEEE/NPSS Symposium Fusion Engineering, San Diego, Ca. Oct 6-10, 1997, p.321.
23. C.H. Skinner et al., "Tritium removal from codeposits on carbon tiles by a scanning laser" Report # PPPL3603 (Sept. 2001) 29pp, submitted to J. Nucl. Mater. 2001 http://www.pppl.gov/pub_report/.
24. http://www.pppl.gov/tritium/page11.html

DIAGNOSTIC FIRST MIRRORS FOR BURNING PLASMA EXPERIMENTS

V.S.Voitsenya, A.F.Bardamid, A.I.Belyaeva, V.N.Bondarenko, A.E.Costley, A.A.Galuza, A.V.Gorshkov, N.Kishimoto, V.G.Konovalov, D.V.Orlinskij, A.V.Rogov, I.V.Ryzhkov, A.H.Shapoval, S.I.Solodovchenko, A.F.Shtan', K.Yu.Vukolov, S.N.Zvonkov*

1. INTRODUCTION

The lifetime of refractive components exposed to reactor grade plasmas will be very short and so all diagnostics which use UV ($\lambda > 5$nm), Visible and IR radiation (λ up to ~100μm) have to view the plasma via a mirror. The diagnostic first mirrors (FM) must survive in extremely hostile conditions and maintain a good optical performance for the duration of reactor operation. In ITER-FEAT[1] the FMs will receive intense UV and X-ray radiation, neutron and gamma fluxes, and particle fluxes (due to charge exchange atoms (CXA)). In addition, they will be subjected to the deposition of material eroded from the divertor and first wall. Of the different kinds of radiation and fluxes only CXA impact will result in direct surface modification of metallic FMs that can lead to degradation of optical properties. The fluxes of all radiation components to the mirror surface depend strongly on the mirror location. For example, the FMs of a wide-angle observation system (endoscope with open architecture) in ITER will be bombarded by CXA fluxes of about the same magnitude as the first wall. On the other hand, the FMs in the LIDAR system[2] are located in a long duct (~2 m in length) and will receive CXA fluxes ~ 10^{-2} of the first wall flux. This corresponds to about the lowest flux received by a FM in ITER.

Only limited knowledge exists of the effect of the potentially damaging radiation and particle fluxes on candidate FM materials and so dedicated experiments are required. Special experiments simulating the environmental impact of a burning plasma on mirror properties are in progress, in particular, at the IPP Kharkov and the Kurchatov Institute, Moscow.

* V.S.Voitsenya, V.N.Bondarenko, V.G.Konovalov, I.V.Ryzhkov, A.N.Shapoval, S.I.Solodovchenko, A.F.Shtan', IPP NSC KIPT, Kharkov, Ukraine; A.F.Bardamid, National Kiev University, Kiev, Ukraine; A.I.Belyaeva, A.A.Galuza, Kharkov Technical University, Kharkov; A.E.Costley, ITER Joint Work Site, Naka, Japan; D.V.Orlinskij, A.V.Rogov, K.Yu.Vukolov, S.N.Zvonkov, Kurchatov Inst., Moscow, Russia; N.Kishimoto, National Institute for Material Study, Tsukuba, Japan

The CXAs not only give rise to the erosion of the FM surface but also to erosion of the material of the diagnostic ducts in which the FMs are located. This material can be re-deposited on the FM surface[3] and build-up a contaminating film. It is probable that the ducts will become coated by the material of the first wall (beryllium) and/or the divertor tiles (carbon) also by erosion and deposition, and so these materials will be the ones that contaminate the FMs. Both processes of FM degradation, that is sputtering leading to erosion and re-deposition will occur at the same time, but in simulation experiments they are investigated separately. In the present paper, the current state of understanding of the FM problem is reviewed. The paper continues and extends earlier work on this topic.[4,5]

2. CHARACTERIZATION OF FMS

Because of the wide wavelength range of FM operation planned in ITER[6], the requirements on the quality of mirror surfaces have to be very different as follows from the well-known formula for specular reflectance R at normal incidence versus surface roughness[7]:

$$R=R_0\cdot\exp\left[-(4\pi d)^2/\lambda^2\right],$$

where R_0 is the reflectivity of the ideally smooth surface, d is the average roughness, and λ is the wavelength of light. For example, a 10% limit on the decrease of reflectivity due to mirror surface roughening gives $d < \lambda/50$. This limitation on the value of d is quite a stringent requirement even for the visible range of the spectrum and probably can not be satisfied for mirrors subjected to high fluxes of CXAs.

In practice, only some of the mirrors will be used for transmission of an image, others will operate as reflectors, transmitting the radiated power into the diagnostic labyrinths. For the latter mirrors the requirements on the surface quality are not so strict and the mean surface roughness can be significantly higher.

The energy distribution and CXA flux on the FMs will also depend on their poloidal location[8] The lowest CXA mean energy will be for mirrors in the divertor region where surface contamination, rather than erosion, will probably be the most important factor affecting mirror operation.

3. EXPERIMENTAL

In a fusion reactor the CXA energy spectrum will be very wide.[8-10] Today there is no single simulation experiment with such a wide particle energy distribution and effects of long-term ion bombardment on mirror properties are investigated mainly with fixed ion energy. For the first time results with a wide energy spectrum of ions of deuterium plasma were presented recently[11]. The ion energy distribution differed from that measured in other work for CXA flux[8,9], nevertheless, the data for copper[11] and stainless steel (SS)[12] mirrors showed that the rate of mirror degradation strongly depends on the particle energy.

In experiments in Kharkov, the measurements of reflectance at normal incidence, R, (in the range $250 < \lambda < 650$ nm) and thickness of layer, h, eroded from the mirror surface due to ion bombardment (by measuring the mirror weight loss) were carried out *ex situ* in many steps. Thus the R(h) dependence was obtained for mirrors of different materials.

The resolving power (RP), i.e., the ability of a mirror to transfer an image was measured, with the highest resolving power achieved in the range of 63 line/mm with a accuracy of ±3 line/mm.

4. EXPERIMENTAL RESULTS OF SPUTTERING TESTS

Mirrors of polycrystalline materials

The surface of polycrystalline (pc) metals has grains with different orientations of the crystalline planes and, as a result, different sputtering rates. Thus, during long-term ion bombardment, the initially smooth mirror surface changes with time into a step structure (tens to hundreds of μm in size in the surface plane). With increasing time of erosion of the pc surface, a small-scale microrelief develops inside the individual grains which depends on the grain orientation and the mirror material. As a result, the pc mirror surface is characterized by a co-existence of strongly eroded roughened areas interspersed with smooth areas. Such microrelief has been observed in many experiments (e.g., papers[11-13]). The step structure has a strong effect on the property of the mirror to transfer an image. To illustrate this, in Fig. 1 the h dependences of R (at λ =400 nm and 650 nm) and RP are shown for a Cu mirror bombarded with ions with a wide energy distribution from a deuterium plasma. As can be seen, the RP dropped by several times after ~1μm thick near-surface layer was eroded, while the reflectance at normal incidence only slightly decreased.

The SS mirrors demonstrated a different behaviour compared to other pc metals that were tested, for example copper[11]. On the surface of SS samples[12] the grains continued to be very smooth plateaus without developing the in-grain microrelief, even after a layer with a mean thickness ~ 4.5μm was eroded. It was found that the higher the ion energy

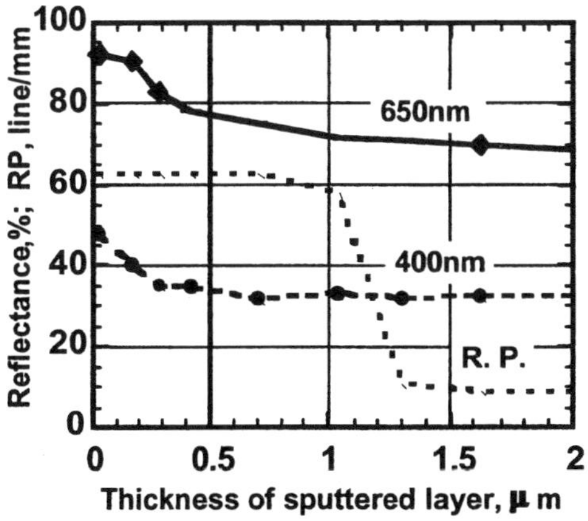

Fig.1. Dependences of reflectance (at λ=650 nm and λ=400 nm) and resolving power of the polycrystalline Cu mirrors on thickness of near-surface layer sputtered by ions of deuterium plasma with wide energy distribution.

the greater the difference between the plateau levels at any sputtered layer thickness, in correlation with faster R(h) degradation when increasing ion energy.[5,12]

It follows from data on long-term erosion of pc metal mirrors that the appearance of a large-scale stepped structure can be suppressed or even avoided by using materials either of an ordered structure (single crystals) or of a small size pc structure (films). Note that surface roughness growth of pc materials with a corresponding degradation of the optical properties is more important for short wavelengths ($\lambda < 1$ μm). For longer wavelengths, for example the far infrared region ($\lambda \sim 100$μm), Mo and W reflectors have shown high reflectance (0.94-0.99) even with the mean roughness of 2-3 μm[14].

Mirrors of monocrystal materials

Mirrors fabricated of monocrystalline (sc) metals have been found to behave differently from pc mirrors when subjected to long-term erosion due to sputtering. For the mirrors of sc Mo and W no decrease of R was observed even after erosion (due to ion bombardment) of a layer ~ 7 μm thick,[5,15] consistent with smooth surfaces seen on scanning electron microscope photographs.

A very similar behaviour was also observed with mirrors fabricated of sc SS and nickel. The tested SS sample mirrors had different orientations of their surface relative to crystallographic planes: (100), (110), and (111), and for the pure Ni mirror the surface was parallel to the (110) face. It was found that the reflectivity of all mirrors degraded slowly. Typically R degraded by only 2- 4% (at $\lambda = 650$ nm) after a ~ 5μm layer was eroded due to ion sputtering. This result is very important because SS monocrystals can be grown up to large size (~ 100 mm) exceeding the maximum size of W (< 30 mm) and Mo (< 50 mm) monocrystals.

The comparison of R(h) dependences for sc mirrors of some materials for the same face (111) is shown in Fig.2 together with R(h) graphs for pc W and Mo mirrors. The symbol "(111) block" denotes the data for a mirror fabricated of pc W consisting of (111) subgrains with an orientation within the limit 1-2 degrees relative to the main direction.

Metal film on metal substrate mirrors

Film mirrors on a metal substrate have several advantages over the sc mirrors: they permit much wider possibilities for choosing the mirror material and can be fabricated of large size. The material of the film can be chosen in accordance with requirements on the optical properties and at significantly lower cost (rhodium is one example of such a material[2]). The grain size in pc metal films is much less (up to ~ two orders in magnitude) than that in solid pc metals and the stepped microrelief developing under ion bombardment is also of small scale. In addition, the technology of film deposition is being continuously developed[16,17] and it is probable that a high adhesion of a substantial thick film (1-10 μm) will be available in the near future.

Several metal film on metal substrate mirrors have been tested in simulation experiments. In one case, a Be film was deposited on a pc Cu substrate by magnetron sputtering and occupied about half of the copper sample surface (22x22 mm)[18]. As distinct from the Cu part of the mirror where RP dropped after erosion ~ 1μm (Fig.1), the Be film mirror continued to have a low-scale microrelief and maintained the R and RP

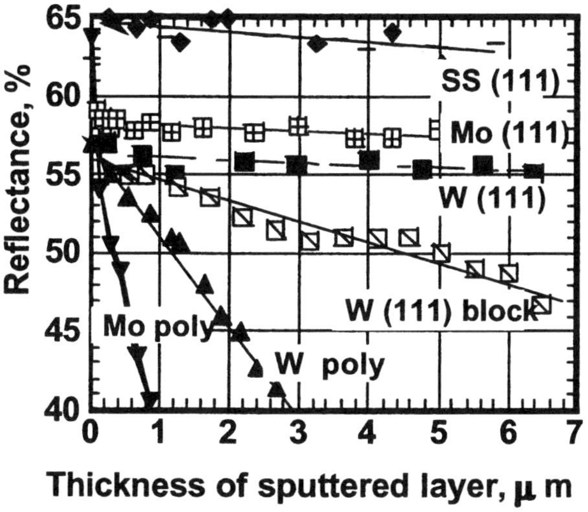

Fig.2. Dependences of reflectance (at λ=650 nm) on thickness of near-surface layer sputtered by ions of deuterium plasma with wide energy distribution for monocrystal and polycrystal mirrors of indicated metals.

values up to an eroded layer ~ 2μm thick. This corresponds to the erosion of most of the the initial thickness (~ 2.5μm) of the layer.

A very similar behaviour was also observed under long-term ion bombardment of a Cu film deposited on a pc Cu substrate. The part of the specimen initially coated with a Cu film (of ~ 1.5μm thick) maintained the mirror-like surface quality until the film was fully eroded.

Rhodium is a candidate material for some of the mirrors used in the LIDAR system[2], and so the ability of Rh film to maintain its optical properties under ion bombardment was recently investigated. In a simulation experiment Rh films deposited on Cu, Ni, SS and V substrates were tested. The thinnest film was on Ni (~0.15 μm), and the thickest ones on Cu substrates (5, 10, and 14 μm) were prepared on recently developed technology[16]. Some details on the laser damage test of film mirrors are presented in the satellite paper[17].

5. CONTAMINATION OF THE MIRROR SURFACE

Tungsten, beryllium and carbon-based composite materials are under consideration for protecting the first wall and the divertor in ITER. The eroded beryllium and carbon will be transported in the vacuum chamber and will accumulate on surfaces remote from plasma, including diagnostic ducts where some mirrors are located. As has been shown in present day fusion devices, the deposited layer can reach several microns in thickness, thus for ITER conditions deposits on the surfaces of FMs can be expected leading to a degradation of optical performance.

The spectral reflectance of SS mirrors has been shown to degrade significantly when coated with arc-evaporated carbon film of 11 and 21 nm in thickness[5]. A similar strong effect of carbon film deposition was observed for Mo mirror, as Fig. 3 demonstrates. The spectral dependences of optical conctants (i.e., the refractive index, n, and the extinction index, k) for the clean Mo surface were taken from paper[19] and for the carbon film the n and k values measured by ellipsometry[5] were used. One can see a quite reasonable qualitative agreement between the calculated values (using formula in reference[20]) and the measured R(λ) values for light at normal incidence.

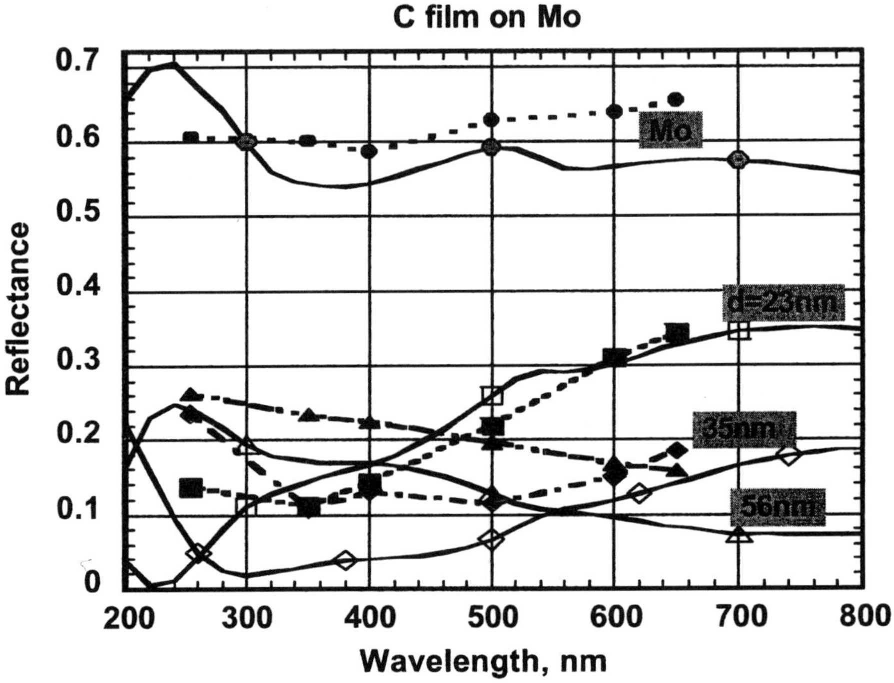

Fig.3. Effect of carbon film coating on spectral reflectance of Mo mirror. Solid lines with open symbols are results of calculation by using formula for reflectance of metal film mirror with semi-transparent contaminating film[20], dashed lines with closed symbols – results of measurements. The carbon film thickness is marked near every group of points.

The graphs of Fig. 3, as well as data for SS mirrors with boron[15] and carbon[5] films, indicate that even thin a contaminating film (a few nm thick) can strongly modify the reflectance of a metal mirror. There is therefore a need to investigate the possibilities of *in situ* cleaning of the mirror surface. Currently there are two possible approaches: in one case the contaminated surface is exposed to a low-temperature deuterium plasma produced near the mirror. In a second method, a not high power laser is used to ablate the surface contaminant. The recovery of reflectance (at λ=600nm) of SS mirror coated with a carbon film exposed to a deuterium plasma has been measured[5]. During that experiment with grounded SS mirror the ion energy was in the range 10-15 eV and estimation shows an efficiency of film cleaning ~0.03 carbon atom per every D atom in agreement with other experiments[21]. These preliminary data give some optimism for the prospect of plasma of local ECR discharge to clean the C film. The efficiency of C film removal by low-temperature plasma can be increased if the discharge is made in H_2 (D_2) with the addition of noble gases (Ne, Ar, Kr, Xe)[22], which may be acceptable for ITER.

The laser ablation method of C film from the SS mirror surface was tested for films of $\geq 0.1\mu m$ thick (the detail description of this experiment can be found in reference[17]). The laser wavelength was 308 nm, pulse duration 20 ns, pulse frequency 10 Hz, and total pulse energy - in the range 20-50 mJ. Within the laser spot (6x2 mm^2) the energy density of the laser beam varied in these experiments from 0.03 to 0.25 J/cm^2 per laser pulse. It was found that the cleaning process has a threshold character depending on the laser beam energy density, and the threshold is higher for thicker carbon film. With energy exceeding the threshold, the full cleaning of the illuminated part of the SS mirror surface and, correspondingly, the full restoration of the mirror reflectance was reached after 10 - 15 laser pulses.

6. THE PROSPECT OF BE AS THE FM MATERIAL

Deposits of Be on a mirror will lead to strong modification of R. Calculations shows that a mirror of any metal with a Be film of ~ 20 nm thickness will tend to have the optical properties close to those of a bulk Be mirror. Therefore, an experiment has been initiated in Kharkov to investigate the prospect for Be to be the FM material with the aim of preparing for the situation where Be deposition dominates over CXA sputtering. Be mirrors have a definite advantage in comparison to mirrors of some other metals: R is almost independent of λ in the range λ=300-700 nm[19]. The reflectance depends on the method of Be mirror preparation, and for the different bulk processed Be mirrors R ≈ 45%, i.e., in the range typical for tungsten.[19]

It was found that after a short time of exposure to ions of deuterium plasma, when the ion erosion can be fully neglected, the Be reflectance drops 5-20 %, depending on λ of reflecting light ($220 < \lambda < 650$ nm) and the pre-history of a given sample. This effect is probably due to the transformation of the BeO film into a $Be(OD)_2$ film due to bombardment of the mirror by deuterium ions as described in reference[23]. By ellipsometry at $\lambda = 632.8$ nm the extinction index of Be hydroxide film was found to be ~ 0.1, i.e., it is not negligible in comparison to that of a BeO film[24]. The annealing of the mirror after deuterium ion bombardment at 200-300 °C during a period of 1-3 hours resulted in the decomposition of $Be(OD)_2$ on the surface of the mirror and restoration of the BeO film with practically full restoration of the initial mirror reflectance in the visible range. These initial results with Be mirrors indicate that to prevent a similar degradation of Be FMs they should be: (i) fabricated as Be film of necessary thickness (to avoid a development of the step structure) and (ii) maintained at an elevated temperature (to avoid an accumulation of Be hydroxide). The range of temperature should be determined in the course of further experiments.

7. MIRRORS IN DIVERTOR

The environmental conditions for mirrors in the divertor region will be very different from those in the main chamber. The plasma particles will have an energy < 10 eV, i.e., below the threshold energy for physical sputtering of any material which is the prospect one for mirror fabrication. Thus, only chemical erosion of C-C composition will be important resulting in the production of volatile hydrocarbon molecules. The second distinctive feature of mirror operation in the divertor is that substantial quantities of dust may be produced due to erosion of the target plates and deposited locally, probably on the mirror surface also.

Thus, for the FMs in the divertor both the possibility of carbon film deposition and accumulation of dust on the mirror surface will have to be taken into account. However, while some progress has been made in carbon film cleaning (see above), investigations on the role of dust in the divertor region on FM optical properties are at a very early stage. The investigation of important dust characteristics, such as dust dimension and volume, dust distribution along the divertor region, dust adhesion to metal surfaces, etc., and their influence on FM properties, have yet to be carried out. In a recent paper[25] it was suggested that laser irradiation resulting in evaporation of the dust without destruction of the surface itself could be used for cleaning. The method was tested for several materials of dust and metal surfaces and looks quite promising.

A very useful step in the understanding of the role of environment effects for operation of FMs in divertor would be the installation of mirrors in current large-scale fusion devices with divertors, for example on JET, JT-60 or ASDEX, and the experimental investigation of their properties. This would aid the development of a model of contaminating film deposition, principally similar to the model described in paper[26].

8. DISCUSSION

The investigation of the long-term sputtering effects on the reflectance of metal mirrors shows that a step structure can be avoided by using materials either with an ordered structure (single crystals) or with a structure based on small-scale crystals (films). In both cases the maintenance of mirror reflectivity is possible even after the erosion of a layer of ~ 7 μm thick from the mirror surface. It has been estimated that for ITER the CXA atom flux (T^0 and D^0) to the first wall will be $\sim 2 \cdot 10^{19}$at/m^{2}s and mean atom energy < 250 eV[10]. Taking into account the characteristics of CXA flux and using known sputtering yields[27], the sc W and Mo mirrors would have a lifetime sufficient for the planned ITER operation ($\sim 10^4$ plasma pulses of $\leq 10^3$ s duration) even for mirrors at the first wall. By locating the mirror in a duct it is possible to reduce the CXA flux by an order of magnitude or more. In such a location the ≤ 10 μm thick Rh mirrors would have a sufficient lifetime. Thinner films (1-2 μm) could be used in a deeper duct or for mirrors protected by an aperture with higher degree of CXA flux attenuation.

The existing technologies permit the manufacturing of monocrystalline mirrors of limited size. For the film mirrors, there is no practical limit on the mirror size, e.g., in Kurchatov Institute a specially developed magnetron produces a flux of sputtered metal atoms and ions with quite homogeneity at a diameter ~ 30 cm. The measured reflectivity of mirrors fabricated in ENEA[28] and at the Kurchatov Institute with Rh film of ~1- 2 μm thickness on V, SS and Cu substrates, is very close to that which is calculated from the optical constants of these materials recommended in the literature[19]. Thus, it appears that the problems associated with the erosion of FMs due to the CXA flux can be overcome on the basis of existing technologies of mirror manufacture.

On the other hand, a deposited film of only 10-20 nm can substantially degrade the reflectance of metal mirrors. Therefore, the investigation of mechanisms of film growth on the FM surfaces is very important. Up to now only one process, potentially responsible for the deposit observed on the diagnostic windows of current fusion devices and which would probably give rise to deposits on FMs in ITER and similar machines, has been discussed in the literature[3]. It involves the redeposition of material of the diagnostic duct walls eroded due to physical sputtering by CXA. It could, therefore, be important if the duct wall is maintained as a metal surface of either the original composition or with beryllium deposited. The use of materials with low sputtering yield (W, Ta) for the protection of diagnostic duct walls could significantly decrease the role of this process in the contamination of FMs. However, such a modification will only have a minor impact on the deposit problem if the erosion products of components with a high sputtering yield diffuse inside the duct due to the creation of volatile molecules (hydrocarbons), i.e., when chemical erosion is predominant over physical sputtering. In the latter case, effective methods of FM cleaning from contaminants *in situ* have to be developed. It is known that the most effective method of removing carbon films from metal surfaces is to use a discharge in oxygen[29]: the efficiency of carbon atom cleaning reaches 0.2-0.3 per every oxygen atom (ion). However, the use of this gas is probably impossible in ITER and similar machines. Therefore, a significant improvement of existing cleaning methods is needed, and could possibly be achieved by using either a low temperature deuterium plasma with addition of the noble gases[22] or a laser ablation method[17].

9. CONCLUSIONS

The simulation work has established several options for the fabrication of diagnostic FMs. For FMs which have to withstand the highest CXA flux, for example those with a wide solid angle of observation located close to the first wall, mirrors fabricated of sc metals with low sputtering yield (W, Mo) are probably the best solution. FMs installed in diagnostic labyrinths where the CXA flux levels are attenuated by at least an order of magnitude can probably be fabricated from thick (~ 10 μm) metal film (e.g., Rh) on metal substrates. Thinner film (~ 1μm) will be sufficient in the case of ~ 100 times attenuation of CXA flux.

Further simulation tests are required. The performance of mirrors fabricated from metal films on metal substrates has yet to be examined with long-term bombardment by sub-keV energy ions of deuterium plasma. Moreover, the model of deposition of eroded material on FM surfaces needs to be developed and checked experimentally at fusion devices under operation so that quantitative predictions of the effect of deposition can be made. The behaviour of Be FMs should also be investigated for possible application in situations where the dominant contaminant is expected to be a redeposited Be from the first wall.

In-situ methods of cleaning the FM surfaces from carbon film and dust have to be improved and methods of calibration of the FM reflectance should be developed.

10. REFERENCES

1. Outline Design of the ITER-FEAT. To appear in the IAEA documentation series.
2. P.Nielsen, et al. LIDAR Thomson scattering for the ITER core plasma. In: Diagnostics for Experimental Thermonuclear Reactors 2, edited by P.E.Stott et al. (Plenum Press, 1998), p. 217.
3. V.S.Voitsenya. Rev.Sci.Instr. **70**, 787 (1999).
4. A.F.Bardamid, et al. Rev.Sci.Instr. **70**, 790 (1999).
5. V.S.Voitsenya et al. Rev.Sci.Instr. **72**, 475 (2001).
6. A.E.Costley, et al. Overview of the ITER Diagnostic SYSTEM. Ibid 2, P.41.
7. H.E.Bennett. JOSA, **53**, 1389 (1963).
8. H. Verbeek, et al. Nuclear Fusion **38**, 1789 (1998).
9. Y.Miura and JFT-2M Team. Nuclear Fusion, **37**, 175 (1997).
10. M.Mayer, et al. Ibid [2], p. 279.
11. A.F.Bardamid, et al. Surface Coatings & Technol. **103-104**, 365 (1998).
12. A.F.Bardamid, et al. Vacuum **58** (2000) 10.
13. R.Behrisch, et al. J. Nucl. Mater. **60**, 321 (1976).
14. M. Nagatsu. Private communication.
15. V.S.Voitsenya et al. J. Nucl. Mater. **290-293** (2001) 336.
16. N.V.Klassen, et al. "Materialovedenie", No 2, 1997, p.47 (in Russian).
17. K.Yu.Vukolov et al. This Proceedings.
18. A.M.Zimin, et al. Plasma Devices and Operations. **8**, 15 (1999).
19. Handbook of Optical Constants of Solids, E.D.Palik Editor, Acad.Press, 1985 and 1991.
20. G.Hass. JOSA. **47**, 1070 (1957).
21. V.S.Voitsenya. Sov. J. Plasma Phys. **17** (1991) 135.
22. Lianghua Yao, Shouqi Sun, Donghai Yan et al. J. Nucl. Mater. **176-177** (1990) 645.
23. V.M.Sharapov, V.Kh.Alimov, L.E.Gavrilov. J. Nucl. Mater. **258-263** (1998) 803.
24. D.F.Edwards and R.H.White. Beryllium Oxide. – Ibid [19], p.805.
25. A.G.Razdobarin, G.Federici, V.M.Kozhevin et al. Submitted to Fusion Science and Technology.
26. Y.Hirooka. J. Nucl. Mater., **258-263**, 1045 (1998).
27. Y.Yamamura and H.Tawara. Report NIFS-DATA-23, March 1995.
28. F.P.Orsitto, et al. Rev. Sci Instr. **72** (2001) 540.
29. B.Landkammer, A. von Keudell, W.Jacob. J.Nucl.Mater. **264**, (1999) 48.

RADIATION HARDNESS TEST OF MICA BOLOMETERS FOR ITER IN JMTR[1]

R.Reichle, T. Nishitani, E.R. Hodgson, L.C. Ingesson, E. Ishitsuka, S. Kasai, K.F. Mast, T. Shikama, J.C. Vallet, S.Yamamoto*

INTRODUCTION

Maximum radiation loss without disruption is essential to extend the maintenance intervals of ITER and following fusion reactors. The strong radiation distribution at the divertor must be known in fine detail to control such discharges. The corresponding reference diagnostic uses metal resistor bolometer cameras for detailed tomographic reconstructions of the radiation distribution[2]. The radiation hardness of the detector had been identified as the main question concerning the reference design. We report on the results of a Japanese/European collaboration (under ITER task T492) achieving the first in situ neutron irradiation tests of present day bolometers. A neutron dose equivalent to 0.1 dpa was reached, corresponding to the fluence between two maintenance periods at the bolometer locations behind the first wall elements or in the divertor under full load conditions of ITER-FEAT.

EXPERIMENTAL SET-UP

We used a former JET high temperature bolometer[3], derived from the ASDEX-Upgrade[4] version by replacing the kapton foil with a Muscovite mica foil of 20 µm

* R.Reichle, J.C. Vallet, DRFC, Association Euratom CEA sur la Fusion, CE Cadarache, F13108 Saint-Paul-lez Durance, France, T. Nishitani, S. Kasai, JAERI, Department of Fusion Engineering, Tokai-mura, Naka-gun, Ibaraki-ken 319-1195, Japan, E.R. Hodgson, Euratom / CIEMAT Fusion Association, Avenida Complutense 22, 28040 Madrid, Spain, L.C. Ingesson, FOM-Instituut voor Plasmafysica, Association Euratom-FOM, Nieuwegein, The Netherlands, E. Ishitsuka, JAERI, JMTR, Oarai Reserch Establishment, Oarai-machi, Higashi-ibaraki-gun, Ibaraki-ken 311-1394, Japan, K.F. Mast, Max-Planck-Institut fuer Plasma-physik, Euratom-IPP Association, Boltzmannstr.2, 85748 Garching, Germany, T. Shikama, Oarai Branch, Institute for Materials Research, Tohoku University, Oarai, Ibaraki, 311-1313, Japan S.Yamamoto, ITER-JCT, ITER Joint Work Site, Max-Planck-Institut fuer Plasmaphysik, Boltzmannstr.2, 85748 Garching, Germany

Advanced Diagnostics for Magnetic and Inertial Fusion, Edited by Stott *et al.*
Kluwer Academic/Plenum Publishers, New York, 2002

thickness and reducing the resistance to 200 Ω. The bolometer was connected with gold plated spring contacts to external connection wires arranged in twisted pairs and inserted in a metal tube to be installed into the Japan Materials Testing Reactor (JMTR) in Oarai. This reactor has a maximum thermal power of 50 MW and a maximum fluence of thermal neutrons of $4{\cdot}10^{18}$ n/(m^{2}s). The bolometer was placed inside the reactor for three 25 day reactor cycles to a total dose of 0.1 dpa (10^{24} n/m^2, E> 1 MeV) at a thermal neutron fluence of $1.5{\cdot}10^{18}$ n/(m^{2}s). A thermocouple was installed next to the bolometer to monitor the temperature. The tube was filled with helium at 0.3 Mpa. During the irradiation, resistance measurements were performed as well as electrical calibration measurements, to determine the sensitivity and the time constant of the bolometer. The principle of the electrical calibration (fig. 1a) is to step up the supply voltage to the bolometer (DC or 1 kHz AC) and monitor the temporal evolution of the consequent signal voltage while the additional electrical power heats up the bolometer. The DC calibration of the sensitivity was also done by measuring the current to voltage characteristic of the bolometer. The sensitivity is here defined as bridge output voltage per bridge supply voltage per incident power.

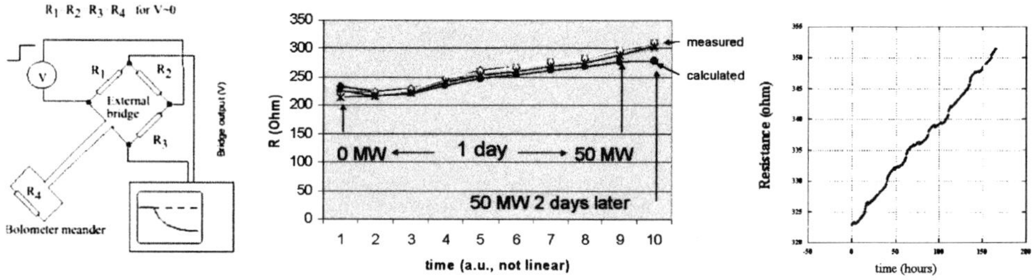

Figure 1. a) experimental set-up; b) resistance during ramp-up; c) resistance rise at constant reactor power.

EXPERIMENTAL RESULTS

The resistance of the gold meanders increased significantly during the experiment. During the ramp-up of the power in one day, the resistance follows closely the evolution expected from the heating up to 100°C (fig. 1b). Only after the full power of 50MW is established for 2 more days a clear additional rise of the resistance of 5% is noticed. This trend continued (fig. 1c) with an almost linear rise to 390 Ω at the end of the first cycle. During cool-down of the first cycle one meander broke in a process which showed at first a gradual small increase of the resistance during 1 minute and then a sudden increase to double the value indicating a rupture of the circuitry on the foil itself. Four hours later and two hours after reactor shutdown all connections went open circuit after the resistances had decreased to 315 Ω leaving a residual resistance increase of 90 Ω. Some resistance measurements recovered on start-up of the second cycle, allowing to continue with the measurements during the second cycle. Later in the second cycle some more connections were re-established to the meanders but one connection remained open, although not the one which had failed first. After 2/3 of the second cycle all contacts were definitively lost. The total permanent resistance rise had been about 75% of the initial value for a dose of 0.055 dpa. During the ramp-up phase of the reactor, the time constant rose from 25 ms to 32 ms, and then stabilized around 30 ms where it stayed throughout the rest of the experiment (fig. 2a). This value is considerably lower than the time constant in vacuum (160 ...230 ms). The sensitivity measurements showed some fluctuations during the startup phase (fig. 2b), but soon settled to a stable value of 0.1 1/W. To investigate the influence of high pressure He atmosphere on the bolometer, laboratory measurements were performed at Tore Supra with a similar bolometer at various pressures (fig. 3) which quantitatively confirmed the reduction of the time constant and also agree well with the sensitivity data from the irradiation experiment.

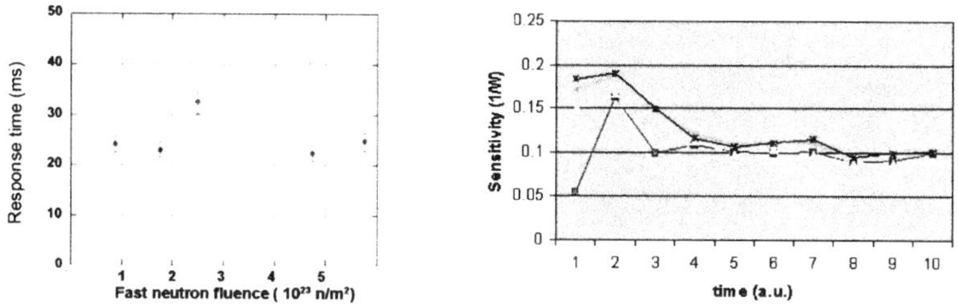

Figure 2. In situ measurements of : a) time constants, b) sensitivity.

DISCUSSION

One plausible explanation for the large permanent increase in resistance is, that the substrate becomes distorted due to radiation swelling, and hence a strain gauge effect causes the resistance to rise. This seems to be excluded by the previous JMTR investigations [5], which indicated an expansion of the mica substrate of less than 0.3% for 0.01 dpa, but seems possible in the light of other investigations[6] reporting an appreciable distortion after a fission reactor irradiation at 10^{-3} dpa. During the latter experiment water was present and the foil was supported in a frame, whereas in the former experiment this was not the case. It is however also known that H_2O may form in mica under irradiation [7]. Another candidate to increase the resistance is in principle the formation of point defects by atomic displacements. But the experience of JMTR is that Cu which has similar defect properties as gold, does not show an appreciable increase of resistance under similar conditions. Finally the resistance may increase due to nuclear transmutation of Au to Hg, which is a few % in 1 cycle from fast neutrons. Loss of contact at cool-down may be related to radiation induced embrittlement causing the gold meander to be less ductile on cool-down and possibly breaking. The observed variations of the time constant and the sensitivity at start-up may reflect changes of the conductivity of the surrounding atmosphere during start-up. Since the helium does not have a forced flow in and out of the tube but only past the tube in which the bolometer is held, a change of the composition, pressure and temperature is possible, all of which may influence the thermal conductivity and thus the time constant and sensitivity. Other explanations are also possible, e.g. the initial loss of sensitivity can partly be explained by the normal loss of sensitivity with rising temperature. The following plateau should therefore probably be somewhat lower than actually found. The difference might be explained by a degradation of the heat conductivity in the insulator and the metallic heat conduction layer.

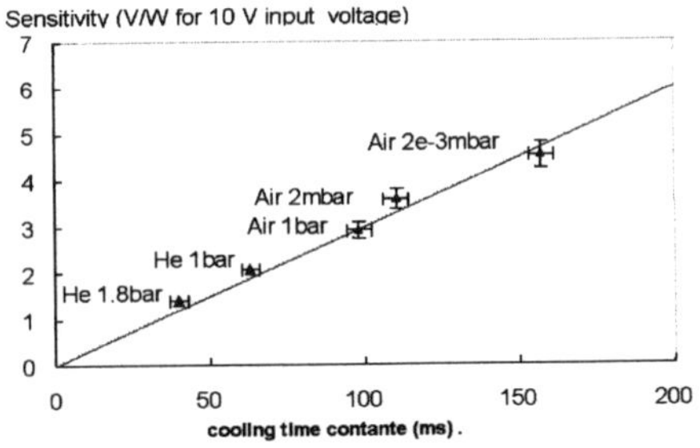

Figure 3. Sensitivity and time constant measurements at Tore Supra.

CONCLUSIONS

Until Post Irradiation Examination (PIE) has been performed[8], the causes for the problems with the resistance of the meanders and their stability are speculative. From the present understanding it may be better to replace the gold meanders with platinum ones. Mica remains for the moment a valid substrate candidate. However it is considered advisable to continue the development of alternative substrates using radiation resistant materials such as Al_2O_3 or AlN at least until a similar experiment has been done under vacuum conditions. Today's standard bolometers seem suitable for ITER start-up, and with reasonable further development should become usable also for the highly active phase.

REFERENCES

1. R. Reichle, T.Nishitani, E.R. Hodgson, L.C. Ingesson, E. Ishitsuka, S. Kasai, K.F. Mast, T. Shikama, J.C. Vallet, S.Yamamoto, Radiation hardness test of mica bolometers for ITER in JMTR, contr. 28th EPS Contr. Fus. Plasma Physics, Madeira, Funchal, Portugal 18-22 June 2001
2. R. Reichle, M Di Maio, L. C. Ingesson, Progress of the reference design for ITER bolometers and development of a high performance alternative, in: *Diagnostics for Experimental Thermonuclear Fusion Reactors 2*, eds. P.E. Stott, G. Gorini, P; Prandoni and E. Sindoni (Plenum Press, New York, 1998) pp. 399
3. R. Reichle, J.C. Fuchs, R.M. Giannella, N.A.C. Gottardi, H.J. Jaeckel, K.F. Mast, P.R. Thomas, 'Bolometer for ITER', in: *Diagnostics for Experimental Thermonuclear Fusion Reactors*, eds. P.E. Stott, G. Gorini and E. Sindoni (Plenum Press, New York, 1996) pp. 559
4. E.R. Mueller, G. Weber, K.F. Mast, G. Schramm, E. Buchelt, C. Andelfinger, IPP 1/224, Max-Planck Institut für Plasmaphysik, Garching, Oct 1985
5. M. Narui et al. to be presented at ICFRM-10, Oct. 2001 Baden-Baden, Germany
6. R. Heidinger, ITER diagnostics progress meeting 3 Feb 2000, Madrid, Spain
7. O.Sisman et al., Solid State Ann.Prog. Rep. Aug. 31, 1957, ORNL-2413, p80 University Press, 1969
8. T. Nishitani, T. Shikama, R.Reichle, E.R. Hodgson, L.C. Ingesson, E. Ishitsuka, S. Kasai, K.F. Mast, J.C. Vallet, and S.Yamamoto, Neutron irradiation effects on mica substrate bolometers, to be presented at ICFRM (Int. Conf. Fusion Reactor Materials)-10, Oct 2001, Baden-Baden, Germany

EXPERIMENTAL STUDY OF FIRST MIRROR DEGRADATION UNDER INFLUENCE OF ITER RELEVANT CONDITIONS

K. Vukolov, A. Bardamid, A. Gorshkov, V. Konovalov, A. Rogov, V. Sannikov, S. Solodovchenko, V. Voitsenya, S. Zvonkov*

INTRODUCTION

Being placed in a straight view of burning plasma the First Mirror (FM) is one of most critical elements of ITER diagnostics. The sputtering by charge exchange atoms (CXA) and the deposition of contaminating films and dust on a mirror surface will be the main reasons of FM reflectivity degradation under ITER conditions, and for laser diagnostics the thermomechanical stresses occurring at every laser shot may accelerate the mirror destruction [1]. The materials traditionally used for optical mirrors are not suitable for building-up the ITER diagnostic mirrors. The use of mirror with reflective coating from metals with low sputtering yield and sufficiently high reflectivity (e.g., Rh, Mo, W, Ag) on metallic substrate is one of the possible variant of FM choice for a survival in extreme ITER environments [2]. The production practice of mirrors with Rh and Mo coatings on SS and Mo substrates (Rh/SS and Mo/Mo, correspondingly) by means of sputter deposition of reflective coating on polished substrates is elaborated using a dc planar magnetron [3]. The influence of main damage factors on FM prototypes and experiments on a remote cleaning of mirrors from carbon films by pulsed laser radiation are considered in the paper.

CXA SPUTTERING

Rh/SS and Mo/Mo mirrors with thickness of coating 1-2 μm were prepared for tests. The dynamics of mirror reflectivity change depending on thickness of layer sputtered by ions of deuterium plasma was studied and the reflective coating structure of mirrors was

* K. Vukolov, A. Gorshkov, A. Rogov, V. Sannikov, S. Zvonkov, RRC "Kurchatov Institute", Moscow, Russia; A. Bardamid, Taras Shevchenko Kiev University, Ukraine; V. Konovalov, S. Solodovchenko, V.Voitsenya, IPP NSC Kharkov Institute of Physics and Technology, Ukraine.

Advanced Diagnostics for Magnetic and Inertial Fusion, Edited by Stott *et al.*
Kluwer Academic/Plenum Publishers, New York, 2002

investigated [3]. These tests have shown that properties of substrate material and a structure of coating have the determining influence on the mirror behavior under ion bombarding. In particular, the reflectivity degradation of Mo/Mo mirrors was determined by a quality of substrates fabricated of a TZM type Mo alloy. In the case of Rh/SS mirror a relatively fast degradation of reflectivity was the result of difference of the Rh layer structure from necessary for stability under long-term sputtering a needle-shaped structure. Fig.1 shows the results on sputtering of a 2 μm rhodium reflecting coating.

CONTAMINATION OF MIRROR SURFACE AND LASER CLEANING

The deposition of contaminating films, the second damage factor, will be especially dangerous for diagnostic mirrors located in the divertor area where hydrocarbon films may be deposited as a result of chemical erosion of divertor tiles fabricated of carbon-based material. To model this process the carbon films of various thickness (10, 20, 50 nm) were deposited on a surface of SS mirrors by means of magnetron sputtering of a graphite cathode by Ar ions at 10^{-3} Torr gas pressure. The influence of mirror temperature (20 and 200°C) and additional letting-to-hydrogen on reflectivity degradation were studied. The specular reflectivity (k_{ref}) as function of film thickness was measured for λ=650 nm at the 45° angle of incidence. Results are shown in Fig.2.

Besides, three SS mirrors were exposed during 6 months inside the T-10 tokamak (2×10^{3} plasma pulses, ~1 s duration each) in the diagnostic duct near to the carbon limiter and at different distances (5 cm, 30 cm and 55 cm) from the plasma edge. The reflectivity degradation of the mirror located on a distance of 5 cm from plasma was small (Δk_{ref}= 4%) and was caused by erosion of mirror surface due to sputtering during tokamak chamber cleaning by means of induction or glow discharges practically every night in ~10 hours. Two other mirrors were covered by brown (3 μm of thickness - Δk_{ref}=54%) and yellow (0.9 μm - Δk_{ref}=9%) sandwich-type hydrocarbon films with a rather smooth surface. This contaminating material was removed from the walls and limiter during the discharge cleaning procedure and redeposited on mirror surfaces.

These mirrors were used to study the contamination ablation under impact of eximer laser radiation (λ=308 nm, repetition rate F=10 Hz, shot duration τ=20 ns). The eximer laser beam was focused on a surface of tested mirror in rectangular spot 6×2 mm^2 and energy density (ρ) in the centre of laser spot on mirror surface could vary from 0.03 up to 0.25 J/cm^2. The beam of continuous diode laser (λ_{max}=650 nm) was directed to the centre of eximer laser spot on the mirror surface and the intensity of its specular reflection and diffusive scattering were recorded during the laser ablation process.

The cleaning process has a threshold character relatively to the energy density of laser radiation depending on thickness and type of contaminating film. The experimental results are summarised in Fig.3, where the dependences of laser energy density threshold on film thickness are presented. The destruction of carbon film takes place with gradual removing of its micro pieces when the energy density is higher than a certain threshold. The mirror cleaning is going faster with higher energy density of a laser beam. But only scanning of laser beam gives the complete cleaning of mirror and for effective cleaning the laser energy density should be about 50% higher than the damage threshold of the carbon film. It was shown that the cleaning does not make worse the initial mirror

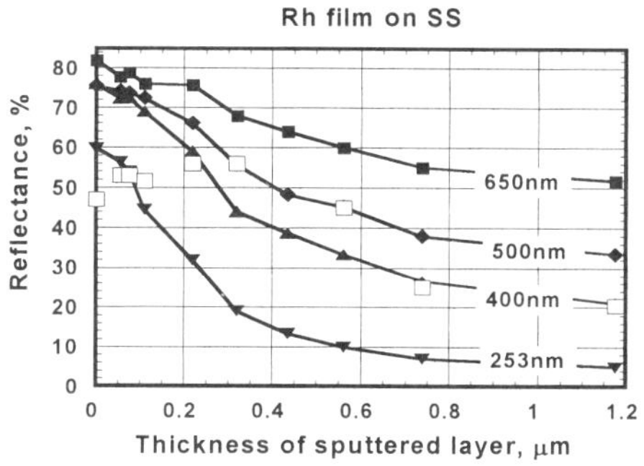

Figure 1. The long-term sputtering effect on reflectance of Rh coating on SS substrate for various wavelengths. The change of mirror's resolving power is shown as opened squares

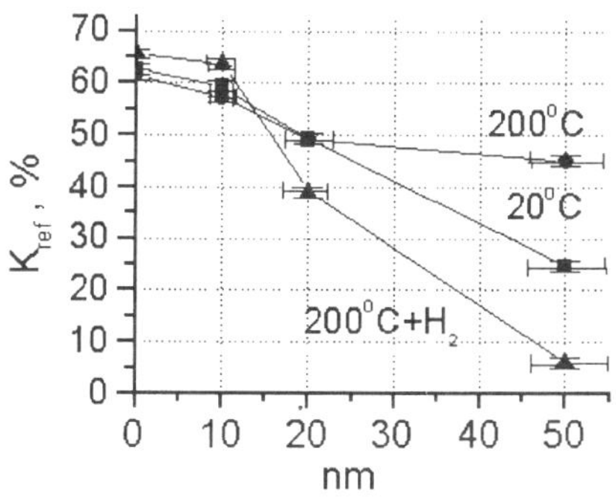

Figure 2. Specular reflectivity on λ=650 nm as function of carbon film thickness (X axis, nm)

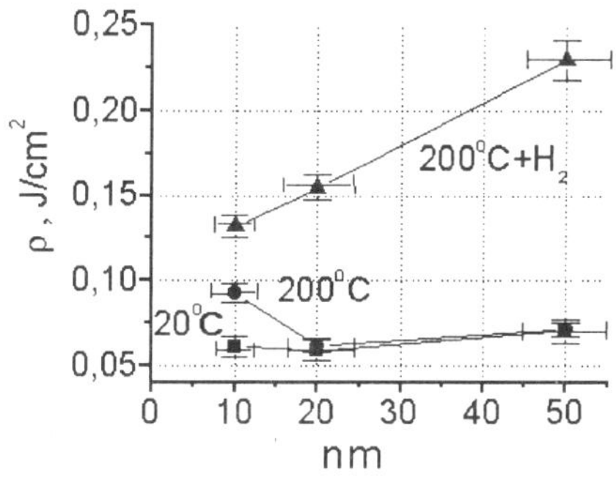

Figure 3. Laser energy density threshold (ρ) depending on carbon film thickness (X axis, nm)

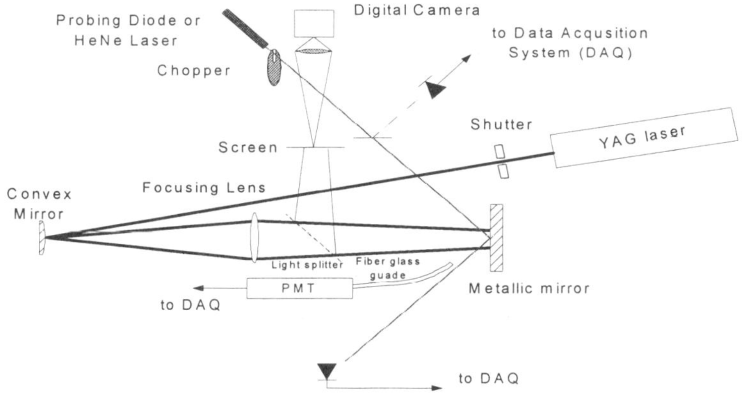

Figure 4. Lay-out of laser test bench for measurements of laser damage thresholds

quality and, in particular, the eximer laser damage threshold of SS mirror (0.8±0.1 J/cm^2) was not changed after cleaning by the scanning laser beam.

LASER DAMAGE

First data on laser damage thresholds under the influence of the frequency-operated pulsed YAG laser (λ=1.06 μm, τ=12 ns, energy - 30 mJ, F=12.5 Hz) were obtained for single laser shot and after about of $4 \cdot 10^4$ laser shots. The set up of used equipment is shown in Fig.4. The output beam of the laser operating in the TEM_{00} mode has a Gaussian profile. Due to moveable focusing lens the diameter of laser spot and, in that way, the laser energy density on the mirror surface can be changed.

At first, the single shot damage thresholds were measured and they were equal to 1.3 J/cm^2 for SS, 1.7 J/cm^2 for Mo/Mo and 2.2 J/cm^2 for Rh/SS mirrors. After that a long-term effect of frequency-operated YAG laser on mirror reflectivity was studied. At the latter tests an average laser energy density was several times lower than the single shot damage threshold. Nevertheless under the SS mirror test the degradation of mirror optical quality started after ~10^3 laser shots at ρ=0.4 J/cm^2 and after $4 \cdot 10^4$ laser shots the specular reflectivity decreased by 6% and the diffusive scattering became seven times over its initial value. But it turned out, that Mo/Mo and Rh/SS mirrors are much more stable. In fact, the Mo/Mo mirror did not degrade at 0.7 J/cm^2 and the reflectivity of Rh/SS mirror was not changed even at 1.8 J/cm^2, both samples having been subjected to 40000 laser shots.

CONCLUSION

The tests of FM prototypes with reflective coatings were performed, that has allowed to find out their stability under ITER conditions and also to reveal a few imperfections of technology. The initial quality of mirrors was quite satisfactory and some defects were discovered only during long-term sputtering test. The efforts on an improvement of the sputter deposition technology proceed at the moment.

The experiments on modelling of contaminating film's deposition have shown an importance of creation of conditions that would be as much as possible similar to the ITER environments, because the properties of mirrors with contamination on a surface do very strongly depend on films chemical composition and structure, which are wholly determined by deposition conditions. The successful tests of a technique of laser cleaning of carbon and hydrocarbon films from the mirror surface were carried out.

The diagnostic laser test bench was assembled and the first results on impact of a multi-shot laser radiation on metallic mirrors have been obtained.

REFERENCES

1. V.S. Voitsenya et al. Rev. Sci. Instr. 70, 2016 (1999).
2. V.S. Voitsenya et al. Rev. Sci. Instr. 72, 475 (2001).
3. K.Yu. Vukolov et al. Final report of ITER R&D T481, wbs No. N 55 TT 10 FR

THOMSON SCATTERING IN INERTIAL CONFINEMENT FUSION RESEARCH

S. H. Glenzer*

INTRODUCTION

Thomson scattering [1] using short-wavelength probe lasers [2] is the standard diagnostic to characterize high-density inertial confinement fusion (ICF) plasmas [3]. Its unique capability to measure plasma parameters such as temperatures, densities, plasma flow, and ionization stages with high accuracy together with possible measurements of plasma wave fluctuations is now widely recognized in the area of laser-produced plasma research. Initially, many basic Thomson scattering experiments with applications to ICF have been performed [4] on the Nova laser facility at the Lawrence Livermore National Laboratory that produced hot mm-scale plasmas with laser energies of up to 30 kJ. More recently, we have also begun to perform Thomson scattering experiments at a second large facility namely the Omega laser facility at the University of Rochester [5].

One of the most important applications of Thomson scattering in ICF is the characterization of closed-geometry hohlraums that will be employed on future indirect drive ignition experiments to implode DT-filled fusion capsules to ignite and to create a burning plasma [3]. These hohlraums consist of multi-ion-species plasmas that can show steep gradients on a scale of 100 μm at electron densities in excess of 10^{21} cm^{-3}. Because Thomson scattering measurements have not been routinely performed in these hostile conditions [6], we have also applied Thomson scattering in open-geometry plasmas to investigate fundamental properties of ion acoustic or electron plasma (Langmuir) waves in two-ion species plasmas [7] and in highly ionized Au plasmas with steep gradients [8]. When implementing a high-energy ultraviolet probe laser on Nova [9] that operated at 263 nm (4ω), accurate measurements of the electron and ion temperature became possible in high-density gas-filled hohlraums. These measurements have provided us with a unique data set that we used to test integrated radiation-hydrodynamic modeling using the code LASNEX [10] which is presently being applied to design future fusion experiments, e.g, on the National Ignition Facility. Our Thomson scattering experiments have shown the existence of steep electron temperature gradients

---

* S. H. Glenzer, L-399, Lawrence Livermore National Laboratory, University of California P.O. Box 808, Livermore, CA 94551, U.S.A.

Advanced Diagnostics for Magnetic and Inertial Fusion, Edited by Stott *et al.*
Kluwer Academic/Plenum Publishers, New York, 2002

in the hohlraum that can only be modeled when including a self-consistent calculation of magnetic fields [4]. These temperature measurements have been applied to test atomic kinetics calculations of x-ray spectra observed from these hohlraum plasmas [11,12] and were critical for calculations of stimulated scattering losses by parametric laser-plasma instabilities [13] inside the hohlraum. The latter occur when ion acoustic or electron plasma waves are excited to large amplitudes by the laser producing stimulated Brillouin and Raman scattering [14,15,16], respectively. Recently, we also applied Thomson scattering to directly investigate large amplitude fluctuations [17] as well as these large amplitude waves [18] by matching the **k**-vectors of the Thomson scattering experiment and the plasma waves. These measurements have shown that ion waves are saturated in ICF plasmas providing us with a better understanding of the laser plasma coupling in high-temperature laser-fusion plasmas.

EXPERIMENTAL SETUP

The experiments discussed here were performed on the Nova laser facility [19], a Nd:glass laser that operated at 1.055 μm (1ω) and that could be frequency converted to 2ω and 3ω. The cylindrical hohlraums of 2.75 mm length and of 1.6 mm diameter were filled with gas (methane or propane) giving electron densities of $n_e > 10^{21}$ cm^{-3} when fully ionized by the laser heater beams and compressed by the inward-moving high-Z wall plasma. These hohlraums are the standard size for capsule implosions [20] on present large scale laser facilities. For the heating of the hohlraums we employed ten 3ω heater beams that were smoothed with kinoform phase plates giving a total energy of 27 kJ in these experiments. The beams were arranged in cones on either side of the hohlraum so that each beam forms an angle of 50° to the hohlraum axis. The heater beams penetrate the hohlraum at both ends through laser entrance holes that were covered with 0.35 μm thick polyimide membranes. We applied shaped laser pulses of 2.4 ns duration which rise from 0.6 to 1.8 TW per beam reaching peak power at 1.5 ns.

The Thomson scattering experiments were performed with Nova's 4ω probe laser [9]. In these experiments, the probe was operating at 4ω (λ_0 = 263.3 nm) with a 2 ns long pulse providing 50 J (at 4ω) on target. This energy is negligible compared to that of the heater beams and is therefore producing little perturbation to the plasma. We focussed the probe beam to a 100 μm spot inside the hohlraum into the gold blow off plasma at a distance of 300 μm from the hohlraum wall and into the gas plasma on the hohlraum axis. The scattered light has been measured at a scattering angle of $\theta = 90°$ through a diagnostic window cut in the side of the hohlraum. Figure 1 shows a schematic of the Thomson scattering experiment. The plasma composition at the location of the scattering volumes were monitored with temporal-resolved 2-D x-ray images (E > 2 keV) that observed the hohlraum plasma emission though the laser entrance hole. For the Thomson scattering measurements we employed f/10 optics with a 1.5x magnification and a 1m spectrometer equipped with an optical streak camera to record spectra with a wavelength resolution of 0.1 nm and a temporal resolution of 50 ps. Collective Thomson scattering is expected for the parameters of the experiments (electron density, temperature, scattering angle, and probe laser wavelength). Typical scattering parameters are $\alpha = 1/k\,\lambda_D > 2$, and light is predominantly scattered into the narrow ion feature of the Thomson scattering spectrum.

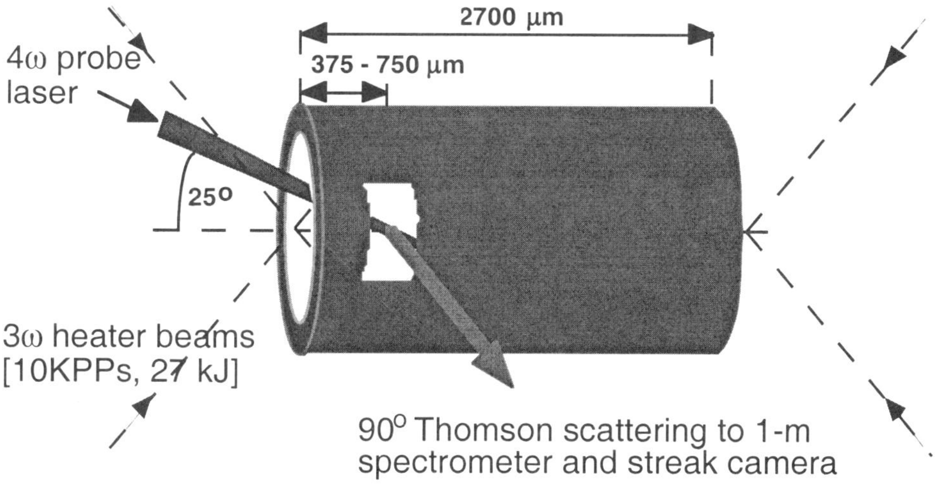

Figure 1. Schematic of the experiment.

RESULTS AND DISCUSSIONS

Figure 2 shows an example of Thomson scattering data along with fits of the theoretical Thomson scattering form factor giving the electron temperature, T_e, in case of the (pure) gold plasma from the hohlraum wall, and T_e and T_i in case of the Thomson scattering data from the gas plasma. The Thomson scattering spectra from the low-Z gas plasma show two broad ion acoustic peaks. This is because the two ion acoustic waves belonging to C and H give rise to four unresolved ion acoustic peaks on the Thomson scattering spectrum that measure the co- and counter-propagating wave fluctuations along the Thomson scattering wave vector **k**. Furthermore, ion Landau damping is efficient for the low-Z species causing an additional T_i-dependent broadening of the peaks. From the spectra we infer the plasma parameters with high accuracy where error bars (shown in Figs. 3 and 4) are obtained by varying the fits within the noise of the experimental data. Electron temperature gradients are also taken into account although they primarily affect the measurement of the ion temperature. A small uncertainty in T_e of about 5% is added due to this effect resulting in a total error bar for T_e of 15%. The error bar for T_i is larger because noise, and plasma gradients are more important, resulting in an uncertainty of up to 25%. Finally, the macroscopic plasma flow velocity along the scattering vector **k** is inferred from the Doppler shift of the whole Thomson scattering spectrum away from the incident laser wavelength λ_0. In the data shown in Fig. 2, we observe a small red-shift indicating that the plasma is moving away from the hohlraum center towards the laser entrance hole consistent with x-ray imaging and with modeling.

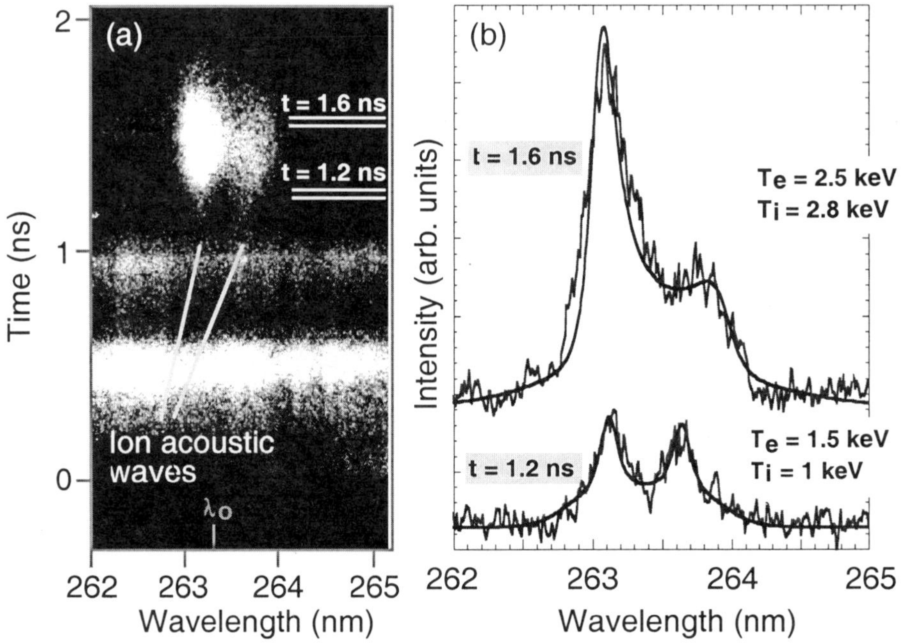

Figure 2. (a) Example of Thomson scattering streak spectra from the low-Z plasma inside a hohlraum that was filled with methane. (b) Example of line-outs at t = 1.2 ns and t = 1.6 ns together with fits of the theoretical form factor $S(\mathbf{k},\omega)$ to the data giving T_e, T_i and macroscopic plasma flow.

Figure 3 shows the experimental electron temperatures measured on the axis at various distances from the hohlraum center. Data are shown at t = 1.3 ns from methane-filled hohlraums together with results from two-dimensional radiation-hydrodynamic modeling. We applied the code LASNEX with two approximations: The first model uses a flux-limited diffusion model for heat transport where the heat flow per unit area in regions of large classical heat flow, $-\kappa$ grad T_e, has an upper bound of $f\,(n_e\,T_e\,v_e)$. Here, $(n_e\,T_e\,v_e)$ is the so-called free streaming value of heat transport and f is the flux limit chosen to be a constant value of f = 0.05. The second model includes large-scale magnetic fields [4].

The comparison in Fig. 3 clearly shows that the inclusion of magnetic fields into the modeling results in excellent agreement between the Thomson scattering data and the calculations. At early times, magnetic fields of the order of 1MG are produced at the hohlraum walls where the laser beams initially deposit their energy. In addition, the hohlraum window is also magnetized. These fields convect into the low-Z plasma of the hohlraum causing inhibition of electron thermal conduction since the product of the electron cyclotron frequency and electron-ion collision time, $\omega_c\,\tau_e$, approaches values of ten (for $\omega_c\,\tau_e \sim 1$, thermal conduction inhibition by magnetic fields become important). In particular, hot plasma regions develop in the path of the laser beams which are surrounded by magnetic fields of order 100kG. Furthermore, temperature gradients develop along the axis of the hohlraum and are clearly correlated to the presence of the magnetic fields. Our Thomson scattering measurements have observed both regions of hot plasma, the hot off-axis regions [6] and the axial temperature gradient (Fig. 3). The good agreement of this model with the Thomson scattering data is very encouraging as its demonstrates our level of theoretical understanding of hohlraum plasma conditions and electron thermal transport as well as the importance of magnetic fields in ICF hohlraums.

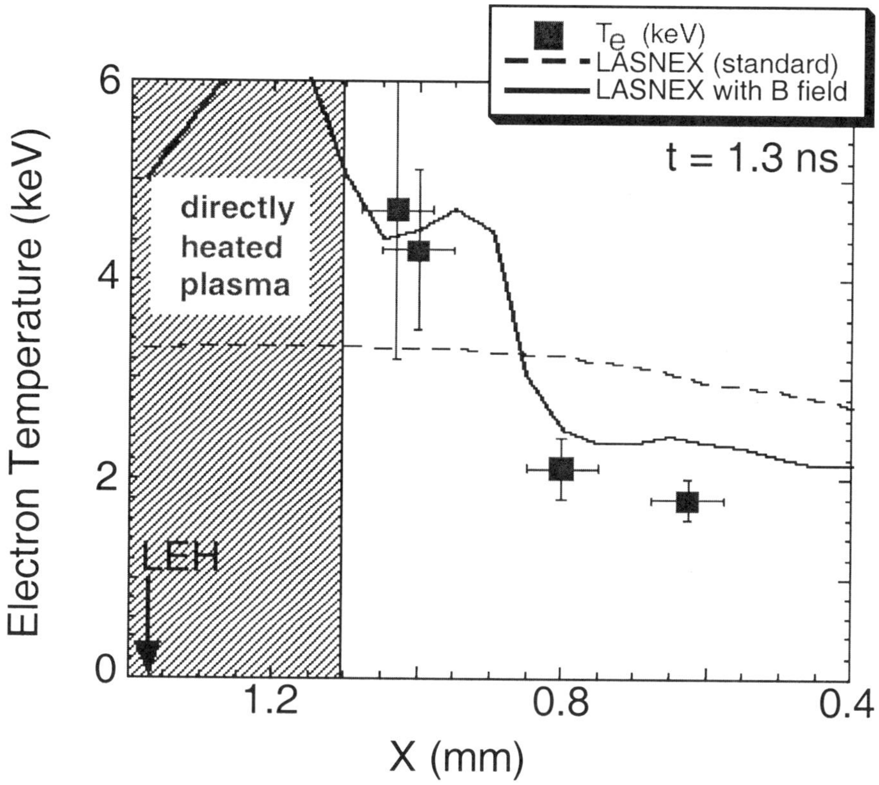

Figure 3. Results of the Thomson scattering measurements at t = 1.3 ns showing the electron temperature along the axis of a gas-filled hohlraum plasma (CH_4) together with results from two-dimensional radiation-hydrodynamic modeling. The laser entrance hole (LEH) is pointed out at x = 1.375 mm.

Figure 4 shows the results of the electron and ion temperature measurements as a function of time inside the hohlraum at a distance of x = 800 μm on the axis from the hohlraum center. We find that T_i is insensitive to the heat transport model used in the simulations. The ion temperature is determined by p dV-work when the low-Z plasma is compressed on the axis by the inward blowing gold plasma from the hohlraum walls. The experimental ion temperatures rise rapidly at t = 1.2 ns to peak values of T_i = 3 keV. The hydrodynamic simulations confirm that the rise of T_i is due to stagnation of the compressed low-Z plasma on the axis of the hohlraum where 90° particle scattering occurs on a scale length of 30 μm. When increasing the gas fill density by more than a factor of two using higher-density gas fills, Thomson scattering shows that the stagnation time is delayed by 0.3 ns [4]. This finding is in good agreement with the radiation-hydrodynamic calculations verifying that a high-density gas fill in ignition target designs will be beneficial in preventing early axial stagnation. Consistent with our findings in Fig. 3 we find that the electron temperature is fairly well modeled at this location. The dashed curve uses the code LASNEX with the magnetic field model.

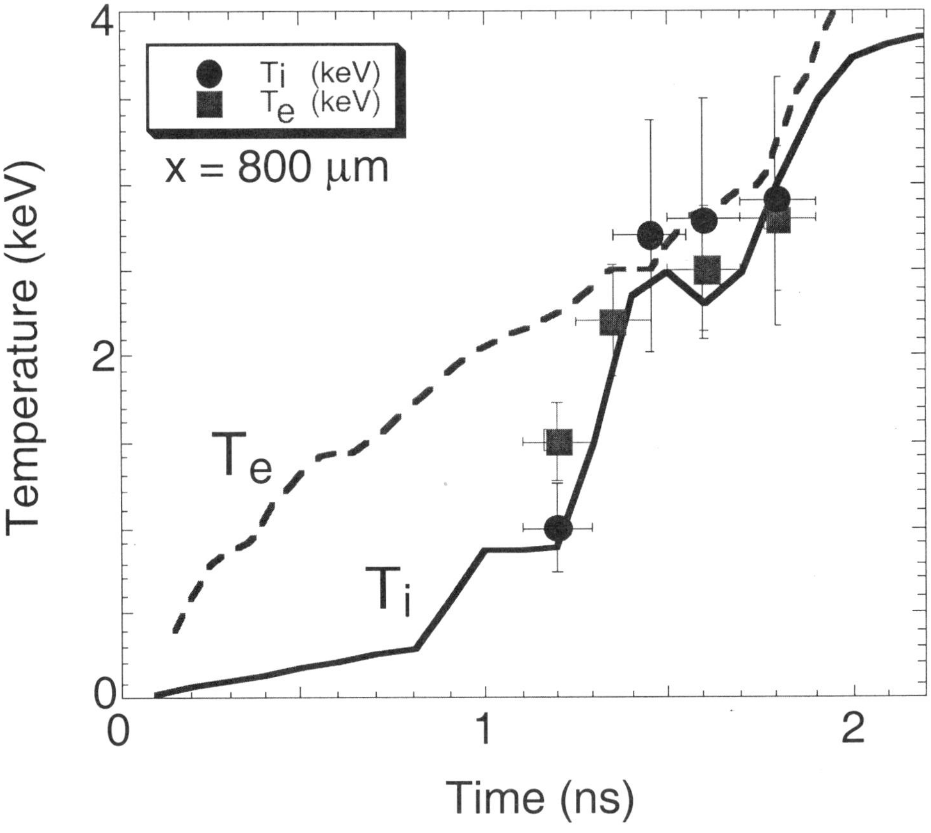

Figure 4. Results of the Thomson scattering measurements at x = 800 μm showing the electron and ion temperature as function of time of a gas-filled hohlraum plasma (CH_4) together with results from two-dimensional radiation-hydrodynamic modeling.

CONCLUSION

We have shown that Thomson scattering is the key diagnostic for accurately measuring the plasma conditions in laser plasmas, in particular inside inertial confinement fusion hohlraums. The comparison of the radiation-hydrodynamic simulations with the results of the Thomson scattering measurements have shown that including magnetic fields into the heat transport modeling is important for predicting the plasma conditions in inertial confinement fusion research. This finding is important for a number of applications. For example, we find that stimulated Raman scattering spectra calculated from a convective gain model, with LASNEX calculations providing the input plasma parameters, show improved agreement with measurements when including magnetic fields. For ignition experiments, the presence of electron temperature gradients as seen with the Thomson scattering measurements will spatially de-tune the stimulated Raman scattering resonance which might lead to a smaller scale length for the growth of this unwanted parametric laser-plasma instability. Verifying these plasma wave phenomena in future laser-plasma interaction experiments on Omega [21] will be an important part of our research program in preparing ignition experiments on NIF.

ACKNOWLEDGEMENT

This work was performed under the auspices of the U.S. Department of Energy by University of California Lawrence Livermore National Laboratory under contract No. W-7405-ENG-48.

REFERENCES

1. H.-J. Kunze, The laser as a tool for plasma diagnostics, in *Plasma Diagnostics*, edited by W. Lochte-Holtgreven (North- Holland, Amsterdam, 1968) pp. 550-616.
2. H. A. Baldis, D. M. Villeneuve, and C. J. Walsh, Plasma Waves in Laser Fusion Plasmas, *Can. J. Phys.* **64**, 961-968 (1986).
3. J. D. Lindl, *Inertial Confinement Fusion*, (Springer, New York, 1998).
4. S. H. Glenzer, W. E. Alley, K. G. Estabrook, J. D. De Groot, M. Haines, J. H. Hammer, J.-P. Jaddaud, B. J. MacGowan, J. D. Moody, W. Rozmus, L. J. Suter, T. L. Weiland, and E. A. Williams, Thomson scattering from laser plasmas, *Phys. Plasmas* **6**, 2117-2128 (1999).
5. O. L. Landen, S.H. Glenzer, M.J. Edwards, R.W. Lee, G.W. Collins, R.C. Cauble, W.W. Hsing and B.A. Hammel, Dense Matter Characterization by X-ray Thomson Scattering, *J. Quant. Spectrosc. Rad. Transfer* **71**, 465-478 (2001).
6. S. H. Glenzer, C. A. Back, L. J. Suter, M. A. Blain, O. L. Landen, B. J. MacGowan, G. F. Stone, R. E. Turner, and B. H. Wilde, Thomson Scattering from Inertial Confinement Fusion hohlraum plasmas, *Phys. Rev. Lett.* **79**, 1277-1280 (1997).
7. S. H. Glenzer, C. A. Back, K. G. Estabrook, R. Wallace, K. Baker, B. J. MacGowan, B. A. Hammel, R. E. Cid, and J. S. De Groot, Observation of Two Ion Acoustic Waves in a Two-Species Laser-Produced Plasma with Thomson Scattering, *Phys. Rev. Lett.* **77**, 1496-1499 (1996).
8. S. H. Glenzer, W. Rozmus, B. J. MacGowan, K. G. Estabrook, J. D. De Groot, G. B. Zimmerman, H. A. Baldis, J. A. Harte, R. W. Lee, E. A. Williams, and B. G. Wilson, Thomson scattering from high-Z laser-produced plasmas, *Phys. Rev. Lett.* **82**, 97-100 (1999).
9. S. H. Glenzer, T. L. Weiland, J. Bower, A. J. MacKinnon, and B. J. MacGowan, High-Energy 4ω Probe Laser for laser-plasma experiments at Nova, *Rev. Sci. Instrum.* **70**, 1089-1092 (1999).
10. G. B. Zimmerman and W. L. Kruer, *Comments Plasma Phys. Controlled Fusion* **2**, 85 (1975).

11. S. H. Glenzer, F. B. Rosmej, R. W. Lee, C. A. Back, K. G. Estabrook, B. J. MacGowan, T. D. Shepard, and R. E. Turner, Measurements of suprathermal electrons in hohlraum plasmas with x-ray spectroscopy, *Phys. Rev. Lett.* **81**, 365-368 (1998).
12. S. H. Glenzer, K. B. Fournier, B. G. Wilson, R. W. Lee, and L. J. Suter, Ionization Balance in Inertial Confinement Fusion Hohlraums, *Phys. Rev. Lett.* **87**, 5002-5004 (2001).
13. S. H. Glenzer, R. L. Berger, L. M. Divol, R. K. Kirkwood, B. J. MacGowan, J. D. Moody, J. E. Rothenberg, L. J. Suter, and E. A. Williams, Reduction of stimulated scattering losses from Hohlraum Plasmas with Laser Beam Smoothing, *Phys. Plasmas* **8**, 1692-1697 (2001).
14. W. L. Kruer, *The Physics of Laser Plasma Interactions* (Addison-Wesley, New York, 1988).
15. B. J. MacGowan, B. B. Afeyan, C. A. Back, R. L. Berger, G. Bonnaud, M. Casanova, B. I. Cohen, D. E. Desenne, D. F. Dubois, A. G. Dulieu, K. G. Estabrook, J. C. Fernandez, S. H. Glenzer, D. E. Hinkel, D. H. Kalanter, R. L. Kauffman, R. K. Kirkwood, W. L. Kruer, A. B. Langdon, B. F. Lasinski, D. S. Montgomery, J. D. Moody, D. H. Munro, L. V. Powers, H. A. Rose, C. Rouseaux, R. E. Turner, B. H. Wilde, S. C. Wilks, and E. A. Williams, Laser Plasma Interactions in ignition-scale hohlraum plasmas, Phys. Plasmas **3**, 2029-2040 (1996).
16. R. K. Kirkwood, D. S. Montgomery, B. B. Afeyan, J. D. Moody, B. J. MacGowan, S. H. Glenzer, W. L. Kruer, K. G. Estabrook, K. B. Wharton, E. A. Williams, and R. L. Berger, Observation of the nonlinear Saturation of Langmuir Waves Driven by Ponderomotive Force in a Large Scale Plasma, *Phys. Rev. Lett.* **83**, 2965-2968 (1999).
17. J. D. Moody, B. J. MacGowan, S. H. Glenzer, R. K. Kirkwood, W. L. Kruer, D. S. Montgomery, A. J. Schmitt, E. A. Williams, and G. F. Stone, First measurements of short scale-length density fluctuations in a large laser-plasma, *Phys. Rev. Lett.* **83**, 1783-1786 (1999).
18. S. H. Glenzer, L. M. Divol, R. L. Berger, C. Geddes, R. K. Kirkwood, J. D. Moody, E. A. Williams, and P. E. Young, Thomson scattering measurements of saturated ion waves in laser fusion plasmas, *Phys. Rev. Lett.* **86**, 2565-2568 (2001).
19. E. M. Campbell, Recent results from the Nova Program at LLNL, *Laser Part. Beams* **9**, 209-231 (1991).
20. B. A. Hammel, C. J. Keane, M. D. Cable. D. R. Kania, J. D. Kilkenny, R. W. Lee, and R. Pasha, X-ray spectroscopic measurements of high-densities and temperatures from indirectly driven inertial confinement fusion capsules, *Phys. Rev. Lett.* **70**, 1263-1266 (1993).
21. W. Seka, R. E. Bahr, R. W. Short, A. Simon, R. S. Craxton, D. S. Montgomery, A.E. Rubenchik, Nonlinear laser-matter interaction processes in long-scale-length plasmas, *Phys. Fluids* **4**, 2232-2240, (1992).

FAST ION DYNAMICS MEASURED BY COLLECTIVE THOMSON SCATTERING

H Bindslev, L Porte, J A Hoekzema, P Woskov, D Van Eester*

1. FAST IONS IN FUSION PLASMAS

Magnetically confined fusion plasmas contain highly non-thermal populations of fast ions resulting from fusion reactions and plasma heating. With energies in the MeV range, two to three orders of magnitude above the bulk ion and electron energies, the fast ions typically carry 1/3 of the plasma kinetic energy and even more of the free energy. It is essential that these energetic ions remain confined while they slow down and heat the thermal bulk plasma.

Non-linear wave particle interaction can significantly affect both bulk plasma and fast ion dynamics, and can lead to catastrophic loss of confinement. Wave particle interaction is also the basis of Ion Cyclotron Resonance Heating (ICRH); one of the main plasma heating schemes relying on the absorption of radio waves. In both cases the wave particle interaction depends critically on the phase space distribution of the energetic ions, and the interaction manifests itself in the dynamics of this distribution.

Understanding the plasma dynamics involving fast ions is essential for optimizing fusion performance. Experimental data on the phase space distribution of energetic ions would be particularly valuable in challenging and improving our understanding here, but fast ions are notoriously elusive to experimental observation.

2. COLLECTIVE THOMSON SCATTERING

Fast ions draw a wake in the electron distribution, detectable by Collective Thomson Scattering (CTS)[1]. From this the velocity distributions can be inferred. The electrons also draw wakes, but unlike the case for ions, the electrons are actually largely hidden by their wakes when viewed on scale lengths larger than the Debey length, λ_D. At scales larger

* H Bindslev, Association EURATOM-FOM, Netherlands, now Association EURATOM-Risø, Denmark, L Porte and P Woskov, MIT plasma science and fusion center, USA; J A Hoekzema, Association EURATOM FZ-Jülich, Germany; D Van Eester, ERM/KMS Association EURATOM Belgian state.

than the Debey length ion wakes are thus the dominant cause of microscopic fluctuations in the plasma[2]. It is these fluctuations, which are detected by CTS. Modeling the ensemble of wakes drawn by the ions and electrons, inference is made about the velocity distribution of the ions.

In CTS, spatial Fourier components of the microscopic fluctuations are selected by the Bragg condition, $k^{\delta} = k^{s} - k^{i}$, where k^{δ}, k^{i} and k^{s} are respectively the wave vectors of the resolved fluctuations, the incident probing radiation and the received scattered radiation. The directions of k^{i} and k^{s} are essentially determined by the orientations and locations of the probe and receiver antennae, while the magnitudes are mainly functions of frequency, though all are affected by refraction. Selecting fluctuations with scale lengths larger than the Debey length, $k^{\delta} \lambda_D < 1$, implies an upper limit on the frequency of the probing radiation and an upper limit on the scattering angle $\angle (k^{i}, k^{s})$.

CTS has been successfully applied to the diagnosis of the bulk ion temperature using a D_2O laser (385 μm)[5] and using millimetre waves generated by gyrotrons.[6-10] For mm waves produced by gyrotrons the frequency of the probing radiation is sufficiently low that there is little or no limitation on the range of usable scattering angles. This facilitates flexible scattering geometry with good spatial localisation.[11]

The spectrum of microscopic fluctuations resolved for a range of frequencies and one direction of the wave vector, $\boldsymbol{k}^{\delta}$, provides information on the distribution of the velocity component along $\boldsymbol{k}^{\delta}$. Resolving anisotropy in the fast ion distribution thus requires two or more CTS spectra with $\boldsymbol{k}^{\delta}$ pointing in different directions. Designs permitting variable scattering geometry are thus attractive.

Refraction, and electron cyclotron emission (ECE) and absorption (ECA) constrain and encumber CTS in the mm wave range.[9] Below the electron cyclotron frequency there can be a usable range depending on the ratio of the plasma frequency to cyclotron frequency. At the low frequency end refraction associated with the approach of the left-hand cut-off is the limiting factor. This range was exploited in the TFTR fast ion CTS and shows the most promise for a mm CTS for JET-EP[12] and ITER. The second frequency range is between the fundamental and second harmonic of the electron cyclotron frequency. This range is available if there is little harmonic overlap, which sets a lower limit on the aspect ratio (major radius/minor radius) and an upper limit on the temperature (Doppler broadening). This range is accessible and was exploited in JET, except at the highest temperatures reached with full heating at modest densities. It is also accessible in TEXTOR, where it is currently being exploited.

The feasibility of measuring MeV range fast ion velocity distributions in fusion plasmas using collective Thomson scattering was demonstrated experimentally at JET [13]. The results were obtained with the fast ion CTS at JET[7] using 140 GHz, 100-400 kW gyrotrons for the probing radiation. To discriminate the weak fast ion CTS signal (1-4 eV spectral density) from the much stronger ECE background (200-2000 eV) the probing radiation was modulated. The probing radiation was scattered in plasmas with energetic proton minority populations (deuteron bulk) heated by 4-8 MW ICRH. The velocity distribution of these ions was inferred from the scattered radiation. The forward model for fast ion CTS used in fitting was based on a fully kinetic description of the scattering, including scattering from electron density, flux and field fluctuations, and with a fully electromagnetic description of the microscopic fluctuations.[3,4] For the inverse problem of estimating the fast ion distributions, a generalized least squares fitting method (LSN) was used, which accounts for uncertainties in plasma and system parameters (nuisance parameters) inferred from other diagnostics.[14].The JET CTS was closed in 1998, but experience gathered there formed the basis for a new CTS project at TEXTOR[15].

3. THE TEXTOR CTS

The probing radiation is provided by a gyrotron delivering 100 kW for up to 200 ms at 110 GHz. The probing beam is launched at the mid-plane on the low field side, while scattered radiation is received at the location labelled "Rec. 1" in Figure 1.

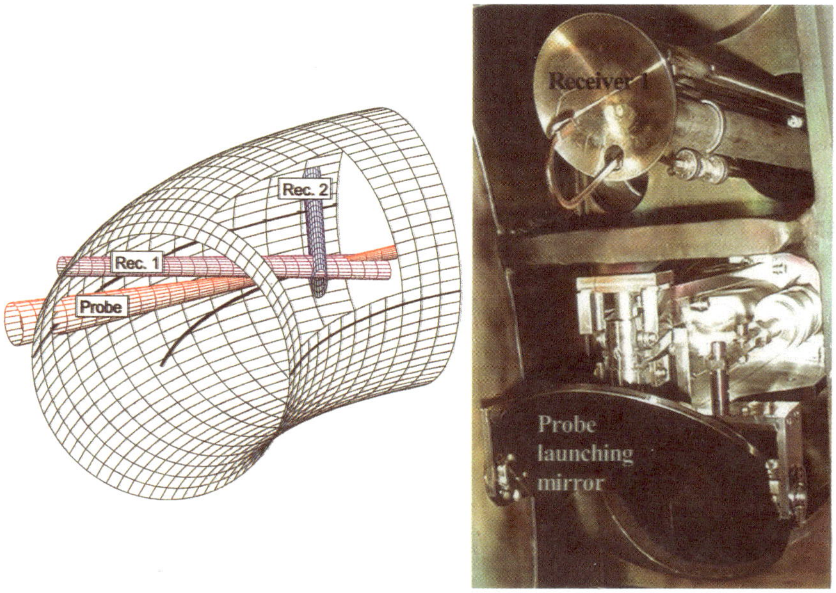

Figure 1 Right Geometry of the TEXTOR fast ion CTS diagnostic. Receiver 1 is placed at nearly the same toroidal location as the launcher of the probe, and at a poloidal angle of 24.2° above the mid-plane. Receiver 2 will be located at a toroidal angle of –42.7° relative to the launcher and a poloidal angle of 19.1° above the mid-plane. **Left** Photo taken inside TEXTOR 17 April 2000.

For the initial operation (till February 2001) only receiver 1 was operational. For this phase a compact Cassegrain antenna design, depicted in Figure 1, was used. This design makes use of fundamental rectangular wave-guide with a rectangular horn as feed, which implies that only horizontal linear polarisation is received. Good coupling to O-mode will thus only be achieved for geometries where the beam plane is near the poloidal plane. With significant toroidal components, part of the received power will have propagated to the receiver in X-mode, which in general had no overlap with the probe beam. Reduced CTS signal strength is the consequence.

The diagnostic became operational August 2000. In 7 dedicated shot days just over 200 useful shots were recorded in pursuit of both generic fast ion physics such as fast ion sawtooth sensitivity and confinement time, and specific Ion Cyclotron Resonance Heating (ICRH) physics. Operating with a magnetic field of 2.6 Tesla the electron cyclotron emission and other plasma noise in the relevant spectral range is on the order of 100 eV. The CTS signal is distinguished from the background radiation by modulating the probe radiation as illustrated in Figure 2. The signal to noise ratio and system constraints currently permit the recording of CTS spectra in each shot at up to 100 time slices, 4 ms apart, each spectrum representing a 2 ms temporal average.

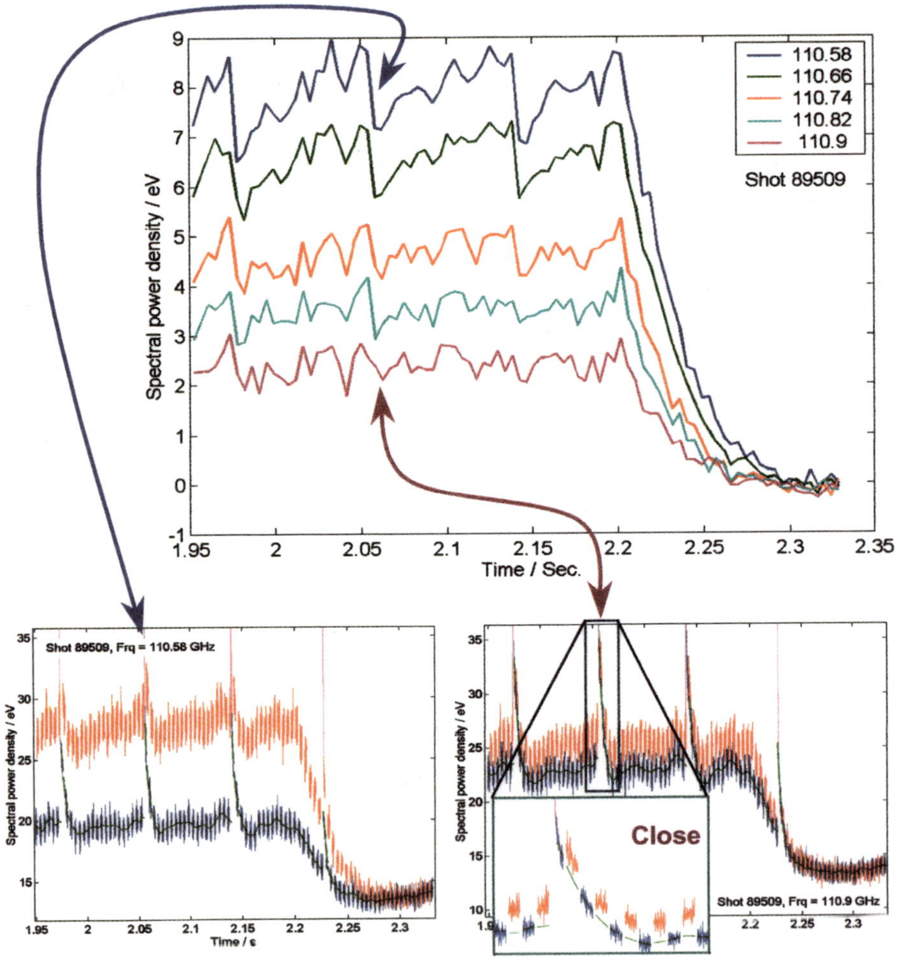

Figure 2 Shot 89509. **Top**. Spectral power density (in eV) versus time (in seconds) estimated for a range of discrete spectral channels centred at 110.58 to 110.90 GHz. **Bottom**. Calibrated samples recorded in two channels (109.58 and 110.90 GHz). Upper (lower) samples are recorded when the probe radiation from the gyrotron is on (off). The background radiation level is estimated from the data recorded when probe is off (see fit in *close up*) and subtracted from the radiation recorded when the probe is on to estimate the spectral power density of the scattered radiation. The latter is plotted in the top plot.

The spectra are detected in 40 discrete channels, 23 of which could be selected for recording. A typical set of spectral data is shown in Figure 3, where the effects of sawteeth and fast ion population decay at the end of the heating period, are apparent.

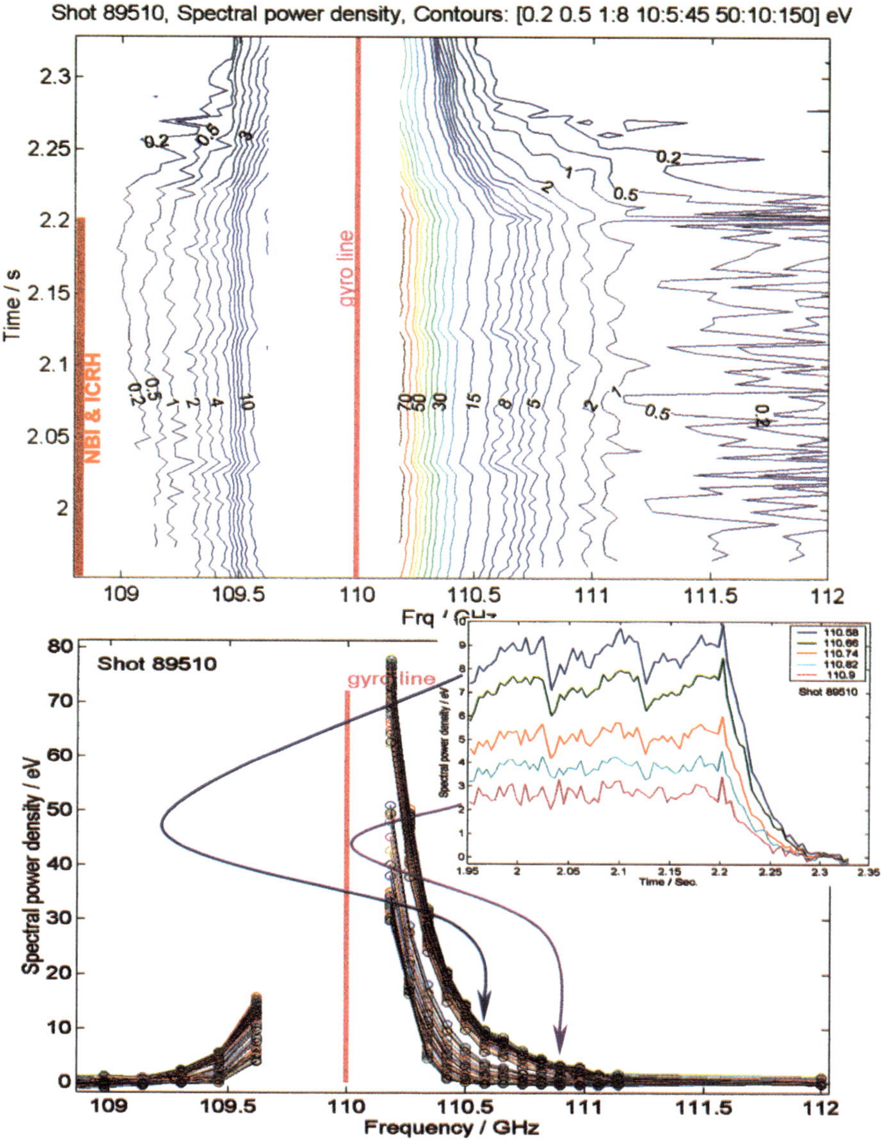

Figure 3 Spectral power density of received scattered radiation as functions of frequency and time. 90 spectra are recorded 4 ms apart, each spectrum representing a 2 ms temporal integration with probe on. **Bottom** the spectra are overlaid. Left inset shows time traces from selected channels. **Top** a contour plot of the spectral power density. Auxiliary heating (NBI and ICRH) was on till 2.2 seconds.

From the CTS data shown in Figure 3 the localized one dimensional ion velocity distribution is inferred. The result is shown in Figure 4 . The auxiliary heating (NBI and ICRH) was switched off at 2.2 seconds, after which the fast ion population is seen to decay to nothing in approximately 100 ms.

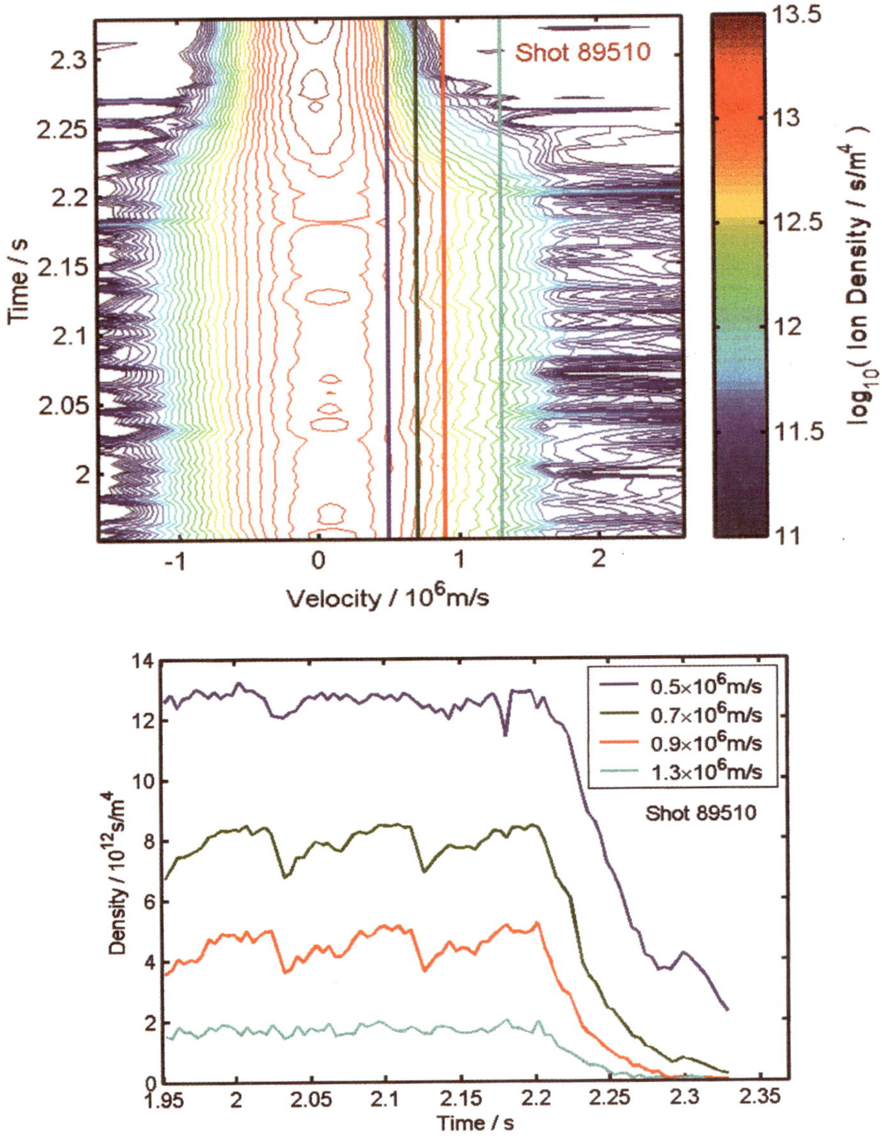

Figure 4 Shot 89510. Ion phase space densities (particles per unit volume and unit velocity along resolved velocity component) as functions of velocity and time. **Top** log contour plot of phase space density, **bottom** time traces of density at selected velocities.

The data shown in Figures Figure 2, Figure 3 and Figure 4 were recorded with a scattering geometry, which locates the measuring volume on the high field side of the plasma center, near the electron density sawtooth inversion radius, and resolves a velocity component near perpendicular to the magnetic field, as illustrated in Figure 5.

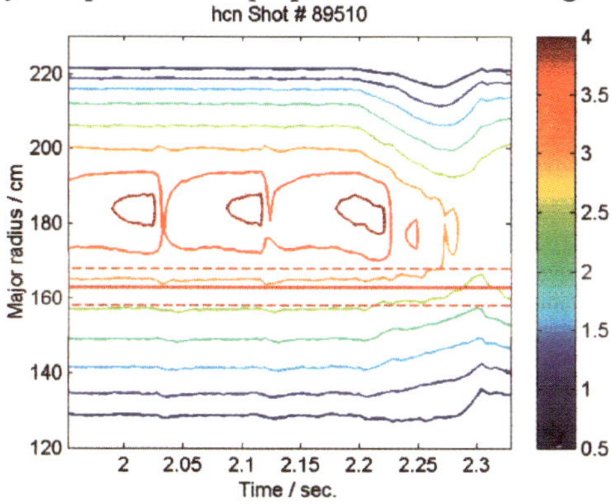

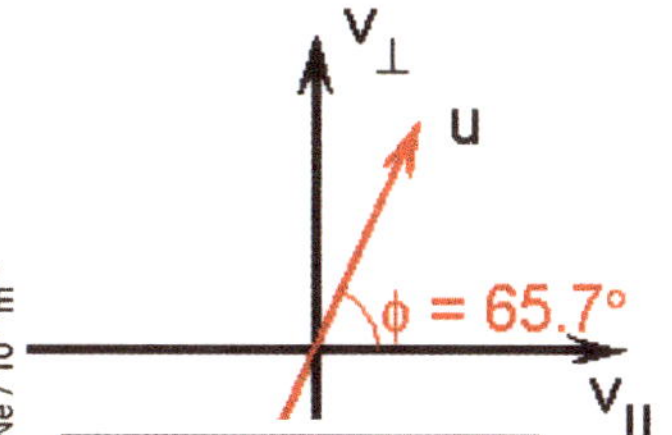

Shot #	R/cm	φ
89509	163	65.7°
89510	163	65.7°
89511	199	71.3°
89519	171	12.1°

Figure 5 (shot 89510) **Left**, contour plot of electron density in plasma midplane as functions of major radius, and time. Solid straight line indicates centre of CTS measuring volume and parallel dashed lines its radial extent. Here the volume is inside but near the electron density inversion radius. The **right** plot shows the direction of the resolved velocity component, **u**. for this scattering geometry. $v_{\|}$ is parallel to the magnetic field, positive in the direction of the plasma current. $\phi = \angle(u, v_{\|})$.The table gives major radius of scattering volume centre and direction of resolved velocity component for a range of selected shots.

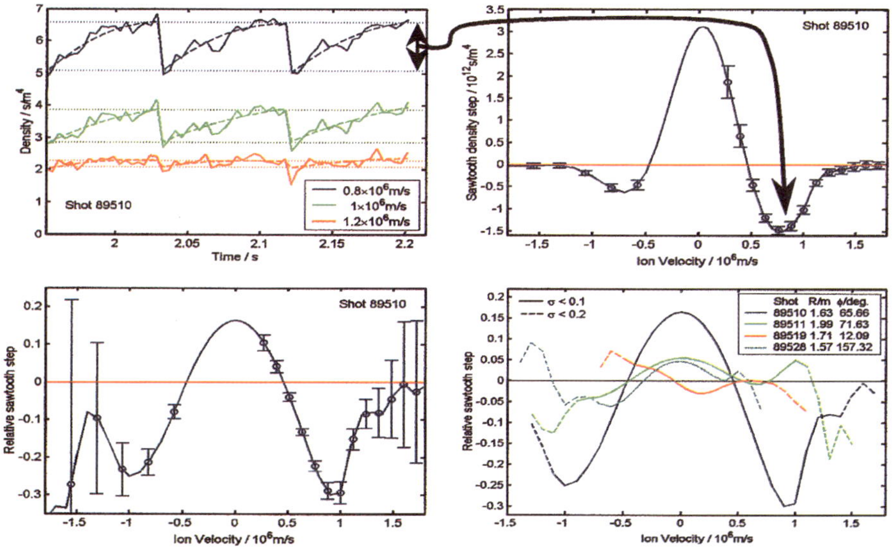

Figure 6 Sawteething of the ion phase space distribution. Shot 89510. Top left, polynomial fit to extract sawtooth step, negative for loss. Top right, sawtooth step as a function of velocity. Bottom left, sawtooth step normalised by phase space density to give relative sawtooth step. Bottom right, relative sawtooth step for four shots resolving different velocity components and at different radial locations, see Figure 7.

Before the heating is switched off we note sawtoothing activity in the ion phase space density. Extracting the change in population density at each sawtooth, as illustrated in Figure 6, we find that the ion sawtooth activity depends on the ion velocity, with a loss in local phase space density of 25 to 30 % for velocities in the range 0.7 to1.0×10^6 m/s.

Varying the scattering geometry we explore anisotropy and inhomogeneity in the ion distributions and in their dynamics. The lower right plot in Figure 6 shows how this is used to investigate the link between ion sawtooth sensitivity to ion drift orbit class. Shots 89510, 89519 and 89528 explored the same spatial location in the plasma but resolved respectively velocity components near perpendicular and near parallel to the magnetic field. In contrast to the perpendicular distribution, the near parallel distributions showed no or very little sawteething. This suggests that only ions in a narrow range of pitch angles near perpendicular are lost from the measuring volume at sawteeth. On the low field side near the inversion radius (shot 89511) the near perpendicular ion distribution displays no sawteething, also in stark contrast to the situation on the high field side.

In conclusion, fast ion diagnosis by collective Thomson scattering is now providing access to the fast ion phase space distribution, resolved in space, velocity and time simultaneously.

REFERENCES

1. For a review of developments in CTS theory see H. Bindslev, *Proc. 3rd Int Workshop Strong Microwaves in Plasmas*, Volga, Russia, p109, IAP, Nizhny Novgorod (1996). (http://www.risoe.dk/euratom/cts/publications/SMIP96_2.pdf); See also Sitenko and Yu. Kirochkin, *Soviet Physics Uspekhi* **9**, 430 (1966); R. Aamodt, D. Russell, *Nuclear Fusion* **32**, 745 (1992) and references 2, 3 and 4 below. For current activities see http://www.risoe.dk/euratom/cts.
2. E. E. Salpeter. *Physical Review* **120**, 1528, (1960).
3. H. Bindslev, *J. Atmospheric and Terrestrial Physics* **58**, 983 (1996)
4. H. Bindslev, *Plasma Physics and Controlled Fusion*, **35**, 1615 (1993)
5. R. Behn, *et al. Phys. Rev. Lett.* **62**, 2833 (1989)
6. J.A. Hoekzema, *et al. 22nd EPS Conference on Controlled Fusion and Plasma Physics*, Bournemouth, UK, Vol. 19C, part II, p445, EPS JET Joint Undertaking, UK (1995)
7. J. A. Hoekzema, *et al. Rev. Sci. Instrum.*, **68**, 275 (1997).
8. J.A. Hoekzema, H. Bindslev, J. Egedal, J.A. Fessey, T.P. Hughes, *Proc. 7th Int. Symp. on Laser-Aided Plasma Diagnostics.*, Fukuoka, Japan, p166, Kyushu Univ. (1995)
9. H. Bindslev, J. Egedal, J.A. Fessey, J.A. Hoekzema, T.P. Hughes, *Proc. 3rd Int Workshop Strong Microwaves in Plasmas*, Volga, Russia, p142, IAP, Nizhny Novgorod (1996) (http://www.risoe.dk/euratom/cts/publications/SMIP96_1.pdf)
10. E. V. Suvorov, *et al., Plasma Physics and Controlled Fusion* **37**, 1207 (1995)
11. P. Woskoboinikow, *Rev. Sci. Instrum.* **57**, 2113 (1986)
12. http://www.risoe.dk/euratom/cts/jet
13. H. Bindslev, J. A. Hoekzema, J. Egedal, J. A. Fessey, T. P. Hughes, and J. S. Machuzak, *Phys. Rev. Lett.* **83**, 3206 (1999).
14. H. Bindslev, *Rev. Sci. Instrum.* **70**, 1093 (1999).
15. H. Bindslev, L. Porte, A. Hoekzema, J. Machuzak, P. Woskov, D. van Eester, 26th EPS Conf. Contr. Fusion and Plasma Physics, Maastricht, Netherlands, Vol. 23J, part II, p765, EPS (1999) (http://epsppd.epfl.ch/pdf/p2049.pdf); H. Bindslev, L. Porte, A. Hoekzema, J. Machuzak, P. Woskov, D. van Eester, J. Egedal, J. Fessey, T. Hughes, *Fusion Engineering and Design*, **53**, 105 (2001); H. Bindslev, Journal of Plasma and Fusion Research. 76, 878 (2000); L. Porte, H. Bindslev, F. Hoekzema, J. Machuzak, P. Woskov, and D. Van Eester, *Rev. Sci. Instrum.* **72**, 1148 (2001); (http://www.risoe.dk/euratom/cts/TEXTOR)

UTILIZATION OF LIBEAM POLARIMETRY FOR EDGE CURRENT DETERMINATION ON DIII–D

D.M. Thomas,[*] D.K. Finkenthal,[†] T.N. Carlstrom,[*] D.H. Kellman,[*] G. Brewis,[‡] T.H. Osborne,[*] J.I. Robinson,[*] A.S. Bozek,[*] K.H. Burrell,[*] M.A. Makowski[Δ], S.G.E. Pronko,[*] and S. Delaware[*]

ABSTRACT

Local details of the edge current profile can have profound effects on the stability and ultimate performance of many advanced tokamak (AT) modes.[1–3] This is true for both bootstrap and externally driven currents that may be used to tailor the edge shear. Absent a direct local measurement of j(r), the best alternative is a determination of the poloidal field. On the DIII–D tokamak[4] we are using precision polarimetry of an injected lithium beam (LIBEAM) to make measurements of the necessary precision in the region of interest (ρ = 0.7–1.1).[5] Combined polarimetry/spectroscopy of the various Zeeman-split 2S-2P lithium resonance line components gives direct information on the local magnetic field components. Because of the negligible Stark mixing of the relevant atomic levels, this method of determining j(r) is insensitive to the large local electric fields typically found in enhanced confinement (H–mode) edges, and thus avoids an ambiguity present in motional Stark effect (MSE) measurements of B.[6,7] Beam intensities and energy are suitable for measurements in the edge region of even high density discharges. Key issues for utilizing this technique include good beam quality, an optimum viewing geometry, and a suitable optical pre-filter to isolate the polarized emission line. Details of the existing DIII–D diagnostic are presented. We also discuss details of the analysis necessary to determine the field component from these measurements, and our plans for future improvements to the system.

1. INTRODUCTION

At present, the performance of "advanced" modes of tokamak operation is set more by the stability limits of various modes (peeling, ballooning, and kink) than by transport considerations.[8,9]. The avoidance and extension of these limits to further improve the plasma performance requires careful tailoring of the plasma pressure and current profiles. This in turn places a premium on precise measurement of these parameters. Measurement of the current density profile through the use of hydrogenic beams and the motional Stark effect (MSE)[10] has played a key role in the understanding and development of these

[*]General Atomics, P.O. Box 85608, San Diego, California 92186-5608. [†]Palomar Scientific Instruments. [‡]Allied Optical. [Δ]Lawrence Livermore National Laboratory.

Advanced Diagnostics for Magnetic and Inertial Fusion, Edited by Stott *et al.*
Kluwer Academic/Plenum Publishers, New York, 2002

advanced operational modes. However, since the polarization of the Stark-split emission manifold is determined by the electric field in the rest frame of the beam, complication occurs for most advanced tokamak modes, where substantial intrinsic radial electric fields are found to be coexistent with the high gradients that are often formed by the transport barriers.[11,12] In this case it is quite difficult for a simple MSE-style measurement to distinguish between the intrinsic (E_T) and motional (v×B) Stark terms without such complications as multiple views, multiple beam injection angles, multiple beam energy components, etc. This is of particular import for studies of the edge bootstrap current on DIII–D where small poloidal magnetic field changes (~1%) need to be resolved in the presence of large (~ 100 kV/m) electric fields associated with high-performance H–modes.[13] As an alternative measurement for this edge region, we have deployed a diagnostic based on the Zeeman effect in the lithium 2S-2P resonance transition. In this case, because of the wide separation of the atomic levels there is no Stark mixing and the polarization and splitting of the resonance emission is strictly due to the local magnetic field.[14] To make the measurement we take advantage of (1) the known polarization state of emission of the various Zeeman components as well as (2) the known angular distribution of fluorescence emission for various views relative to B.

2. NEUTRAL LITHIUM BEAM AND POLARIZATION OPTICS

Figure 1 shows the deployed beam system, which is basically the resurrected LIBEAM system.[5] Details of the system and installation are reviewed in Reference 8. To optimize radial resolution and sensitivity to the poloidal field, we have chosen radial injection and vertical viewing. This implies we must analyze both the circular (equivalent to emission along the field direction) and linear (emission perpendicular to the field direction) polarization in order to precisely determine the magnetic field direction and components.

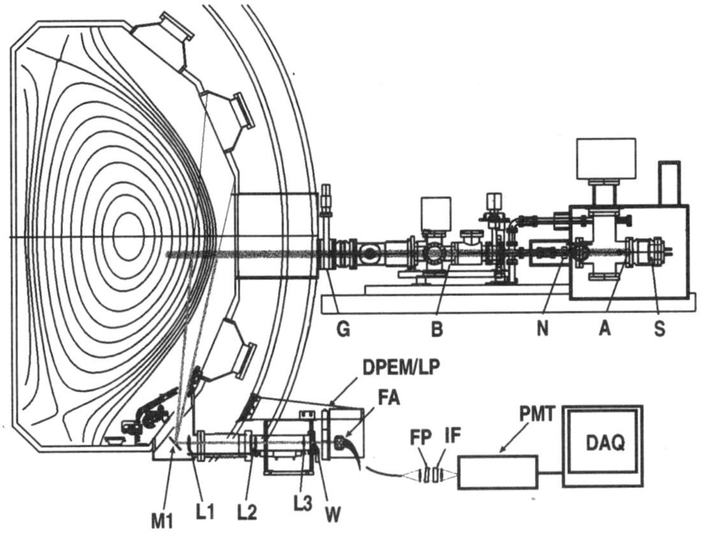

Figure 1. Neutral lithium beam system and viewing optics system on DIII–D for the Zeeman polarization diagnostic. Beam parameters: 30 keV, 10 mA, 1–2 cm diameter. Radial resolution 5 mm. All lenses are SFL6 (low Verdet constant) glass and the polarization-maintaining mirror has a special coating compatible with extended vacuum exposure and high (350°C) temperature. Dual PEM fundamental frequencies are 23 and 20 kHz. Fiber array allows up to 32 radial channels to be observed simultaneously.

A substantial amount of effort went into the design of the optical system. Design features include low Verdet constant glasses, polarization maintaining mirrors, and an insertable diagnostic polarizer. A dual photoelastic modulator (PEM) is mounted immediately outside of the vacuum window and is used to modulate the polarization state of the emitted light. Lock-in detection is then used to determine the input polarization state. In-vessel tests of the optics after the 2000 experimental campaign revealed no large systematic variations of the measured polarization due to the various elements, and only minor (~1 mm) spatial shifts of the 32 viewing locations. We believe we have a robust set of optics with good fidelity for polarization determination.

3. DETECTION AND POLARIZATION ANALYSIS

Due to the relatively small separation of the Zeeman states for typical DIII–D discharge conditions,a premium is set on efficient narrow band spectral filtering of the line profile to isolate one of the σ components. A prototype filterscope employing a tilt-tuned Fabry-Perot etalon and lithium line filter serves to isolate the proper component, depending on the line broadening that is actually present due to beam thermal spread, finite viewing angle, etc. To accurately determine the ratio of circular to linear polarization using modulation techniques requires lock-in measurements at four different frequencies (fundamental and first harmonic of each of the two PEM resonant frequencies). To avoid deploying and tuning 128 separate analog lockin amplifiers, increases in computer speed and memory have allowed us to pursue true digital lock-in techniques to extract all four required frequency components in near real-time. This technique relies on off the shelf PC/PCI hardware and is scaleable to many channels. Two big advantages of this approach are (1) the elimination of analog noise during the reference mixing and (2) digitizing the PEM reference frequencies allows us to avoid error due to frequency drift. The system is presently capable of doing eight separate channels simultaneously using a single PC and PCI DAQ board at digitization rates up to 1 MHz for the full DIII–D shot length.

4. INITIAL RESULTS

Figure 2 shows the beam fluorescence collected at several radial locations inside the plasma, spanning the approximate range $0.7<\rho< 1.05$. The signal levels are reasonably consistent with beam penetration modeling. These signals represent the total line intensity and are not polarization resolved. Figure 3 shows the measured circular polarization for a channel looking near the edge (ρ=0.91) after the Fabry-Perot was tuned to the wavelength appropriate for the doppler shifted $\sigma+$ component. The overall polarization was small (~1%). We believe this is due to excessive energy spread in the beam, which increases the overlap of the doppler-broadened π and σ states. Despite the small level of polarization, the digital lock-in technique is still capable of accurately extracting the polarization components. The ratio, which is roughly proportional to the poloidal field, is seen to increase with the current ramp, then decrease possibly due to current diffusion.

5. SUMMARY AND FUTURE WORK

Deployment of the neutral beam accelerator and the associated polarization optics on DIII–D is essentially complete, and we have tested a few channels of the digital lock-in/data acquisition system. In the near future we plan to improve the beam energy resolution to increase the measurable polarization fraction. We will increase the number of detection channels to the full complement of 32. In addition, we will continue to search for improved spectral resolution elements such as Lyot filters to improve on the Fabry

Perot performance, as well as improving the computer control of various aspects of the experiment including PEM, detector, and beam control. Our ultimate goal is to have the resolved field components input directly as constraints to EFIT, similar to the existing MSE data.

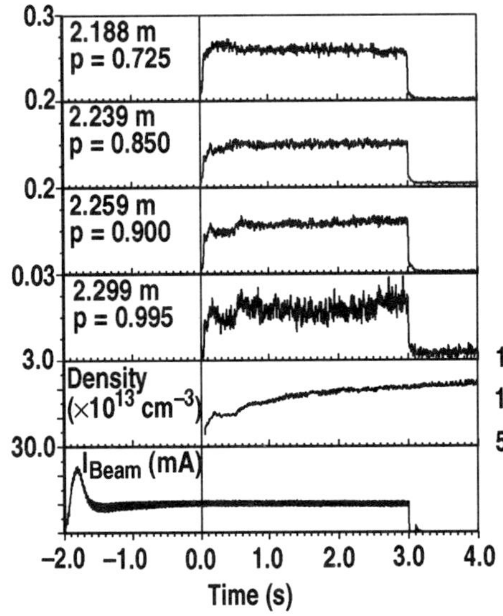

Figure 2. Lithium beam fluorescence versus radius for DIII–D shot 107208. Plasma density was ~2.5×10^{13} cm^{-3}. Beam is turned off at 3.0s.

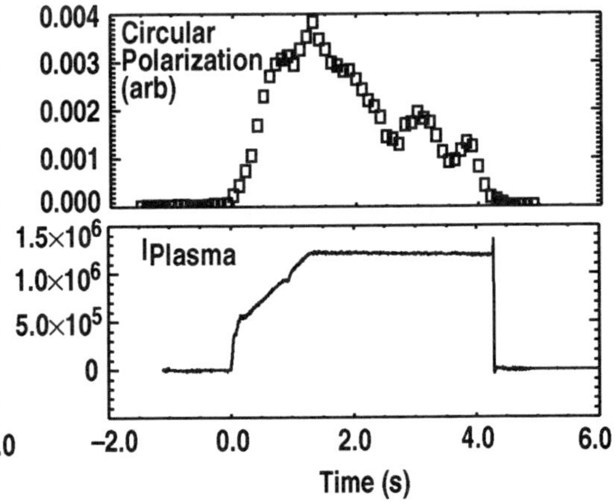

Figure 3. Polarization data for a single channel at ρ=0.91, for DIII–D shot 107805. Also shown is the total plasma current. Beam is turned off at 4.0 s.

6. ACKNOWLEDGEMENTS

The authors have benefited from numerous stimulating and informative discussions with T.S. Taylor, L.L. Lao, J. Jayakumar, and A.D. Turnbull. Work supported by U.S. Department of Energy under Contracts DE-AC03-99ER54463 and W-7405-ENG-48.

7. REFERENCES

1. H.R. Wilson et al., Phys. Plasmas **6**, **5**, 1925 (1998).
2. J.R. Ferron et al., Nucl. Fusion **40**, **7**,1411 (2000).
3. M.R. Wade et al., Phys. Plasmas **8**, **5**, 2208 (2001).
4. J.L. Luxon, L.G. Davis, Fusion Technol. **8**, 873, (1985).
5. D.M. Thomas, Rev. Sci. Instrum. **66**, 806 (1995).
6. B.W. Rice et al., Nucl. Fusion **37**, 577 (1997).
7. D.M. Thomas, Rev. Sci. Instrum. **66**, 806 (1995).
8. L.L. Lao et al., Nuc. Fusion **39**, 11Y, 1785 (1999).
9. C.M. Greenfield, et.al, Phys. Plasmas **7**, 5, 1959, (2000).
10. F. Levinton et al, Phys. Rev. Lett. **63**, 2060 (1989).
11. B.W. Rice et al., Phys. Rev. Lett. **79**, 14, 2694 (1997).
12. M.C. Zarnstorff et al., Phys. Plasma **4**,1097 (1997).
15. K.H. Burrell, et al, Phys. Plasmas **8**, 5, 1499 (1997).
14. H.A. Bethe and E.E. Salpeter, Quantum Mechanics of One- and Two-Electron Atoms, (Plenum, 1977) 205 ff.

ROBUST LINE DENSITY MEASUREMENTS AT W7-AS USING A COTTON-MOUTON POLARIMETER

T. Bindemann, Ch. Fuchs, H. J. Hartfuß, M. Hirsch[1]

INTRODUCTION

A magnetized plasma is birefringent. The wave which is propagating in the plasma can be regarded as superposition of two characteristic waves which propagate with different phase velocities, giving rise to a phase shift between them. The refractive indices depend on the probing beam frequency, f, the direction and absolute value of the magnetic field, $\vec{B}$, and the electron density n_e. For an arbitrary angle between the magnetic field $\vec{B}$ and the wavevector $\vec{k}$, the characteristic waves are elliptically polarized. For the special cases $\vec{B} \parallel \vec{k}$ and $\vec{B} \perp \vec{k}$ the Faraday- and the Cotton-Mouton effect can be observed, respectively. For the Cotton-Mouton effect, the characteristic waves are linear polarized, the o-mode ($\vec{E} \parallel \vec{B}$) and the x-mode ($\vec{E} \perp \vec{B}$). A wave composed of an o- and an x-mode changes its ellipticity when passing the plasma. By measuring this change, the line integrated electron density ρ_e can be determined. An unambiguous measurement of the line density is especially important for future long pulse machines. In this context a Cotton-Mouton polarimeter was proposed for ITER [1] and for W7-X. A proof of principle experiment was performed successfully at W7-AS [2, 3].

MATHEMATICAL TREATMENT OF POLARIZATION CHANGES

The polarization state is described by the three Stokes parameters s_1, s_2 and s_3, the components of the Stokes vector $\vec{s}$. The evolution along the line of sight (z-direction) is

$$\frac{d\,\vec{s}\,(z)}{dz} = \vec{\Omega}\,(z) \times \vec{s}\,(z)\ , \qquad (1)$$

the vector $\vec{\Omega}(z)$ describing the plasma wave interaction.

[1]Max-Planck-Institut für Plasmaphysik, EURATOM Ass., D-17491 Greifswald, Germany

Advanced Diagnostics for Magnetic and Inertial Fusion, Edited by Stott *et al.*
Kluwer Academic/Plenum Publishers, New York, 2002

With ω the probing frequency and ω_p and ω_c are the plasma- and the electron cyclotron angular frequencies for $\omega^2 \gg \omega_p^2$ and $\omega^2 \gg \omega_c^2$ the approximation

$$\vec{\Omega} = \frac{\omega_p^2}{2c\,\omega^3} \begin{pmatrix} \frac{e^2}{m^2}(B_x^2 - B_y^2) \\ \frac{e^2}{m^2}\,2\,B_x B_y \\ 2\,\omega \frac{e}{m} B_z \end{pmatrix} \tag{2}$$

can be made [4]. The components Ω_1 and Ω_2 represent the Cotton-Mouton, while Ω_3 describes the Faraday effect.

Equation (1) describes a sequence of infinitesimal rotations of the vector $\vec{s}$. The state of polarization after passing the plasma is given by

$$\vec{s}\,(z_1) = \mathbf{M}(z_1) \cdot \vec{s}(z_0)\,. \tag{3}$$

The Müller matrix $\mathbf{M}$ describes the birefringent properties of the plasma along the line of sight [4, 5].

In W7-AS sigthlines can be found, where $B_z = 0$, and B_x and B_y are almost constant. Thus the Müller matrix in 0th order approximation is:

$$\mathbf{M}(z_1) \simeq \begin{pmatrix} 1 & 0 & 0 \\ 0 & 1 & -W_1 \\ 0 & W_1 & 1 \end{pmatrix}\,. \tag{4}$$

With Ω from eqn. (2), W_1 is given by:

$$W_1 = \int_{z_0}^{z_1} \Omega_1(z)dz = \frac{e^2}{2\,c\,\omega^3\,m_e^2} \int_{z_0}^{z_1} \omega_p^2(z)\,(B_x^2(z) - B_y^2(z))dz \propto B_{eff} \int_{z_0}^{z_1} \omega_p^2\,dz\,. \tag{5}$$

For z-independent B_x and B_y, W_1 only depends on the line density $\rho_e \propto \omega_p^2$.

ELLIPTICITY MODULATION

By introducing a periodic phase shift with frequency ω_m between the characteristic waves of the probing wave its ellipticity is modulated. With identical amplitudes of the probing wave components one yields for the Stokes component s_2 at the beginning (z_0) and at the end (z_1) of the line of sight respectively from equation (3) and (4):

$$\begin{aligned} s_2(z_0) &= \cos(\omega_m\,t) \\ s_2(z_1) &= \cos(\omega_m t + W_1)\,. \end{aligned}$$

W_1 appears just as the phase difference of the Stokes parameter s_2 of the probing wave before entry and after leaving the plasma. If B_{eff} is known and constant, ρ_e can be evaluated.

EXPERIMENTAL SET-UP

The total experimental set-up is sketched in figure 1. A backwardwave oscillator (BWO) is used as a tunable radiation source at a frequency of about 600 GHz. Oversized waveguides ($\phi = 24\,\text{mm}$) are used throughout in the guided wave sections. The modulation of the wave ellipticity is accomplished in a way as described in [6]. A polarizing wire grid reflects the y-polarized component of the wave, whereas the x-component is guided to a delay line (DL). After reflection at two mirrors which rotate the polarization plane by 90^0 this component is added to the reflected one. The phase difference of the two components depends on the wave frequency and the length of the delay line. A BWO-frequency modulation with $\omega_m/(2\pi) = 47\,\text{kHz}$ and $\Delta f = 300\,\text{MHz}$ then transforms to an ellipticity modulation of the probing wave. Detectors D_1 (reference detector) and D_2 (signal detector) with 45^0 analyzers (A) deliver signals proportional to $s_2(z_0)$ and $s_2(z_1)$ respectively. Their phase difference is the quantity of interest. The detectors are corner cube detectors with a long wire antenna contacting a low capacitance Schottky barrier diode.

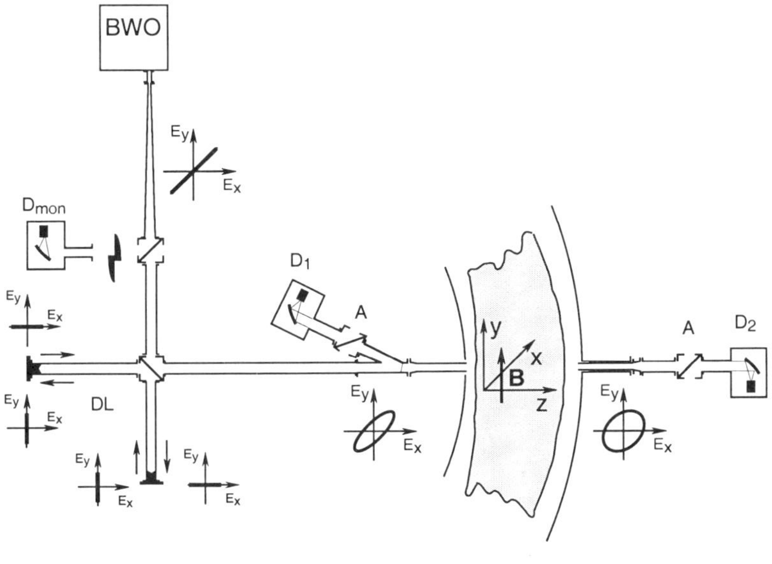

Figure 1: Experimental setup at the W7-AS

RECENT EXPERIMENTAL RESULTS

The resolution $\Delta\rho_e$ of a line density measurement for the Cotton-Mouton polarimeter depends on B_{eff}, f and the signal-to-noise ratio. For the same signal-to-noise ratio, $\Delta\rho_e$ is worse compared with typical interferometers because of the lower phase differences involved.

However, the stability and robustness is much higher because no reference beam is needed and phase differences remain below 360^0 so that keeping track of fringe jumps is not necessary. This is demonstrated in figure 2. The line density measurement of two different conventional interferometers used for comparision and the Cotton-Mouton polarimeter respectively, are disturbed by transient instabilities in the plasma in the time interval 200ms$< t <$380 ms. Only the polarimeter delivers reliable information in the silent phase of the discharge ($t > 550$ ms).

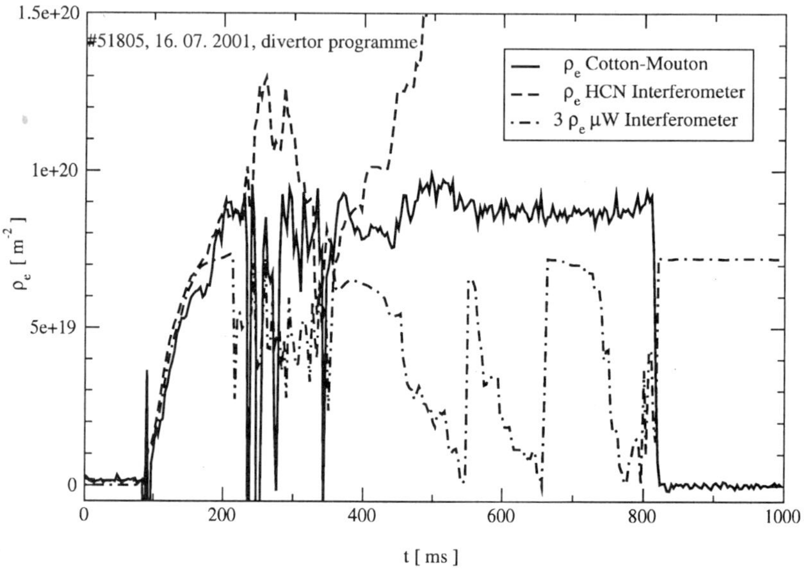

Figure 2: Line densities: Cotton-Mouton polarimeter at 600 GHz, microwave interferometer at 160 GHz, HCN interferometer at 890 GHz, B=2.48 T

There is good agreement between the Cotton-Mouton polarimeter and the microwave interferometer at low densities ($\rho_e < 8 \cdot 10^{19} \mathrm{m}^{-2}$). In divertor experiments under high density and high β conditions clear deviations are observed which are being studied. Several reasons could cause the observations amongst which B_{eff} changes, due to β effects in connection with Shafranov shifts might be the most probable ones.

REFERENCES

1. P. Buratti, proposed and discussed at meetings with the European Home Team on microwave diagnostics for ITER (1995).
2. Ch. Fuchs, H.J. Hartfuß, Phys. Rev. Lett. **81**, 1626(1998).
3. Ch. Fuchs, H.J. Hartfuß, Rev. Sci. Instr. Sci. Instr.**70** 722 (1998).
4. S. E. Segre, Phys. Plasmas **2**, 2908 (1995).
5. S. E. Segre, Phys. Plasmas **3**, 1182 (1996).
6. V. F. Shevchenko, A. A. Petrov, V. G. Petrov, and Y. A. Chaplygin, Plasma Physics Reports **22**, 28 (1996).

DETERMINATION OF THE ELECTRON DENSITY AND MEAN ENERGY FOR ANISOTROPIC VELOCITY DISTRIBUTIONS BY THOMSON SCATTERING

Bernd Kurzan, Heindieter Murmann, Denis Bolshukhin, Martin Jakobi, and the ASDEX Upgrade Team*

1. INTRODUCTION

Thomson scattering is a standard diagnostic for measuring electron temperature T_e and density n_e robustly and routinely on nearly every fusion device. When evaluating the experimentally determined scattering spectra, the assumption of a Maxwellian velocity distribution function (VDF) for the electrons, is usually a good approximation. For some H-mode discharges with early application of Neutral Beam Injection (NBI) the electron temperatures and densities evaluated for two different Thomson scattering systems on the ASDEX Upgrade tokamak, however, differ quite substantially. Here the assumption of a Maxwellian VDF clearly fails. In the following an attempt is made to reconstruct the real VDF from the two Thomson scattering diagnostics to determine the mean energy and the density of the electrons.

2. DIAGNOSTICS

On ASDEX Upgrade two different Thomson scattering diagnostics are now in operation: (1) The "Vertical Thomson Scattering" (VTS) system is constructed for 90 deg scattering and its cluster of 6 laser beams is aligned perpendicular to the toroidal magnetic field. Here the electrons with high poloidal-radial velocities generate the strongest Doppler shift (fig. 1a). Thus with this system the mean perpendicular energy of the electrons is measured. (2) The newly installed "Horizontal Thomson Scattering" (HTS) system has approximatly 45 deg scattering and its cluster of 5 laser beams encloses with the toroidal magnetic field an angle of about 45 deg. In this geometry the electrons moving with a pitch angle of approximately 135 deg generate strongly blue shifted

* Max-Planck-Institut für Plasmaphysik, EURATOM Association, Boltzmannstr. 2, Garching, Germany D-85748.

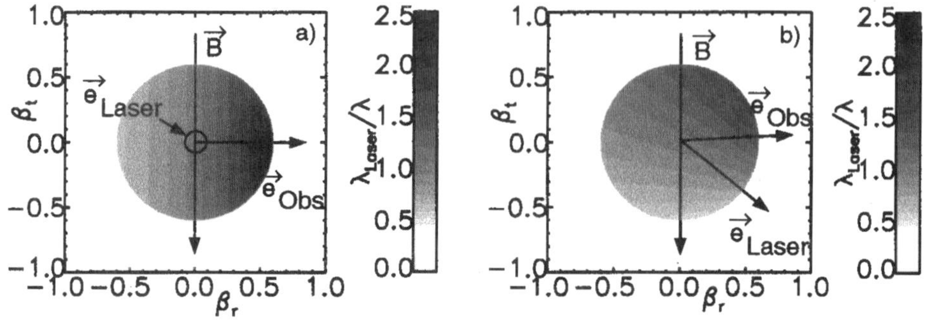

Figure 1. Doppler shift for the VTS (a) and HTS (b) system. For $\beta_r = \beta_t = 0$ there is no Doppler shift. β_t, β_r are the radial and toroidal electron velocities normalized to the speed of light. $\vec{e}_{Laser}$, $\vec{e}_{Obs}$ indicate the directions of the laser beam and of the observation of the scattered light, and $\vec{B}$ is the toroidal magnetic field. λ_{Laser} and λ are the wavelengths of the laser and of the scattered light.

scattering photons, while the electrons carrying the inductively driven current having pitch angle 0 deg lead to strongly red shifted photons (fig. 1b). In both diagnostics only the blue-shifted spectra are analyzed by four (VTS) and three (HTS) spectral channels.

Suprathermal electrons generated in the inductively driven toroidal electric field appear in the HTS system in the red-shifted spectrum and therefore cannot be observed.

3. EXPERIMENTAL RESULTS

In H-mode discharges with NBI as the only additional heating source, switched on in the plasma current ramp up phase, or shortly after it, and followed by a density plateau phase, differences in the electron densities and temperatures, derived with the VTS and HTS systems, develop. They are assigned to anisotropic VDFs. The measured signal amplitudes of the spectral channels are well described by the light spectrum scattered from a Maxwellian plasma in the directions of observation $\vec{e}_{Obs}$ of the VTS and HTS diagnostics: The χ^2–distribution of the fits of electron density and temperature to the spectral channels[1] follows closely the theoretically expected one for normally distributed and independent signals of two degrees of freedom for the VTS system. For the HTS system, however, discrepancies between the measured χ^2–distribution and the theoretically expected one for one degree of freedom appear at large values of $\chi^2 \approx 5$ indicating a non-Maxwellian character of the VDF. In a typical discharge of this kind where NBI was switched on at the time t = 1 s the anisotropy develops with a relaxation time constant of about 3 s. The anisotropy in the temperature has its maximum in the plasma center and decreases towards the plasma edge (fig. 2a and b). The effect of the anisotropic VDF on the derived density is most strongly pronounced in the plasma edge and decreases towards the plasma center where it practically vanishes (fig. 2c and d). In the Soft X-ray spectra obtained for a line of sight through the plasma center a high energy tail develops with time, which is fitted by a Maxwellian distribution of temperature 18 keV confirming the high electron temperatures of about 12 keV measured with the HTS system in the plasma center. The electron temperatures determined from X mode

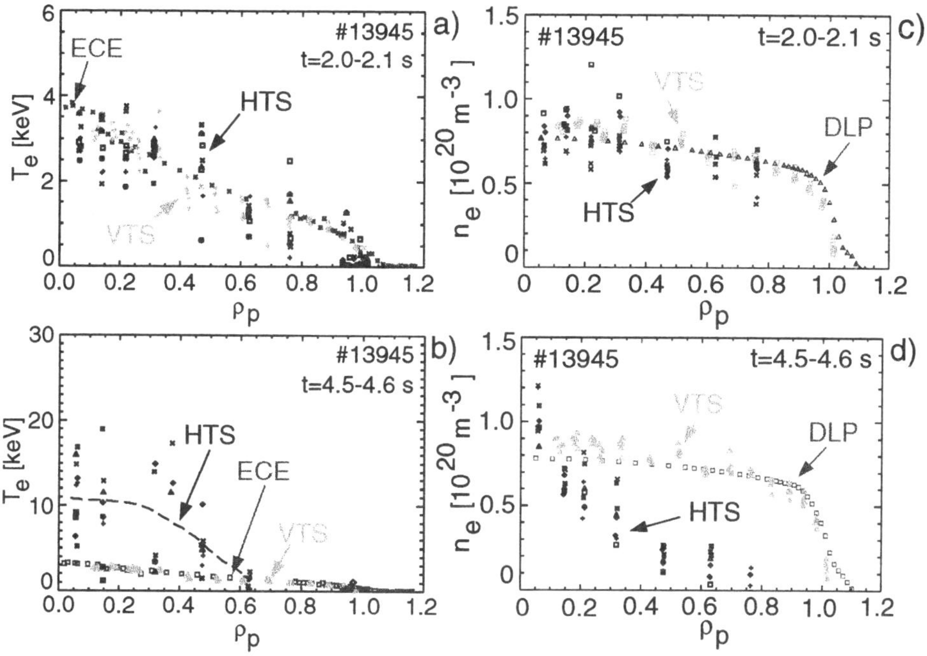

Figure 2. Anisotropies in the electron temperatures and deviations in the electron density develop after $t = 2.1$ s. The density profiles determined from a deconvolution of interferometric signals and the Li-beam diagnostic are indicated by DLP.

Electron Cyclotron Emission (ECE) agree with the temperatures measured with the VTS system, because both diagnostics are sensitive to the electron velocity perpendicular to the magnetic field.

The observed maximum apparent temperature anisotropy in the center of the plasma is in the range $[T_e(\mathrm{HTS}) - T_e(\mathrm{VTS})]/[T_e(\mathrm{HTS}) - T_e(\mathrm{VTS})] = 0.55 - 0.80$.

4. RECONSTRUCTED VELOCITY DISTRIBUTION FUNCTION

A simple reconstruction of the electron VDF is possible in the center of the plasma where both the HTS and VTS diagnostics observe the scattered light from volumes which are equivalent in terms of the behaviour of the trapped particles. As an ansatz for the VDF an expansion into Legendre polynomials,

$$f(v,\mu) = f_M\left(v, T_e\right)\left(1 + 3\mu\frac{n_b}{n_e}\frac{v_b v}{v_{Te}^2}\right) + \alpha f_M\left(v, T_{e2}\right)\frac{3\mu^2 - 1}{2}$$

is used with v: electron velocity, μ: cosine of pitch angle, n_b: ion beam density, v_b: ion beam velocity. The temperatures T_e and T_{e2} in the Maxwellian distributions

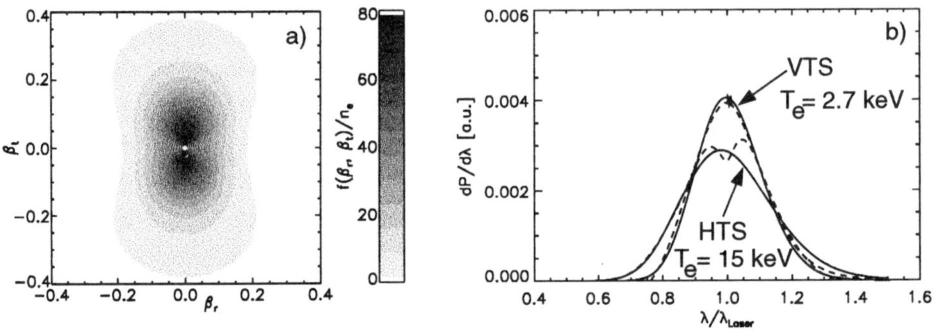

Figure 3. Reconstructed VDF (a) and scattering spectra (b). The solid lines correspond to the spectra of Maxwellian VDFs for the measured temperatures and the dotted lines show the spectra due to the reconstructed VDF shown in a).

f_M and the parameter α are adjusted for best agreement between the scattering spectra belonging to the temperatures observed with the VTS and HTS systems and the scattering spectra calculated with the distribution function. The NBI current drive term[2] $\propto \mu v$ does not change the distribution function significantly for the realistic beam densities of $n_b = \left(10^{17} - 10^{18}\right) m^{-3}$. For the case of maximum anisotropy the parameters $T_e = 4.5\,\text{keV}$, $T_{e2} = 13\,\text{keV}$ and $\alpha = 2.2$ are found (fig. 3) where T_e is also the mean energy of the electrons.

5. CONCLUSION

For strongly additionally heated plasmas non-Maxwellian electron VDFs have to be expected. In these cases the mean energy contained in the electrons can be obtained by at least two different Thomson scattering diagnostics probing different segments of the VDF.

REFERENCES

1. B. Kurzan, H. Murmann, H. Salzmann, ASDEX Upgrade Team, Improvements in the evaluation of Thomson scattering data on ASDEX upgrade, *Rev. Sci. Instrum.* **72**(1), 1111–1114 (2001).
2. J. G. Cordey, E. M. Jones, D. F. H. Start, A. R. Curtis, I. P. Jones, A kinetic theory of beam-induced currents, *Nucl. Fus.* **19**(2), 249–259 (1979).

HIGH RESOLUTION THOMSON SCATTERING ON JET – AN ASSESSMENT OF THE FEASIBILITY

C Gowers, C J Barth, R Behn, M Beurskens, B Kurzan, P Nielsen, R Pasqualotto, M Peres Alonso, F Orsitto, J Sanchez, M Walsh, D Wilson and contributors to the EFDA-JET work programme*

1. INTRODUCTION

From discussions with the EFDA-JET Task Forces in early 2000, a range of requirements emerged for higher spatial resolution T_e and n_e Thomson scattering measurements in various specific regions of JET plasmas. These requirements are summarized in table 1 and seek significantly better spatial resolution than is achieved or is achievable with the two LIDAR Thomson scattering systems[1,2] currently working on JET (4Hz core LIDAR ~12cm; 1Hz Edge LIDAR 2-3cm, outer ~10cm only). Two High Resolution Thomson Scattering (HRTS) proposals were received from the Associations, one from C J Barth, (FOM)[3] and one from P Nielsen and R Pasqualotto, (RFX)[4]. These, in their final form, used almost identical "90 degree" scattering geometry (see figure1).

The feasibility of making HRTS measurements is dependent upon successfully resolving several key issues, which are common to both schemes. For example: achieving a sufficiently low stray laser light level; achieving a sufficiently low background plasma light level; achieving relatively high collection and detection efficiency. Estimates of the first two of these items based on measurements taken on JET, are presented and discussed below. The issue of the collection window is also discussed briefly.

* C Gowers, D Wilson Euratom-UKAEA Association, Culham Science Centre, Abingdon, UK; C J Barth, M Beurskens, EURATOM-FOM Association, Nieuwegein,NL; R Behn, EURATOM-Suisse Association, CRPP Lausanne, CH; B Kurzan, EURATOM-IPP Association, Garching, D; P Nielsen, R Pasqualotto, EURATOM—ENEA Association, RFX, Padova, I; M Peres Alonso, EURATOM-IST Association, Lisbon, P; F Orsitto, EURATOM-ENEA Frascati, I and EFDA-CSU, Culham Science Centre, UK; J Sanchez, EURATOM-CIEMAT, Madrid, E; M Walsh, EURATOM-DCU Association, Dublin, EI;

Table 1. Specification by the task forces (those marked with ✓ are met by the proposals, **X** are not met, ? are maybe met in some modes of operation but not simultaneously i.e. 10% accuracy can be achieved but not at a density of 5e18 m-3.)

Region	Spat. Resol.	Coverage	Accuracy	Frequency	Min. density
Edge (G Matthews)	2 mm in midplane X	2 cm region ✓	5-10% ?	10 pulses/ discharge ✓	1e18 m-3 ?
Edge (W Suttrop)	1 cm in midplane ✓	(-5, +15cm) at LCFS ✓	10% ?	10 Hz ✓	5e18 m-3 ✓
ITB (A Becoulet)	< 2 cm ✓	R=3.3 till 3.7m ✓	5-10% ✓	10 Hz ✓	1e19 m-3 ✓
MHD (T Hender)	< 2 cm ✓	Full chord? ✓	<<10% X	10 kHz ?	1e19 m-3 ✓

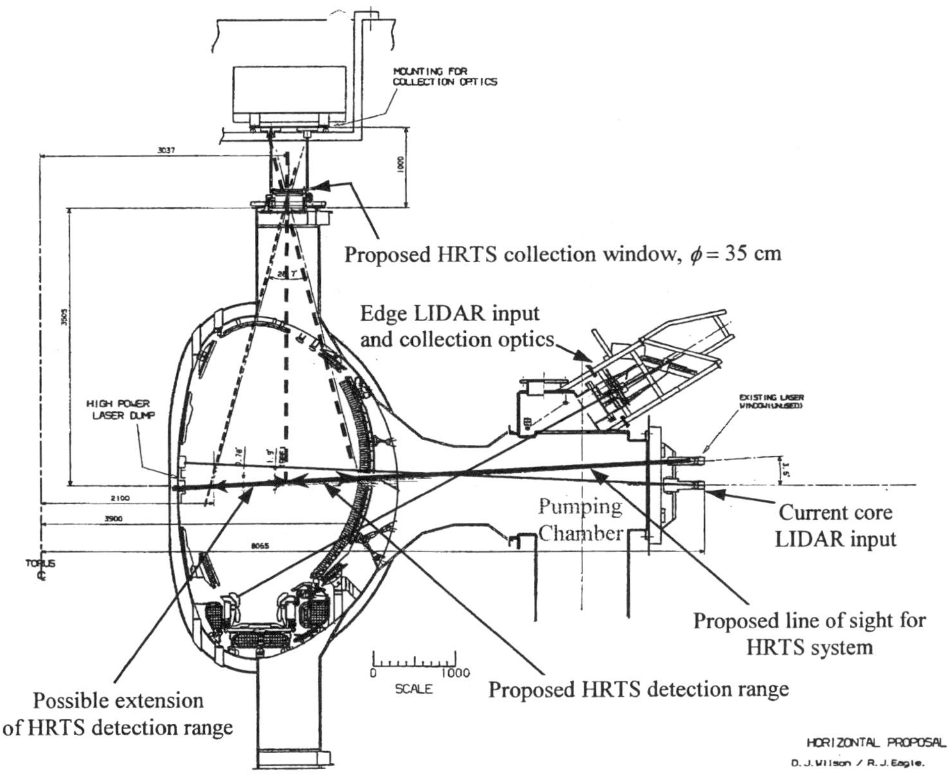

Figure 1. The proposed scattering geometry for the HRTS superimposed on the existing LIDAR systems

2. STRAY LASER LIGHT

As usual with a Thomson scattering system, stray laser light could swamp the scattered light signals making measurements difficult or impossible to interpret. In addition on JET it is difficult to make all the refinements used on smaller machines to minimise stray light levels. The LIDAR Thomson scattering advantage of being able to discriminate against stray light signals by their time of arrival at the detector and gate them out, is not available for the conventional imaging systems proposed. Assessment of the expected stray laser light levels is best undertaken experimentally and we report here the results from such an experimental investigation.

The proposed HRTS laser path was simulated by using the existing Core LIDAR ruby laser, figure 1. This enters the vessel via a window on the pumping chamber door, as shown and hits a simple grooved carbon beam dump fixed to the inner wall. The collection system in this test experiment was a 50mm diam, 100mm focal length lens mounted just above a window on the upper vertical port. The light collected was imaged on to an Avalanche Photodiode (APD)[4]. The geometry is such that the detection system cannot see the laser beam dump directly, which reduces the level considerably. These stray light results were scaled with laser energy and collection solid angle giving an expected n_e equivalent level with the new system of $\sim 2 \times 10^{23} m^{-3}$. This indicates that, despite the simplicity of the scheme, with the addition of notch filters (atten. $\sim 5 \times 10^4$) in the collection path the stray light levels should be tolerable without additional measures.

3. PLASMA LIGHT BACKGROUND

High resolution Thomson scattering measurements require very good signal-to-noise ratio for the scattered signals. The level of plasma light background could be so intense that the ratio of Thomson scattered signals to plasma light background renders high resolution measurements impossible. Estimates of the expected plasma light levels have been made by measuring the level of plasma light on the signals of the Core LIDAR system on representative plasmas. For a high performance Deuterium-Tritium pulse (#42676) with $T_e(0)$ of 12keV, $n_e(0)$ 4.e19m^{-3} the plasma light intensity equated to about 20 times that expected for pure H-H bremmstrahlung whereas for an optimised shear pulse of similar $T_e(0)$ and $n_e(0)$ (#47413) a value of about 40 times was obtained. These numbers are essential input to simulation codes so that accurate estimates of the expected signal–to-noise ratio for the new system (e.g. reference 4) can be made.

However, since the HRTS system uses a vertical collection line of sight, some parts of the field of view can see the divertor. Intense plasma light from the divertor is potentially an additional serious problem. Measurements of plasma light using different vertical sight lines of the visible spectroscopy diagnostic have been carried out. Sight lines with and without a divertor view have been compared, figure 2, for representative plasmas. These indicate that for the outer half of the profile (R=3.0 - 3.9 m) where the image does not include the divertor, the plasma light levels should be tolerable without degrading the signal-to-noise ratio significantly. For measurements at R<3 m the signal-to-noise ratio could be significantly degraded, particularly by ELM's. However, the region of the plasma called for in table 1 is still adequately covered.

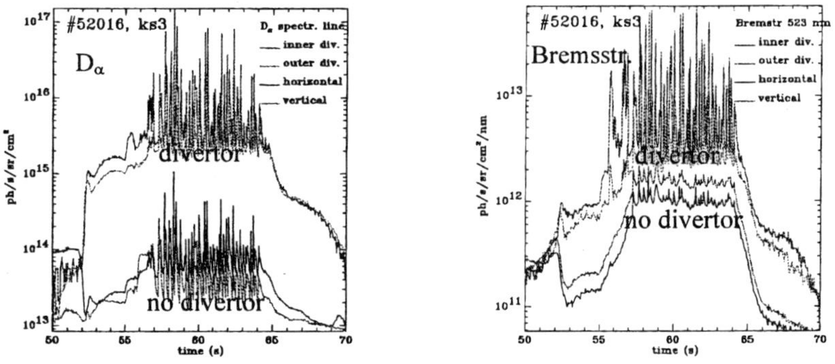

Figure 2. Time traces of plasma D_α and bremsstrahlung in a neutral-beam heated ELMy H-mode along four lines of sight. The horizontal and vertical l.o.s. without divertor closely correspond to the current core LIDAR l.o.s. and the proposed HRTS collection l.o.s. Clearly, inside 3.0 m (Fig. 1) the increased plasma light level will disturb the TS measurement.

4. COLLECTION WINDOW

To have sufficiently strong scattered signals, a high collection efficiency is required for the HRTS system. To achieve the required collection F/no of 16 for measurements near the plasma axis and 25 for the edge region, a new large window is required on the upper vertical port. The development of such a window is a key element in the development of the system. This window is approximately twice the diameter of the largest window currently in use on JET and is therefore identified as a key technological development for the HRTS system. However, since the clear aperture through the vertical port is not in fact circular, it has been suggested[5] that it may be possible to utilise two adjacent standard large windows and still achieve almost as high a collection solid angle.

5. CONCLUSION

The experimental assessment of the stray laser light and background plasma light levels reported here indicates that, in the geometry proposed, High Resolution Thomson scattering measurements are feasible on JET. This geometry requires the development of a large double vacuum window to achieve the collection solid angle required (but see reference 4 for further comments on this).

Acknowledgements
This work has been conducted under the European Fusion Development Agreement and is partly funded by Euratom and the UK Dept of Trade and Industry

References

1. C Gowers et al, Rev. Sci. Instrum. 66 471(1995)
2. M.Beurskens et al, Proc 27th EPS conf. on Controlled Fusion and Plasma Physics, Budapest, 2000, p273.
3. C J Barth, H J van der Meiden, T Oyevaar, N J Lopes Cardozo, Rev. Sci. Instrum. 72 (2001), 1138
4. P Nielsen, R Pasquelotto (this conference)

HIGH RESOLUTION THOMSON SCATTERING FOR JET

Roberto Pasqualotto and Per Nielsen *

1. INTRODUCTION

LIDAR Thomson scattering diagnostics have proven successful at JET to measure electron temperature and density profiles along a diameter in the equatorial plane and in the divertor region. A hard limit of the LIDAR approach is the low spatial resolution (~ 12 cm), given by the response time of the system. This resolution is difficult to significantly improve, based on the available technology. Two regions of the plasma require a higher resolution: gradient zones at the ITBs (Internal Transport Barriers) and the edge pedestal.

A conventional TS system is being developed for JET-EP, which will provide 1.5 cm resolution for the ITBs and edge pedestal, with a predicted accuracy of 15% at a density of 1.0×10^{19} m^{-3}.

2. LAYOUT

The laser is located in the roof lab. The laser beam enters the Torus Hall through the same penetration as the Divertor LIDAR (KE9) beam. A mirror in front of the original KE9 laser window on the horizontal port sends the laser beam horizontally to a beam dump on the inner wall. It might be preferable to steer the laser beam to an existing lower beam dump should this be required to better view the edge, but in this case the beam passes about 25 cm below the plasma centre.

The system is a conventional 90 degree system viewing the laser chord through a main vertical port. The distance of this port to the scattering volume is such that a window of ~35 cm diameter is required. A window of this size has not been installed on JET in the past. The window needs to be developed and tested. The design effort for this is in progress. As a backup solution we also consider the use of two standard 17 cm diameter windows. The port is quite narrow and these windows would cover roughly the same area. However, the longer

* Consorzio RFX, Associazione EURATOM-ENEA sulla Fusione, corso Stati Uniti 4, Padova 35127, Italy

Advanced Diagnostics for Magnetic and Inertial Fusion, Edited by Stott *et al.*
Kluwer Academic/Plenum Publishers, New York, 2002

tubes associated with these windows cause additional vignetting. By tilting the innermost window tube towards the outside edge we get the same throughput in this region as with a full size window. Figure 1 compares the solid angles available with the two solutions.

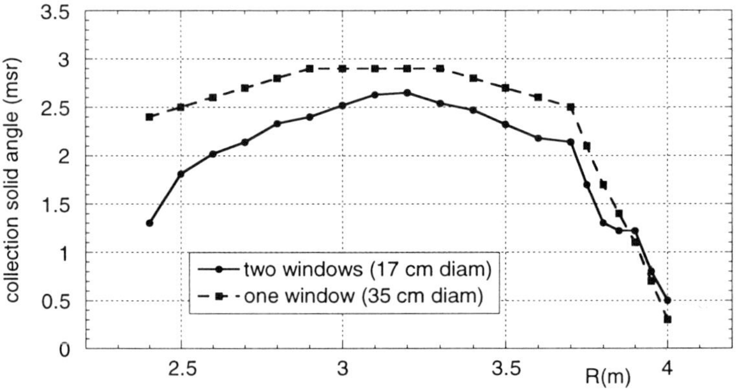

Figure 1. Collection solid angles through the two proposed windows solutions

The collection optics outside the window has to be able to withstand the radiation level from the D-T campaign. This excludes elaborate multi-element lenses in the system. Two solutions are considered for the imaging of the scattering volumes. The most straightforward solution images the scattering volumes from R=3.0 m to R=3.9 m onto a single row of 1 mm diameter Si/Si fibres giving a spatial resolution ~15 mm. The fibre lengths are substantial and contribute significantly to the cost of the system.

An alternative approach is to relay the images to the torus hall wall. Two singlet quartz lenses and a flat mirror provide sufficiently good imaging for better than 1 cm resolution. The scattering volumes are imaged onto aspherical lenses who in turn image the two quartz lenses onto each their fibre. The fibres at this location can be Si/silicone and the lengths are in any case much shorter.

Two locations along the path will have two fibres side by side to allow alignment. The signal in these four fibres will not be fed into the spectrometers but directly to detectors fitted with fixed filters to eliminate laser stray light. The signal in each pair should be balanced for proper alignment.

The viewing port allows observation of the plasma inside the quoted R = 3.0 m as illustrated in Figure 1. The system will be prepared to allow additional observation channels to be fitted. Scattering inside this radius will be limited by two factors. The laser beam is focused midway on the outer half to ensure a sufficiently large beam size on the dump. The beam diameter on the inside may therefore exceed the size of the detector image. The locations on the inside are viewing into the divertor region and the background light may make measurements impossible.

3. SPECTROMETERS

The fibres are brought to a cubicle outside the biological wall containing a number of filter spectrometers. Filter spectrometers generally stand out in terms of optical throughput. Groups of a few adjacent scattering positions are fed into each their spectrometer. This makes it easy to match and change the dynamic temperature range of the individual spectrometers to expectations and experience.

To reduce cost we propose to combine three scattering locations in each spectrometer: two fibres from each position plus one for calibration will fill the aperture of the proposed spectrometers. By using transient recorders it is possible to combine time delayed signals into one recording channel. The signals from the second and third location are delayed by an optical delay fibre of 10 m and 20 m length respectively giving sufficient separation between the scattered signals.

Each spectrometer has four spectral channels (spectral bands are illustrated in Figure 2). Their global transmission is about 80% and stray light rejection is at least 10^{-5}. Detectors are avalanche photodiodes: model C30956 from Perkin Elmer, 3 mm diameter, which have been chosen for their high quantum efficiency at 1064 nm (about 45%). Effective quantum efficiency is reduced because of excess noise factor F and dark current /amplifier noise at low light level. The spectrometers recently built at RFX have a noise figure of F=3 and the minimum detectable signal is 90 photoelectrons at an internal gain of 100. Characterisation of spectrometers and detectors is reported in [1].

4. ACQUISITION SYSTEM

With the advent of "cheap" transient recorders at GHz speed we plan to use the recording system prepared at RFX. The APDs amplifiers have two outputs, an AC-output which is fed into the transient recorders and a DC output for recording the background level for reference. The AC output has a lower frequency cut-off at a time constant of 5 μs allowing a high level of background signal without affecting dynamic range. The AC output when recorded by a transient recorder is sufficient for our analysis since we can directly determine the noise of the background signal. As the spectral channels are recorded simultaneously the accuracy of the signal level may be improved using correlation techniques. The DC output is recorded by a slower recording system to ensure that the system is within dynamic range.

5. LASERS

We propose to use Nd:YAG lasers as they are the most developed lasers for a high repetition rate system. Typically you can have 2J/pulse @ 10 Hz or 1.5 J/pulse @ 30 Hz. This energy level may be insufficient in some scenarios but adequate in others. To get sufficient energy in all situations you need to combine 2 or 3 lasers as e.g. done at ASDEX. This gives us the additional freedom of either combining the lasers for high energy or using the higher repetition rate resulting from several lasers.

We could opt for an Alexandrite laser which would give us 7J/pulse @ 10 Hz. Even though this would result in slightly better performance at the edge because of the drop in quantum efficiency of the detectors towards the YAG wavelength we consider this option to be inferior and less flexible than the Nd:YAG option.

6. PERFORMANCE SIMULATION

Error calculations are made to illustrate the potential performance of the proposed system. The scattering angle is set to 90 degrees. The overall transmission of laser and collection is set to 0.2. The laser is assumed to be 3 Nd:YAG lasers of 1.5 J/pulse @ 30 Hz.

Figure 2 shows the result for a 4.5 Joule Nd:YAG laser with a filter spectrometer, assuming F/# = 25 and a density of 1.0×10^{19} m^{-3}, typical of the edge region. The data for the detector, i.e. excess noise factor F and NEP are measured on actual instruments at RFX. We have also assumed that we combine two 1 mm fibres to one scattering volume. The background noise is the sum of the contributions from the three combined scattering volumes. The left graph is the used slitfunction. The right graph shows the relative error in % of the fitted temperature and density respectively. The curve with the largest error is the temperature uncertainty.

The decrease in accuracy at high temperature is a result of the filter selection made. This can easily be changed with another set of filters: 4 filters can cover the range of 100 eV – 15 keV.

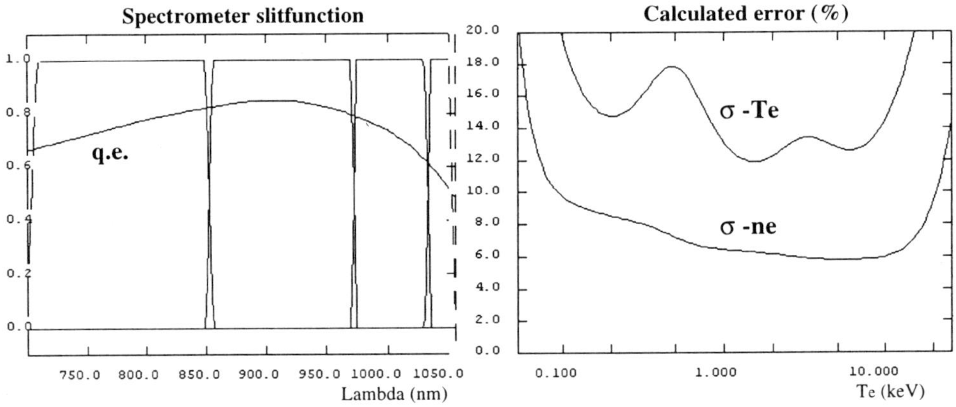

Figure 2. Spectral configuration and expected performance for the edge region

7. REFERENCES

1. R.Pasqualotto, P.Nielsen, L.Giudicotti, The new RFX Thomson scattering system, Rev. Sci. Instr. **72**(1), 1134-1137 (2001)

RECENT PROGRESS IN MICROWAVE IMAGING DIAGNOSTICS

N.C. Luhmann, Jr., C.C. Chang, B.H. Deng, C.W. Domier, C. Liang, X.P. Liang, H.J. Lu, J. Wang, and Z.G. Xia[1]; E. Mazzucato, T. Munsat, and H. Park[2]; A.J.H. Donné and M.J. van de Pol[3]; A. Mase[4]; K. Kawahata and Y. Nagayama[5]; K. Mizuno[6]; H. Matsuura and A. Miura[7]

1. INTRODUCTION

The anomalous transport in magnetized plasmas remains to be an outstanding issue, which in large part due to the difficulty in diagnosing the small amplitude fluctuations in the high temperature plasma core[1]. Recent progress in fusion studies has motivated the development of advanced microwave imaging diagnostics, both to visualize the elegant three-dimensional (3-D) structures of plasma turbulence revealed by extensive computer simulations[2] and to perform high spatial and temporal resolution measurement of turbulent fluctuations in the plasma core, such as T_e fluctuation measurements by ECE imaging. A third motivation is the recent discovery of the problem with conventional reflectometry and the need for imaging reflectometry as an essential tool for density fluctuation measurements[3]. The fourth is the need to measure fine scale structures in profiles of transport significance, such as filamentation and internal transport barriers[4]. The focus and emphasis of this paper will therefore primarily be on the dramatic breakthroughs occurring in electron cyclotron emission imaging (ECEI) and microwave imaging reflectometry (MIR), which will ultimately permit the imaging and visualization of complicated 2-D and 3-D structures of both electron temperature and density.

---

1. Department of Applied Science, University of California at Davis, Davis, California 95616, U.S.A.
2. Plasma Physics Laboratory, Princeton University, Princeton, New Jersey 08543, U.S.A.
3. FOM-Instituut voor Plasmafysica Rijnhuizen, Associatie EURATOM-FOM, Trilateral Euregio Cluster, 3430 BE, Nieuwegein, The Netherlands
4. Advanced Science and Technology Center for Cooperative Research, Kyushu University Kasuga 816-8580, Japan
5. National Institute for Fusion Science, Toki 509-5292, Japan
6. Research Institute of Electrical Communication, Tohoku University, Sendai 980- 0812, Japan
7. Development Project Center, Corporate R & D Center, Yokogawa Electric Corporation, Musashino 180-8750, Japan

Advanced Diagnostics for Magnetic and Inertial Fusion, Edited by Stott *et al.*
Kluwer Academic/Plenum Publishers, New York, 2002

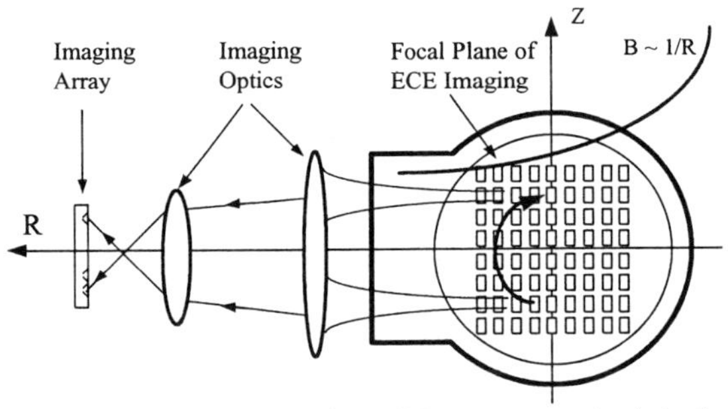

Figure 1. Schematic diagram illustrating the ECEI system principle (reprinted from Ref. 5 with permission from AIP).

2. ELECTRON CYCLOTRON EMISSION IMAGING (ECEI)

ECE imaging is a novel 2-D ECE diagnostic[5-7], in which a millimeter wave antenna/mixer array is utilized as the receiver, as shown in Fig. 1. Second harmonic, extraordinary mode electron cyclotron radiation is focused onto the imaging array by specially designed imaging optics. The elements of the array are aligned along the E field (vertical) direction to collect ECE radiation from vertically aligned plasma volumes. It is seen that with a 1-D imaging array, this new diagnostic has 2-D capability, as the horizontal position of the vertically aligned sample volumes can be scanned by varying the magnetic field strength from shot to shot while keeping the LO frequency fixed, or by sweeping the LO frequency at a constant magnetic field during the shot. In the various ECEI systems, the sample volumes were essentially 1.2 cm cubes (diameters at 1/e power) and were capable of resolving temperature fluctuations with a wavenumber range of $k < 2.8\ cm^{-1}$ [7]. The advantages of the ECE imaging diagnostic over conventional ECE radiometers were first demonstrated through experiments on the TEXT-U tokamak, in which the poloidal wavenumbers, statistical dispersion relations, and poloidal correlation lengths of electron temperature fluctuations in the plasma core were measured for the first time[8]. This high spatial resolution ECEI system was also utilized in plasma transport studies to measure the fine scale structures of T_e profiles due to plasma filamentation in ECRH plasmas[9]. The power of ECEI diagnostics was again demonstrated in comprehensive fluctuation and transport studies in the RTP tokamak[6]. An example is shown in Fig. 2, which is a 2D image of the fluctuation amplitude, measured on a shot to shot basis in RTP[6]. It is seen that turbulence is not uniformly distributed over the plasma cross-section. Instead, prominent peaks are seen in the plasma confinement region, which may be related to the coherent structures predicted by theory[10]. Of particular interest are the peaks of both low frequency (< 30 kHz) coherent fluctuations and high frequency (> 30 kHz) turbulence in the tokamak midplane, outside the q = 2 radius, in both the high and low field side[5, 6, 11]. This experimental finding is not expected from current theory.

More recently, a prototype ECEI system was developed and implemented on TEXTOR-94 [12]. Figure 3 illustrates initial ECEI data obtained on TEXTOR-94. In Fig. 3(a), it is seen that the island grows rapidly during the sawtooth ramp phase. The detailed island structure before the sawtooth crash is shown in Fig. 3(b), in which the second harmonic is clearly seen indicating the complex structure of the magnetic island.

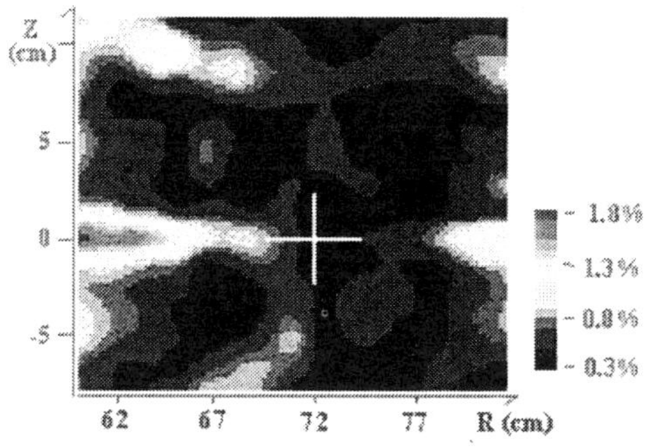

Figure 2. 2-D image of the relative fluctuation level ($\tilde{T}_{e,rms}/T_e$) of the turbulent fluctuations (30-170 kHz) measured at I_p = 60 kA, and $\bar{n}_e$ = 1.5×10^{19} m^{-3}. The statistical noise level is about 0.3%. *R* is the plasma horizontal position (72 cm corresponds to the plasma axis) *z* is the plasma vertical position, and the yellow-cross indicates the center of the vacuum chamber. The data are from about fifty identical discharges (reprinted from Ref. 5 with permission from AIP).

In the ECEI system under development for TEXTOR-94, a multi-frequency heterodyne receiver will be constructed for each of the imaging channels, so that 2-D maps of T_e profiles and fluctuations can be obtained in a single discharge, unlike the previous 1-D systems implemented on TEXT-U and RTP, in which successive discharges with varying toroidal *B*-field conditions or LO frequency were required[5-7]. For the study of $\tilde{T}_e$ turbulence, this new instrument makes it possible to unambiguously measure k_r and k_θ simultaneously, which is essential for mode identification. In the case of T_e profiles and coherent mode phenomena such as MHD and sawteeth, this will allow direct observation of island/filament rotation without the assumption of rigid body rotation. The 2-D mapping also makes it considerably easier to overlap with MIR sample volumes for correlation studies between $\tilde{T}_e$ and $\tilde{n}_e$ fluctuations.

ECE imaging systems have now also been developed and installed on mirror machines and stellarators[13, 14]. In the ECEI system developed for the large helical device (LHD) at the National Institute for Fusion Science (NIFS) in Japan, an ellipsoidal mirror and a plane mirror are located inside the vacuum vessel, converging the ECE signals to

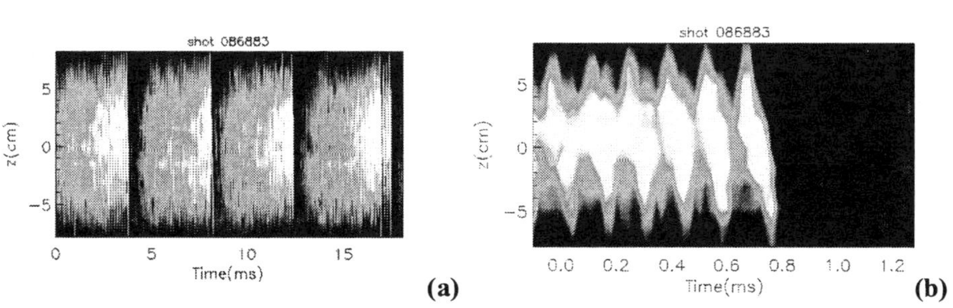

Figure 3. Observation of temperature profiles on TEXTOR-94 with ECEI, showing complex island structure during high power ECRH. Time starts from 1.566 sec. Sensitivity calibrated with Thomson scattering.

pass through a 192 mm diameter fused-quartz vacuum window. These in-vacuum optics make it possible to image an object plane located at the plasma center which is 2.7 m in front of the ellipsoidal mirror.

3. MICROWAVE IMAGING REFLECTOMETRY (MIR)

Microwave reflectometry has been widely utilized to diagnose electron density fluctuations in a magnetized plasma[15]. In a conventional reflectometer, usually a pair of horn antennas are utilized to transmit the probe beam and to receive the reflected wave, without any imaging optics. Density fluctuations are inferred from the measured phase fluctuations of reflected waves. It was recently discovered that, due to the complicated fluctuation structure of plasmas, the phase of the reflected waves measured in this simple point measurement configuration becomes scrambled, making it impossible to deduce density fluctuation from phase data [3, 15]. This has been verified from experimental and numerical studies, and the latter indicates that imaging techniques are essential for density fluctuation measurements via reflectometry[3,15]. This is because the reflected wave "appears" to originate from a 'virtual' cutoff layer ($r = r_{min}$), where its amplitude fluctuations are minimum, and the density fluctuation spectrum can be recovered from the phase measurements. As this 'virtual cutoff layer' is behind the actual cutoff layer ($r = r_c$), imaging optics are required to map it onto the detector plane. Therefore, a microwave imaging reflectometer has been developed and tested in TEXTOR-94.

The system configuration of MIR is similar to that of ECEI shown in Fig. 1, except that an MIR system has the added complication of the probing beam, which is separated from the reflected wave by a beam splitter. To maximize the reflection power focused onto the detector array, the imaging optics are designed so that the wave front curvature matches that of the curved plasma cutoff layer. From our first experience, it was found that spurious reflections from lens surfaces may contaminate the reflectometer data, and therefore, mirrors are preferred for the imaging optics. Shown in Fig. 4 is a 3D AutoCAD

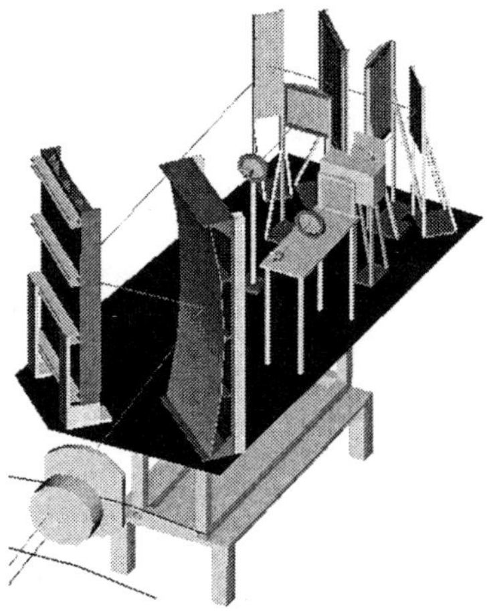

Figure 4. System configuration of a microwave imaging reflectometer developed for TEXTOR-94.

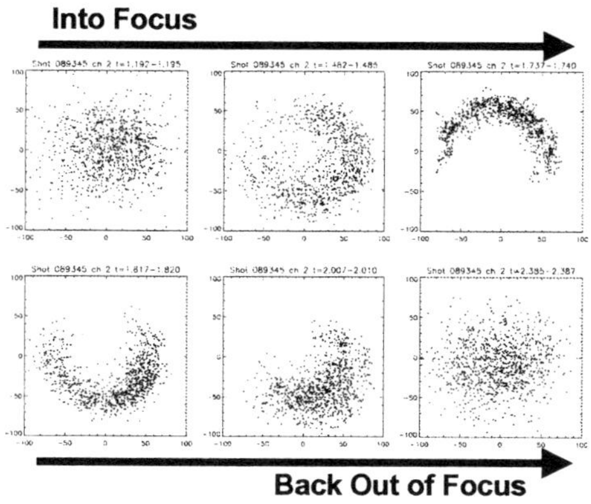

Figure 5. Complex field amplitude from the prototype TEXTOR-94 MIR system as the cutoff layer is swept through the focal plane of the imaging optics. Each frame represents a 3 ms time window.

design of the mirror based system briefly implemented and tested in TEXTOR-94, which confirmed the feasibility of the imaging scheme.

One of the most significant results from the prototype MIR system is the demonstration of the virtual cutoff surface. In this experiment, both the probing frequency and the focal plane of the optics were held fixed, and the electron density was ramped over the course of the shot to bring the cutoff surface through and beyond the focal plane of the optics. Figure 5 shows the complex signal recorded over several 3 ms duration time windows as the cutoff surface moves outward through and beyond the optical focal plane. Both the characteristic crescent-shaped phase modulation when the cutoff is in-focus and the fill-in of the complex plot as the cutoff moves out of focus are clearly identifiable.

Standard reflectometry fails due to the complicated interference wave front pattern in the detector plane. Similarly, good Gaussian beam quality from the power source is required, and the optics should be fabricated with high quality to maintain a good Gaussian beam during propagation to minimize the wave front interference. For the new TEXTOR-94 MIR system, a specially designed horn antenna will be utilized to launch the probe beam.

4. SIMULTANEOUS ECEI/MIR

To understand the critical physics issues confronting magnetic fusion; for example, to resolve the relation between anomalous transport and microturbulence, there is a need for simultaneous n_e and T_e fluctuation measurements. Figure 6 illustrates the concept of such a combined ECEI and MIR system using refractive optics and dichroic plates to separate the ECEI and MIR signals in frequency space. The primary optical components are shared between ECE and reflectometer subsystems, with each subsystem using a dedicated imaging array. Currently, a combined ECEI and MIR system is under

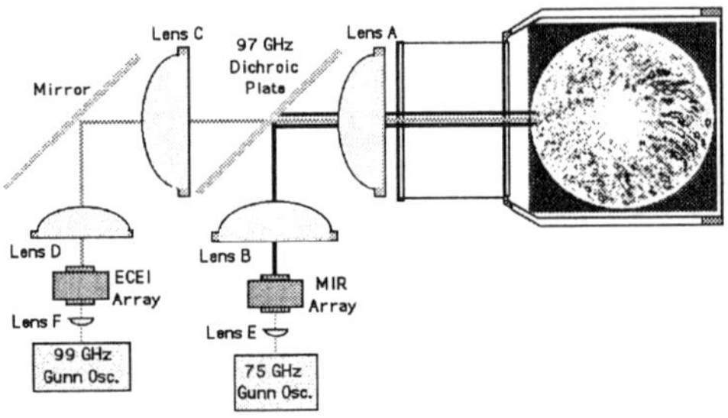

Figure 6. Combined 3 D ECEI and MIR system for simultaneous measurement of density and temperature fluctuations.

development for the TEXTOR-94 tokamak using large mirrors, which will be installed on TEXTOR-94 in 2002.

5. IMAGING TECHNOLOGY DEVELOPMENT

Millimeter wave imaging arrays are the key component for plasma imaging diagnostics. Both hybrid and monolithic imaging arrays have been fabricated at UC Davis for use in plasma imaging diagnostics[16, 17]. Slot bowtie antennas are utilized due to the advantages of low cost, ease of fabrication, compact size, and wide bandwidth. Comparing to hybrid arrays, monolithic arrays have improved response in terms of sensitivity, reproducibility, compactness and reduced LO power requirement. Shown in Fig. 7(a) is a photograph of monolithic slot bow tie array assembly developed by UCD. Shown in Fig. 8(a) is a 2-D imaging array developed by the University of Tsukuba and Tohoku University for imaging diagnostics on the GAMMA 10 tandem mirror device[18]. Shown in Fig. 8(b) is a bowtie array integrated with IF electronics, developed by Kyushu University and the Teratec Corp. for LHD[19]. More recently, a different class of imaging arrays is under development, which utilizes dual dipole antennas [Fig. 7(b)]. The initial

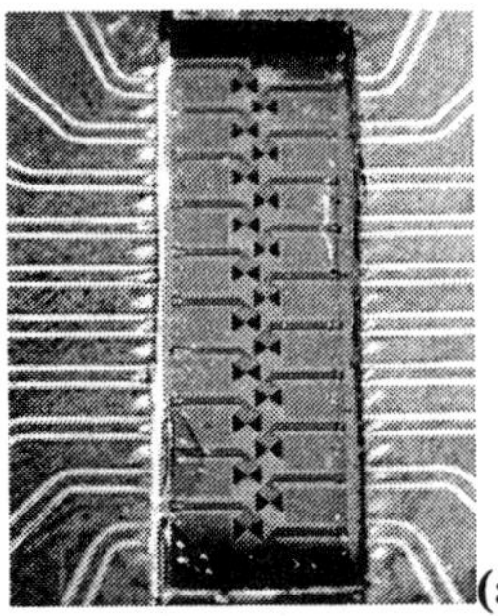
(a)

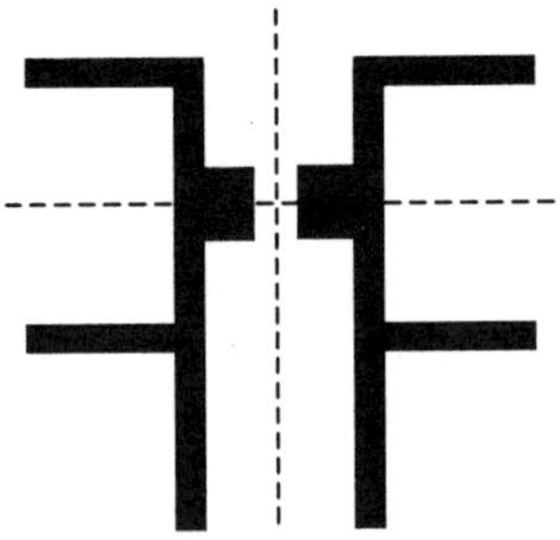
(b)

Figure 7. Photograph of a monolithic slot bow tie antenna/mixer array assembled with printed circuit board (a) and the diagram of the new dual dipole antenna array under development (b).

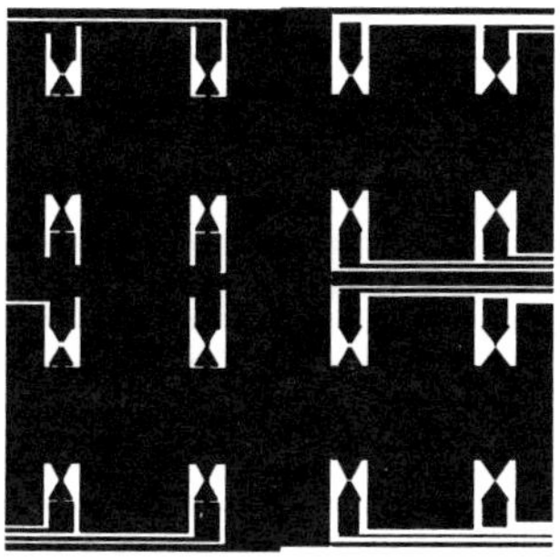

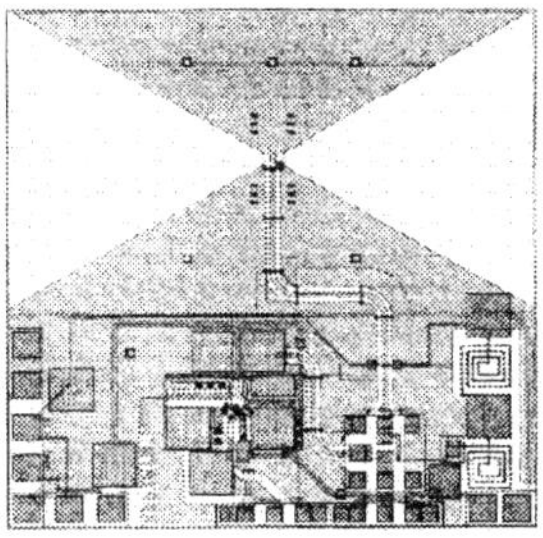

Figure 8. (a). Schematic of the 2-D imaging array. (b). Mask pattern of the monolithic detectors integrated with bow tie antennas, Schottky diodes, and HBT amplifiers. (Reprinted from Fusion Engineering and Design, Ref. 13, Copyright (2001), with permission from Elsevier Science)

test results of hybrid dual dipole antennas are very promising, and their monolithic version will be fabricated and tested soon.

To sufficiently pump the mixers in the imaging array, an LO power of about 100 mW or more is required. Therefore, subharmonic mixers are utilized at frequencies where high power LO source is not available[20]. Another approach is the passive video camera, which eliminates the need of an LO source[21].

The massive channel numbers of an imaging diagnostic system drive the need to develop low cost post detection electronics, especially for ECE imaging IF detection circuits. Currently, a prototype 2 to 18 GHz multichannel IF detection system using printed circuit board components is being developed at UCD. Such low cost IF electronics make it practical to construct real time 2-D ECEI systems, which will be invaluable in visualization of both coherent and turbulent T_e fluctuations (see Sec. 2).

For the MIR diagnostic system, a critical issue is to match the wave front curvature of the illumination beam to that of the plasma cutoff layer. To account for the variation of density profiles and thus the curvature of cutoff layers, a phased antenna array (PAA) based on nonlinear delay lines (NDLs)[22] is being developed for electronic control of the wave front. As shown in Fig. 9, NDLs and MMIC amplifiers are inserted between the source and the antennas. By controlling the phase and amplitude profile across the antenna array, focusing or defocusing beams can be achieved. A prototype beam control array has been fabricated and successfully tested in the 6-18 GHz region[23].

(a) Defocusing PAAs

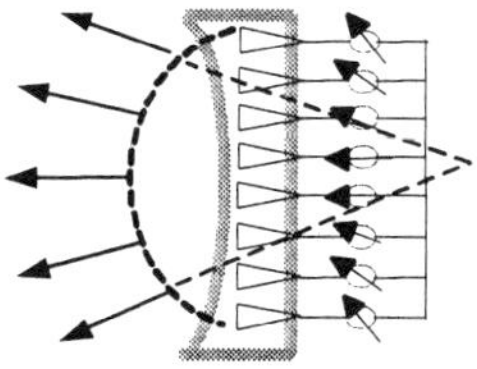

(b) Focusing PAAs

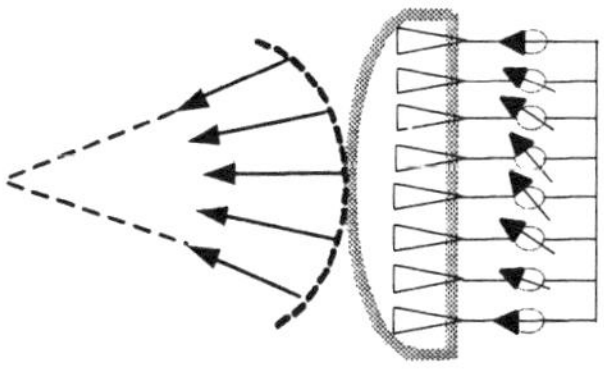

Figure 9. (a) Convex wave fronts generated by defocusing PAAs (b) Concave wave fronts generated by focusing PAAs.

6. ACKNOWLEDGMENTS

This work is supported by the U.S. Department of Energy under contract Nos. DE-FG03-95ER-54295 and W-7405-ENG-48, by NWO and EURATOM, and by the Japanese Ministry of Education, Culture, Sports, Science, and Technology. Assistance from S.R. Burns, R. Hsia, T. Hillyer, and the RTP team is gratefully acknowledged.

7. REFERENCES

1. N. Bretz, "Diagnostic Instrumentation for Microturbulence in Tokamaks", Rev. Sci. Instrum. **68**, 2927-2964, (1997).
2. W. Horton, "Drift Waves and Transport", Rev. Modern Phys. **71**, 735 (1999).
3. E. Mazzucato, "Microwave Imaging Reflectometry for the Visualization of Turbulence in Tokamaks", Nuclear Fusion **41**, pg.203-213 (2001).
4. N.J. Lopes Cardozo, *et al.*, "Plasma Filamentation in the Rijnhuizen Tokamak RTP", Phys. Rev. Lett. **73**, 256 (1994).
5. B.H.Deng, et al., "ECE Imaging of plasma T_e profiles and fluctuations", Fusion Engineering and Design **53**, pp77-85 (2001).
6. B.H. Deng, et al., "Electron Cyclotron Emission Imaging Diagnostic of T_e Profiles and Fluctuations", Phys. Plasmas **8**, no. 5, pp. 2163-2169, 2001.
7. B.H. Deng, et al., "ECE imaging of electron temperature and electron temperature fluctuations (invited)", Rev. Sci. Instrum. **72**, 301 (2001).
8. B.H. Deng, et al., "Mode Structure of Turbulent Electron Temperature Fluctuations in the Texas Experimental Tokamak Upgrade", Phys. Plasmas **5**, pp. 4117-4120 (1998).
9. G. Cima, et al., "Electron Heat Diffusivity in the Sawtoothing Tokamak Core", Plasma Phys. Control. Fusion **40** (no. 6), pp. 1149-58, (1998).
10. W. Horton, *et al.*, "Coherent Drift-wave Structures in Toroidal Plasmas", *J. Plasma Phys.* **56**, part 3, pp. 605-613, (1996).
11. B.H. Deng, et al., "Imaging of Core Electron Temperature Fluctuations of Tokamak Plasmas", submitted for publication in the 3rd Triennial Special Issue of the IEEE Transactions on Plasma Science, Images in Plasma Science.
12. B.H. Deng, et al., "ECE Imaging on TEXTOR-94", Rev. Sci. Instrum. **72**, pp. 368-70 (2001).
13. A. Mase, et al., "ECE-Imaging Work on GAMMA 10 and LHD", Fusion Eng. Design **53**, pp. 87-95, (2001).
14. A. Mase, et al., "Application of Millimeter-Wave Imaging System to LHD", Rev. Sci. Instrum. **72**, no. 1, pp. 375-378, 2001.
15. E. Mazzucato, "Microwave Reflectometry for Magnetically Confined Plasmas," Rev. Sci. Instrum. **69**, pp. 2201-2217 (1998).
16. R.P. Hsia, et al., "ECE Imaging Array Diagnostic Development of TEXT-U," Rev. Sci. Instrum. **66**, pp. 834-836, (1995).
17. Shelley Xiao-Lei Cheng, "Monolithic Schottky-Diode Imaging Array", Master thesis, UCLA, (1993).
18. N. Oyama, et al., "Millimeter-Wave Two-Dimensional Imaging Array for theGAMMA 10 Tandem Mirror", Rev. Sci. Instrum. **68**, no. 1, pp. 500-503, (1997).
19. H. Negishi, et al., "A Fully Monolithic Millimeter-Wave Detector for Plasma Diagnostics", Conf. Digest 24th Int. Conf. on Infrared (1995).
20 S. Baumel, et al., "ECE Imaging Development at W7-AS", 26th EPS Conf. on Contr. Fusion and Plasma Physics, Maastricht, 14-18 June 1999. ECA **23J**, pp. 1581-1584 (1999).
21. P. Moffa, *et al.*, "Passive Millimeter Wave Camera Flight Tests." Proc. of the SPIE - The International Society for Optical Engineering **4032**, p.14, (2000).
22. W. M. Zhang, et al., "Novel Low-Loss Delay Line for Broadband Phased Antenna Array Applications", IEEE Microwave Guided Wave Letters 6(11), pp. 182-184, (1996).
23. C. Liang, et al., "Wide Band True Time Delay Line Controlled Phased Antenna Arrays " 24th International Conference on Infrared and Millimeter Waves, U.S.A., p. PM-B5 (1999).

ELECTRON BERNSTEIN EMISSION DIAGNOSTIC OF ELECTRON TEMPERATURE PROFILES AT W7-AS STELLARATOR

Francesco Volpe[1], Heinrich P. Laqua[2] and the W7-AS Team

1 INTRODUCTION

Magnetically confined plasmas do not necessarily fulfill the accessibility criteria for electromagnetic waves in the electron cyclotron (EC) frequency range, polarized in the ordinary (O) or extraordinary (X) mode. This may hold especially in high density stellarator plasmas, not affected by the Greenwald limit, and in low magnetic field spherical tokamaks. In such cases the electron Bernstein (B) mode, the third electron cyclotron mode in a *hot* magnetized plasma is the "natural" candidate for EC heating, current-drive and diagnostic, since it propagates without upper density limit. On the other hand, it is an electrostatic mode unable to propagate in vacuum, which has to be excited and detected within the plasma, e.g. by mode conversion or direct coupling [5]. Overdense plasma heating by electron Bernstein waves (EBWs) was demonstrated at W7-AS [1] taking advantage of the O-X-B mode conversion mechanism [4]. The coupling of O and slow X-mode occurs at the cutoff layer for the O-mode: if it is approached with a suitable inclination defined by [4]

$$k_y = 0, \qquad k_z = \omega\omega_c/(\omega + \omega_c)c \tag{1}$$

in a frame with the z axis parallel to the magnetic field $\mathbf{B}$ and the x axis along the density gradient, the two modes coalesce, locally. In the phase space, then, one branch is the prolongation of the other without crossing a region of evanescence ($k_x^2 < 0$). In the physical space this means that 100% of input O-mode energy flows into X-mode and (being unaffected by the O-mode cutoff) goes ahead. Then it turns back to the upper hybrid layer and converts completely into the B-mode. The reversed scheme, BXO, was used at W7-AS for the first spectral measurements of electron Bernstein emission (EBE) [2]. Since emission reachs blackbody

[1]Max-Planck-Institut für Plasmaphysik, EURATOM Association, Boltzmannstr.2, D-85748 Garching, Germany, e-mail: `frv@ipp.mpg.de`

[2]Max-Planck-Institut für Plasmaphysik, Teilinstitut Greifswald, EURATOM Association, Wendelsteinstr.1, D-17491 Greifswald, Germany

Advanced Diagnostics for Magnetic and Inertial Fusion, Edited by Stott *et al.*
Kluwer Academic/Plenum Publishers, New York, 2002

levels, more recently was possible even to measure the first temperature profile [3]. This paper describes the first routinely operating dedicated diagnostic and its results, with the focus on the features distinguishing EBE from electron cyclotron emission (ECE) of underdense plasmas. The article is organized as follows. The Sec.2 is devoted to the ray tracing code for mode-converted electron Bernstein waves (EBWs), developed to design a dedicated antenna and to interpret the experimental results. The experimental setup is described in Sec.3 with special regard to the broadband phase shifter allowing for pure O-mode detection. In Sec.4 we present a typical profile, reconstructed from the spectral measurements by means of the above mentioned ray tracing code. Heat waves measurements combining EBW emission diagnostic with modulated EBW heating at very high density are presented in Sec.5. Finally Sec.6 shows applications of the good EBE time resolution compared to Thomson scattering, which also access high density but cannot resolve, for example, ELMs or profile oscillations observed by EBE before radiative collapse.

2 RAY TRACING CODE AND ANTENNA OPTIMIZATION

The drawback of the BXO scheme is the oblique sightline, resulting in a Doppler shift not negligible in both the antenna design and data interpretation. A ray tracing code was developed keeping these goals in mind. Unlike standard codes, the hot dielectric tensor and the three-dimensional (3D) magnetic topology of W7-AS stellarator were implemented, for the following reasons.

The electron Bernstein waves are electrostatic modes sustained by electron gyration in a *hot magnetized* plasma. Electrons exhibit fronts of rarefaction and compression perpendicular to the magnetic field with wavelength $\lambda \leq \rho$, where the Larmor radius ρ is supposed to be finite. As a consequence also the thermal velocity has to be finite, i.e. the hot plasma dielectric tensor must be used.

By virtue of the different restoring force (Lorentz rather than electrostatic) and of the different relevant frequency (ω_c instead of ω_p), the electron cyclotron motion can "propel" a space charge perturbation even on longer wavelengths than Langmuir plasma waves, $\lambda > \lambda_D$ (but $\lambda \leq \rho$), i.e. in the overdense part of the plasma, where $\omega_p > \omega \geq \omega_c$. Since the Lorentz force acts as restoring force, it is clear that EBW propagation depends strongly on the magnetic field. This explains why the tracing code was interfaced to the TRANS code, for an accurate knowledge of the W7-AS magnetic field.

The ray tracing code takes into account also the mode conversions, in one case just as by-product of the hot dielectric tensor: in the hot plasma dispersion relation the B-mode connects to the slow X-mode at the upper hybrid resonance, thus including automatically the B-X conversion. The other mode conversion occurs at the O-mode cutoff, where the WKB approximation breaks and the ray tracing approach is no longer valid. The problem is by-passed by tracing the O-mode as close to the cutoff as permitted by the WKB approximation, then restarting the X-mode on the other side as soon as the approximation applies again. Following Hansen and Lynov [6], the initial conditions for the X-mode are matched to the last point of the O-mode solution by symmetry arguments in the (thin) plasma slab lying in between. This slab is thin for the strong density gradient typical of W7-AS.

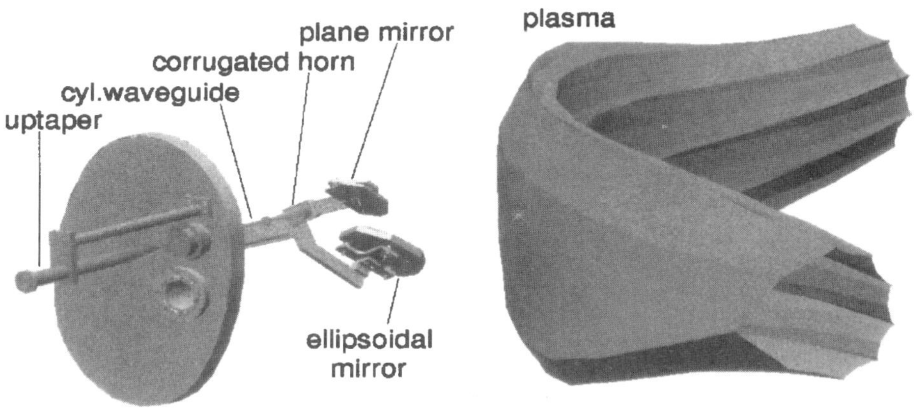

Figure 1: In-vessel part of the diagnostic

2.1 Antenna optimization

In the nonaxisymmetric device W7-AS the initial toroidal angle ϕ influences the evolution of the component of EBW refractive index parallel to $\mathbf{B}$, $N_{\|}$. A finite $N_{\|}$ in the emission region broads the resonance. The high optical thickness for EBWs (a factor c^2/v_{th}^2 greater than ECE), delocalizes the emission to the very outskirt of the broadened resonance layer, any internal emission being reabsorbed. The mentioned Doppler broadening results thus in temperature-dependent, inhomogenous, shift of the spectrum. This undesired spectrum deformation was minimised minimising $N_{\|}$ in a toroidal scan in 3D ray tracing calculations in full stellarator geometry [3].
At the same time the transmittivity was maximised according to eq.(1) by varying the other free parameters, like the inclination with respect to the equatorial plane.

3 EXPERIMENTAL SETUP

3.1 Antenna

A gaussian optic antenna consisting of an ellipsoidal mirror and a corrugated horn was installed in the W7-AS vessel (fig.1). The antenna collects the mode-converted EBE escaping the overdense plasma along the best sightline (best transmittivity and minimal Doppler shift), with the focus at the OX conversion layer. The beam is focused almost at the diffraction limit (waist radius $w = 9mm$, wavelength $\lambda \simeq 4mm$) because within few centimeters converts in B-mode and gets high refractive index perpendicular to $\mathbf{B}$, $N_{\perp} \approx c/v_{th} = 15 - 30$, thus diverges $15 - 30$ times slower than in vacuum, conserving a transverse size similar to that at the waist. A cylindrical mono-mode waveguide in-vessel, an uptaper and an oversized transmission line out-vessel conserve the polarization before it is processed by a polarization rotator and a phase-shifter.

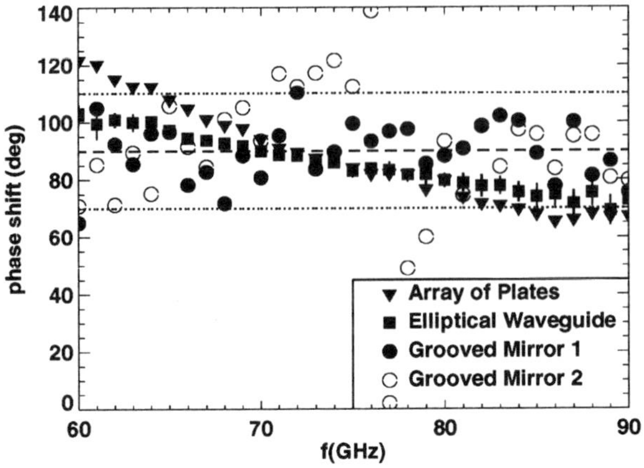

Figure 2: Frequency response of tested phase-shifters. Dotted lines represent *min* and *max* allowed phase shift.

3.2 Broadband phase-shifter

The O-mode polarization detected along the oblique sightline is elliptical, but for proper optical axes orientation it is changed into a linear one by a $\lambda/4$ phase-shifter. After comparison of several devices, summarized in fig.2, an elliptical waveguide resulted to be broadband enough for the diagnostic purposes: the frequency range $f = 66 - 78GHz$ is covered with a systematic error in phase shift smaller than 10% [3], corresponding to 2% in intensity which can be neglected vs. errors of about 10% associated to the radiometer calibration with an hot/cold source. The phase shift originates simply from different phase velocity for orthogonal modes in the elliptical waveguide.

3.3 Radiometer

The linearly polarized output of the phase-shifter is spectrum-analysed in the 70 GHz ECE heterodyne radiometer [7], with 8 extra-channels added for the occasion. Recent coupling with the "zoom" device [8], a 16 narrow filter section with variable intermediate frequency, raises to 48 the total channel number. The Doppler shift for the new antenna is so low that the EBE spectrum matches the band $f = 63 - 78GHz$ of the radiometer without necessity to reduce the magnetic field as in ref.2 or to replace the local oscillator in the radiometer.

4 PROFILES

Fig.3 shows an EBE spectrum in a NBI-heated discharge, rescaled to take into account the transmittivity $< 100\%$. In fact, in spite of antenna optimisation for penetration through a *laminar* cutoff layer, a realistic *turbulent* layer is corrugated and oscillates, therefore locally and transiently can form a non-optimal angle with the ray. By the way the effect is small and a transmittivity $75 - 85\%$ is found, thanks to the short length scale L_n of density profile in the gradient region. Typical

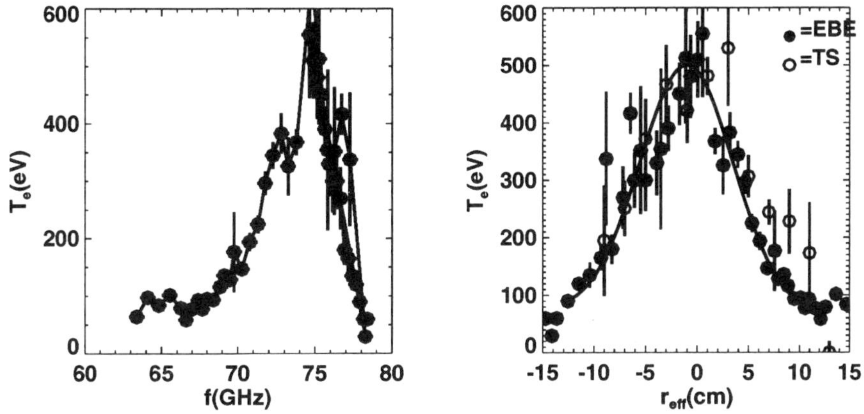

Figure 3: EBE spectrum (left) and profile (right), Thomson scattering (TS) profile

W7-AS values of $L_n = 1 - 4cm$ yield in fact effective OX coupling in a cone of aperture $7^0 - 16^0$ around the best direction [1, 3]. Turbulence-induced deviations from the optimal angle are negligible if smaller than this aperture. The finite $N_{||}$ intrinsic in the oblique detection explains the peculiar asymmetric shape of the spectrum, characterized by a larger frequency upshift in the center, where the plasma is hotter. As a consequence the high field side of the plasma emits in a narrower band, where most of new channels have been concentrated. Due to the complicated ray paths and to the shift of emitting layer with respect to the $\omega = \omega_c$ one, the conversion of EBE spectra in profiles is not as straightforward as ECE [3]. Experimental density data from Thomson scattering and Lithium beam diagnostic and magnetic data from the TRANS code are assigned as input to the parallel version of the ray tracing code. This maps the 48 channel frequencies into as many emitting locations in plasma, thus allowing profile reconstruction. Indeed, being the temperature profile also an input for calculations, which involve the *hot* dielectric tensor, a "guess" is necessary at the beginning, but in 2-4 iterations the code converges, within error bars, to the definitive profile.
The vertical error bars in the spectrum depend on the precision of the calibration with an hot/cold source, and with some exceptions are of order 10%. Horizontal error bars of $0.5 - 1cm$ in profiles are assigned according to emission region widths computed by the ray tracing code.

5 HEAT WAVES

The heat wave method has been extended for the first time to overdense plasmas combining EBWs heating and diagnostic. The plasma cannot be diagnosed and heated at the same harmonic because the heating beam drives a parametric instability at the UHR layer, where the XB conversion occurs. Such instability results in high non-thermal peaks overlapping the thermal spectrum [1].
The problem has been overcome combining EBE measurements at the first harmonic ($66 - 78GHz$) with EBWs heating at the second harmonic ($140GHz$), mod-

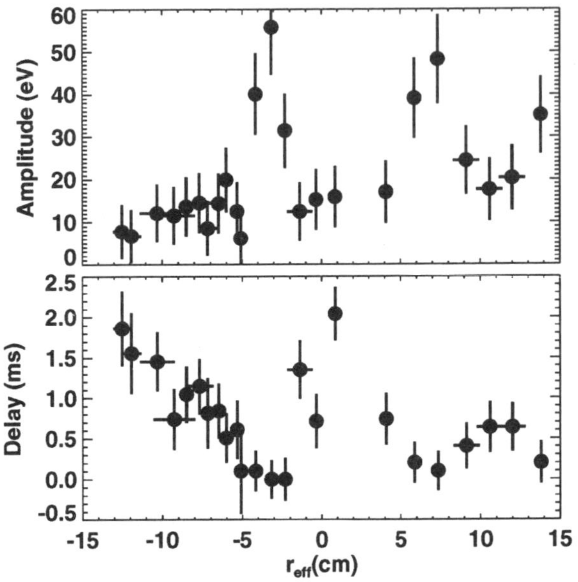

Figure 4: Heat wave amplitude and phase for $n_{e0} = 3 \cdot 10^{20} m^{-3}$

ulated in this case with a frequency of $184Hz$.
It should be emphasized that for this the plasma had to be overdense up to the second harmonic O-mode, i.e. $n_e > 2.4 \cdot 10^{20} m^{-3}$. So high densities are possible at W7-AS after installation of the island divertor [10]. A cross-correlation analysis of EBE channels response to the periodic temperature perturbation yields the heat waves amplitudes and phases plotted in fig.4. The two amplitude peaks and delay zeros correspond to off-axis power deposition at $r_{eff} \approx 5cm$. High β effects as Shafranov shift are not yet included in the analysis and are expected to remove the systematic $1.5cm$ offset. Heat waves can be recognized in fig.4, propagating from deposition regions inward and outward.

6 FAST MEASUREMENTS

6.1 ELMs

Innovative applications of EBE diagnostic may concern phenomena occurring at too high density for ECE and over too fast time scales for Thomson scattering.
Examples are L/H mode transitions and edge localized modes (ELMs), which at W7-AS often take place at densities out of ECE operational range, $n_e > 1.2 \cdot 10^{20} m^{-3}$ (see fig.1 and 4 in ref. [11]). Fig.5 shows EBE measurements of ELMs at $n_e = 1.4 \cdot 10^{20} m^{-3}$, with $1\mu s$ time resolution. As expected, edge EBE is correlated with H_α, mostly coming from the edge. On the other hand, for any edge temperature spike, the core temperature drops. This is well visible in correspondence with the two biggest spikes in fig.5 and is confirmed for the whole sample by negative correlation with H_α. The time lag for propagation of temperature

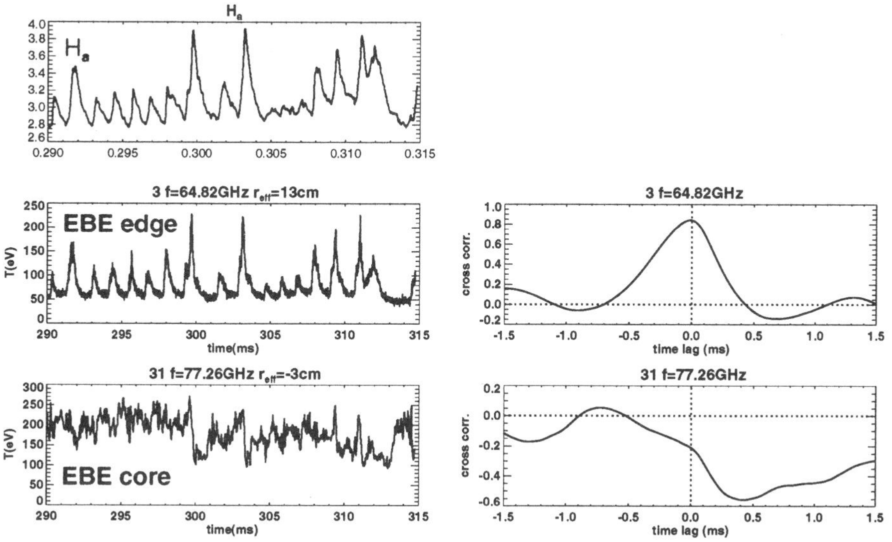

Figure 5: left: ELMs measured by H_α, edge and core EBE; right: EBE-H_α cross-correlations

information from edge to center is $0.4ms$ in the considered discharge with limiter.

6.2 Radiative collapse

Another high density fast phenomenon at W7-AS is the radiative collapse [12]. Fig.6 shows temperature profile oscillations observed by EBE before the definitive collapse. A single collapse is a transition from one stable branch to another of the multistable solution of the time dependent temperature diffusion problem [13]. Observed oscillations can therefore be interpreted as multiple transitions between the two states, like an hysteresis cycle along the S-shaped solution.

7 CONCLUSIONS

In short, ECE radiometry of overdense plasmas was proved by considering, rather than ordinary (O) or extraordinary (X) mode emission perpendicular to the magnetic field, the black body emission of electron Bernstein (B) waves (EBWs) escaping the plasma after B-X-O mode conversion along a special oblique sightline.
The original features of such approach, compared to traditional ECE, are:

- the antenna optimization with respect to transmittivity through the cutoff layer and with respect to the Doppler deformation of the spectrum,
- the broadband phaseshifter, for modal purity measurements along the oblique sightline,
- the mode-converted EBW ray tracing code in 3D hot plasma for profile recovery from spectra.

Temperature profiles and diagnostic applications in a variety of conditions occurring at too high density for ECE (up to $n_e = 3.5 \cdot 10^{20} m^{-3}$) and over too fast time scales for Thomson scattering (as fast as $1\mu s$) have been presented. Moreover, for the

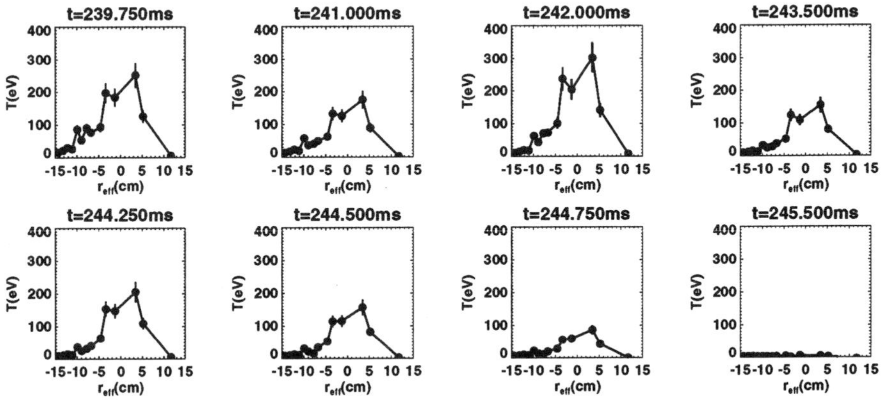

Figure 6: Profile oscillations before radiative collapse, time t in ms

first time the heat waves method has been extended beyond the ECE cutoff density combining EBW emission measurements at the first harmonic ($f = 63 - 78GHz$) with modulated EBW heating at the second harmonic ($140GHz$).

8 REFERENCES

1 H.P.Laqua *et al.*, Phys.Rev.Lett.**78**, 3467 (1997)

2 H.P.Laqua *et al.*, Phys.Rev.Lett.**81**, 2060 (1998)

3 F.Volpe *et al.*, Proc.27th EPS Conf.Contr.Fusion Plasma Phys., 1669 (2000)

4 J.Preinhaelter and V.Kopecký, J.Plasma Phys.**10**, 1 (1973)

5 A.K.Ram and S.D.Schulz, Phys.Plasmas **7**, 4084 (2000)

6 F.R.Hansen, J.P.Lynov and P.Michelsen, Plasma Phys. Control. Fusion **27** 1077 (1985)

7 H.J.Hartfuß *et al.*, Proc.6th Joint Workshop on ECE and ECRH, Oxford 1987, p.281; see also H.J.Hartfuß *et al.* Plasma Phys.Control.Fusion **39**, 1693 (1997)

8 Ch.Fuchs and H.J.Hartfuß, Rev.Sci.Instrum.**72**, 383 (2001)

9 U.Gasparino *et al.* Plasma Phys.Control.Fusion **40**, 233 (1998)

10 P.Grigull *et al.*, to be published in Plasma Phys.Control.Fusion **44** (Jan.2002)

11 M.Hirsch *at al.*, Plasma Phys.Control.Fusion **42**, A231 (2000)

12 L.Giannone *et al.*, Plasma Phys.Control.Fusion **42**, 603 (2000)

13 P.Bachmann et al., Proc.24th EPS Conf.Contr.Fusion and Plasma Phys., Berchtesgaden 1997, p.1817

TARGET DIAGNOSTIC SYSTEM FOR SANDIA NATIONAL LABORATORIES' Z FACILITY

R. J. Leeper[a], J. R. Asay[a], J. E. Bailey[a], G. R. Bennett[b], A. L. Carlson[a], G. A. Chandler[a], G. W. Cooper[c], M. Cuneo[a], C. Deeney[a], G. Dunham[b], D. L. Fehl[a], D. Hanson[a], C. A. Hall[a], M. J. Hurst[a], D. Jobe[b], M. Knudson[a], P. Lake[b], S. E. Lazier[b], J. S. McGurn[a], R. C. Mock[b], T. Moore[b], T. J. Nash[a], D. Nielsen[b], R. E. Olson[a], J. Porter[a], J. Pyle[b], G. A. Rochau[a], L. E. Ruggles[a], C. L. Ruiz[a], T. W. L. Sanford[a], J. F. Seamen[a], J. J. Seamen[b], W. W. Simpson[a], R. Smelser[b], R. B. Spielman[a], W. A. Stygar[a], J. A. Torres[a], and D. F. Wenger[a*]

1. INTRODUCTION

The U. S. Department of Energy is supporting research at Sandia National Laboratories (SNL) to investigate the possibility of using pulsed power driven magnetic implosions (z-pinches) to drive inertial confinement fusion (ICF) targets.[1] This research is being conducted on SNL's Z facility that is capable of delivering peak currents of 20 MA in z-pinch loads producing implosion velocities as high as $7.5X10^7$ cm/s, x-ray energies of 1.6 MJ, and x-ray powers exceeding 200 TW. Several different target concepts that employ DT cryogenic capsules (2-4 mm in diameter) indirectly driven by radiation fields generated by z-pinch hohlraum sources are being pursued.[1]

Many of the diagnostic needs for z-pinch ICF experiments are similar to any other approach to indirect drive ICF. We must measure both the radiation hydrodynamic

* Authors[a], Sandia National Laboratories, Albuquerque, New Mexico, 87185. Authors[b], K-Tech Corporation, Albuquerque, New Mexico 87106. Authors[c], University of New Mexico, Albuquerque, New Mexico 87131

Advanced Diagnostics for Magnetic and Inertial Fusion, Edited by Stott *et al.*
Kluwer Academic/Plenum Publishers, New York, 2002

environment that the capsule is exposed to and the response of the capsule to that environment. However, there are some special demands imposed in z-pinch experiments. The standard ICF measurements mentioned above must be conducted in an intense hard x-ray bremsstrahlung background of some 10^9-10^{10} rads/s, detectors are subjected to intense debris, and delicate instruments are impacted by large ground accelerations as the 11.5 MJ of electrical energy is dissipated throughout the system. In addition, the plasma conditions of interest span a greater range than in laser driven ICF. Since we must investigate all the physical phenomena involved as the current flows into the z-pinch load, the solid wires are transformed into a z-pinch plasma, the z-pinch implodes, and finally the stagnating z-pinch creates radiation. The plasma conditions encompass a wide range of temperatures and densities as well as the plasma formation from solid state materials. This implies that a diagnostic suite capable of studying a wide range of conditions is required.

A comprehensive suite of diagnostics has been developed for application on the Z facility.[2] This diagnostic system can be categorized in five diagnostic areas: x-ray power and fluence, x-ray imaging, optical and x-ray spectroscopy, shock physics, and neutron diagnostics. The system is capable of measuring the power, energy, and temperature of Z pinches and associated hohlraums, gamma-ray emission in Z driven target experiments, neutron emission from inertial confinement fusion capsule experiments, and the mechanical response of materials to both isentropic and shock compression.

Diagnostic lines-of-sights (LOSs) are located in several positions on the Z facility. Eleven LOSs are located around Z at an angle of twelve degrees up from Z's horizontal axis. Nine LOSs are located in an axial package that are located at zero degrees, six degrees, and nine degrees from the vertical axis of Z.[2,3] Fig. 1 is a photograph of the Z facility that shows one of the horizontal LOS in the foreground and the remote screen boxes used to record diagnostic signals. Fig. 2 is a photograph of the Z axial diagnostic package. Other LOS positions that are sometimes used for diagnostics are located at 30 degrees from the vertical and on the bottom of the facility.

Figure 1. Z hoizontal lines-of-sight

Figure 2. Z axial diagnostic package

2. X-RAY POWER AND FLUENCE DIAGNOSTICS

2.1 X-ray diode array diagnostic

Broadband x-ray spectral measurements of soft x-ray emission are performed with a five element x-ray diode detector (XRD) array.[4] The biplanar XRD detectors used consist of a vitreous carbon photocathode and an electroformed nickel anode mesh built into a N-connector housing. The XRDs are electrically coupled to fast, high bandwidth (1 GHz Tektronix TDS 684 or 645 digitizers), through a high bandwidth, Sandia designed insertion unit. This single-ended operating mode for the detectors minimizes the electrical noise and reduces the complexity of the detectors at the expense of slight signal degradation due to the RC droop and the impedance mismatch in the insertion unit. The measured system bandwidth is 600 MHz and is dominated by the digitizers. This diagnostic is designed to measure target temperature from sources in the spectral range 137 eV to 2.3 keV. X-ray filters in front of the detectors provide broad energy cuts ($E/\Delta E$=0.5-1.0). The time resolution of these detectors is approximately 600 ps FWHM which is 6 times shorter than the measured radiation pulse. The five spectral cuts are unfolded to produce a soft x-ray spectrum.

2.2 Transmission grating spectrograph

Three x-ray spectrometers, each with a transmission grating dispersion element, are routinely used to measure the spectrum and temporal history of the absolute soft x-ray power emitted from z-pinch and hohlraum sources.[5] Each of the instruments uses a gold transmission grating to disperse the x-ray spectrum onto an array of 16 subnanosecond-response silicon photodiodes and onto a gated microchannel-plate-intensified x-ray detector. The photodiodes record the continuous temporal history of 16 spectral segments whose $E/\Delta E$ ranges from 8 to 70. The gated microchannel-plate-intensified x-ray detector records a continuous spectrum for 2 ns near the time of peak x-ray intensity. One of the spectrometers, located 26 meters from the z-pinch, diagnoses the source spectral distribution of Plankian sources between 150 eV and 250 eV. This spectrometer has a photon energy range of 250 eV to 2250 eV. The other two instruments are at 8 meters from the z-pinch and are generally used to measure hohlraum wall emissions at blackbody equivalent temperatures of 75 eV to 150 eV. These two instruments have a photon energy range of 100 eV to 900 eV. The accuracy of the spectral power measurements made with these spectrometers is ±30%.

2.3 Bolometer array diagnostic

Nickel and gold x-ray bolometer arrays have been developed to measure the x-ray radiation fluence and flux emitted by Z sources.[6] The resistive element for the nickel bolometer consists of a 1 μm thick Ni film and the resistive element for the gold bolometer employs a 1 μm thick Au film. The resistance of either film increases approximately as a linear function of absorbed x-ray energy. Nickel is used as the resistive element for <1.4 keV X-rays and gold is used for x-rays in the energy range of 1-4 keV. In the operation of either bolometer, a bias current of 40-60 A is pulsed in the thin Ni or Au film and the radiation x-ray source is discharged at the peak of this bias current. The incident x-ray radiation heats the film and the resulting resistive voltage change across the bolometer foil provides a measure of the absorbed x-ray radiation fluence and flux.

2.4 Filtered Diamond Photoconducting Detector Array

Photoconductive detectors (PCDs) fabricated from natural IIa diamonds are employed to measure the x-ray power emitted from a variety of z-pinch sources on the Z facility.[7] The resistance of (or current through) of the diamond detector element is linearly dependent on the incident photon power and the applied bias voltage. In the principle, PCDs are "flat response" detectors. The high photon energy response of a PCD detector is determined by the thickness of the diamond element. The very rapid electron-hole recombination time (measured to be ~ 200 ps) in diamond gives this detector the potential for excellent time response. The diamond detector elements of PCDs used on Z are typically 1 mm wide by 3 mm long by 0.5 mm thick and are mounted in type-N connectors on gold-coated circuit board substrates. A nominal bias of 100 volts applied to the detector yields a sensitivity of 4×10^{-4} A/W.

2.5 Soft x-ray calorimeter diagnostic

A soft x-ray calorimeter has been developed to measure the soft x-ray fluence emitted by Z sources.[8] In the operation of this diagnostic, incident x-rays (< a few keV) are deposited in an aluminum absorber within ~ 1 μm of the surface on a 10 ns time scale. The resulting thermal energy is measured over ~ 100 s interval. Absorbed photon energy is quickly converted into thermal energy δQ, which then slowly diffuse to ambient conditions (in the base of the calorimeter) by thermal conduction through a "thermal sandwich". This structure consists of two outer, electrically insulating plates, separated by a regular, two-dimensional array of posts made of alternating thermoelectric materials. The posts are arranged in a square array and electrically connected in series, forming a thermopile. As thermal energy diffuses across the array, a spatially varying temperature difference ΔT (proportional to the local thermal current $\dot{q}$ is produced across the posts. Such temperature differences generate proportional thermal emfs across pairs of posts (junctions), and the output signal $V_{th}(t)$ of the thermopile is the sum of these local thermal emfs. Thus $V_{th}(t)$ is proportional to the total heat current across the thermopile array, and $\int V_{th}(t)dt$ is proportional to δQ deposited by the x-rays. For calibration purposes, a small ohmic heater is wrapped around the absorbing disk. The calorimeter is fielded on Z at the end of a vacuum line of sight pipe, 2245 cm from the source. 2 MJ of x-rays radiated isotropically into 4π yield fluences of ~ 20 mJ/cm^2 at the detector.

3. X-RAY IMAGING DIAGNOSTICS

3.1 Time-resolved and time-integrated x-ray pinhole camera diagnostics

A schematic of a standard Z time-resolved x-ray pinhole camera is shown in Fig. 3. The cameras developed for application on Z are used to obtain time-resolved images of soft x-ray emission from Z-pinch driven sources in the spectral range of 100 eV to several keV. The cameras operate by forming pinhole images of the soft x-ray emission source onto separate micro-channel plate striplines in the image plane. The image groups formed on each MCP stripline are spectrally resolved by broadband x-ray filters located in front of the camera's pinhole aperture plate. Typical soft x-ray spectral cuts are at 200 eV, 270 eV, 450 eV, and >800 eV. The interframe time of 1-5 ns is set by cable delay and the dwell time of each frame is set by the MCP gate pulse width of 1-5 ns. The output of the MCP is a standard phosphor/fiber optic/film arrangement. The pinhole limited spatial resolution of the cameras is nominally in the range of 100-200 μm. A number of time-integrated soft x-ray and hard x-ray pinhole cameras are also fielded as part of Z target experiments.

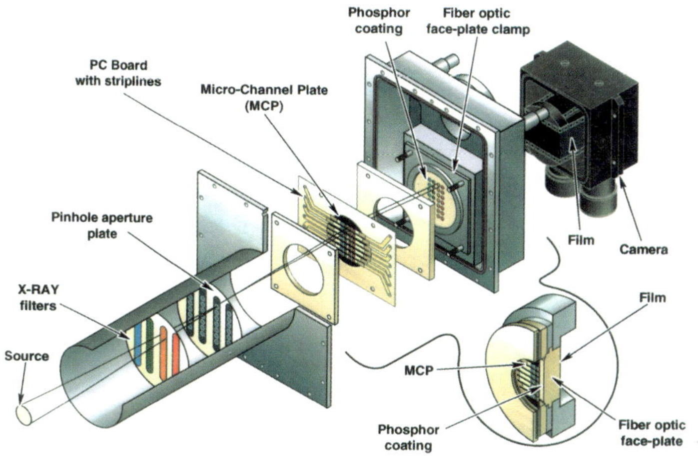

Figure 3. Schematic of a standard Z time-resolved x-ray pinhole framing camera

3.2 Streaked fiber array one-dimensional imaging diagnostic

A streaked fiber array 1-D imaging diagnostic has been developed for Z to obtain temporally continuous images of soft x-ray sources.[9] The diagnostic operates by forming a 1-D slit image of the target onto two separate regions of a scintillator/fiber optic array image plane. The two images formed on the fiber optic array are spectrally resolved by broadband x-ray filters located just in front of the scintillator. The scintillator used is BC-418 or BC-422Q plastic scintillator that is 2 μm thick. The coherent fiber array is 35 m long and consists of 120 fibers per spectral cut that are 100 μm in diameter. The fiber optic bundle is proximity coupled to a fiberoptic face plate that is coated with scintillator. The output of the fiber bundle is coupled onto the photocathode of a streak camera. The output of the streak camera is coupled to a CCD camera.

3.3 Z/Beamlet Laser Backlighter

We have recently finished construction of the Z-Beamlet Laser (ZBL) system for use as an x-ray backlighter for Z experiments. ZBL is a kiloJoule-class Nd:Glass laser capable of peak powers of 2 TW at 2ω (526 nm) with pulse durations variable between 250 ps and 2 ns. ZBL is optimized for use in the point-projection backlighting mode. The laser can be focussed to spots as small as 50 μm producing peak power densities $\geq 10^{16}$ W/cm^2 allowing x-ray probe energies up to 10 keV to be used in backlighting experiments on Z. The timing synchronization between ZBL and the Z radiation pulse is approximately ± 3 ns. A film-based x-ray detector is presently used for backlighting experiments.[10] This x-ray detector can be fielded within 75 cm of the z-pinch radiation source. It uses a fast electomagnetically-driven shutter with peak velocities of 150 m/sec to protect the x-ray film from debris and thick shielding to protect the film from the hard x-ray background generated on typical Z shots.

4. SPECTROSCOPY DIAGNOSTICS

4.1 Optical Spectroscopy

Optical spectroscopic diagnostics have been developed for application on Z that measure spectral line intensities, wavelengths, and line profiles and continuum intensity as a function of wavelength. Space- and time-resolved data are often required to produce un-ambiguous results. These requirements are met in the optical-ultraviolet regime by employing fiber optic coupled streaked spectrograph systems.[11] The light from the fibers is injected into Czerny-Turner spectrometers and recorded by a streak camera located in the exit focal plane. These measurements are useful for investigations of initial wire plasma formation, pulsed power propagation in high-current transmission lines, current diffusion through high current conductors, shock wave physics, and hohlraum preheat.

4.2 X-ray Spectroscopy

For application at higher plasma temperatures, x-ray spectroscopy diagnostics have been developed for Z. The measurement needs are similar to the optical-ultraviolet regime: simultaneous time, space, and spectral resolutions. The spectral dispersion is accomplished using an assortment of crystal spectrometers with crystals bent in both convex and elliptical geometries.[12,13] The elliptical geometry is particularly useful for high power z-pinch experiments because the detector is isolated from the softer photons in the line of sight. The detectors in use at Z are either x-ray film or time-gated microchannel plate framing cameras.

5. SHOCK PHYSICS DIAGNOSTIC TECHNIQUES

5.1 Active Shock Breakout Diagnostic

An active shock breakout diagnostic has been developed to measure hohlraum radiation temperatures on Z.[14] The radiation pressure of the x-rays contained in the hohlraum drives a shock into a stepped aluminum witness plate located on the side of the hohlraum. The back surface of the witness plate is polished to a mirror finish. A laser probe illuminates the rear surface of the witness plate at normal incidence. Before the shock wave reaches the rear surfacc, the witness plate will reflect the laser into a set of collection optics that relays the light to a visible streak camera. When the shock wave reaches the back surface of the witness plate, it will rapidly heat the surface, causing the reflectivity to drop dramatically. The shock will break out of the thinner region of the witness plate first and then later in time break out of the thicker region of the witness plate. Typically steps of 100-200 μm are used. The time difference of the shock breakout over a known thickness of witness plate gives the shock velocity and hence the radiation temperature as interpreted through radiation/hydrodynamics calculations of the hohlraum temperature.

5.2 Conventional and Line VISAR Diagnostics

The Velocity Interferometry System for Any Reflector (VISAR) is a diagnostic widely used for continuously measuring the motion of a surface or interface.[15] The principal behind the diagnostic is the measurement of the Doppler shift of laser light incident on the moving surface. Reflected light collected from the surface is split into two legs, typically using a Wide Angle Michelson Interferometer (WAMI), each having a slightly different transit time. The two legs are then recombined, forming an interference pattern. This interference pattern is stationary provided the surface is at rest or moving at constant velocity. If however the surface accelerates or decelerates, a change in the fringe pattern occurs. A detector is used to monitor the number of fringe intervals during the acceleration or deceleration periods. This fringe information can then be converted to velocity information based upon the velocity per fringe (VPF) constant, which is a function of the transit time difference between the two legs of the interferometer. A slight variant of the conventional VISAR design provides spatially resolved velocity information along a 1-D line segment of the surface. This diagnostic, referred to as a line-imaging VISAR, incorporates two subtle differences from the conventional design. First, a set of lenses are used to create an image path from the target to the detector, typically a streak camera, with the interferometer contained with the image path. Second, the interferometer is purposely misaligned slightly to create a linear interference pattern across the image of the target on a streak camera slit, with the slit perpendicular to the line of interference. The streak camera records the shifting of the interference lines along a 1-D line segment of the target, providing a spatially resolved velocity measurement.

6. NEUTRON DIAGNOSTICS

6.1 Total Neutron Yield Diagnostics

Indium and copper activation are used on Z for the measurement total neutron yield of 2.45 MeV DD neutrons and 14.1 MeV DT neutrons, respectively.[16] The indium diagnostic is based on the inelastic neutron scattering reaction: $^{115}In\ (n, n')^{115m}In$. This reaction has a 336 keV threshold so it is suitable for the measurement of DD neutrons. The ^{115m}In metastable state has a 4.5 hours half-life. The induced activity is measured with a high purity germanium detector. The copper diagnostic is based on the nuclear reaction: $^{63}Cu(n, 2n)^{62}Cu(\beta^{+})$. This reaction has an 11 MeV threshold, so it is sensitive to DT neutrons but not DD neutrons. The induced ^{62}Cu positron activity has a 9.74 minute half-life and is measured using a NaI gamma-gamma coincidence system. Another activation diagnostic that is employed on Z is the lead activation technique that can be used for measuring either DD or DT neutrons.[17]

6.2 Neutron Time-Of-Flight Diagnostics

Scintillation detectors are used on Z to make time-of-flight measurements of the neutron spectra and provide a time-dependent measurement of the neutron yield. These detectors are operated in current-mode and consist of a fast-plastic scintillator (Bicron 418 or 422) coupled to either a photomultiplier tube (PMT) or a microchannel plate (MCP). To provide time-of-flight separation, the detectors are located at distances ranging from 7 to 9 meters from the target. Because Z produces a large, hard-bremsstrahlung background, these detectors must be shielded by 20 cm of lead. The effect of this large amount of lead shielding on the pulse shape of 3 ns wide DD and DT neutron pulses has been measured.[18]

7. SUMMARY AND CONCLUSIONS

A comprehensive suite of diagnostics has been developed for application on the Z facility. Special demands are imposed on these diagnostics due to harsh backgrounds, intense debris, and large ground accelerations.

8. ACKNOWLEDGMENTS

Sandia is a multi-program laboratory operated by the Sandia Corporation, a Lockheed Martin Company, for the U. S. Department of Energy under Contract No. DE-AC04-94AL85000.

9. REFERENCES

1 R. J. Leeper et al., Nuclear Fusion **39**, 1283 (1999).
2 T. J. Nash et al., Rev. Sci. Instrum. **72**, 1167 (2001).
3 M. J. Hurst et al., Rev. Sci. Instrum. **70**, 468 (1999).
4 G. A. Chandler et al., Rev. Sci. Instrum. **70**, 561 (1999).
5 L. E. Ruggles et al., Rev. Sci. Instrum. **72**, 1218 (2001).
6 R. B. Spielman et al., Rev. Sci. Instrum. **70**, 651 (1999).
7 R. B. Spielman et al., Rev. Sci. Instrum. **66**, 867 (1995).
8 D. L. Fehl et al., Rev. Sci. Instrum. **70**, 270 (1999).
9 M. S. Derzon et al., Proc. SPIE **2002**, 31 (1993).
10 G. R. Bennett et al., Rev. Sci. Instrum. **72**, 657 (2001).
11 J. E. Bailey et al., Rev. Sci. Instrum. **68**, 1009 (1997).
12 T. J. Nash et al., Rev. Sci. Instrum. **70**, 302 (1999).
13 B. L. Henke et al., Rev. Sci. Instrum. **54**, 1311 (1983).
14 R. E. Olson et al., Rev. Sci. Instrum. **72**, 1214 (2001).
15 L. M. Barker and R. E. Hollenbach, J. of Appl. Phys. **43**, 4669 (1972).
16 G. W. Cooper and C. L. Ruiz, Rev. Sci. Instrum. **72**, 814 (2001).
17 L. Ruby and J. B. Rechen, Nucl. Instrum. Methods Phys. **53**, 290 (1967).
18 R. J. Leeper and J. Chang, IEEE Trans. On Nucl. Sci. **NS-29**, 798 (1982).

DIAGNOSTICS FOR HELICAL SYSTEMS

Shigeru Sudo, Hans J. Hartfuss, Toru Mizuuchi, Tetsuo Ozaki, Yoshio Nagayama, Tsuyoshi Akiyama, Akihide Fujisawa, Yasuji Hamada, Katsumi Ida, Takeshi Ido, Harukazu Iguchi, Shinichrou Kado, Kazuo Kawahata, Katsumi Kondo, Kazumichi Narihara, Akimitsu Nishizawa, Hiroyuki Okada, Byron J. Peterson, Fumimichi Sano, Mamiko Sasao, Kuninori Sato, Mamoru Shoji, Kenji Tanaka, Shunji Tsuji-Iio, Kiyomasa Watanabe, Arthur Weller, and Ichihiro Yamada*

1. INTRODUCTION

Diagnostics for helical systems are reviewed. A helical system is expected to be a disruption-free steady-state reactor since no plasma current is required. Different types of medium-sized helical systems such as CHS[1], W7-AS[2], Heliotron E[3] and J[4], H-1NF Heliac[5] and HSX[6] have been developed and constructed. Recently, a large helical device (LHD[7]) has been constructed in Japan and another large one (W7-X[8]) is under construction in Germany. These two devices are fully equipped with superconducting magnets, modern diagnostics and high power heating systems. The plasma parameters are expected to be comparable to large tokamaks. While most plasma diagnostics are similar to those for tokamaks, there are also specific features: (1) measurement of vacuum magnetic flux surfaces; (2) intrinsic 3-D structure including divertor; (3) intensive diagnostics for the electric field; (4) diagnostics to verify device design elements; (5) high density diagnostics;

* S. Sudo, T. Ozaki, Y. Nagayama, T. Akiyama, A. Fujisawa, Y. Hamada, K. Ida, T. Ido, H. Iguchi, K. Kawahata, K. Narihara, A. Nishizawa, B. J. Peterson, M. Sasao, K. Sato, M. Shoji, K. Tanaka, K. Watanabe, I. Yamada, National Institute for Fusion Science, Toki, Japan. H. J. Hartfuss, A. Weller, Max-Planck-Institut für Plasmaphysik, Garching and Greifswald, Germany. T. Mizuuchi, K. Kondo, H. Okada, F. Sano, Kyoto University, Uji, Japan. S. Kado, Tokyo University, Bunkyou-ku, Tokyo, Japan. S. Tsuji-Iio, Tokyo Institute of Technology, Meguro-ku, Tokyo, Japan.

Advanced Diagnostics for Magnetic and Inertial Fusion, Edited by Stott *et al.*
Kluwer Academic/Plenum Publishers, New York, 2002

(6) steady state operation. W7-X optimization principles call for sensitive current diagnostics (Pfirsch-Schlüter and bootstrap currents) and tomographic means to verify the reduced Shafranov shift[9, 10] as well as fast particle confinement diagnostics[11]. Steady state or long pulse operation fulfills the demands of correlation diagnostics as T_e fluctuation measurements, it allows for coherent averaging and it enables scans of various diagnostics. Examples are: i) polarization scans of ECE, ii) wavelength scans of spectrometers, and iii) spatial scans of several diagnostics. The special features of diagnostics for helical systems are described in the following chapters.

2. MEASUREMENT OF MAGNETIC FLUX SURFACES

In helical systems the confining field is produced exclusively by the external coil set in contrast to tokamaks. It is the advantage of the helical system to be able to measure the confining magnetic field without plasma current. Instead of using the small electrical probe detection technique[12], the Heliotron group demonstrated a new technique[13], where a mesh screen coated with a fluorescent material was inserted into the confinement region and the image was taken with a camera. Because of the usage of normal conductors, this measurement was only possible at a low magnetic field ($|B| \sim 0.05$ T). In this case, the terrestrial magnetism and residual fields made rather large islands in the confinement region. In LHD, owing to superconducting magnets, the magnetic field mapping experiments can be done at a high magnetic field ($|B| \sim 2.75$ T), and thus, effects such as the terrestrial magnetism can be neglected. The examples of the mapping results are shown in Fig. 1. The widths of the magnetic islands are much decreased by adjusting the currents of the auxiliary coils: local island divertor (LID) coils.

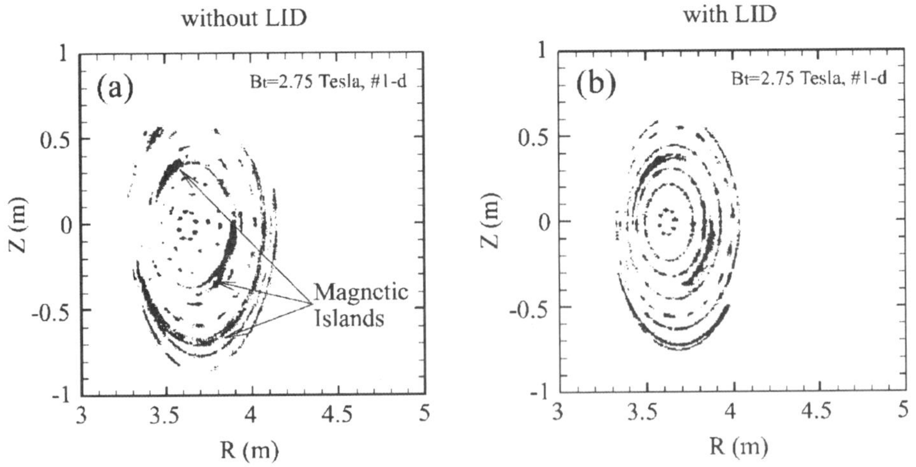

Figure 1. Magnetic surface mapping of LHD at $|B| = 2.75$ T. (a) Without LID, in this case magnetic islands can be seen at $\iota/2\pi \sim 0.5$ and 1, (b) With LID, the width of the magnetic islands is much decreased compared to the case (a).

The method using a fluorescent rod was applied in W7-AS and has been shown to be a very sensitive method of mapping the vacuum flux surfaces[14]. It was possible to identify magnetic islands caused by perturbing fields with amplitudes as low as 10^{-5}.

Based on the stellarator-diode method[15], the Heliotron group (Kyoto University) also has developed a beam-impedance method[16], where an electron beam is used instead of non-directional electron emission.

In Heliotron E, the distributions of the plasma and heat flux were investigated in the divertor field region and at the divertor legs on the vacuum chamber by using several sets of Langmuir probes and calorimeter probes[17]. The experimental results show that the edge plasma is localized only at the position where the divertor fields are running.

First divertor experiments at W7-AS have successfully been conducted to test the island divertor concept[18] which is foreseen for the W7-X device as well[19].

3. DIAGNOSTICS FOR SECONDARY CURRENTS

In helical devices, net toroidal current is not required to confine the plasma. However, several kinds of toroidal currents, such as the bootstrap current, the Pfirsch-Schlüter current, the beam driven current and the current driven by strong microwave beams in the electron cyclotron range of frequencies[20] have been studied. Since the reduction of bootstrap and Pfirsch-Schlüter currents is an important optimization element for W7-X, a number of magnetic loops and saddle coils are foreseen to measure the dipole field emerging from the Pfirsch-Schlüter currents and to verify the low bootstrap current, its parameter dependencies and the flow direction[9].

In LHD, net toroidal currents of up to 100 kA have been observed in NBI plasmas. The currents are directly measured using Rogowski coils that are installed inside the cryostat. The comparison between the experimental results and the theoretical model for bootstrap currents was made under the condition that the Ohkawa current is cancelled out by the NB balanced injection. The dependence of the net toroidal current on W_p is consistent with that of the bootstrap current, and the experimental data are quantitatively consistent with the theoretical prediction by using a SPBSC code[21].

4. PROFILE MEASUREMENTS

4.1 Thomson Scattering

A Thomson scattering (TS) system is important for observing electron temperature and density profiles. In LHD, an oblique back scattering configuration is employed in order to reduce the port requirements: the laser beam is injected in the radial direction on the equatorial plane[22]. The backward scattered light is collected by a mosaic rectangular mirror of 1.5 m x 1.8 m in area and focused onto an array of 200 optical fibers of 2 mm in diameter each. Using this TS system, T_e and n_e can be measured on the whole plasma region along a

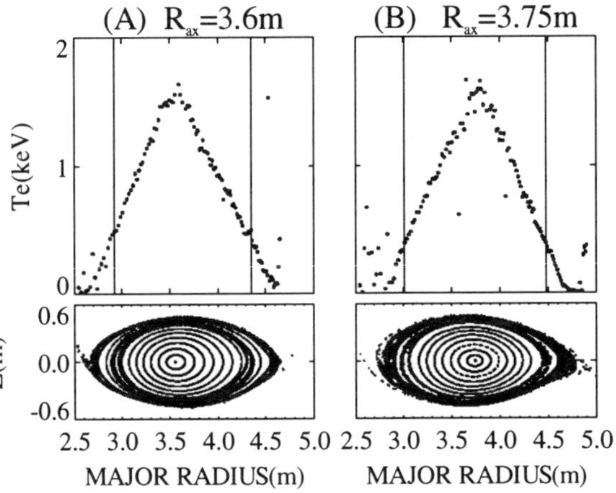

Figure 2. T_e profiles in the R_{ax}=3.6 m and R_{ax}=3.75 m configurations of LHD.

major radius with a repetition rate up to 200 Hz. Typical T_e profiles of the plasmas with the magnetic axis at different positions, R_{ax} =3.6 m and R_{ax}=3.75 m, are shown in Fig. 2 together with the poloidal cross sectional views of the magnetic field lines. The boundary of the plasma, defined by the positions where T_e is below the minimum detectable value (30 eV), is well correlated with the outer limit of the stochastic magnetic region. It is shown that the left and right equal-temperature positions lie on the same magnetic surface within the error of ±2 cm. This is a direct demonstration of the commonly used assumption that in a toroidal plasma T_e is a function of a parameter labeling the nested flux surfaces. The straight lines indicate the positions where the rotational transform $\iota/2\pi \sim 1$.

For local transport analysis, profile data of T_e and n_e with sufficient spatial resolution is important. For the 2-D measurement, a TS system with two ruby lasers (4-5 J) and TV cameras has been used in Heliotron E[23].

4.2 X-ray Tomography

The detailed reconstruction of equilibrium flux surfaces of W7–AS and MHD mode structures by X-ray tomography has been accomplished with an imaging system consisting of 10 compact cameras with a total number of 320 lines of sight[24]. It is installed inside the vacuum vessel of W7–AS close to a plane where the plasma cross section is triangular. The cameras are mounted on a support structure and contain a pinhole aperture, a fixed curved Be-foil of 6 μm thickness, a planar 35 channel pin-photodiode array and preamplifiers for 32 radial channels used in each camera. The radial resolution is about 1 - 2 cm at the plasma center. Neutral beam driven Alfvén eigenmodes are the most prominent MHD instabilities in W7-AS[25], and therefore, continuous efforts are made to study their characteristics and their relevance for the fast particle confinement. The advanced stellarator configuration of W7-AS, characterized by very low shear (~10 %), large aspect

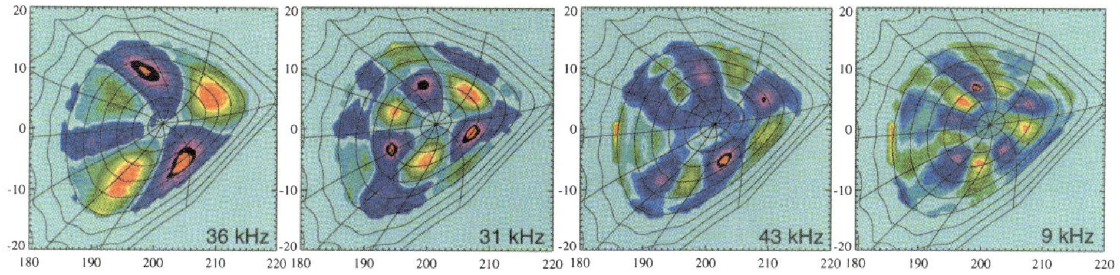

Figure 3. X-ray tomograms of m=3 and m=5 GAE modes of different frequencies. The modes appear simultaneously (W7 #39029) and are separated by Fourier filtering.

ratio (~10) and a particular 3-D field geometry, leads to modifications of the Alfvén spectrum as compared with conventional tokamaks and high shear heliotrons/ torsatrons. In particular, low shear causes TAE gaps to disappear and results in the appearance of weakly damped global Alfvén eigenmodes (GAE) below the Alfvén continua with frequencies $\omega_{GAE} = (k_{\|} \cdot v_A)_{min}$, which are excited through resonances with fast passing particles from NBI. For the comparison of the experimental data with code calculations tomographic reconstructions of the mode structures are required. An example for GAE mode structures is shown in Fig. 3, where several GAE modes of different poloidal and radial structure occurred simultaneously.

4.3 Diagnostics for Transport Study

A unique method for measuring particle diffusion coefficients with a tracer-encapsulated pellet (TESPEL) injection has been developed[26]. The lithium tracer was successfully observed by CXS in CHS, which resulted in the profile of the particle diffusion coefficient. In LHD, the CXS for lithium was not adequate due to the high kinetic energy of NBI (150 keV), but the titanium tracer was successfully observed by a soft X-ray pulse height analyzer.

A CO_2 laser phase contrast imaging method, LPC, for density fluctuations in plasmas was intensively developed at Heliotron E[27], to detect spatially resolved fluctuation spectra, with the feature that the pitch angle of the magnetic field is a function of the spatial position due to a sheared magnetic field (up to 40 degrees). The wavenumber-frequency spectra S(k, f) of fluctuations were successfully obtained through 2D-FFT analysis with time resolution of 1.6 μs and spatial resolution of 2.5 mm.

5. DIAGNOSTICS FOR ELECTRIC FIELD

The electric field plays an important role in the transport of helical plasmas[28]. In order to measure the electrostatic potential directly with a high temporal resolution, a heavy ion beam probe (HIBP) is developed intensively. An active trajectory control method was proposed for the HIBP on helical systems[29]. The method was tested in the CHS before it is applied to the HIBP of LHD. Particularly, the HIBP measurements reveal that the density fluctuations are reduced at the connection layer of the dome pattern, and Thomson scattering measurements show that the electron thermal ITB is established[30].

In LHD, singly charged gold ions (Au^+) are accelerated up to 6 MeV with a tandem accelerator, and the device has already been installed. In the next experimental campaign, this HIBP will be operated, and the ions ionized in a plasma (Au^{2+}) will be detected. The beam is swept with an octupole deflector in order to sweep the observation area in a plasma. In a helical device, the extraction angle of the beam varies widely during the sweep. The change of the angle causes an error in the analysis of the beam energy. Accordingly, an additional octupole deflector has been installed at the extraction port so as to control the beam trajectory and to reduce the change of the angle.

The radial electric field profiles measured with the charge exchange spectroscopy (CXS) method are compared with those measured with the HIBP in CHS[31]. The relation of the two radial electric fields measured with the CXS and HIBP for all radii available is:

$$E_r(\mathrm{CXS}) = (0.89 \pm 0.03) \times E_r(\mathrm{HIBP}) - (0.19 \pm 0.08)\ \mathrm{kV/m}.$$

If we take the ratio of the radial electric field measured with CXS to that measured with HIBP by ignoring the offset, the coefficient is 0.9 ± 0.03.

6. DIAGNOSTICS UNDER HIGH DENSITY OPERATION

It has been shown in W7-AS[18] that divertor operation enables access to a new operational regime with NBI at very high densities up to $3.5\times10^{20}\ \mathrm{m}^{-3}$. This necessitates the development of non-standard heating techniques for central electron heating. The main heating scenario at W7-AS, Electron Cyclotron Resonance Heating at 140 GHz, second harmonic, X-mode is cut off at $1.2\times10^{20}\ \mathrm{m}^{-3}$. Efficient heating of an NBI target plasma by electron Bernstein waves at 140 GHz has been achieved. The method is based on the conversion of a launched O-mode into an X-mode, and finally to an electron Bernstein wave (O-X-B). In addition to the heating, clear evidence for the two steps of mode conversion has been demonstrated[32, 33]. The processes can be inverted to use the thermal emission of electron Bernstein waves as a radiometric tool to measure the electron temperature under conditions where ECE is cutoff. The emission is measured with heterodyne radiometers similar to the standard ECE systems[34]. However, to convert the spectra into temperature profiles, it is necessary to reconstruct the complicated mode-converted ray-paths inside the plasma[35].

7. DIAGNOSTICS UNDER LONG PULSE OPERATION

Since ECE spectrometers[36] have a polarizer, the X-mode and the O-mode can be selected by using the polarization rotator. The experimental results show that the polarization of ECE is aligned at the plasma edge, where n_e sharply drops so that the refractive index of O- and X-modes become unity[37]. This experiment was possible owing to the long pulse operation of LHD.

The time-of-flight neutral particle analyzer (FNA)[38] with the capability of two-dimensional measurement has been installed to study the behavior of the helical trapped particles. Precise spatial distribution of the high energy particles can be measured using the continuous scanning in the long discharge of LHD. The sight lines can be scanned horizontally with the speed of 0.17 degree/second. Fast ion orbit losses are also studied by energy-and-pitch angle-resolved lost ion probes on CHS[39] and W7-AS[40], and resonance particle losses corresponding to both transition and trapped particle orbits were observed during MHD[41].

Long pulse helical devices offer unique conditions for coherent averaging and the application of cross correlation techniques which are of special importance in the field of fluctuation diagnostics. One example is cross correlation radiometry of the electron cyclotron emission. The method has been applied to detect temperature fluctuations in the core of W7-AS plasmas[39, 40].

Polarimeters for density measurements were proposed and demonstrated based on both the Faraday[41-43] and the Cotton-Mouton effects[44]. They have generally poorer density resolution but are almost insensitive to vibrations, a property which makes them especially suited as robust line integrated density diagnostics for long pulse experiments. In addition, by selecting a CO_2 laser it is possible to suppress the refraction of the probing beam and to avoid "fringe jump" errors even in the high density plasma with a simple optical setup.

In terms of application of bolometers to a steady state reactor several factors should be considered: steady-state operation, real-time measurement, durability to neutrons. The infrared imaging bolometers (IRIB)[45] towards 3-D tomography are successful on LHD .

8. SUMMARY

Diagnostics for helical systems are reviewed mainly from the viewpoint of their unique features compared to tokamaks. These unique features are: (1) magnetic flux surfaces in vacuum; (2) multi-dimension structure; (3) electric field; (4) verification of device design elements; (5) high density diagnostics; (6) steady state operation. Steady state operation brought us advantages for diagnostics such as time averaging and temporal scan of diagnostics. The essential diagnostics for helical systems are well established, although diagnostics for detailed studies of turbulent and diffusive transport in two or three dimensions should be developed further.

REFERENCES

1. K.Matsuoka, et al., Plasma Physics and Controlled Nuclear Fusion Research, 1988, Vol.**2**, IAEA, Vienna (1989) 411.
2. H.Renner, et al., Plasma Phys. Control. Fusion, **31**, 1579 (1989).
3. K. Uo, et al., in Controlled Fusion and Plasma Physics (Proc. 10th Int. Conf. Moscow, 1981), EPS(1981)E-1.
4. T.Obiki, et al., Nucl. Fusion, **41**, 833 (2001).
5. J.Harris, et al., The Australian & New Zealand Physics, **35**, 245 (1988).
6. D.T. Anderson, et al., J.Plasma Fusion Res. SERIES **1**, 49 (1998).
7. O. Motojima, et al., Phys. Plasma **6**, 1021 (1999).

8. G. Grieger et al., Phys. Fluids **B 4**, 2081 (1992), F. Wagner, Int. Conf. on Accelerator and Plasma Research, Tae Jeon, Korea, Advanced Fusion Science Research, p. 41 (1995).
9. H.J. Hartfuss et al., Rev. Sci. Instrum. **68**, 1244 (1997), and: Diagnostic System for the W7-X Stellarator, this conference.
10. A. Weller et al., Fusion Engineering and Design **34-45**, 17 (1997).
11. A. Werner et al., Fast Ion Loss Diagnostics for the Wendelstein 7-X Stellarator, this conference.
12. M. S. Jones Jr., et al., Princeton Plasma Phys. Lab. Report., MATT-14 (July, 1959). T. Mizuuchi, et al., Nucl. Fusion **22** (1982) 247.
13. R. Takahashi, et al., Proc. Int. Stellarator/Heliotron Workshop (IAEA Tech. Committee Meeting) (Kyoto, 1986). Plasma Phys. Lab. Report PPLK-5, **1**, 220 (1986).
14. R. Jaenicke et al., Nuclear Fusion **33**, 687 (1993).
15. A. G. Dikiji, et al., Sov. J. Plasma Phys. 14 (1988) 160. J. V. Gutarev, et al., Proc. Int. Stellarator/Heliotron Workshop (IAEA Tech. Committee Meeting) (Kyoto, 1986). Plasma Phys. Lab. Report PPLK-5, Vol. **1** (1986) 233.
16. F. Sano, et al., J. Nucl. Mater., **241-243**, 961 (1977).
17. T. Mizuuchi, et al., J. Nucl. Mater., **162-164**, 105 (1989).
18. P. Grigull et al., First Island Divertor Experiments on W7-AS, **28**th EPS CFPP, Madeira, Portugal, June, 2001, to be published in Plasma Phys. Control. Fusion.
19. H. Renner et al., Nucl. Fusion **40**, 1083 (2000).
20. V. Erckmann et al., Fusion Engineering and Design **53**, 365-375 (2001).
21. K. Watanabe et al, Nuclear Fusion **32**, 1992 (1992).
22. K. Narihara, I. Yamada and T. Minami, Proc. 6th International Symposium on Laser-Aided Plasma Diagnostics, Bar Harbor (1993) p.108.
23. H. Okada et al., Proc. 1996 int. Conf. on Plasma Physics, Vol. **2** (Nagoya, 1996) pp.1434-1435.
24. A. Weller et al., Rev. Sci. Instr. **70**, 484 (1999).
25. A. Weller et al., Physics of Plasmas **8**, 931 (2001).
26. S. Sudo et al., **17**th Conf. Proceedings of Fusion Energy, IAEA, Yokohama, Japan, EXP1/18 (1998).
27. S. Kado et al., Jpn. J. Appl. Phys. **34**, 6492 (1995).
28. R. Brakel et al., Plasma Phys. Control. Fusion **39**, B237 (1997).
29. A. Fujisawa, H. Iguchi, M. Sasao et al., Rev. Sci. Instrum. **68**, 3393(1992).
30. A. Fujisawa, H. Iguchi, T. Minami et al. Physics of Plasmas **7**, 4152(2000).
31. K.Ida, A.Fujisawa, et al., Phys. Plasmas, **8**, 1 (2001).
32. H. P. Laqua et al., Phys. Rev. Lett. **78**, 3467 (1997).
33. H.P. Laqua and H.J. Hartfuss, Phys. Rev. Lett. **81**, 2060 (1998).
34. Ch. Fuchs and H.J. Hartfuss, Rev. Sci. Instrum. **72**, 383 (2001).
35. F. Volpe and H.P. Laqua, Development of an electron Bernstein emission diagnostic for temperature profiles above the EC-cutoff density at Wendelstein 7-AS Stellarator, 28th EPS CFPP, Madeira, Portugal, June, 2001. See also paper by F. Volpe, Electron Bernstein Wave Diagnostics, at this conference.
36. Y. Nagayama, et al., Rev. Sci. Instrum. **70**, 1012 (1999).
37. P. de Vries, et al., Phys. Plasma. **7**, 3707 (2000).
38. T. Ozaki, et al., Rev. Sci. Instrum. **72**, 2698 (2000).
39. M.Isobe et al., Rev. Sci. Instrum. **70** (1999) 827.
40. A. Werner et al., Rev. Sci. Instrum. **72** (2001) 780.
41. T. Kondo et al. Nuclear Fusion **40** (2000) 1575.
42. S. Sattler and H. Hartfuss, Phys. Rev. Lett. **72**, 653 (1994).
43. H. Hartfuss et al., Plasma Phys. Control. Fusion **38**, A227 (1996).
44. F. C. Jobes, D. K. Mansfield: Rev. Sci. Instrum. **63**, 5154 (1992).
45. T. Akiyama, E. Sato, T. Nozawa *et al*: Rev. Sci. Instrum. **72**, 1073 (2001).
46. Ch. Fuchs and H. Hartfuss, Phys. Rev. Lett. **81**, 1626 (1998).
47. B.J. Peterson *et al.*, Rev. Sci. Instrum. **72**, 923 (2001).

DIAGNOSTIC SYSTEM FOR THE W7-X STELLARATOR

H.J. Hartfuss and the W7-X Diagnostics Team*

INTRODUCTION

A full set of diagnostics for the optimised stellarator W7-X being under construction in Greifswald, Germany, has been defined following the strategy as published earlier[1]. W7-X[2] is a superconducting modular device equipped with a divertor and heated by high power ECRH, NBI, and ICRH. The diagnostics definition was oriented towards the aims of W7-X, considering the neoclassical optimization criteria, the divertor operation, and the possibilities opened up by the long pulse, high density, and steady state operation of the device[1,3]. The whole work is organized in nine specialist's groups which are either subject or method oriented. Diagnostics proposed by the various groups are briefly summarized in the following sections. Details can be found in the W7-X diagnostics documentation system: http://tuovi.rzg.mpg.de/KRONO

FLUCTUATIONS

A large number of the proposed diagnostics for W7-X are capable of delivering data on plasma turbulence. The most important amongst them are Langmuir probe arrays, magnetic Mirnov probes, $H_\alpha$detectors, reflectometers, ECE correlation radiometers, and SX cameras, as well as some spectroscopic and atomic beam diagnostics.

New developments have been started and improvements of existing systems are being tested at the W7-AS device. Main work in this group concentrates on the identification of the topics to be investigated for which the proposed diagnostics must be adapted for. Investigations on the basis of turbulence theory on one hand and on stellarator specific properties like the variable magnetic well depth, on favourable and unfavourable magnetic curvature, the influence of trapped particles, the 3D structure, as well as on the influence of pressure and rotational transform are being conducted.

---

* Max-Planck-Institut für Plasmaphysik, EURATOM Ass., 17491 Greifswald, Germany

PLASMA EDGE

The main aim is to define and to locate diagnostics capable to characterize the polymorphic plasma edge of W7-X i.e. the 3D SOL and the 2D deposition zones at the divertor targets. The problem to describe the plasma edge in non-axisymmetric stellarators is very complex in general[4]. Nevertheless, an interplay of various diagnostics at different magnetic field periods of W7-X but at topologically equivalent positions allows to do it with a certain degree of redundancy if the 5-fold stellarator symmetry is preserved. With the symmetry broken due to misalignement of targets, SOL bifurcation etc. it becomes almost impossible. This underlines the importance of tools for supervision and control of the SOL symmetry. In the conceptual design of the edge diagnostics it is foreseen that the SOL symmetry is supervised by 2D imaging cameras (thermography, spectroscopy) viewing the targets in each divertor module. Target integrated pop-up probe arrays and various H_α diagnostics will allow for the comparison of different divertor modules. The other edge diagnostics, Thomson scattering, fast reciprocating Langmuir probes, divertor bolometry, SOL and divertor spectroscopy and atomic beam spectroscopy are distributed around the torus according to the availability of appropriate ports and space.

MICROWAVES

It is foreseen to install a ten-channel vibration compensated infrared CO_2 laser interferometer. It will be complemented by a robust single channel, submillimeter Cotton-Mouton effect polarimeter specially suited for long pulse operation[5]. For ECE measurements an extremely wideband multichannel ECE radiometer with zoom device[6] will be installed. Various mm-wave reflectometers operating in x-mode are proposed both for high and low field side. Later extensions of the ECE radiometric system to correlation[7] and imaging radiometers and the completion of the reflectometry system by extremely fast swept reflectometers will especially concentrate on fluctuation measurements. High density operation demands for new diagnostics. Succesful developments have been conducted using mode conversion from electron Bernstein waves[8,9].

SPECTROSCOPY

An extensive number of spectroscopic diagnostics is foreseen, for plasma core and plasma edge investigations. A CXRS system with a dedicated diagnostic neutral beam will be employed for the determination of ion temperature, low-Z impurity density, and poloidal and toroidal rotation velocity profiles. From visible bremsstrahlung Z_{eff} will be deduced. For medium and high-Z impurities in the bulk plasma a number of VUV spectrometers is proposed. A boron, carbon, oxygen monitor and two low resolution SPRED systems, covering the wavelength range from about 10 to 200 nm, will serve as impurity monitors and for transport studies. These are complemented by two high resolution grazing incidence spectrometers allowing for detailed studies in the range from 1 to 250 nm. A high spectral resolution X-ray imaging crystal spectrometer[10] will provide information on impurity content and impurity transport as well as on electron and ion temperatures and on plasma rotation, while a ten-channel Bragg spectrometer fan will be utilized to measure radial density profile evolutions of different ionisation states of various impurity species. The island divertor plasma will spectroscopically be diagnosed mainly in the visible. Initially a number of imaging systems consisting of 2D CCD-cameras equipped with H_α filters will be used to look for possible toroidal asymmetries. Some of these systems will later be equipped with beam splitters and further interference filters for CII and/or CIII and H_γ. High spectral and spatial resolution of the visible spectroscopy will give detailed information on the interaction zones of the islands with the target plates, complemented by thermal He-beam and laser induced fluorescence systems. A large number of video cameras distributed around the machine are foreseen for operational purposes (plasma position, configuration, interaction with in-vessel components). The total radiated power will be measured by two bolometer arrays on the basis of metal resistor detectors.

THOMSON SCATTERING

Three different Thomson scattering systems are being planned. For the bulk plasma, a Nd-YAG laser system is foreseen to deliver profiles in real time allowing for online plasma monitoring and profile shape event recognition. A second system is especially dedicated to the divertor region. The laser beam will be movable in radial direction to investigate 2D structures in front of the divertor surfaces i.e. island structures in the divertor region. The third system uses a ruby laser. It will be placed in a local minimum of the B-field close to the ECRH deposition planes. The scattering distribution function will be measured with high spectral and spatial resolution. The motivation is to look for small spatial structures as well as for distortions in the electron distribution function.

SOFT X-RAY AND ELECTROMAGNETIC DIAGNOSTICS

The X-ray measurements will basically provide information on plasma equilibrium, MHD instabilities, electron temperature, impurity content, and impurity transport. To achieve images of the 2D X-ray emissivity and hence the magnetic topology, a multi-camera tomography system is proposed installed inside the vacuum vessel[11]. A second conventional flexible camera system is proposed with the possibility to adjust the spectral range by interchangeable energy filters enhancing the sensitivity for particular impurity species to tailor the radial emissivity profile for equilibrium and MHD mode analysis and to derive the electron temperature according to the 2-foil absorption method. In addition pulse height analysis systems are being planned.
Electromagnetic diagnostics measure the magnetic fields produced by currents inside the plasma. Various types of coil arrangements like Rogowski coils, diamagnetic loop, saddle coils, flux loops, and poloidal magnetic field coils will exist to measure different

aspects of the plasma current distribution such as total current, diamagnetic current, or dipole moment of the parallel currents. The whole set of measurements will be used for equilibrium identification. In addition Mirnov coils will be installed to investigate instabilities. Challenges due to properties of W7-X are the reduced parallel currents and the planned long-time discharges.

HEAVY ION BEAM PROBE AND FAST PARTICLE DETECTORS

The heavy ion beam probe is a unique diagnostic to measure the local plasma potential. On the basis of 3D orbit calculations, the port geometry of the beam injection and detector ports has been determined for a primary beam species of Cs^+ ions with 3 MeV energy. An optimization with respect to best access, minimum necessary beam energy, and maximum detectable plasma region has been performed. Fast particle confinement is a major issue for W7-X. Fast ion loss detectors are being developed. Successful measurements have been conducted at W7-AS[12].

NEUTRON DIAGNOSTICS

W7-X deuterium plasmas are estimated to generate 10^{12} to 10^{16} neutrons per second depending on the heating scenarios allowing in principle to evaluate information on plasma parameters like ion velocity distribution, deuterium density as well as MHD activity and the thermal reaction fractional contribution in the neutron emission rate and the energy distribution. In the first priority level one concentrates on the development of 6 neutron counters with sufficient sensitivity, high dynamic range, and with good accuracy and stability. One will be arranged in the center and the other five on equivalent positions in the outer region of the torus. Ports are foreseen to complete the neutron diagnostics by appropriate spectrometers in a later stage.

REFERENCES

1. H.J. Hartfuss et al., Rev. Sci. Instrum. **68**, 1244 (1997).
2. G. Grieger et al., Phys. Fluids **B 4**, 2081 (1992).
3. S. Sudo et al., Diagnostics for Helical Systems, this conference.
4. P. Grigull et al., First Island Divertor Experiments on W7-AS, EPS Funchal, Portugal (2001).
5. Ch. Fuchs and H. Hartfuss, Phys. Rev. Lett. **81**, 1626 (1998).
6. Ch. Fuchs and H. Hartfuss, Rev. Sci. Instrum. **72**, 383 (2001).
7. H. Hartfuss et al., Plasma Phys. Control. Fusion **38**, A227 (1996).
8. H.P. Laqua and H.J. Hartfuss, Phys. Rev. Lett. **81**, 2060 (1998).
9. F. Volpe et al., Electron Bernstein emission diagnostic, this conference.
10. F. Bertschinger, High Resolution X-Ray Imaging Spectrometer, this conference.
11. S. Klose et al., X-Ray Multi Camera Tomography System for W7-X, this conference.
12. A. Werner et al., Fast Ion Loss Diagnostic for the Wendelstein 7-X Stellarator, this conference.

DIAGNOSIS OF ST PLASMAS IN NSTX

Challenges and Opportunities

D. Johnson, P. Efthimion, J. Foley, B. Jones, E. Mazzucato, H. Park, G. Taylor, F. Levinton, and N. Luhmann*

1. INTRODUCTION

During the past decade, considerable experimental and theoretical progress has been made in the development of the Spherical Torus (ST) configuration. A second generation of 'proof of principal' devices, including GLOBUS-M, MAST and NSTX, are now operating with plasma currents in the 1-2 MA range. The ST has already been the basis of a reactor study, and the concept has also been considered in the context of a volume neutron source.

As the aspect ratio of a conventional circular tokamak is reduced towards one, the cross-section naturally assumes a shape with high elongation and triangularity, with only modest requirements for shaping fields. This configuration can sustain high plasma current with high values of edge q, meaning that the toroidal field can be much lower than that in a conventional tokamak. Compared to a standard tokamak, the magnetic field pitch can be very large (pitch angles > 50° at outer midplane). High values of toroidal beta can reached, and the combination of high β, high κ, and high q(a) leads to the potential for a large bootstrap current sustainment.

The National Spherical Torus Experiment (NSTX) is a low-aspect-ratio torus ($R/a \leq 1.3$) designed to produce and study high β_T (25-40%) plasmas that are non-inductively sustained with high bootstrap fraction (≤70%). Heating and current drive is done with a 3-source neutral beam (5 MW) and a 12 strap, high-harmonic-fast-wave (HHFW) antenna (6 MW). In addition, coaxial-helicity-injection (CHI) is used for non-inductive plasma current initiation. Basic device capabilities are R=0.85 m, a=0.68 m, $I_P \leq 1.4$ MA, $B_T \leq 0.45$ T with elongation ≤ 2.0 and triangularity ≤ 0.5. Central plasma parameters in the range of $n_e = 4\text{-}5 \times 10^{19}\ m^{-3}$ and $T_e, T_i = 3\text{-}4$ keV are obtained for high power heating experiments, which began in late 2000.

* D. Johnson, P. Efthimion, J. Foley, B. Jones, E. Mazzucato, H. Park, G. Taylor, Princeton Plasma Physics Lab, P.O. Box 451, Princeton, NJ, 08540. F. Levinton, Nova Photonics, 1 Oak Place, Princeton, NJ, 08540. N. Luhmann, University of California, Davis, 228 Walker Hall, Davis, California 95616

Advanced Diagnostics for Magnetic and Inertial Fusion, Edited by Stott *et al.*
Kluwer Academic/Plenum Publishers, New York, 2002

This paper will highlight some of the challenges and opportunities present in the diagnosis of ST plasmas on NSTX, and discuss the corresponding diagnostic development that is currently underway. After a brief description of diagnostic systems currently installed, examples of ST-specific diagnostic challenges will be highlighted, as will another case, where the ST configuration offers opportunities for new measurements.

2. PRESENT DIAGNOSTICS

The evolving NSTX diagnostic set currently has a set of magnetic sensors adequate for plasma control and between-shot equilibrium reconstruction. Filtered, fast visible cameras provide valuable information on overall discharge evolution, with focussed views on edge turbulence and divertor region. IR cameras view areas such as the divertor region which have significant power efflux. Profile diagnostics include a multi-pulse Nd:YAG Thomson scattering system (T_e and n_e), edge reflectometry (n_e), charge exchange spectroscopy (T_I and v_ϕ), bolometer arrays (P_{rad}), and VB arrays (Z_{eff}), Impurities are monitored by visible and UV spectrometers, filtered fiberscopes, and soft x-ray detectors. A scanning neutral particle analyzer and fixed fast ion loss probe measure fast ion behavior. MHD activity is monitored by Mirnov arrays, soft x-ray arrays, and correlation reflectometry. Measurement developments in the next period will emphasize diagnostics for the divertor, for q(R), and for turbulence.

3. ELECTRON BERNSTEIN WAVE RADIOMETRY

At the low toroidal field and moderate densities on NSTX, $\Omega_{ce} << \omega_{pe}$, and electromagnetic radiation is not generated for the first several cyclotron harmonics, making it impossible to use standard electron cyclotron emission as a fast profile diagnostic for electron temperature. Although multipoint Thomson system provides periodic snapshots, these are generally not adequate for MHD studies and detailed studies of other fast transients in $T_e(R)$. A promising new development for diagnosing electron temperature is electron Bernstein wave (EBW) radiometry[1], which is currently being pursued at PPPL on NSTX as well as the smaller ST, CDX-U.

EBW's are electrostatic waves that can propagate in ST plasmas and exhibit high optical thickness at the electron cyclotron resonances. These waves can only exist within the upper hybrid resonance layer typically located near the outside edge of the plasma. If the density scale length L_n^{UHR} at this layer is short enough, these waves will mode convert and tunnel to fast X-mode electromagnetic waves, which can be easily detected and provide information on $T_e(R,t)$. Studies are underway at PPPL to measure the mode conversion efficiency and its strong dependence on L_n^{UHR}. On NSTX, there is a sudden increase in the measured core emission as the density scale length shortens at the onset of the H-mode transition, and the inferred $T_{rad}/T_e(0)$ increases from 3% to 13%, in agreement with model calculations on the scaling with L_n^{UHR}. On CDX-U, a local, radially-scannable limiter has been used to modify L_n^{UHR} in the vicinity of the receiving horn, and thereby maximize the conversion efficiency at near 100%.

4. MOTIONAL STARK EFFECT MEASUREMENTS OF q(R) AND E_r

Since non-inductive current initiation and sustainment are top level NSTX research goals, measurements of the current profile J(R) are essential to many planned experiments. On tokamaks with neutral beams, the standard technique for this task is motional-stark-effect polarimetry (MSE),[2] which recently has also been crucial in measurements of $E_r(R)$. The standard implementation of this technique would be impossible on NSTX, due to the difficulty of spectrally isolating the polarization components at the low magnetic field.

Several modifications are planned to adapt the MSE technique to lower field.[3] First, the Doppler broadening of the individual components will be minimized, by reducing the spread in angles between the beam and the viewing aperture. This will be done by narrowing the viewing aperture in the horizontal dimension, and to compensate, vertically elongating the aperture. Second, a narrower bandwidth birefringent filter is being developed, based on a Lyot design currently utilized in solar spectroscopy and dye laser tuning. The goal of this development is an effective passband width of 0.075 nm, compared to 0.7 nm width of standard interference filters. Figure 1 shows the transmission of a single stage of this multistage filter. To block all but the central peak, it will be used in series with two subsequent stages with spectral frequency 1/2 and 1/4 that shown, in addition to a standard interference filter. Modeling the performance of a system, with the new aperture shape and a 0.075 nm filter passband, one obtains polarization fractions of P_f = 0.2-0.4 for the central σ_0 component. Typical MSE systems operate with a $P_f \sim 0.8$. To compensate, the optical throughput on the NSTX system will be much larger than on previous systems. The optics is designed to image the beam onto 19 fiber optic bundles, providing spatial resolution of 2.5 cm at the edge and 3.0 cm in the core.

The emission used in MSE systems results from collisionally-induced-fluorescence (CIF) arising from collisions between injected D^0 atoms and the background plasma. Using a tunable laser to induce this emission can have significant advantages. On a somewhat delayed but parallel path to the MSE work described above, development is underway on laser-induced-fluorescence (LIF) MSE. The plan is to inject a hydrogen diagnostic neutral beam with a coaxial tunable dye laser resonant with the Doppler shifted H_α transition. The linear polarization of the laser is rotated, and the phase between the laser polarization and the induced fluorescence is used to determine the pitch angle. With a narrow spectral width, polarization fractions near unity are predicted even at the low fields of NSTX. In a lab test with no B field, initial LIF signals have been observed with a beam operating at 32 keV and 25 mA, with a 1.2 cm diameter. If the CIF and LIF MSE systems described above are successful on NSTX, determination of both q(R), and E_r will be possible.

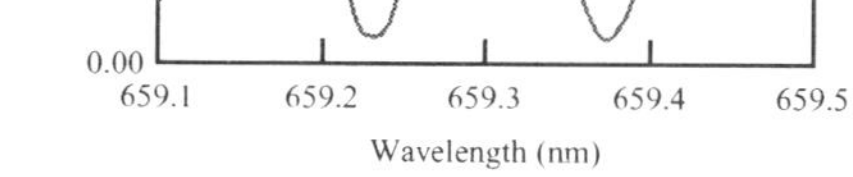

Figure 2 Transmission of a single stage birefringent filter in arbitrary units.

5. MICROWAVE SCATTERING STUDIES OF HIGH-k TURBULENCE

In recent years, experimental and theoretical studies of internal transport barriers in tokamaks have indicated that different turbulent mechanisms drive electron and ion transport. Numerical studies have shown that on NSTX, fluctuations with $k_{\perp}\rho_{i} < 1$ are absent or completely suppressed by **ExB** velocity shear, but also that Electron Temperature Gradient (ETG) modes with $k_{\perp}\rho_{e} \sim 0.1\text{-}0.3$ ($k_{\perp} \sim 10\text{-}40\ cm^{-1}$) are unstable over most of the plasma cross-section.[4]

Mazzucato has proposed a microwave scattering diagnostic, which takes advantage of the large pitch angles on an ST, to provide spatially and spectrally resolved measurements of $k_{\perp}$ for this turbulence. The scattering geometry is shown in Figure 2.

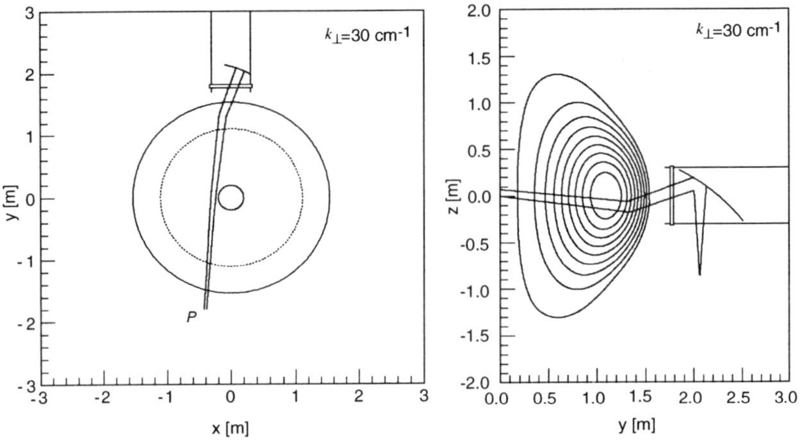

Figure 2. Equatorial and poloidal projections of scattering geometry to probe ETG turbulence.

A 290 GHz beam with a waist of 2 cm is launched through a window ~ 30 cm above the midplane and traverses the plasma near the center stack to a scattering region. The scattered light is imaged by a large mirror through a window to a detector plane, where light from different values of R and $k_{\perp}$ are dispersed in 2 dimensions. Radial resolution of 1-3cm for the outer 90% of the plasma, and spectral resolution of ~ 1 cm^{-1} for $k_{\perp} \sim 10\text{-}40\ cm^{-1}$ appear feasible. Using available sources and detectors, such a system is predicted to easily have adequate sensitivity to detect the fluctuation amplitudes ($\delta n/n \sim 10^{-3}$) predicted from a mixing length criterion for NSTX parameters

6. REFERENCES

1. G.Taylor, P.C. Efthimion, B. Jones, *et al.*, "Electron Bernstein Wave Research on CDX-U and NSTX", Proceedings of 14th Topical RF Conf., (Oxnard, Calif., May 2001) To be published in proceedings by American Institute of Physics
2. F. M. Levinton, Rev. Sci. Instrum., **70**, 810 (1999)
3. F. M. Levinton, private communication.
4. C. Bourdelle, *et al.*, Joint US-EU Transport Task Force Mtg. Fairbanks, Alaska, May 16, 2001

DIAGNOSTIC DEVELOPMENTS FOR THE MAST SPHERICAL TOKAMAK

A. W. Morris and the MAST team*

INTRODUCTION

MAST[1] is a relatively new spherical tokamak (ST) with plasma cross-section and parameters comparable to those of ASDEX-U and DIII-D but at much lower field (0.5T) and aspect ratio (~1.3). The plasma is heated with neutral beams and two 60GHz RF schemes (ECRH and, under development, electron Bernstein wave heating). The aims of MAST are to contribute to mainstream tokamak physics, for ITER, and to test the potential of the ST in its own right. Diagnostics for MAST combine new and conventional approaches to match the plasma and machine layout. Systems must accommodate a flexible physics programme and an unusually wide range of plasma configurations.

CORE AND PROFILE MEASUREMENTS

MAST has a large vacuum tank, with internal poloidal field coils. This restricts views from the top of the vessel but allows excellent imaging of the whole plasma cross-section when tangential midplane views are used. An active/passive multi-chord charge-exchange spectroscopy system for $T_i(r,t)$, $v_{\theta,\phi}(r,t)$ views the neutral beams and the background plasma. It is presently tuned to CVI (n=8-7, 529nm). The background view has the added advantage of reducing the need to notch the beams. To assist all optical systems the midplane region of the MAST vessel has been blackened with colloidal graphite. Furthermore the neutral beams do not interact with the thin centre column.

Electron temperature measurements in STs are challenging – ECE is not an option due to the high density (typically $\omega_p^2/\omega_{ce}^2 >>1$), and the large number of harmonics present in ST plasmas. Thus although electron Bernstein waves (EBW) are being

*EURATOM/UKAEA Fusion Association, Culham Science Centre, Abingdon, Oxfordshire, OX14 3DB, UK

considered for STs, multi-pulse Thomson scattering is the only proven option. A two-laser 100Hz, 19 point NdYAG system is under construction, to complement the existing 40 spatial point single pulse ruby system (being upgraded to 300 spatial points) from START.[2] The ruby and NdYAG lasers will use the same input window, graphite beam dump and collection lens (designed to work at both wavelength ranges), merely having the laser beams and collection fibre arrays displaced. The fibre array for the ruby system is composed from ordered bundles rather than randomised to allow the CCD detector to be used for alignment checks on each pulse. The NdYAG system uses compact 5 channel filter spectrometers, and time delay from differing fibre lengths will allow two spatial channels per spectrometer, reducing calibration effort and uncertainty as well as hardware (3-filter single fibre spectrometers are also used). The two NdYAG lasers can be triggered exactly synchronously to provide higher intensity, at 50Hz, or with any desired separation of the pulse trains.

For edge transport barrier studies, the high resolution Thomson scattering system is to be augmented by a D_α camera[3], and shared optics for HeI line ratio and Doppler spectroscopy. This will provide good space-time synchronous data on edge T_e, n_e, $n_{neutral}$, v_i and hence electric field profiles. On COMPASS-D this system gave unique data on the development of the H-mode transport barrier.[4,5]

PLASMA DENSITY AND PELLET MEASUREMENTS

MAST plasmas move rapidly during formation, and are fuelled by pellets (with an 8-pellet injector from FOM) so systems robust to fast transients are required. There are 5 systems for density profiles and fluctuations in use or under development. The basic system is a midplane tangential view using a low-cost CO_2 interferometer with He-Ne laser for vibration compensation. Both inboard and outboard parts of the plasma are probed without the need for in-vessel retro-reflectors. This is supplemented, especially for low density operation and plasma start-up, by a novel unambiguous-phase microwave system (132GHz, ~2mm) viewing along a major radius, which is essentially immune to fringe jumps. The two Thomson scattering systems provide full profile information. A fast 4-band (18-59GHz) swept-frequency reflectometer system has been constructed by IST (Lisbon) and will provide the outboard profile and data on fluctuations, extending the existing simple 33, 38, 43GHz X-mode prototype turbulence reflectometer. Reflectometry in STs is complicated by the large fieldline angle on the outboard side, so the polarisation mix of launched and received waves needs to be considered carefully.

A key physics issue for pellet injection is improved penetration. The model used to explain the effectiveness of inboard launch (ExB drift from grad-B) can in some cases provide inward drift in an ST for outboard launch allowing both a test of the model and hopefully deep fuelling from outboard launch. The 3-D pellet trajectory (and hence penetration depth) can be constructed from a position sensitive light detector system. A strip-masked CCD camera will provide some information on the fieldline pitch and hence q-profile. Both systems are on loan from RFX.

PLASMA BOUNDARY AND HALO DIAGNOSTICS

Much information is available from whole-plasma images from a fast wide-angle CCD camera – for example helical filamentary structures outside the separatrix that may affect SOL transport – and this is also a key operational diagnostic. The natural sweeping of the divertor strike-points (by up to 30cm) and the large range of plasma configurations means that studies of the divertor target need large arrays of probes. Low-cost high spatial resolution Langmuir probe arrays with real-time processing and multiplexing have been developed, as has an outboard reciprocating probe.

MAST was designed after the issue of halo currents arose in the fusion community, so it was possible to implement a system[6] where essentially all current paths are monitored, together with extensive measurements of toroidal asymmetries. The result is perhaps the most comprehensive set of halo diagnostics on any tokamak, and already a substantial halo-current database has been generated, showing both magnitude and toroidal asymmetry to be much lower than at large aspect ratio.[7]

FAST PARTICLE AND CURRENT PROFILE MEASUREMENTS

Fast ion physics is a key area for MAST: neutral beams are used for main heating and current drive, and the low toroidal field means the beam ions are super-Alfvénic. This makes STs like MAST natural laboratories for testing the physics of energetic particle modes relevant to ITER as well as burning ST devices . To determine the fast ion energy content and spectrum a sophisticated large-orbit code (LOCUST) suited to the ST has been developed and is used in conjunction with a 2-species neutral particle analyser on loan from PPPL. Excellent agreement has been obtained between the energy content from the thermal and fast particles and the energy deduced from magnetic reconstruction (EFIT), by iterative use of EFIT and LOCUST.[8]

For the current profile, the low magnetic field makes conventional MSE very challenging, and it is likely that a combination of EBW emission[9], optical imaging of field-line pitch, instability locations and reconstruction codes such as EFIT will be used. Mode conversion of EBWs to free-space waves has already been observed on MAST in H-mode (and acts as an H-mode signature[11]). This is also being used to assist the design an EBW heating antenna. An important contribution will be accurate measurements of the total edge pressure gradient which will allow the edge bootstrap current to be calculated. A key issue is the necessary accuracy of $J(r)$, which is determined by the physics issues to be resolved (edge stability, magnitude of non-inductive currents etc).

SUMMARY

The diagnostic capability of MAST is still being developed, but already the existing systems are revealing new physics, and some novel systems and approaches are being implemented. The table below summarises some of the diagnostic advances and results arising from the development of diagnostics for MAST.

DIAGNOSTIC	SPECIAL FEATURES RELATED TO MAST
Langmuir probes	Ultra compact arrays, signal multiplexing, analogue processing.
Thomson scattering	Ordered fibre bundles for pulse-by-pulse alignment checks; shared optics and dumps for both ruby and NdYAG systems.
EBW emission	H-mode signature, and possible measurement of the q-profile.
Edge profiles	Shared optics for n_e, T_e, T_i, v_i ensure accurate spatial alignment for ELMs and L-H transitions (following COMPASS-D).
Interferometry	Novel two-colour fringe-free microwave system.
Plasma viewing	Full views of the whole torus are possible, allowing the entirety of helical structures and filaments to be measured.
Halo currents	Inclusion of halo diagnostics from the outset to provide a comprehensive measurement set including toroidal asymmetries.

ACKNOWLEDGEMENTS

This work is jointly funded by the UK Department of Trade & Industry and Euratom. Collaborations with DCU, ENEA (Frascati and Padua), FOM, IST, GA, PPPL, Ioffe, and others are gratefully acknowledged. MAST NBI & NPA equipment is on loan from US DOE, ORNL and PPPL, and EFIT was provided by GA.

REFERENCES

1. A. Sykes et al, First physics results from the MAST Mega-Amp Spherical Tokamak, *Physics of Plasmas*, **8**, 2101 (2001)
2. M. J. Walsh et al, Interactive optical design and realization of an optimized charge coupled device Thomson scattering system for the spherical tokamak START, *Review of Scientific Instruments*, **70**, 742-746 (1999)
3. M. R. Tournianski, Use of a 1D D_α camera to measure edge neutral and electron densities evolutions, *this conference.*
4. P. G. Carolan et al, Synergy of multiviewing spectroscopic diagnostics on COMPASS-D (invited), *Review of Scientific Instruments*, **72**, 881-887 (2001)
5. N. J. Conway et al, Integrated spectroscopic edge diagnostics on COMPASS-D, *this conference.*
6. T. Edlington et al, MAST magnetic diagnostics, *Review of Scientific Instruments*, **72**, 421-425 (2001)
7. R. Martin et al, MHD activity, disruptions and halo currents on MAST, *Proc. 27th EPS Conference on Controlled Fusion and Plasma Physics, Budapest*, p221 (2000)
8. R. J. Akers et al, Confinement in L and H mode MAST plasmas, *Proc. 27th EPS Conference on Controlled Fusion and Plasma Physics, Madeira*, paper P2.026 (2001).
9. V. F. Shevchenko, ECE measurements via B-X-O mode conversion: A proposal to diagnose the q profile in spherical tokamaks, *Plasma Physics Reports*, **26**, 1000-1004 (2000)
10. V. F. Shevchenko et al, EBW emission observations on COMPASS-D and MAST, *Proc. 27th EPS Conference on Controlled Fusion and Plasma Physics, Madeira*, paper P3.101 (2001)
11. R. J. Akers et al, L – H transition in the Mega-Amp Spherical Tokamak, submitted to *Physical Review Letters* (2001).

ADVANCES IN THERMAL IMAGING

R. Reichle for the TORE SUPRA Infra-Red-team*

INTRODUCTION

Tore Supra's (TS) present mission is to study high energy long duration discharges. The last step in this direction has been to upgrade its power exhaust capability to run 1000 sec discharges with power densities in the 10 MW/m^2 range on actively cooled target plates. High performance water cooling very closely beneath the carbon surface of the targets allows to keep the surface temperature at an acceptable level of 1000°C. This condition needs to be carefully monitored since even only localised overheating can rapidly lead to a water-leak. Responding to this challenge, a new approach to the use of thermal imaging in tokamaks has been taken[1] to ensure safe operation. Endoscopes and detectors needed to be adapted to the environmental conditions while fulfilling the spatial and temporal resolution requirements. Part of the heavy flux area is inaccessible to direct view, requiring the use of optical fibres, which prompted a detour from the standard infrared (IR) range into the near infrared (NIR) region where unexpected spectral features were encountered[2]. The 100% coverage of the target areas and the long duration of the discharges lead to an unusually large amount of data to handle both by the security system and the human supervisor.

PHYSICAL AND TECHNICAL BACKGROUND

The spectral luminance $L_{bb}(\lambda,T)$ of an ideal black body radiator at temperature T and wavelength λ is described by Planck's well known radiation formula. For real objects with large emittance values $\varepsilon(\lambda,T)$ near 1.0 one can often make correct

* R. Reichle, D. Guilhem, C. Balorin, E. Delchambre, C. Desgrange, L. Ducobu, M. Jouve, M. Lipa, P. Messina, R. Mitteau, D. Moulin, C. Pocheau, H. Roche, A. Saille, DRFC Association CEA sur la Fusion, CE Cadarache, F13108 Saint Paul lez Durance, France; Z. Wang: present address: Electronic and Science Dept., East Campus USTC, Hefei, Anhui, 230026, P. R. China; G. Zhuang : present address: CRPP–EPFL PPB-Ecublens CH-1015 Lausanne, Switzerland.

Advanced Diagnostics for Magnetic and Inertial Fusion, Edited by Stott *et al.*
Kluwer Academic/Plenum Publishers, New York, 2002

temperature measurements by assuming that the measured luminance L_{meas} is the product (eq. 1).

$$L_{meas}(\lambda,T) = \varepsilon(\lambda,T) \cdot L_{bb}(\lambda,T) \quad (1)$$

For the carbon fibre reinforced carbon (CFC) used on the target elements of TS this seems to be well justified for the whole wavelength range between 0.9 and 5 μm, both when heated by high energy (125 keV) electrons at 10 MW/m^2 to 1000°C [3], or when heated at 0.1 MW/m^2 by 50 eV electrons. The electron irradiation did not change the appearance of the surface. Deviations from eq. 1 were observed under conditions where TS scrape off layer (SOL) plasma of 10 to 100 eV and a few MW/m^2 was heating a target zone which consisted of pyrolytic graphite covered with amorphous carbon flakes. There, the deduced NIR temperature was significantly higher than the IR temperature (see fig.1) [2]. Pyrolytic graphite is known to have directional, wavelength and temperature dependant emittance coefficients, generally above 0.5, only descending for temperatures above 1700 K to lower values [5], but the observed effects would imply a difference of a factor 5 to 10 of the emittance in the NIR and the IR.

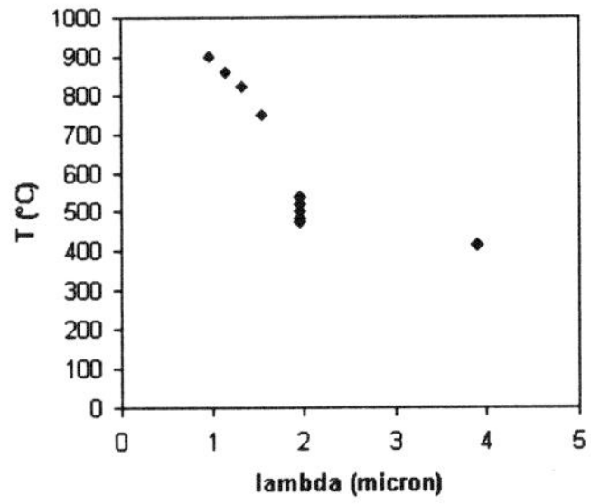

Figure 1. Temperatures determined from the spectral luminance at different wavelengths.

The same effect of additional radiation in the NIR range, albeit with smaller magnitude, was also noted under conditions of 250 keV ions of a few MW/m^2 heating up a B_4C surface, that was also somewhat modified under impact[5]. Bremsstrahlung from the plasma was estimated to be too small to cause this effect and a model assuming an extra radiation of hotter dust particles of sizes of 1 μm or less - omni-present in tokamaks[6] - are a possibility to explain the observed spectra due to the fact that the thermal emission from dust at wavelengths λ longer than the dust size falls of with a $1/\lambda^2$ dependency[7]. The increasing transparency of carbon with wavelength should also play a role, both in terms of wavelength dependent viewing depth on the target as in wavelength dependent absorption strength of carbon films deposited on the observation windows[8].

Assuming that one will find a good spectral window to apply equation (1) one can address properly the technological and physical questions which need large area target observations. The technological question is the uniformity of the target plate elements. The targets have a top layer of 5 mm CFC and a body of CuCrZr cooled by water flowing inside. Between the two is a bond layer, which may deteriorate locally. The worst implications of this is known as critical flux condition, where the cooling properties of the water are locally lost, when the water vaporises. As a consequence, runaway heating will start, leading within 0.2 secs to a melting of the cooling block if the power-flux is not cut earlier, which is the main object of all the installations described here.

IMAGING ENDO SCOPES

Six endoscopes will be installed in regular distribution around the upper ports of TS. They will look down on the toroidal limiter and sideways onto the antennae, covering the totality of their surfaces exposed to plasma. The exterior envelope of the endoscopes will be kept at 150°C by the same water cooling circuit at 30 bar as all the other first wall components. This cooling loop will have to absorb the diffuse radiative heat load of about 0.3 MW/m^2 and locally on a specially cooled protection plate (fig. 2) local power deposition by electronic ripple losses[9] of some MW/m^2 possibly amounting up to 50 kW per endoscope. Due to the high pressure of the cooling water one has chosen CuCrZr as material for the ripple plate and the front end of the endoscopes. The objects to observe will be between 2 and 5 m away and the required optical resolution is in the range of 1 cm. Since the viewing lines will cross the plasma, there is some parasitic radiation from bremsstrahlung and possibly molecular radiation. At 1000°C both are considered negligible in comparison with the thermal radiation. At lower temperatures however they may become visible as well as other radiative effects caused by the presence of strong plasma wall interaction in front of the target region. The absorption of light caused by carbon deposits on the first optical element is relatively small due to the wavelength dependant absorption of carbon, but metallic depositions may degrade the transmission. A window used for about a year at Tore Supra showed e.g. at 5 μm 5% absorption due to carbon and 50% due to metals.

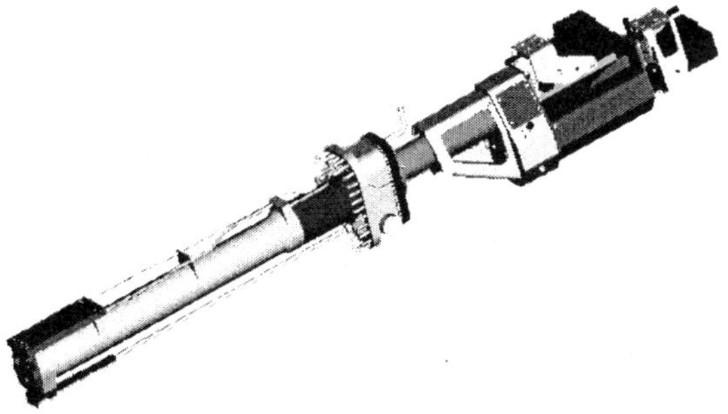

Figure 2. Overview over construction of endoscope.

Each endoscope is designed to have three individual optical transfer lines bridging a distance of about 1.5 m between the front end optics and the camera. Two of these transfer lines recombine half-images of the limiter (fig. 3) onto one camera. The third optic is equipped with a camera of its own looking either at an antenna or procuring a close up of an important zone of the limiter.

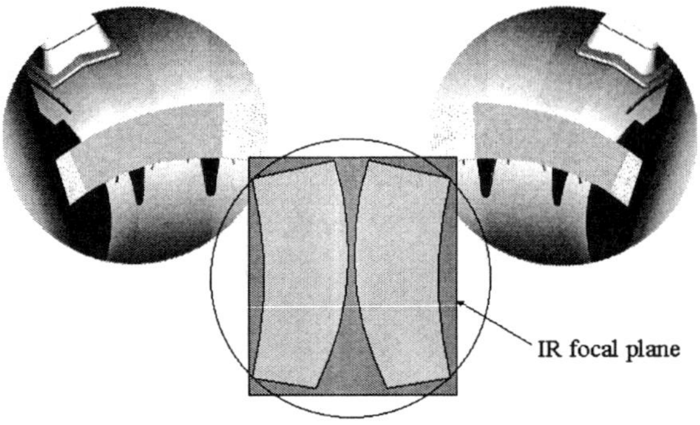

Figure 3. Combination of two half images onto the camera chip.

To keep the parasitic thermal radiation of the 40 lenses in each subassembly as low as possible and to stand a chance to measure as low as 150°C an internal cooling loop at 25°C has been implemented which envelops the lenses. The optical calculation takes the different thermal expansion between the outer skin and inner optic assembly into account. The hottest part will be the front end sapphire window which is made as small as possible (2 cm diameter) and which forms the barrier between machine vacuum and secondary vacuum. Tests in an IR irradiation facility have shown that the windows and their connection to the CuCrZr head will stand up to the required radiative flux load even when well covered with radiation absorbing material. To protect this window against deposition of absorbing materials, a carbon shutter has been put in front of it, which can be used also as uniform reference light source when closed and the tokamak is at its working temperature of 150°C. As another calibration feature, small blackbody sources have been incorporated into the endoscopes themselves. The front-end optic in the endoscope head contains a prism for selecting the viewing angle and sets of small lenses. At the other end of the transfer line, where the two images of the limiter are combined by a prism, the lenses, now outside the tokamak, are considerably larger. The aperture stop of the whole arrangement is F/3. The optical resolution is designed exceed those of the cameras (320*256 pixel). The choice of cameras from which the present candidate was chosen included InSb, micro-bolometer, quantum well and Mercury-Cadmium-Tellurid detector focal plane arrays. The micro-bolometer cameras were excluded due to their incompatibility with the sapphire windows, chosen for their thermo-mechanical properties and proven soldering techniques. Amongst the remaining candidates the InSb chips convinced with their high quantum efficiency and associated low detection limits. The final choice amongst a number of such cameras was made on the basis of the flexibility of the manufacturer to undertake a special development. The aim of this development is to recombine in the camera three consecutive images taken with different exposure times at a rate of 150 Hz such that a significantly enhanced dynamic can be achieved. This should give pictures at 50 Hz covering the range from 100°C to 1200°C with an precision close to 2% of the indicated value over the whole range. The present planning is to install all endoscopes and cameras beginning 2002 and close the security feedback loop 2003.

FIBRE PERISCOPES

Twelve actively cooled pump assemblies will be installed below the toroidal limiter of TS (fig. 4a). They will have leading edges exposed to 14 MW/m^2 under nominal conditions. These target plates, called neutralisers, are of a similar construction as the elements of the limiter with even smaller safety margins with respect to a critical surface temperature of 1000°C. Due to the impossibility to have a direct line a sight onto these objects, the project to observe them individually via 12 fibre periscopes (fig. 4b) was lanced.

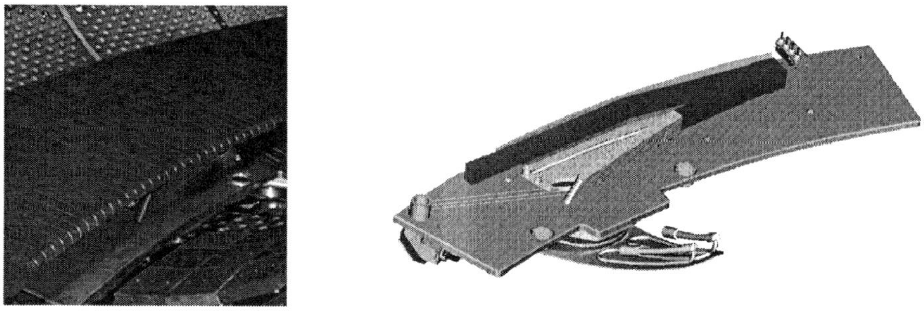

Figure 4. The toroidal limiter with a tip of the neutraliser below visible and schema of periscope observation.

The power deposition pattern on their front face reflects the SOL power flux profile (fig. 5) with its maximum at the top of the neutraliser target tile. Due to cooling efficiency variation one expects to find apart from the hot spot corresponding to the primary target zone at the top of the neutraliser tile also a second hot zone at the bottom of it. In order to observe both areas in detail and to have some redundancy in the system, three fibre optics are aimed at the neutraliser. Two of them observe spots of 1 cm diameter in the expected high temperature region. The third one covers the whole area of interest with a 5 cm spot size (fig 6b). Sapphire balls are used as vacuum feed-through between two fibres. It is foreseen to replace in one of the 12 locations where such installations are planned the wide angle view with a spatially resolving fibre bundle of 19 individual fibres in hexagonal close pack arrangement.

Figure 5. Power-flux profile on actively cooled neutraliser [10].

The first experimental campaign with a neutraliser in place started in August 2001 with an inertial cooled neutraliser entirely fabricated from CFC. For this configuration, the expected temperature profile is governed mainly by the power deposition profile. A different line of sight arrangement is used, which allows to test a ZrFl fibre transparent up to 4 μm (Reflex Anaytical) looking at the same hot point as a NIR silica fiber (Oxford Electronics) (fig 6a). Another advantage of this preliminary configuration is the possibility to exploit an internal fibre as reference (fig. 6c). The latter is inserted in the CFC and approaches the point of interest from the interior thus avoiding the potential spectral complication faced by the fibres looking at the outside.

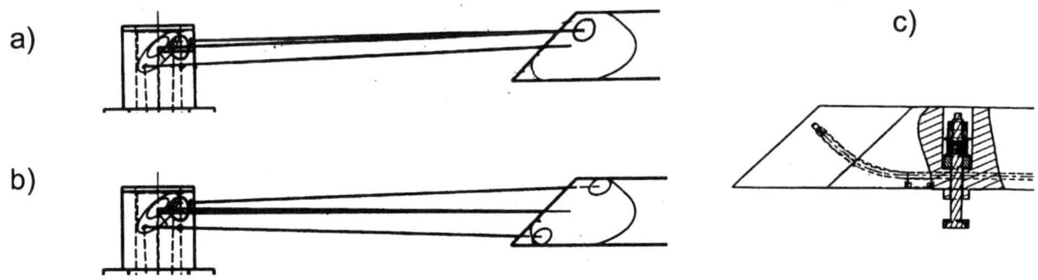

Figure 6 . Viewing lines for inertially (a) and actively cooled (b) neutraliser and interieur fibre (c).

There is a risk of disturbance by carbon contamination of the front-end optics of the periscope due to its proximity to the target plate. The first approach to this question had been to consider 2 colour pyrometry using NIR silica fibers, but due to the problems to reconcile the form of the spectra with black body emission spectra in the NIR range, the strategy now is to investigate these spectra in more detail by covering also a larger spectral range up to 4 μm, using ZrFl fibres. The aim is to search for extrapolation rules - probably based on data similar to those of fig. 1 – in order to deduce sufficiently correct temperatures even from measurements with silica fibres, which have the better thermo-mechanical properties. If one really needs to use the proper IR range to have good temperature measurements, chalcogenide and silverhalide fibres may become also interesting[11]. Not only the fibres, but also the dispersive system and its detector have evolved with the need to cover a larger wavelength range. The first experiments were made with Peltier element cooled InGaAs linear arrays (Lot Oriel) with a sensitivity range from 0.9 to 1.6 μm installed behind a high trough-put (F/2.9) low resolution spectrometre (Jobin Yvon CP200). For the coming experiments a prism spectrometer is under construction which should allow to measure simultaneously with 39 fibres spectra between 1 and 2 μm with a resolution of more than 10 significant points and between 1 and 4 μm with a resolution of more than 20 significant points (fig. 7). The detector is practically identical with the cameras used at the endoscopes except for the differences, that the spectral range of sensitivity is from 1 to 5 μm and not from 3 to 5 μm and the F-number is 2.5 and not 3. The sensitivity of the NIR system previously used allowed measurements down to 300°C with 50 Hz temporal resolution. With the prism spectrometer system in front of the IR camera one expects to measure from 450°C at 1.5 μm and from 350°C at 4 μm. The reason for the lower sensitivity lies in the low F-stop of the first lens of the prism-spectrometer (F/7.5) and the lower quantum efficiency of InSb in the NIR Range. The reasons to chose this solution were that it covers the whole spectral range with a single dispersive system and reunites all measurements onto the same detector. To calibrate the system with the help of the thermal background temperature of 150°C of the tokamak, one will have to replace temporally the prism by a mirror. Another calibration feature is a spatially uniform light source integrated into the dispersive system in order to compensate the in-homogeneity of the detector. For the experimental campaign 2001 it is foreseen to make first measurements with the prism system using the aforementioned 4 fibres and for the following experimental campaign 39 fibres should enter the detection system.

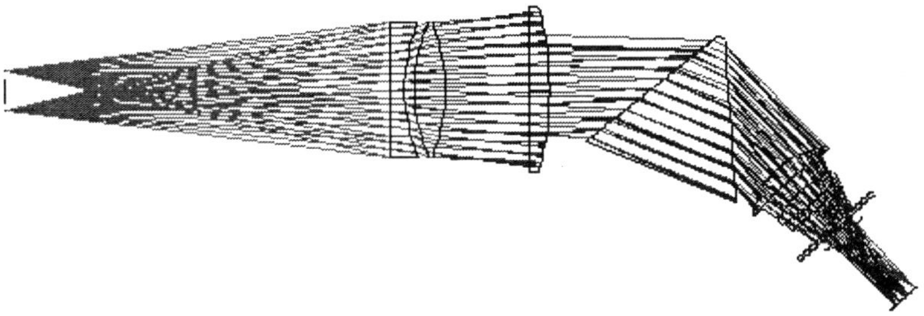

Figure 7 . Layout of dispersive prism system: 3 lines of 13 fibres at the left, produce 3 groups of 13 spectra at the right on the 2 dimensional camera chip.

DATA ACQUISITION

All cameras are integrated into the same data-acquisition system. This system has recently been described[12]. It is designed for 13 camera running at 50 Hz for 1000 sec with pictures of 320*256 pixel at 16 bit resolution. This data-flow will be simultaneously directed into a real-time control system and a dedicated high volume storage system which can be accessed over its own network by a limited number of terminals. The real-time system is based on in-house developed VME cards with field programmable gate arrays which allow the selection of up to 12 arbitrarily shaped analysis zones per camera to detect dangerous temperature excursions. Their output is a selection of temperatures on one line and alarms on another line. The alarm output will be triggering a soft stop of the tokamak, whereas the temperature signals are transferred by reflective memory into the control computer for the tokamak operation. Apart from the real-time section, the rest of the acquisition is PC based. It is geared up not only to store the data but also to process them to provide useful temperature traces. To allow human survey of the group of 12 views of the limiter, it is envisaged to recombine the 12 camera images to a single image. After an initial phase of loss-less compression, more powerful compression will be employed.

DISCUSSION, CONCLUSIONS, OUTLOOK

The challenge of TS to implement safety systems based on thermography that take on the problems of 1000 sec 100% coverage on actively cooled components close to their critical load condition has spurred a number of investigations, solutions and new research as exposed in the article. In particular the NIR spectra and their possible connection with dust and flakes merit interest, since dust tends to capture large amounts of Tritium[13] and may levitate and recycle in the SOL thus becoming an important SOL-parameter[14]. The spectral investigations which are continuing may also shed some light on recently inferred temperature over-estimation in the IR range in the presence of strong chemical erosion of the target[15, 16] and might influence design proposals assuming perfect blackbody spectra from target plates from the NIR to the IR range[17]. The concepts may be useful not only for other large fusion devices facing similar challenges, but also to other IR applications in industry.

ACKNOWLEDGEMENT

We thank the C. Arnas, V. Kaeppelin, and M. Carrere of the LPIIM of the University St. Jerome in Marseille and M. Febvre of the FE200 installation of FRAMATOME in Le Creusot for the help with laboratory measurements at their premises.

REFERENCES

1. D. Guilhem et al., Infrared surface temperature measurement for long pulse real time feed-back control in an actively cooled machine, Rev. Sci. Instr. 1999, Vol 70,1, 427-430
2. R. Reichle et al., Thermography of target plates with near-infrared optical fibres at Tore Supra, *Journal of Nuclear Materials*, Vols. 290-293 (2001) 701-705
3. R. Reichle et al., contr. 28th EPS Contr. Fus. Plasma Physics, Madeira, Funchal, Portugal 18-22 June 2001
4. Y.S Touloukian and D.P. Witt, Thermal radiative properties: nonmetallic solids, Vol 8,1972
5. E.Delchambre et al, 7th Congrès de la Division Plasmas de la SFP, 28-30 Mars 2001 Piriac sur mer, France
6. Ph. Chappuis et al, Dust characterization and analysis in Tore Supra, *J. Nuc Mat* Vols. 290-293 (2001) 245
7. V. Ossenkopf & Th. Henning, Astron. Astrophys. 291, 943-959 (1994)
8. A. H. Lettington, 1991,Application of Diamond-like (Hard Carbon) Films in *Diamond and Diamond-like Films and Coatings*, R.E. Clausing, L. L. Horton, J.C. Angus, P. Koidl, Eds, NATO ASI Series B, 266, Plenum Press, 481, New York
9. V. Basiuk et al . 27th EPS Contr. Fus. Plasma Physics, Prague, 2000
10. E. Tsitrone, internal note of Tore Supra: CFP/NTT-1999.03
11. Proceedings of SPIE, Volume 3849, *Infrared Optical Fibers and their Applications*, M. Saad and J. A. Harrington edit., 21-22 Sept 1999, Boston, Massachusetts, ISBN 0-8194-3442-6
12. D. Moulin et al, proc. 12th IEEE-NPSS Real Time conference (Valencia 4-8 june 2001)
13 J..P. Coad et al., Erosion/deposition issues at JET, *J. Nuclear Materials*, Vols. 290-293 (2001), 224-230
14 J. Winter et al;, Radioactive dust levitation and its consequences for fusion devices, *Journal of Nuclear Materials*, Vols. 290-293 (2001) 509 - 512
15. Th Eich et contr. 28th EPS Contr. Fus. Plasma Physics, Madeira, Funchal, Portugal 18-22 June 2001
16. V. Philipps et al., contr. 28th EPS Contr. Fus. Plasma Physics, Madeira, Funchal, Portugal 18-22 June 2001
17. A. Herrmann and H. Salzmann, High resolution IR Temperature monitoring system for the ITER target plates, in *Diagnostics for experimental thermonuclear fusion reactors 2*, ed P. E. Stott, G. Gorini, O. Prandoni, E; Sindoni, Plenum Press New York, 1998

FAST MEASUREMENTS OF LIMITER SURFACE TEMPERATURE AND HEAT FLUX VARIATIONS ON THE T-11M TOKAMAK BY IR RADIOMETER

V.B. Lazarev, A.G. Alekseyev, A.M. Belov, S.V. Mirnov *

1. INTRODUCTION

Fast variations of limiter surface temperature and heat flux from plasma through the surface of a limiter or divertor plates in a tokamak are important parameters for small experimental installations and also are especially critical for a tokamak-reactor for the control of the plasma edge.

Liquid lithium technology is actively developed now for the tokamak applications requiring recoverable low-Z plasma facing material protecting high-Z vessel components. Earlier within the framework of this activity on the T-11M, the experiments with the lithium limiter were carried out at various initial temperatures of lithium in he range from 20 to 400^{0}C, in helium and hydrogen discharges [1]. Strong dependence of the lithium desorption versus temperature was revealed. Therefore the necessity to measure fast variations of the limiter surface temperature had been recognised, in order to clear up the physical processes responsible for the lithium erosion due to plasma - surface interaction.

The work was performed on the tokamak T-11M within the framework of lithium limiter experiments [2,3]. It had the following purposes: first - the study of lithium surface temperature variations with enhanced temporal resolution, and second - the recovery of the heating power variations loading the limiter surface.

2. SURFACE TEMPERATURE MEASUREMENT BY IR RADIATION

An infrared (IR) radiometer was designed with the use of liquid nitrogen cooled Ge:Au photoconductive detector for the limiter surface temperature measurements. With Ge input window this detector is sensitive ($R_v \sim 10^3$V/W) in a spectral range $\lambda = 2 \div 6\mu$m. An appropriate optical sketch is represented in the Fig. 1.

The detector unit with LN_2 dewar in the magnetic shield was located above an upper port of a tokamak vessel watching 1.5x1.5 см^2 surface area in the central part of the Li limiter installed in the lower port. The BaF_2 vacuum window and NaCl lens were used for the limiter surface IR image transmission with 7.5^x size reduction related to the detector sensitive aperture. Low noise preamplifier with a battery power supply was located in the same shield.

* V.B. Lazarev, A.G. Alekseyev, A.M Belov, S.V. Mirnov TRINITI, Troitsk, Moscow region, 142190 Russia. E-mail: lazarev@triniti.ru

Hence the output voltage of the preamplifier U(t) was proportional to IR radiation power P(t) from the 1.5x1.5см^2 area on the limiter surface. The temporal resolution ~100μsec of was limited by the preamplifier.

The temperature resolution δT_s is limited by the detector noise and depends on the temperature of the limiter surface. In our case it was approximately $\delta T_s \approx 1$K at $T_s \approx 100^0$C. Also it depends on the limiter material and for a graphite limiter it will be better due to higher albedo.

Since the radiometer "watches" the limiter surface through the plasma volume, there is a question, what fraction of the total IR power incident to the detector is contributed by the plasma radiation? We estimated the ratio of plasma radiation to IR radiation of the limiter in λ= 2÷6μm spectral range for the most unfavorable conditions (Z_{eff}=3, T_e=10эВ, $n_e = 10^{13} \div 10^{14}$ cm^{-3}). It was assumed that the main plasma contribution is provided by the bremsstrahlung radiation. The intensity of IR radiation from the limiter surface was found to be approximately 10^5 times higher then bremsstrahlung plasma radiation in the same spectral range. A negligibility of plasma contribution was confirmed experimentally by removing the limiter from the vessel into the vacuum port. The radiometer signal has fallen down to a noise level.

3. ABSOLUTE CALIBRATION OF IR RADIOMITER

The recalculation of the measured signal U(t) into the temperature T(t) was carried out with the use of calibration data, obtained during slow heating of the limiter by internal electric heater and temperature control by thermopile measurements. The calibration curve U(T) for the whole temperature range (20÷420^{0}C) was obtained by the fourth order polynomial approximation of the calibration data. Within a numerical factor it almost coincides with the blackbody radiation in the same spectral range. Non-linearity of U(T) curve results from the selected IR spectral range, where $h\nu > kT$. The drift of radiometer «dark» level due to the IR background variations caused the main error of calibration procedure. It should be noted also, that the device measures an average value of temperature over the surface area 1.5x1.5 cm^2.

4. HEAT POWER VARIATIONS ON THE LIMITER

For the evaluation of a specific incident heat flux on the limiter surface Q(t) [W/cm^2], from the measured T(t) data, it is necessary to solve an inverse task using the recovery procedure of a heat transfer equation taking into account an actual limiter structure. However, within the required accuracy, the task could be essentially simplified.

Let's consider a construction of the rail type limiter, which we used on the T-11M. So-called lithium-filled capillar-porous structure (CPS) [1,2] covering the limiter surface had 5-mm thickness. Lithium surface temperature during the discharge is determined by a specific heating power Q(t) on the surface, initial temperature T_0 and intrinsic parameters ρ, c, λ of thermal sink material. Duration of heating τ and thermal propagation factor $\chi = \lambda/(\rho \cdot c)$ of a limiter material determine the temperature half-value thickness Δ characterising the propagation of temperature wave:

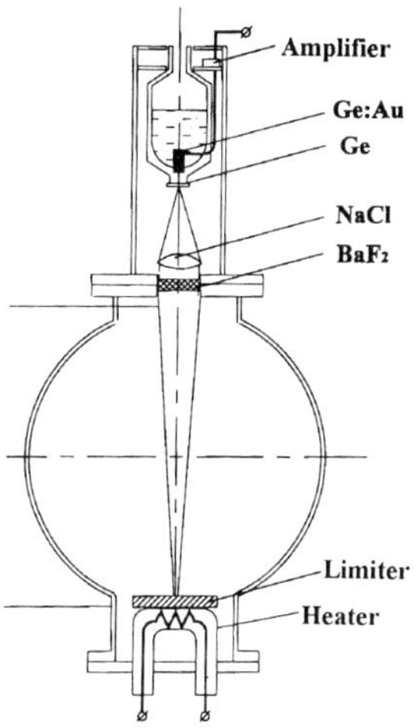

Fig. 1. IR radiometer on the T-11M.

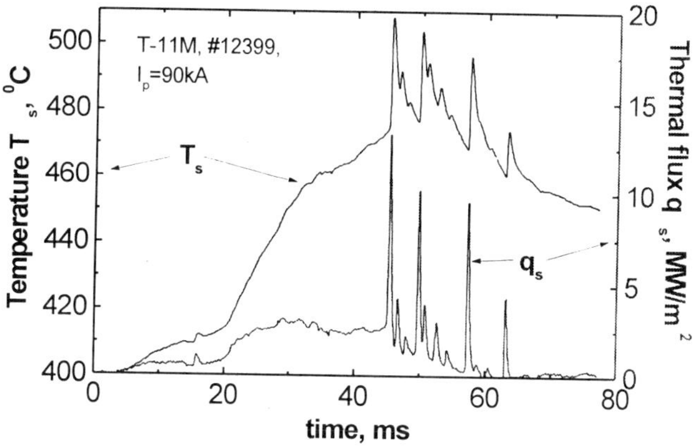

Fig.2. The evolution of Li limiter surface temperature and relevant heat flux during the disruptions.

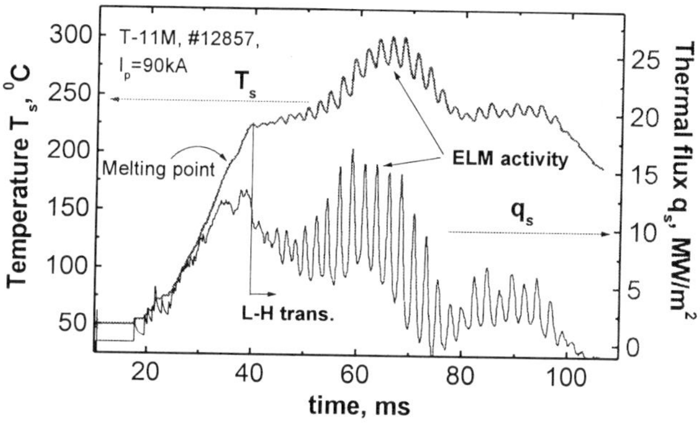

Fig.3. The evolution of Li limiter surface temperature and relevant heat flux during the ohmic H-mode.

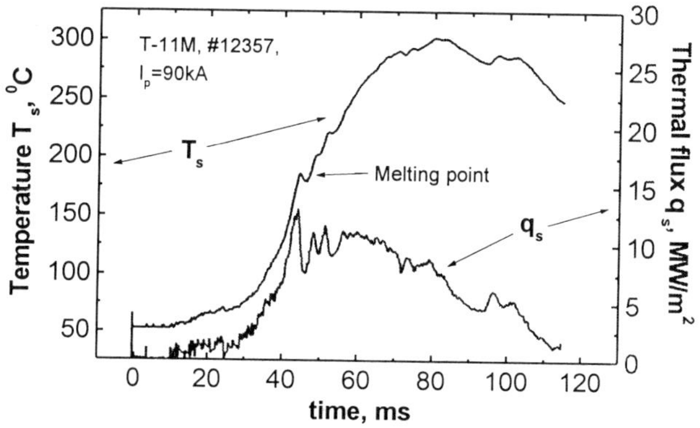

Fig.4. The evolution of Li limiter surface temperature temperature and relevant heat flux during the ohmic L-mode.

$$\Delta \approx (\chi \cdot \tau)^{1/2} = (\lambda\, \tau / \rho \cdot c)^{1/2} \qquad (1)$$

where ρ, c, λ are CPS density, heat capacity and thermal conductivity, accordingly.

Appropriate estimations of the warmed-up CPS depth during the plasma discharge $\tau \approx 0.1$s results in the value $\Delta \approx 1$ mm, being ~5 times lower than CPS thickness. Therefore an infinitive semi-space approach could be applied, providing a simple solution for the recovery of incident power evolution:

$$Q(t) = \frac{\lambda}{\sqrt{\pi \chi}} \cdot \int_{-\infty}^{t} \frac{dT(\tau)}{d\tau} \cdot \frac{dt}{\sqrt{t-\tau}} \qquad (2)$$

Note, that the functions Q(t) and T(t) are related by the formula similar to well known Abel inversion. An example of actual heating power Q(t) recovered numerically from the experimental data T(t) is represented in Fig. 2,3,4. The total value of Q(t) could be attributed to plasma-to-limiter power deposition since other sources of limiter surface heating or cooling (radiation and Li evaporation) are negligible.

5. CONCLUSIONS

The results obtained with IR radiometer on the T-11M tokamak with lithium limiter were found to be quite useful for the evaluation of the following parameters:

- **Surface temperature** evolution of a limiter or a divertor plate;
- **Specific thermal flux** to the limiter or divertor surface;
- **Electron temperature** in the vicinity of a limiter or divertor surface.

The IR radiometer provides an opportunity for the quantitative studies of MHD-related plasma-wall interaction events, e.g. a disruptive instability or ELM activities in the H-mode.

The response time of the device could be improved to the detector limit ~10 μsec with the replacement of preamplifier by more fast one. Further enhancement in both temporal and spatial resolution could be achieved by the application of CdHgTe photodiode array with ~1 μsec response required for studying the localisation of power deposition to the limiter surface during the development of disruptive instability.

REFERENCES

1. V. Lazarev, E. Azizov, et al., Compatibility of the Lithium Capillary Limiter with Plasma in T-11M, 26th EPS Conf. on Contr. Fusion Plasma Physics, 1999, vol. 231, pp. 845-848.
2. V.A. Evtikhin, I.E. Lyublinski, A.V. Vertkov, S.V. Mirnov and V.B. Lazarev, Technology Aspects Of Lithium Capillary-pore Systems Application In Tokamak Device, SOFT-21 (Madrid), A-37.
3. V.A. Evtikhin, I.E. Lyublinski, A.V. Vertkov, E.A. Azizov, S.V. Mirnov, V.B. Lazarev, S.M. Sotnikov, Experimental Study of the Tokamak plasma interaction with Lithium Capillary-Pore Systems", 18-th IAEA FEC 2000 (October, Sorrento, Italy), Exp4/21.

PERISCOPE SYSTEMS FOR INFRARED IMAGING ON KOREA SUPERCONDUCTING TOKAMAK ADVANCED RESEARCH (KSTAR)

S. J. Yoo and M. Kwon*

1. INTRODUCTION

The KSTAR (Korea Superconducting Tokamak Advanced Research) is under development to establish a scientific and technological basis for an attractive fusion reactor [1] and will pursue steady state advanced tokamak mode of operation, aiming at especially a long pulse operation up to 300 s.[2] The KSTAR is equipped with fully superconducting magnets and thus the vacuum vessel is covered with a large cryostat, which results in long view ports around the vacuum vessel.

The long view ports (longer than 2 m from the end flange of the cryostat to the vacuum vessel) make a periscope inevitable to get a field of view for diagnosing the internal space of the vacuum vessel and better observation of plasma discharge on KSTAR. The periscopes has also been used in conventional copper magnet tokamaks such as TFTR [3] and Alcator C-Mod [4] just for specific purposes, for instance, to view plasma discharge phenomena or to inspect vacuum vessel internal structures in more detail, etc. In KSTAR diagnostic systems,[5] however, periscopes are required for routine operation due to the long view ports.

In this report, a design concept of periscopes will be described, which are under development for the infrared imaging system in order to monitor the thermal load of the inner walls and protecting armors of the heating neutral beam, the antennas for ICRH and LHH system, and divertor plates.

2. PERISCOPE DESIGN

Reflective optics and refractive optics are available for periscope design. The reflective optics is preferable in many respects: The reflective optics is free of chromatic aberration and can be made in a large size, and furthermore, it can be made of metallic

* S. J. Yoo and M. Kwon, Korea Basic Science Institute, Taejeon, 305-303, Korea

Advanced Diagnostics for Magnetic and Inertial Fusion, Edited by Stott *et al.*
Kluwer Academic/Plenum Publishers, New York, 2002

material and thus cooled more effectively in high radiation fields. A periscope design study using reflective optics can be referred in reference 6. In spite of the many attractive aspects of the reflective optics, however, the reflective optics has great difficulty in producing images over a wide field of view (FOV).[7] With regard to performance, the periscope is required to have a wide FOV of the interior of the vacuum vessel with resolutions on the order of a few millimeters to centimeters. So, for KSTAR, standard refractive optics is used in designing periscopes to improve the imaging quality with a wide FOV.

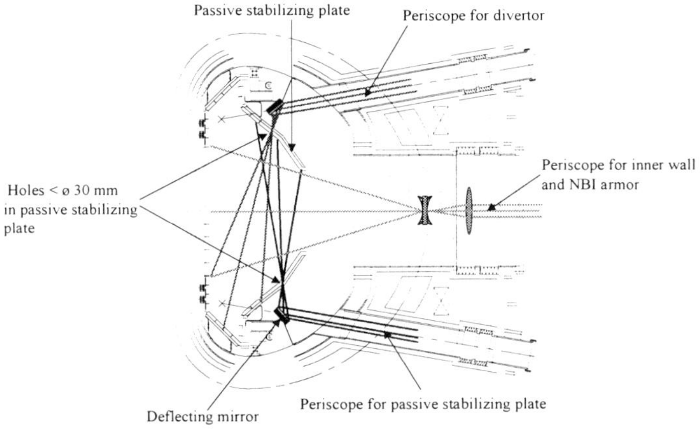

Figure 1. An installation layout of periscopes on KSTAR.

2.1. Periscope for Monitoring Inner Wall and Antennas

An installation layout of the periscope is shown in Fig. 1. For monitoring the inner wall, four periscopes with a FOV of 40° are required in order to provide essentially complete viewing coverage of the inner wall and a periscope with a FOV of 20° is enough for monitoring antennas of ICRH and LHH. The periscope is composed of three sections (collection optics, horizontal relay optics, and vertical relay optics sections) and is located in the radial horizontal ports. The total track of the optical path is over 10 m. The optical path is optimized using a ray tracing code (ZEMAX). A negative lens is inserted into the collection optics section in order to obtain a wide FOV of 40°. A band pass filter of which the spectral range is 3.94 – 4.06 μm is added to a aperture stop following the negative lens in order to reduce the chromatic aberration. The collected images are transferred through the horizontal and vertical relay optics to an infrared TV system which is set up on the experiment room floor and shielded against radiations.

2.2. Periscope for Monitoring Divertor and Passive Stabilizing Plate

For monitoring the divertor heat load and passive stabilizing plate, the periscope has to be equipped with a deflecting mirror system as shown in Fig.1. Since the passive stabilizing plate blocks the line of sight from the port, a hole with a diameter smaller than 3 cm has to be made in the passive stabilizing plate. The divertor or passive stabilizing plate surfaces will be monitored through this hole from the 10-degree-slanted port located at the opposite side. The assembled structure of the periscope is shown in Fig. 2 (a). The mirror has to be controlled from the outside of the vacuum vessel in order

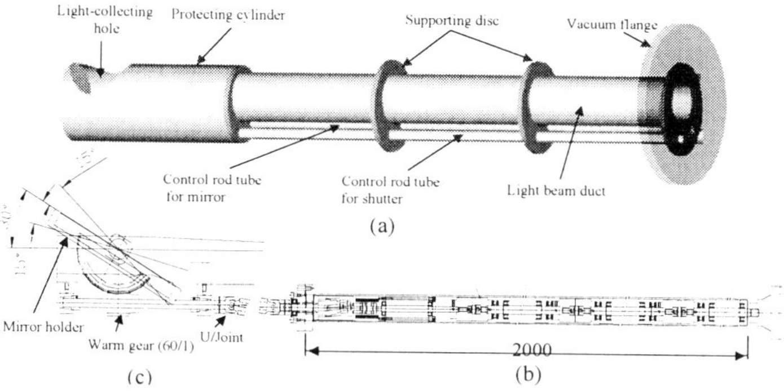

Figure 2. The periscope structure with a deflecting mirror: schematic assembled structure (a), control rod composed of multi-sectors (b), mirror holder driven by a warm gear (c).

to scan different locations in the divertor or passive stabilizing plate. The mirror deflection is controlled with a rod extended to the end flange of the cryostat through multi-sectors as shown in Fig. 2 (b). The rotational motion is transferred to the mirror holder through the rod and a vacuum feedthrough. The rotation is converted into linear motion through a worm gear and then the mirror holder is tilted through a half circle disc gear driven by the worm gear as shown in Fig. 2 (c). The mirror and window have to be protected from electromagnetic radiation and high-energy particles when not used and the shutter has to be a good shield against the rf wave since the plasma may be produced in the location of the deflecting mirror. The shutter system has been designed to cover both the deflecting mirror and the window by closing the light-collecting hole.

2.3. Zoom Lens Set

A zoom lens set is needed when local area can be observed in more detail. A concept of zoom lens set is schematically shown in Fig. 3. A zoom lens set consists of four components, i.e., two moving components and two fixed components. The first moving component (b) sets the magnification and the second one (c) is used to stabilize the focal position. The component (b) is a positive lens with a focal length f_b and the component (c) is a negative lens with a focal length f_c. The fixed components (a) and (d) make an infinite conjugate, i.e., collimated ray trace and are all positive lenses with focal lengths of f_a and f_d, respectively. In case $f = f_b = -f_c$, the magnification M is defined by $M = f^2/(f_d d)$ where d is the distance between the component (b) and (c). The length L of the zoom lens set has a lower limit:

$$L \geq (M + 1/M - 1)f + f_d \tag{1}$$

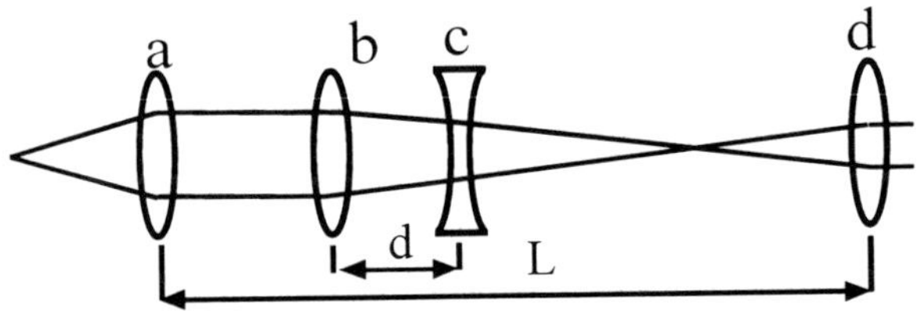

Figure 3. Concept of zoom lens set

3. CONCLUSION

The first step periscope design has been mainly focussed on the optics and mechanical functions. Thermal load of all components of the periscope during the steady state operation of the KSTAR will be analyzed further in the next time. Thus, if necessary, a cooling system has to be added to the periscope according to the result of this analysis.

ACKNOWLEDGEMENT

This work was supported by the Korea Ministry of Science and Technology under the KSTAR Project Contract.

REFERENCES

1. G. S. Lee, et al., The design of the KSTAR tokamak, Fusion Engineering and Design, **46**, 405 (1999).
2. The KSTAR Team, The KSTAR Project: An advanced steady state superconducting tokamak experiment, *Nuclear Fusion* **40** (3Y), 575-582 (2000).
3. S. S. Medley, et al., Periscope camera system for visible and infrared imaging diagnostics on TFTR, *Rev. Sci. Instrum.* **56**(10), 1873-1884 (1985).
4. R. J. Maqueda, G. A. Wurden, J. L. Terry, J. A. Stillerman, New infrared imaging system on Alcator C-Mod, *Rev. Sci. Instrum.* **70**(1), 734-737 (1999).
5. M. Kwon, Diagnotics for the KSTAR: Korea Super-conducting Tokamak Advanced Research, these Proceedings.
6. S. S. Medley, Design study of a visible/infrared periscope for intense radiation applications using reflective optics, *Rev. Sci. Instrum.* **70**(1), 794-797 (1999).
7. R. R. Shannon, *The Art and Science of Optical Design* (Cambridge University Press, 1997).

DIAGNOSTICS FOR THE KSTAR: KOREA SUPER-CONDUCTING TOKAMAK ADVANCED RESEARCH

M. Kwon, J. G. Bak, S. H. Jeong, Y. S. Jung, B. C. Kim, H. G. Lee, S. G. Lee, H. K. Na, H. L. Yang, S. J. Yoo, and KSTAR Team*

INTRODUCTION

The KSTAR (Korea Super-conducting Tokamak Advanced Research) [1,2] is aiming at the successful performance of the advanced tokamak (AT) mode and steady state operations. Though the AT mode can be realized in present tokamaks, it has usually been achieved as a transient phenomenon, shorter than the characteristic time scales. The KSTAR, however, will pursue steady state AT operation with fully super-conducting magnets and with state-of-the-art plasma control schemes. After achieving the first plasma milestone, the KSTAR operations for testing AT modes will be performed in order according to the availability of heating and other ancillary systems. KSTAR diagnostics are also installed in a phased manner, prioritized by the order of the operation scenarios, the availability of related technologies and the required time for development. The whole diagnostics set is classified in three categories as the Machine and Control, Core, and Divertor and Edge. The whole set is also listed in three phases by the order of development as Basic, Baseline and Mission-oriented. The Basic set consists of a minimal diagnostics for the Day One plasma operation of KSTAR. Most of magnetic diagnostics and relatively simple diagnostic sets for measuring basic parameters such as electron temperature and density, ion temperature, impurity behavior, neutral pressure, and run-away electron detection are included. Monitoring TV viewing inside vacuum vessel in the visible and IR region will also be included in this phase of development. As other ancillary systems are installed to KSTAR, relevant diagnostics is going to be added in parallel. Thus for the Baseline campaign when important physics experiments will be performed with intensive heating capabilities with a divertor configuration, profile measurements of important parameters should be provided. Baseline diagnostics are prepared for this purpose. Some diagnostics are still in consideration for development for KSTAR since they may not be urgent or technologies

---

* M. Kwon, J. G. Bak, Y. S. Jung, H. G. Lee, S. G. Lee, H. K. Na, H. L. Yang, S. J. Yoo, Korea Basic Science Institute, Yusung-Ku, Daejeon, Korea 305-333. S. H. Jeong, Korea Atomic Energy Research Institute, Yusung-Ku, Daejeon, Korea 305-353.

Advanced Diagnostics for Magnetic and Inertial Fusion, Edited by Stott *et al.*
Kluwer Academic/Plenum Publishers, New York, 2002

for their realization have not been evolved enough. The time for starting development of such diagnostics will be determined as the demands and necessities are raised along the way of KSTAR operation or as the technical problems are to be solved.

Even though the developments are delayed for certain diagnostic system, the port allocation was done for all specified systems together to minimize the space interference due to the phased development. Figure 1 shows all the diagnostic systems allocated around KSTAR including other ancillary systems such as heating and vacuum systems.

TECHNICAL CHALLENGES

The mission of KSTAR is to explore the advanced tokamak operation regime and to achieve steady-state operation with high performances. Though the operation parameters of KSTAR are not exceeded those of present machines, the planned long pulse (>300 s) operation will require new specifications for technologies currently applied to the various diagnostics. ITER also made great efforts to resolve similar problems during their R&D phase such as radiation-induced damages on the mineral insulated cables and cassette-type port access. For KSTAR diagnostic systems, similar criterion and concepts will be taken for radiation damage assessments and the efficient port access, respectively.

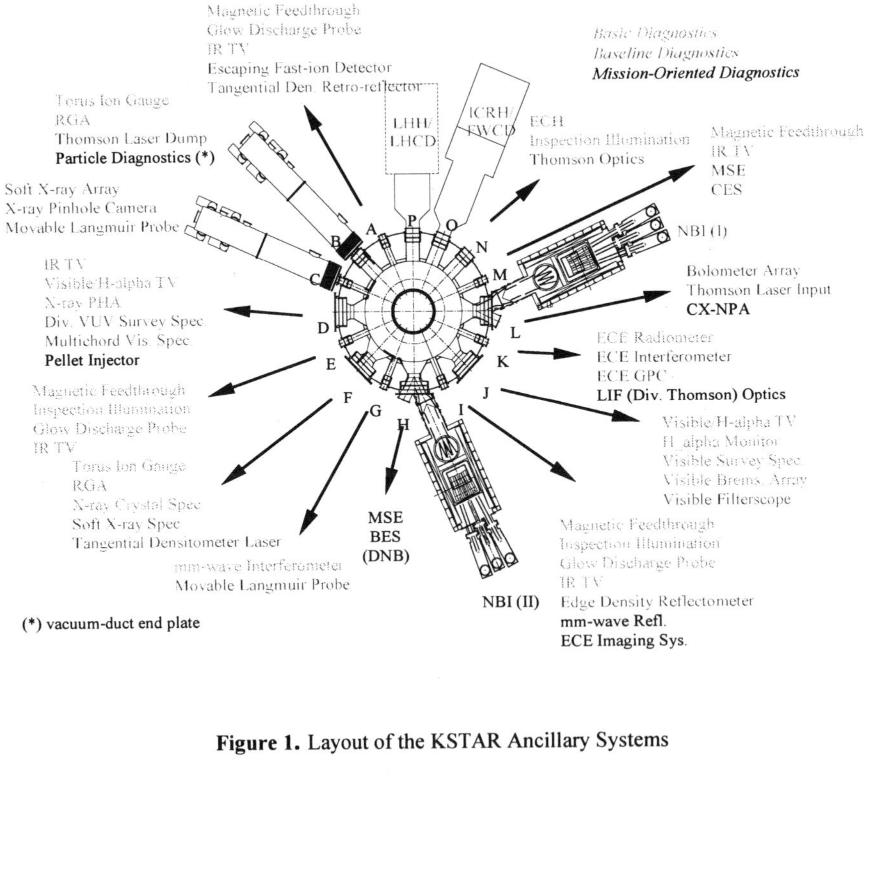

Figure 1. Layout of the KSTAR Ancillary Systems

In KSTAR magnetic diagnostics, due to the high temperature environments MgO insulated cable will be used. The outer sheath is made of stainless steel and the inner conductor is Ni-clad copper. The radiation effect due to long pulse operation on the insulating voltage and the temperature effect on cable inductance and resistance should be examined. The port access to view the plasma is another problem in a super-conducting tokamak because of the long port length through the cryostat. Though its level is not severe as much as ITER, viewing plasma through diagnostic ports is sure to be a puzzle because of various inner components routing around inside vessel and the neutral beam armor tiles for protecting ports from the beam shine-through. A cassette concept was adopted and its usefulness was illustrated. However, maintainability, alignment over many cycles of baking-cool down-operation, and the effect of the disruptions through electromagnetic forces on the cassettes should be examined in detail. Devise of calibration techniques covering wide range of dynamic ranges, assessment of particle bombardments on the optical components, methods to cover full spectral and spatial ranges, and new requirements of data acquisition, storage and retrieval schemes are a few of other challenges that KSTAR is facing with.

STATUS OF THE KSTAR DIAGNOSTIC SYSTEMS

The AT operations imply high beta and high bootstrap current fraction obtained by optimized pressure and current density profiles. To support this mission, KSTAR diagnostics will be concentrated on an accurate and reliable measurement of profiles of related plasma parameters [3]. At the KSTAR Advanced Physics Workshop held in 2000, physicists also recommended that it would be very important to develop some advanced diagnostics for performing important physics experiments during the early phase of KSTAR operations. For the current density measurements, two polarimetries are under consideration; the motional Stark effect (MSE) and the Faraday rotation. The Faraday rotation technique requires a full view of plasma and information from other diagnostics to overcome difficulties caused by uncertainty of information near edge and on axis. However, it provides excellent temporal resolution. The MSE technique requires either diagnostic or heating beam line, which is a major drawback. However, this technique can measure the current distribution with a good time resolution and relatively good accuracy. Simultaneous use of both systems is being considered for KSTAR to cover from the core to the edge region of the plasmas. Since there is no test facility in Korea with neutral beam injector, a 35 kV, 1 A diagnostic neutral beam injector is being developed for HANBIT mirror machine and few beam-aided diagnostic systems will be tested[4]. For measuring the pressure profile, many parameters are involved such as plasma density and temperature, impurity and fast beam ions. Thomson scattering is a conventional tool for electron temperature and density measurement and the electron cyclotron emission (ECE) measurement complements it for the transient phenomena. Interferometry will fulfill the requirements for the electron density measurements. Densitometry based on Faraday rotation angle measurement is also being prepared. Profile diagnostic systems such as Thomson scattering system and ECE, interferometer and reflectometer are also installed and operated in HANBIT for developing measurement techniques for the AT operations in KSTAR.

For obtaining high beta in AT operations, MHD stability needs to be sustained. A critical issue for AT operation in a steady state tokamak is thorough understanding and extending the MHD stability limits through active feedback control of plasma configurations. To support this goal, an extensive and reliable set of magnetic diagnostics as well as ECE and reflectometry is being prepared. Since magnetic sensors will be placed near the plasma, radiation resistance is a crucial requirement for the cable. A cable insulated with MgO was chosen and a prototype of each sensor was fabricated and tested for assessment of the effect of thermal expansion due to baking and magnet cool-down. A Mirnov coil was also designed to have a wide frequency response up to 1 MHz according to the test results. A vacuum feedthrough for sensors was also designed and tested. Due to the long pulse operation, long term drift of the integrator should be compensated. A compensation technique is under development as a part of the design work of a reliable integrator itself.

The plasma boundary and divertor regions are multiply-coupled to the performance of the AT operations and there still remains uncertainty of the compatibility of the AT operation with edge particle and power handling capabilities. More diagnosis of current and density profiles and flow in the edge region is demanded for a better understanding of the core and edge coupling. To resolve the uncertainty of compatibility and to extend boundary control techniques, radiation monitoring, spectroscopy and optical techniques will be used. However, accessibility is highly limited for the divertor region due to passive plates, armor, and toroidally and poloidally distributed piping and support structures. More elaborate technical solutions are being actively pursued.

The machine protection is a critical issue for the advanced tokamaks that will usually be operated in long pulse with high beta conditions. Fast plasma position detection and visible/IR plasma TV are important in this sense. Also a disruption prevention or mitigation provision should be included with the plasma control system that maneuvers plasma position and profiles to find better operation conditions. For visible/IR TV system, a periscope should be used for KSTAR for viewing a large area of inside surface of vacuum vessel and for better observation of plasma discharges. Because of the long distance from the port to the plasma owing to the cryostat region, cassettes for most of diagnostic systems will be needed for better access to the plasma region. Physics and engineering prototypes are being fabricated and will be tested in HANBIT.

ACKNOWLEDGEMENT

This work was supported from the Korean Ministry of Science and Technology.

REFERENCES

1. G. S. Lee, et al., Fusion Engineering and Design **46** (1999) 405.
2. The KSTAR Team, Nuclear Fusion **40** (2000) 575.
3. S. G. Lee, et al., Fusion Technology **33** (1996) 861.
4. S. M. Kwon, et al., Transactions of Fusion Technology **39** (2001) 10.

HELIUM PARTIAL PRESSURE MEASUREMENTS USING A PENNING GAUGE: A NEW APPROACH

P D Morgan and contributors to the EFDA-JET work programme*

1. INTRODUCTION

Future magnetic-confinement devices which produce significant power by the fusion of D-T nuclei will require an adequate removal rate of helium ash from the plasma core, in order to maintain core purity and the level of power production[1]. Hence, the role of the divertor in removing helium is of crucial importance. At JET, several divertor configurations have been studied to characterise their efficacy at this task, under various operating conditions[2].

More recently, a number of plasmas have been produced with high concentrations of He, up to 100%, to study their characteristics. Such plasmas are seen as relevant to ITER during its early phase of operation, where there are concerns about keeping activation of the vessel and its components to a minimum.

The partial pressures of the deuterium and helium in the sub-divertor volume are measured using a Penning gauge, which is a well-established technique. In this paper, a new approach to the measurement is described, which is particularly useful for plasmas with high and varying concentrations of helium.

2. STANDARD METHOD

The use of a Penning gauge as excitation source for the measurement by optical spectroscopy of the partial pressures of tokamak exhaust gases is well known[3]. A detailed description of the JET system is to be found elsewhere[4]. The Penning gauge is mounted about 2.8m below the divertor, by means of a vacuum tube of suitable length. Light from the Penning discharge is collected by a lens and relayed using an optical fibre to a place outside the biological shield. At this location, the light intensity is measured at two transition wavelengths, those of D_α and HeI at 656.1 and 587.6 nm, respectively, using two photomultipliers, each equipped with an interference filter of the appropriate wavelength. In the range of 10^{-5} to 5.10^{-3} mbar, approximately, the intensity in each channel is proportional to partial pressure.

* Euratom/UKAEA Fusion Association, Culham Science Centre, Abingdon, Oxfordshire OX14 3EA, UK.

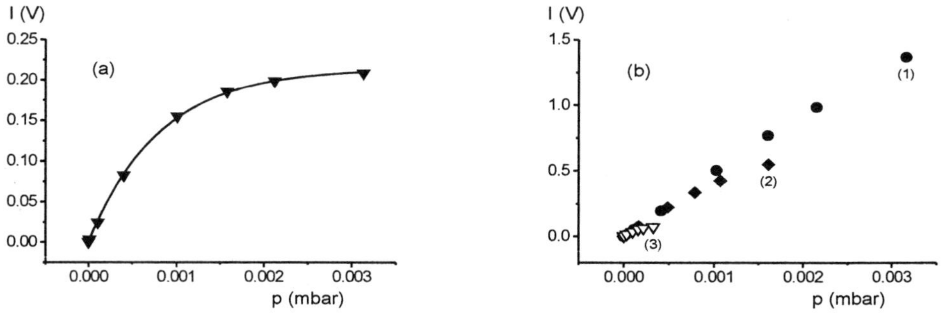

Figure 1. Variation of signal in each channel with partial pressure. **a)** D_α signal for pure deuterium. **b)** HeI signal at three concentrations of helium in deuterium: 1) 100%, 2) 50% and 3) 10%.

In figure 1(a) the triangular points show the variation of the D_α signal with pressure, for pure D_2. (The fitted curve is discussed in section 3.) The response of the HeI channel to helium is similar, but more linear with pressure. However, a complication is that the He signal is polluted by molecular emission from the D_2. This is taken into account through system calibration. Figure 1(b) shows the variation of the HeI signal with partial pressure for three different helium concentrations, having subtracted any cross talk.

The relationship between the D_2 and He partial pressures, p, and the intensities, I, in the two channels is given by:

$$p(D_2) = A_{11}.I(D_2), \quad (1)$$

$$p(He) = A_{21}.I(D_2) + A_{22}.I(He). \quad (2)$$

In the above equations, the coefficients A_{11} and A_{22} relate D_2 and He signal intensities to the corresponding partial pressures, whilst coefficient A_{21} represents the D_2 cross talk in the He channel. During calibration, without plasma, using pure D_2 the intensity in both channels is recorded versus pressure, giving the response to this gas and the cross talk. The process is repeated with a known concentration of He (typically ~ 10%), to yield the system response to that gas. From these data, by taking ratios of intensities versus pressure, as described in detail by Finken et al.[3], the coefficients A_{11} to A_{22} are derived. These are subsequently used to analyse signals obtained with plasma.

In plasmas where the concentration of He varies between 0 - 15% the method works well, although at low pressures the time traces of the calculated results may show large fluctuations. These arise from the method used to calculate the coefficients, which will be poorly defined when taking the ratio of two small signal intensities with significant noise content. However, in plasmas where the He concentration changes markedly during a discharge, and rises to many tens of percent, for all pressures the results become inaccurate whenever the concentration becomes significantly different from that used for calibration. This arises because the sensitivity of each channel varies with the gas concentration, due to changes in the characteristics of the Penning discharge with mixture. Figure 1(b) illustrates this dependence for helium; it can be seen that the response at a given pressure increases with increasing concentration.

3. NEW METHOD

At JET, a new approach to calibration and analysis overcomes both problems described in the previous section. System calibration is performed for a range of concentrations of He in D_2 from 0 to 100%. For a finite concentration of either gas, the variation of intensity, I, with partial pressure, p, is well fitted by:

$$I = A[1-\exp(-p/p_0)], \tag{3}$$

where the value of A and p_0 depend on the gas species and its concentration, C. A curve of the form described by Eq. (3) is fitted to the data points in figure 1(a).

Equation (3) tends smoothly to zero as p decreases, avoiding the possibility of spurious fluctuations at low pressure. In turn, A and p_0 vary monotonically with gas concentration, C. The triangular points in figures 2(a) and 2(b) show the variation of A with concentration, for D_2 and He respectively. The data are well fitted by simple exponential functions of C, as shown in the figures. For deuterium, the variation of both A and p_0 with C is well described by an equation analogous to Eq. (3). In the case of helium, the variation of the two parameters with C follows an offset exponential growth.

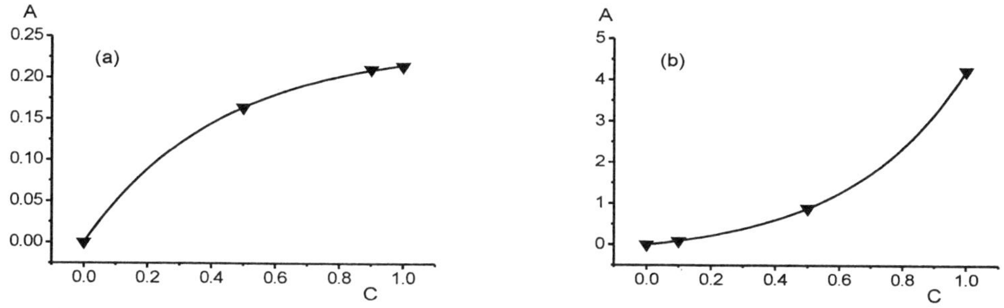

Figure 2. Variation of parameter A with concentration C of relevant gas. **a)** Deuterium. **b)** Helium.

To process signals obtained with plasma, the following analysis is performed on the data at each time point. Starting with values of C of unity, initial values of A and p_0 are calculated using the relationships established from calibration. The pressures, p, are then adjusted iteratively until the intensities, I, calculated from Eq. (3), match those observed. These pressures yield new values of C, enabling better values of A and p_0 to be derived. The process is repeated until convergence to the required precision is obtained.

4. RESULTS

The new method has been verified by its application to an analysis of calibration data. Figure 3(a) shows three traces of the temporal evolution of total pressure for a He concentration of 50%. Trace 1 is the absolute pressure measured using a capacitance manometer. The pressure derived using the new method is shown in trace 2, whilst trace 3 is that derived using the standard method, with coefficients from a calibration with pure helium. Clearly, trace 2 is in good agreement with the absolute pressure measurement, whilst trace 3 exhibits a discrepancy which rises to over 30%.

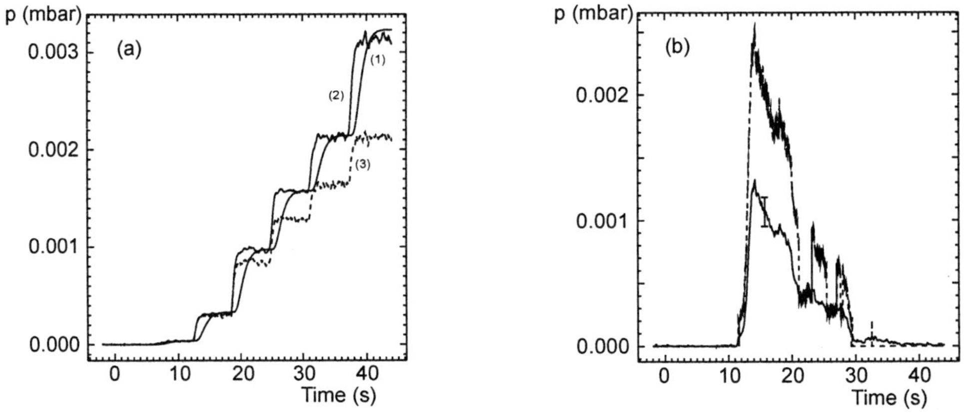

Figure 3. Comparison of results derived using the two analysis methods. **a)** Total pressure in a 50:50 D_2 / He mixture used for calibration: 1) capacitance manometer, 2) new method and 3) standard method. **b)** Helium pressure during a discharge: solid line - new method (error bar +/- 12%), dashed line - standard method.

Figure 3(b) illustrates the results obtained by applying the two methods to the analysis of data obtained in a discharge where the He concentration rose to ~95%. The standard method made use of coefficients from a calibration with a He concentration of 50%. The calculated helium pressure is then too high by almost a factor of two.

5. CONCLUSIONS

A new method has been described for using a Penning gauge to measure the partial pressure of helium in a tokamak exhaust. Its advantage is that it can reliably measure partial pressures for changing and high concentrations of He. A disadvantage is that a number of calibrations are required over a range of He concentrations from 0 – 100%.

6. REFERENCES

1. G. Janeschitz, ITER-JCT and Home Teams, Status of ITER, Plasma Phys. Control. Fusion **37**, A19-A35 (1995).
2. M. Groth, J.K. Ehrenberg, H. Guo, D.L. Hillis, L.D. Horton, G.F. Matthews, P.D. Morgan and M.G. von Hellermann, Helium Enrichment Studies in JET Discharges, Proc. 1998 Int. Congress on Plasma Phys. & 25th EPS Conf. on Control. Fusion and Plasma Phys. **22C**, 361-364, Prague (1998).
3. K.H. Finken, K.H. Dippel, W.Y. Baek and A. Hardtke, Measurement of Helium Gas in a Deuterium Environment, Rev. Sci. Instrum. **63**(1), 1-7 (1992).
4. D.L. Hillis, P.D. Morgan, J.K. Ehrenberg, M.F. Stamp, M. von Hellermann and V. Kumar, Tritium Concentration Measurements in the Joint European Torus by Optical Spectroscopy of a Penning Gauge, Rev. Sci. Instrum. **70**(1), 359-362 (1999).

Acknowledgements: This work was performed under the European Fusion Development Agreement and was partly funded by EURATOM and the UK Department of Trade and Industry.

INTERIOR MAGNETIC FIELD FLUCTUATION AND PROFILE MEASUREMENTS ON MST

D. L. Brower, W.X. Ding, S.D. Terry*

INTRODUCTION

Understanding plasma stability and confinement in a reversed-field pinch (RFP), or any toroidal confinement device, requires detailed information on the poloidal magnetic field and current density distribution. The stability of the plasma to tearing modes and other MHD events such as the sawtooth perturbation or dynamo are closely related to the current density profile. In RFP's, resistive MHD instabilities can produce magnetic islands on rational surfaces that are closely spaced in radius causing magnetic stochasticity resulting from island overlap. Energy transport in conventional RFP plasmas is believed to arise mainly from particles streaming along these stochastic magnetic field lines. For the Madison Symmetric Torus (MST), a major goal is to reduce magnetic fluctuations through current density profile control. Elucidating the physics of this interplay requires high-resolution measurement of both the current density profile time evolution and magnetic fluctuations.

Motivated by the need to address these critical measurement needs within the RFP community and the potential application to other toroidal confinement devices, a new high-resolution **Polar**imeter-**I**nterferometer **S**ystem (**POLARIS**) is being developed to measure the plasma poloidal magnetic field profile and magnetic fluctuations, as well as the electron density profile and density fluctuations. The diagnostic being developed will provide a non-perturbative measure of the spatial distribution of the equilibrium and fluctuating quantities. Time resolution of the simultaneous polarimeter-interferometer system will be approximately 1 μsec. This resolution will be sufficient to measure perturbed equilibrium profiles on the time scale of the sawtooth crash and high-frequency small-amplitude turbulent magnetic fluctuations. The fast time resolution and small-amplitude phase resolution have already been demonstrated by the interferometer.[1,2]

* D.L. Brower, W.X. Ding, S.D. Terry, Electrical Engineering Department, University of California, Los Angeles, U.S.A.

POLARIMETRY-INTERFEROMETRY SYSTEM

To date, the maximum bandwidth achieved by polarimetry-interferometry systems in plasma applications has been 5 kHz using amplitude techniques[3] and 1 kHz by phase measurement techniques.[4] This is insufficient to see the magnetic fluctuations of interest in MST. Observations made by external probes show fluctuations at frequencies up to 200 kHz which are dominated by resistive tearing modes in the range of 5-30 kHz with fractional amplitudes of 1%. In order to observe these frequencies, a polarimeter-interferometer system is currently being tested which will employ a 3-wave far-infrared laser. This approach, originally proposed by Dodel and Kunz[5], produces a phase measurement and involves the launching of 2-waves into the plasma that are circularly polarized and counter rotating. Previous phase measurement techniques involved mechanically spinning wave plates which generated a rotating, linearly-polarized beam. In such systems, the bandwidth is limited by the plate rotation speed and phase noise is generated by the mechanical rotator.

The technique proposed here utilizes two distinct but collinear far-infrared beams to probe the plasma. The probing beams are frequency offset and their polarizations are set to counter-rotating circular. Because of plasma birefringence, each beam will experience different value of refractive index upon propagation through the plasma. The difference in refractive index is related to the Faraday rotation angle, Ψ, by the relation

$$\Psi = \frac{\lambda^2 e^3}{8\pi^2 c^3 \varepsilon_0 m_e^2} \int n_e B_{//} dz = 2.62 \times 10^{-13} \lambda^2 \int n_e B_{//} dz \ ,$$

where $B_{//}$ is the component of the magnetic field parallel to the FIR beam, n_e is the electron density and λ is the FIR laser wavelength, all in MKS units. Since n_e is known from the simultaneous interferometer measurement, the poloidal magnetic field can be determined from inversion of the above equation. The toroidal current density is then obtained from Amperes law. Since no mechanical polarization modulation techniques are used, contamination associated with the modulation is removed.

The frequency offsets that allow us to distinguish and separately demodulate the independent carriers are generated by proper tuning of the three lasers. Signals carried by the three primary intermediate frequencies are multiplexed onto a single detector. The individual phases of each of the mixing products can then be determined and compared with those from a reference mixer. This analysis involves the use of a digital phase comparator techniques.[6] Measurement time resolution is limited by the frequency separation between the mixing products and can easily be set to several hundred kilohertz bandwidth, sufficient to see the fluctuations of interest in MST.

First tests of the 11-chord fast **POLARIS** diagnostic have recently been conducted on MST. For these tests, two far-infrared probe beams are frequency offset at 750 kHz (1.4 μs minimum time response) and set to counter-rotating circular polarization. The mixer for each channel then detects the mutual mixing product of the two probe beams which is directly proportional to twice the Faraday rotation angle as described in the equation above. Sample MST data, bandpass filtered at 20 kHz, are shown in Fig. 1, where clear changes in the measured Faraday rotation angle are observed during each sawtooth cycle. All chords show strong correlation with the sawtooth effect, even during the setup phase of the discharge, i.e., 0-10 msec. At 20 kHz bandwith, the fast polarimeter rms noise level is approximately 0.04^0.

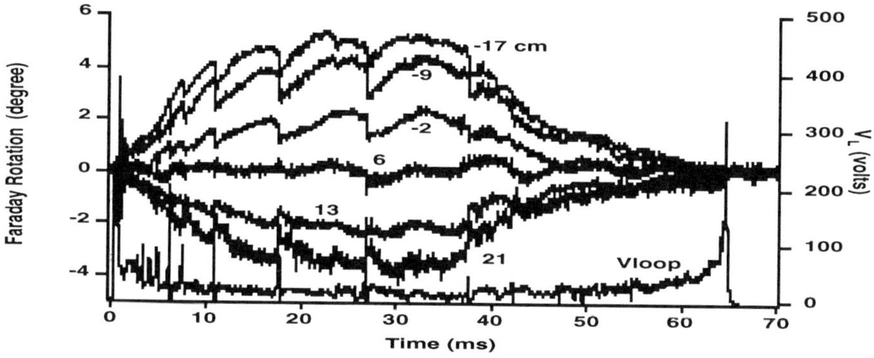

Figure 1. Test of fast-polarimetry measurement on MST.

In Fig. 2(a), fast polarimetry data are shown for a 300 kA MST plasma using an expanded time axis to isolate a sawtooth crash. In these data, a clear coherent modulation is observed on all chords prior to the sawtooth crash at 25 msec. These fluctuations correspond to the dominant m=1 tearing mode in MST and have frequency near 12 kHz. This demonstrates that the fast **POLARIS** can resolve both the sawtooth crash phase of the discharge as well as the dominant core fluctuations. If one computes the coherence between a polarimeter chord and an external magnetic field coil, finite coherence is observed out to 100 kHz as shown in Fig. 2(b). This indicates that not only does the fast polarimeter have sufficient time response, it also has the phase resolution necessary to measure high-frequency small-amplitude magnetic fluctuations.

The polarimeter provides a time-resolved, non-perturbing, chord-averaged measurement of the core and edge magnetic fluctuations. The fluctuating component of the Faraday rotation signal can be written as $\tilde{\Psi} = c_p \int B_0 \tilde{n}\, dl + c_p \int \tilde{B}\, n_0\, dl$, and it is important to note that the fluctuations observed may have contributions both from the fluctuating density and magnetic field. However, for the chord (x=6 cm) passing through the magnetic axis, $B_0 = 0$, thereby making the fluctuating quantity detected a direct measure of the line-integrated fluctuating magnetic field. This component corresponds to $\tilde{B}_r$ and has line-averaged rms amplitude of 33 Gauss (approximately 1%). The polarimeter rms noise level is 10 G. Since the density profile is peaked on axis, the polarimetry measurements are weighted to the plasma core.

CONCLUSION

The fast **POLARIS** diagnostic provides a measure of the equilibrium and fluctuating magnetic field with ≈ 1 μs time response. This permits observation of both fast changes to the equilibrium field and turbulent magnetic fluctuations. The system noise level of 10 G would allow measurement of 5×10^{-4} magnetic fluctuation levels in a tokamak with B = 2 T. Efforts to further increase the system sensitivity are ongoing along with techniques to extract the fluctuating field information from all 11 chords.

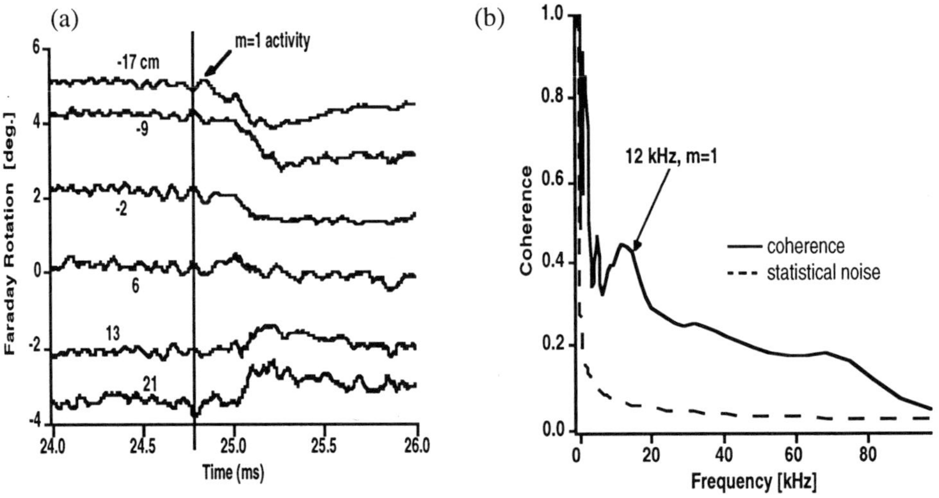

Figure 2. (a) Fast **POLARIS** signals showing m=1 fluctuating quantity. Sawtooth crash occurs at 25 ms. (b) Fast **POLARIS** magnetic fluctuation frequency spectra.

ACKNOWLEDGEMENTS

This work is supported by the U.S. DoE under grant DE-FG03-01ER54615, Task III.

REFERENCES

1. Y. Jiang, D.L. Brower, and N.E. Lanier, *Interferometric measurement of high-frequency density fluctuations in Madison symmetric torus*, Rev. Sci. Instrum. **70**,703-706(1999).
2. N.E. Lanier , D. Craig, T.M. Biewer, B.E. Chapman, D.J. Den Hartog, C.B. Forest, S.C. Prager, D.L. Brower, Y. Jiang, *Control of density fluctuations and electron transport in the reversed-field pinch*, Phys. Rev. Lett. **85**,2120(2000).
3. H. Soltwisch, H.R. Koslowski, Observation of magnetic field perturbations during sawtooth activity in tokamak plasmas, Plasma Phys. Control. Fusion **39**,A341-A349.
4. D.L. Brower, L. Zeng, Y. Jiang, *Multichannel far-infrared polarimeter system on TEXT-Upgrade*, Rev. Sci. Instrum. **68**, 419-421(1997).
5. G. Dodle, W. Kunz, *A far-infrared 'polari-interferometer'for simultaneous electron density and magnetic field measurements in plasmas*, Infrared Phys. **18**, 773-776(1978).
6. Y. Jiang, D.L. Brower, L. Zeng, *Applicatioin of a digital phase comparator technique to interferometer data*, Rev. Sci. Instrum **68**, 902-905(1997).

VACUUM DIODE HIGH VOLTAGE MEASUREMENTS BY MEANS OF PHOTONUCLEAR REACTIONS

S. Dan'ko, G. Dolgachev, Yu. Kalinin,* D. Maslennikov

INTRODUCTION

Plasma opening switches (POS) are well known as very efficient and non-expensive power sharpeners for pulse power technique. The corresponding pulse power generator[1] is an *LC*-contour closed through the POS. The latter is commonly a fragment of a vacuum coaxial line with some part of electrodes connected by means of plasma bridge. One of this line ends is connected with the Marx generator (which is a capacity energy storage) while its another end is joined up to a loading device.

After the generator is switched up, time-increasing current in the line flows through the plasma linker. At this stage POS impedance is a rather small, and generator energy $CU_0^2/2$ is transforming into the magnetic energy $LI^2/2$, with L - a total inductance of the line, generator, and conductors. At proper time POS impedance becomes to be drastically increased with accompanying decreasing of the current. This results in the jump of inductive voltage in the POS, and the magnetic energy stored is evolved as the flows of accelerated particles inside of the POS and in the paralleling load also.

Recently a series of projects was started aimed to the elaboration of pulse power devices for ICF experiments with POS as a final power sharpener, including the "Baikal" machine[2] with expected output voltage up to 10 MV. It was extremely important from the standpoint of this problem solution to perform the most reliable measurement of the POS voltage. However, direct measurement on the base of shunts and voltage dividers in the range of several MV is very difficult and not so reliable. In addition, the POS geometry prevents from proper operations. Normally, they obtain the voltage from calculations based on the current derivative and the values of inductance of the circuit elements. This work presents an independent method of determination of the voltage maximum values for the energy range of 2-4 MV in the disconnected POS.

* Russian Research Center "Kurchatov Institute", NFI, 1, Kurchatov Sq., 123182, Moscow, Russia

Advanced Diagnostics for Magnetic and Inertial Fusion, Edited by Stott *et al.*
Kluwer Academic/Plenum Publishers, New York, 2002

POS VOLTAGE MEASUREMENTS USING PHOTONUCLEAR REACTIONS

The key point of the method proposed is determination of the high-energy boundary of the electron Bremsstrahlung spectrum from the vacuum diode. The method of filters commonly used for the energy range <1 MeV, cannot be used here owing to slight dependence of the γ-radiation absorption coefficient on the energy and its non-monotonic character in the range under study (2÷4 MeV). To overcome this obstacle, photonuclear reactions were applied for our goal achievement.

Two following reaction were used: $\boldsymbol{Be^9(\gamma,n)Be^8}$ and $\boldsymbol{D^2(\gamma,n)p}$ with the thresholds of 1.65 and 2.25 MeV, respectively. It is important that the cross-sections of these reactions are growing very sharp in this range with the photon energy.[3, 4] These features of the reactions selected have provided very high accuracy of the voltage measurements.

The registration of neutrons obtained from each of these reactions was performed using two separated neutron detectors with identical construction, geometry, and sensitivity to the radiation. The operation of neutron detectors was based on the activation method with delayed registration: i) at the beginning, the fast neutrons were slowed down by their passing through the polyethylene cube; ii) further, thermal neutron interaction with six ***In***-plates arranged on the cube surfaces results in the production of active nuclei due to the reaction $\boldsymbol{In^{115}(n,\beta)Sn^{116}}$; iii) finally, the β-particles radiated were registered by the six Geiger counters disposed on the back sides of the ***In***-plates.

The yield of hard γ-radiation from the diode was calculated by means of Monte Carlo method for the real experimental geometry. The neutron yield resulting from the interaction of the radiation with substances ($\boldsymbol{Be, D_2O}$) was estimated as the following convolution: $N_n = \sigma(\gamma n)\cdot N_\gamma(h\nu)\cdot N_{Be,D}d(h\nu)$.

The measurements of (γ, n)-reactions were carried out just after the experiment in a set of sequential time intervals, each of them was equal to the half-decay time of In^{116}-nucleus: $T_{1/2}$=13,4 second. The readings of the neutron counter without the "background" value must decrease twice from each time interval to the next one in the real reaction measurement set. The several measurements of the "background" were performed right before and after the measurements in the experiment. The counters were switched on manually during the first second just after the accelerator run.

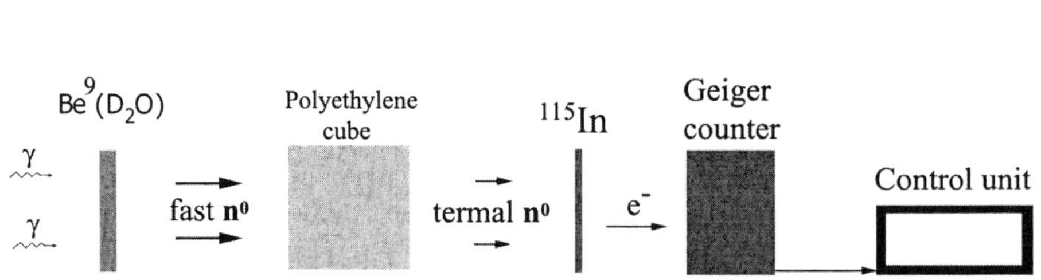

Figure 1. The (γ,n)-reaction measurement technique.

The absolute calibration of the sensitivity of these neutron detectors was not performed, instead, the ratio of the indicated values from two registration channels was used. The number of the available readings from the each registration channel in our condition was calculated for the peak voltage range U_d 2÷5 MV.

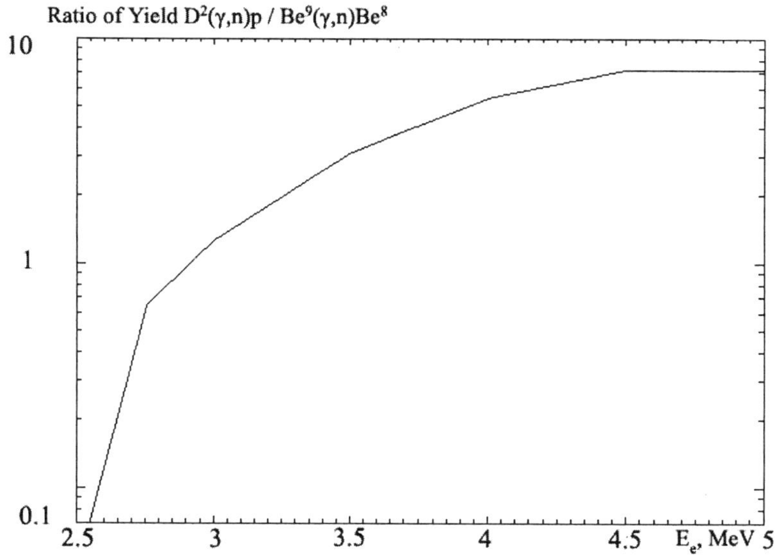

Figure 2. The ratio of neutron yields in the reactions $D^2(\gamma,n)p$ and $Be^9(\gamma,n)Be^8$ as a function of the γ-quantum energy.

Our experiments were carried out on the modified RS-20 facility. Its scheme and sketch of design are presented in Fig. 3.

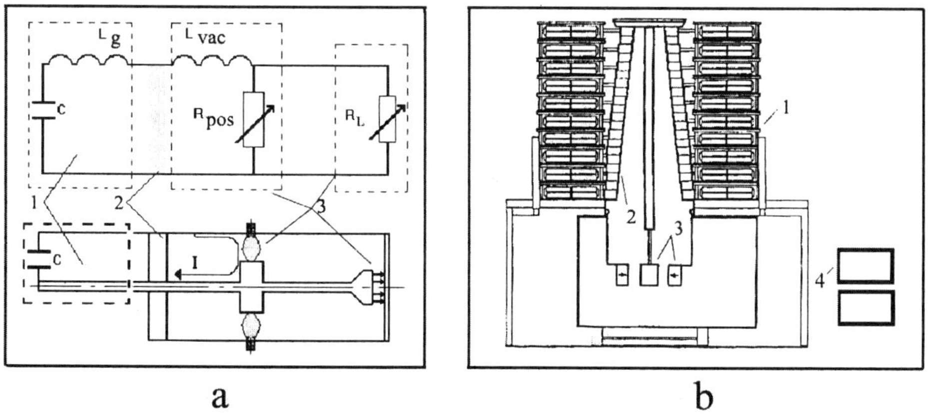

Figure 3. Electrical circuit of the RS-20 facility: a). Electric scheme b). Sketch of the real design. 1 - pulse generator, 2 - insulator, 3 - plasma opening switches - high voltage diode, 4 - detectors for (γ,n)-reaction.

The main electric parameters of our experiments were as follows: 20 Marx cascades, output Marx generator voltage ~0.8 MV, drive current amplitude ~120 kA, and time duration 2 μs / 300 ns. These experiments were performed using POS with the extrinsic longitudinal magnetic field, its strength was equal to 16 kGs. To increase the γ-radiation yield, the base and side surfaces of the cylindrical anode was covered by the ***Ta***-plates.

The results of these measurements including reading from two channels, their ratio, and electron energy calculated are presented in the Table 1.

Table 1. Examples of electron energy determination using photonuclear reactions

Run	Readings		***D/Be*** ratio	E_e, MeV
	from ***Be***-channel	from ***D***-channel		
1	34 ± 11	53 ± 12	1.6	3.05 ± 0.15
2	32 ± 11	33 ± 11	1.0	2.85 ± 0.1

CONCLUSION

We have shown that photonuclear reactions on ***Be*** and ***D*** may serve as a reliable tool for measurement of the electron energy and pulsed voltage in the range of 2.2 ÷ 4 MV. This range may be somewhat broadened by means of the absolute calibration technique.

ACKNOWLEDGEMENTS

This research has been supported by the European INTAS Research Grant 97-0021, and American CRDF grant RP1-2113.

REFERENCES

1. G. I. Dolgachev, L. P. Zakatov, and A. G. Ushakov, Miscrosecond plasma opening switch, *Sov. Journal of Plasma Physics*, **17**(9), 1171-1183 (1991).
2. V. V. Alexandrov, E. A. Azizov, A. V. Branitsky, et al., Multi-wire array implosion physics study. X-ray facility "BAIKAL" project based on 900 MJ inductive store, *Abstracts of the 13th Int. Conf. on High-Power Particle Beams*, BEAMS-2000, Nagaoka, Japan, June 25-30, 2000, p. 214.
3. R. J. Hughes, et al, *Proc. Int. Conf. Photonuclear Reactions and Applications*, Asilomar, 1973, p. 151.
4. G. Breit, *Proc. Int. Conf. Photonuclear Reactions and Applications*, Asilomar, 1973, p. 323.

INDUCTIVE VOLTAGE PROBE IN LINER IMPLOSION EXPERIMENTS ON INSTALLATION ANGARA-5-1

V.V.Alexandrov, E.V.Grabovsky, G.M.Oleinik*

INTRODUCTION

Last years in ICF is widely investigated current driven implosion of liners as powerful x-ray source. Thus one of basic question is the question on effective transfer of power and energy of pulsed power generator to a load. Therefore it is very important to measure current and voltage in the vicinity of load and then to calculate on base of these data the impedance of a load, and power and energy transferred to a load.

Measurements of parameters at vacuum output of pulsed power generator, where the load in most cases is located, are particularly complex. Powerful energy input from generator to the load leads to appearance of intensive x-ray radiation. As result ionization of residual gases takes place and all surfaces of solid bodies become ionized under influence of radiation. Another ionization source - electron fluxes from magnetically insulated transmitting lines which are used in such generators. Plasma appearance at resistive divider surface and in high voltage gap of capacitive divider may provide essential measuring signal distortion.

INDUCTIVE VOLTAGE PROBE DESING

Indicated problems are partially eliminated when inductive divider is used. The design of inductive voltage probe is very simple. In this work the design of inductive voltage probe for voltage measurements at load unit of Angara-5-1 installation and examples of results obtained are presented.

The inductive voltage probe design is presented in Fig.1. The high-voltage part of divider is formed by a metallic rod, which is connected with cathode and anode electrodes. This rod takes on not so large part of total current of the generator, since inductance of a load much less then the inductance of the rod. The low-voltage part of divider is formed by a loop, situated in metallic cavity closely to this rod. To be close is under-

* V.V.Alexandrov, E.V.Grabovsky, G.M.Oleinik. GSC RF TRINITI, 142190, Troitsk, Moscow reg., Russia.

Advanced Diagnostics for Magnetic and Inertial Fusion, Edited by Stott *et al.*
Kluwer Academic/Plenum Publishers, New York, 2002

stood in sense that signal from this loop should depend only of current over the rod. The cavity plays role of a screen, protecting the signal from other currents flowing in the chamber.

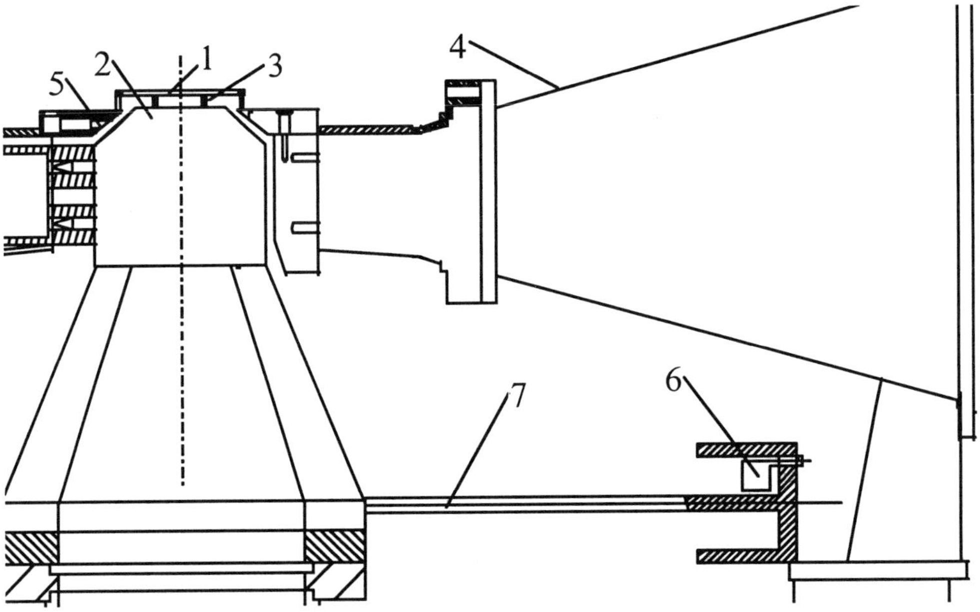

Figure 1. Schematic cross-section of output unit of Angara-5-1:
1 - anode, 2 - cathode, 3 - load (incompressible cylinder or wire array), 4 - vacuum transmission line, 5 - layout of a B-dot current probe, 6 - loop of a voltage probe, 7 - metallic rod of voltage a probe.

The rate of current changing through the rod is proportional to voltage between anode and cathode. As the signal from loop is proportional to the rate of current changing on the rode, the signal from the loop is proportional to voltage between anode and cathode also. Proportionality coefficient can be easily determined at calibration.

Frequency response of inductive divider is determined by parameters of high-voltage part of divider which can be considered as inductance when wave processes in circuit will be finished and there will be established a quasistationary distribution of currents. The temporal resolution for given probe is about 2ns.

EXPERIMENTAL RESULTS

Testing of inductive divider was made in experiments, when a load was a heavy incompressible metallic cylinder, which does not change its geometry at a shot. In this case the rate of current changing in the load and measured voltage should be proportional to each other. In Fig.2 are given signals from inductive probe and a B-dot probe measuring the rate of current changing over incompressible cylinder. The current probe was a loop located at radius of 55mm. It is seen very well that signals are proportional to each other actually.

The voltage behavior was quite another (see Fig.3) when load was the wire array (80 tungsten wires, wire diameter 6μm, initial array diameter 8mm). At first stage of the process these signals are proportional to each other too, but later they become different. The current is diminished, that is visible on change of the sign of a signal «B-dot probe», and the voltage (signal « L-divider ») is increased. Such behavior is connected with liner compression and essential increase of an impedance of a load. In spite of that the voltage measurements were made rather far from the liner, it was received that the voltage probe was sensitive rather well to the compression dynamics and voltage increase connected with the liner implosion.

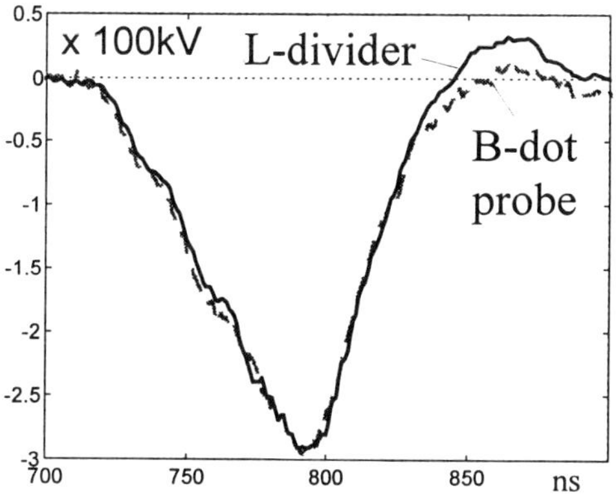

Figure 2. Profiles of rate of current changing over incompressible cylinder (B-dot probe) and voltage (L-divider) measured by inductive probe.

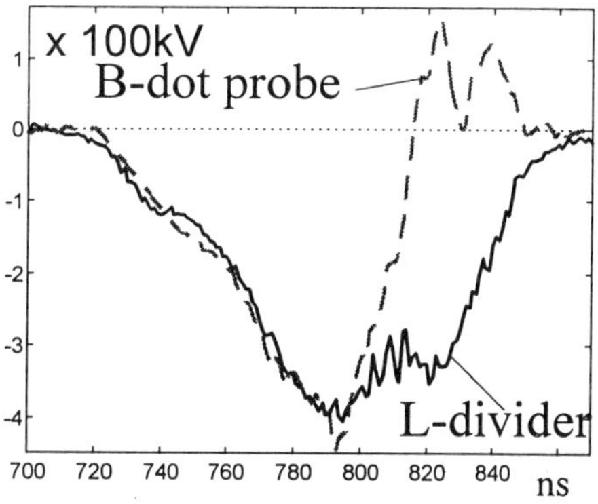

Figure 3. Profiles of rate of current changing over wire array (B-dot probe) and voltage (L-divider) measured by inductive probe. Wire array had 80 tungsten wires (wire diameter 6μm, initial array diameter 8mm).

CONCLUSION

Inductive voltage probe provides non invasive measurement of voltage drop in vicinity of load. Inductive voltage probe has nanosecond temporal resolution. Inductive voltage probe is weakly influenced by surrounding plasma fluxes.

MAGNETIC FIELD MEASUREMENTS INSIDE AND OUTSIDE OF IMPLODING ARRAYS ON ANGARA-5-1

I.Frolov, E.Grabovsky, K.Mitrofanov, G.Oleinik, V.Smirnov, G.Zukakischvili*

INTRODUCTION

The interest to wire arrays as the sources of powerful soft x-ray radiation (SXR) has increased nowadays. The efficiency of transformation of electrical energy into SXR during array implosion is appreciably depended on plasma parameters. It is very important to know the current and magnetic field distributions inside imploding array.

This report is devoted to the investigation of magnetic field distribution in imploding array by magnetic probes on installation Angara-5-1. Evaporation of a probe under influence of SXR and flow of particles, screening of magnetic probes by plasma, plasma flows around the probe limit the using of magnetic probes in the powerful discharges. The elimination of above factors are analyzed in given paper. It is experimentally shown, that at a level of currents up to 4MA and power of SXR up to 3TW, the using of magnetic probes can gives the reliable results.

ARRANGEMENT OF EXPERIMENT AND PROBES CONSTRUCTIONS

The scheme of experiment is presented in Fig.1, the constructions of magnetic probes - in Fig.2 and 3. Anode - cathode gap was 10mm, array radius - 10mm. Array had 40-120 tungsten wire. Wire diameter - 5-10μm.

The azimuthal magnetic field (H_φ) was measured at distance of 2mm from anode. Probes were located both inside the array at radius 5mm and outside at radius 30mm.

Axial magnetic field (H_Z) was measured by probe on the axis of array. These probes were used in experiment of axial magnetic flux compression by imploding array. Probes were situated on the axis both inside the array at distance of 2 mm from anode and outside of it behind a thin metallic foil of great specific resistance, which thickness (15μm) was less than a skin layer of cold metal.

---

* I.Frolov, E.Grabovsky, K.Mitrofanov, G.Oleinik, V.Smirnov, G.Zukakischvili, GSC RF TRINITI, 142190, Troitsk, Moscow reg., Russia.

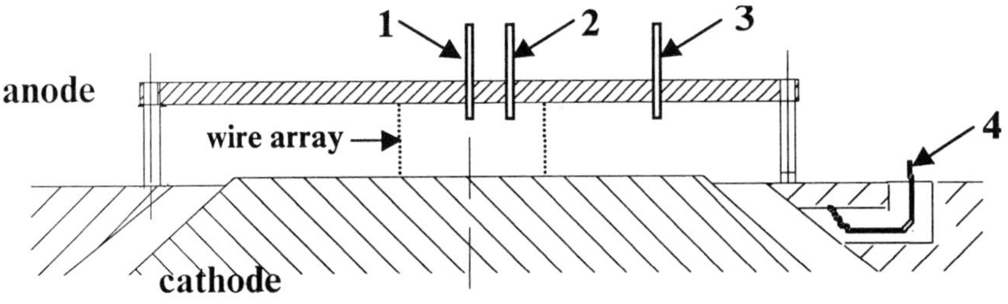

Figure 1. Magnetic probes dispositions: 1- for measurement of axial magnetic field, 2-for measurement of azimuthal field at radius of 5mm, 3- for measurement of azimuthal field at radius 30mm, 4- a system of 8 loops located at radius 55mm (one of 8 magnetic loops is shown in this figure).

Each probe consisted of two miniature coils winded in different directions. Probe diameters for H_z field measurements were 880μm (4-10 turns) and 300μm (4 turns). Probe coil diameter for $H\varphi$ field measurements was 260μm (1 turn). The coils were situated in ceramic tubes. The ceramic tubes were situated in copper tubes of thickness 0.14 mm (more than skin depth). Tubes and coils were filled by epoxy glue.

The ground wires of coils were soldered to the cylindrical screen, and two signal wires passed together through this screen tube of length 250mm and further were connected with two coaxial cables: each wire was soldered to central conductor of cable. Screen was connected with outer conductors of cables. The value of additional signal due to mutual inductance of this system is less than ~1% of main signal.

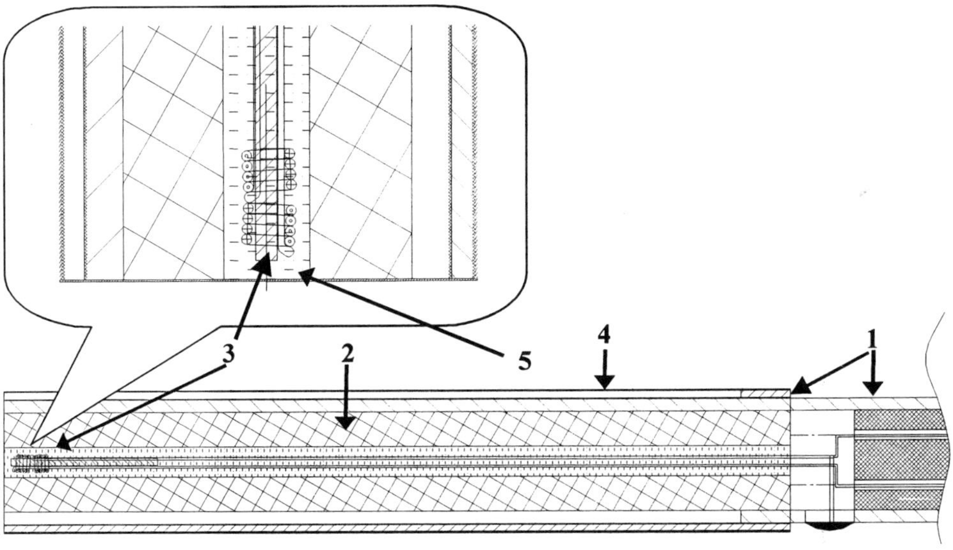

Figure 2. A design of an axial magnetic probe. 1 - copper tubes (thickness of screens is 0.29 mm and 0.14 mm) with cut for penetration of a axial magnetic flux (width of cut is 0.3 mm); 2 - ceramic tube (diameter of tube is 0.7 mm); 3 - two miniature coils (diameter of coils is about 300μm) winded in different directions (clockwise and counter clockwise). 4 - NbTi foil screen (thickness of foil is 15 μm). 5 - epoxy glue.

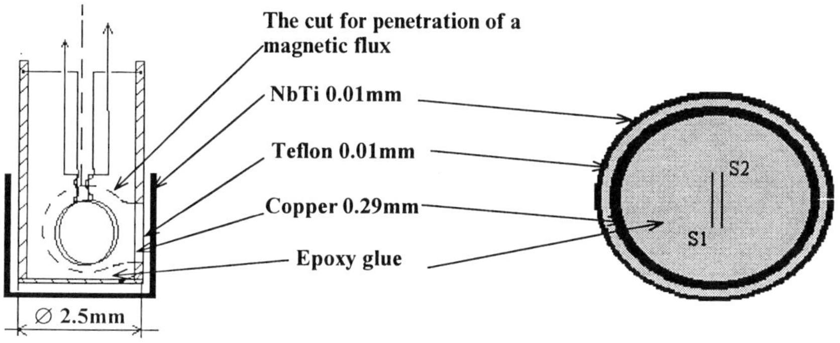

Figure 3. A design of azimuthal magnetic probes.S1, S2-two coils.

A probe immersed inside studied plasma will give at the same time with inductive also capacitive signal. Capacitive coupling between plasma and coil can be removed by using ground electrostatic screen. In this case the value of a capacitive signal is very small (~ 0.1% of main signal). The screens insured mechanical solidity of probes and screening from electron beams generated in anode–cathode gap.

For H_Z measurement the tubes and screens had narrow slots (~0.3 mm) in axial directions. For $H\varphi$ measurement the cut-out from tube were produced. That permitted reliable separation of axial and azimuthal magnetic field components. To avoid the filling of slits by plasma, the foil screens were used with thickness less than that of skin (~1.5 ns at 15μm foil thickness NbTi).

Double screen cables were used for screening of probe signal paths. Noise level was about 0.25V, signal level -1kV, voltage level in the facility -1MV.

The factors limiting usage of magnetic probes

The main obstacles which hinder measurements by magnetic probes are: evaporation of screen of a probe under influence of SXR and flow of particles, probes destruction, penetration of electron beams into probes, probes screening by surrounding plasma, and perturbation of the surrounding plasma by probes.

The estimation of vaporization of a probe screen under influence of X-radiation of plasma was made. The heat conduction equation for NbTi screen was decided for definition of depth of a probe screen sublimated by influence of SXR. The obtained depth of the sublimated layer was 0.8μm. The generated plasma has an electron temperature ~1-3 eV, Z~1, and resistivity about 20 Om·cm. Such plasma can not screen the magnetic probe during discharge time (~100 ns).

The condition of correct measurements is small dimension of magnetic probe in comparison with skin layer thickness of hot plasma. According estimation the temperature in a plasma shell at the compression stage is 20-30 eV and the skin layer thickness ~1mm. So the characteristic dimension of a probe positioned in plasma was chosen about 300μm.

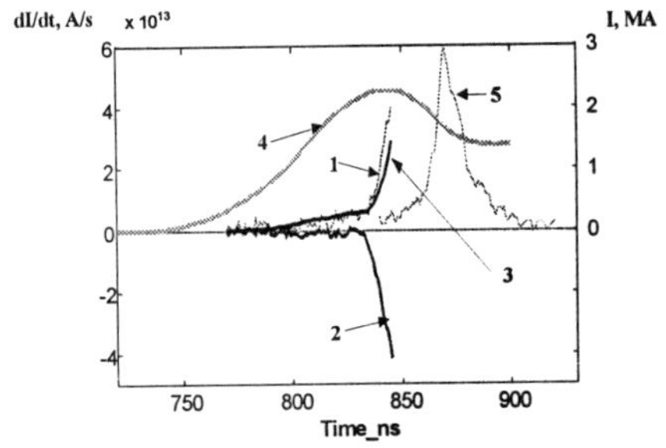

Figure 4. Results of azimuthal magnetic field measurements. 1,2- derivatives of current at radius of 5mm (two polarity), 3- current at radius of 5mm, 4- total current at radius of 30mm and 55mm, 5- SXR(a.u.)

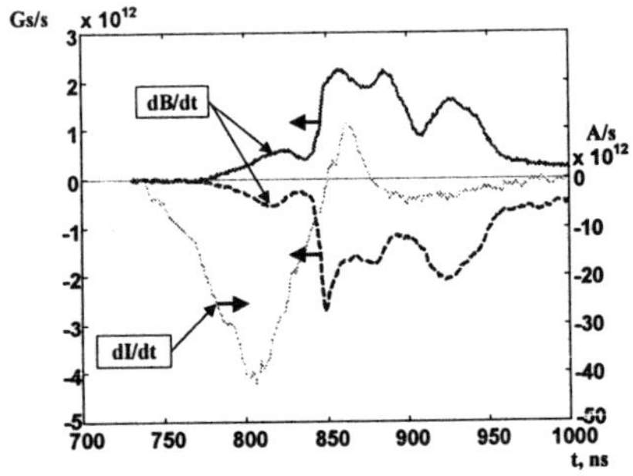

Figure 5. Results of axial magnetic field measurements. dB/dt- derivatives of axial magnetic field on the axis (two polarity), dI/dt- derivative of total current at radius of 55 mm.

EXPERIMENTAL RESULTS

As was mentioned above, each magnetic probe consisted of two coils wounded in opposite directions. Signals from these coils should be similar and with different polarity if they are related to magnetic field changing. The results of measurements are presented in Fig. 4 and 5. Recorded signals are reliable up to the moment when they correspond to each other in form and have opposite sign.

The experiments on measurement of azimuthal magnetic fields inside and outside of wire arrays were carried out (Fig. 4). The magnetic probe, located outside the array, was used for total current measurement at radius of 30mm in the vicinity of the array. It was found that current at radius 30mm is equal to current at radius 55mm up to moment of final pinching. The penetration of an azimuthal magnetic flux inside the array during implosion is fixed. The internal border of a current shell has the high speed ($\sim 10^7$cm/s) and reaches an axis of the array at early stages of the discharge. The plasma precursor carries up to 3% of a total current. The effective thickness of shell of current is determined and is greater than 2mm.

The measurements of axial magnetic fields at the axis were carried out (Fig. 5). In these experiments a compression of a weak axial magnetic flux by a current shell was demonstrated. The dynamics of movement of internal border of current shell was obtained with the help of magnetic probe. It was found that the creation of continuous shell and the beginning of the magnetic flux compression take place in ~30ns from the current start at the distance of 0.3cm from the initial array radius.

CONCLUSION

It is experimentally shown, that at a level of currents up to 4MA and power of SXR up to 3TW, the using of magnetic probes can give the reliable results.

This work was supported by RFBR grands # 99-02-16471 and # 99-02-17952.

MEASUREMENTS OF NEUTRAL DENSITY IN THE ALCATOR C-MOD TOKAMAK

R.L. Boivin, J.W. Hughes, B. LaBombard, D. Mossessian, and J. L. Terry*

1. INTRODUCTION

In magnetically confined plasmas, the study of edge dynamics has many important implications which include overall confinement and performance, and, in particular, calls for a quantitative understanding of the particle inventory and its sources. In fact, when studying the particle, momentum and energy balance of the edge plasma, one must include the contribution of neutral particles. However, direct measurements of the neutral population, i.e. neutral density and associated ionization rate, are particularly arduous in that region. Based on a simple measurement of the excitation of neutral deuterium (and hydrogen), we installed a series of diagnostics in the Alcator C-Mod tokamak, from which we can obtain those two quantities with high radial resolution. In addition we installed a series of modified Penning-type gauges to measure the neutral pressure in close proximity of the discharge itself.

The Alcator C-Mod tokamak is based on a compact design, with high magnetic field, high density and high power density (ohmic and RF heating). Typical densities of ~2×10^{20} m^3 and ~5×10^{19} m^3 are found at the center and at the edge of the plasma, respectively. Measured

*Plasma Science and Fusion Center, MIT, Cambridge, MA, USA

Advanced Diagnostics for Magnetic and Inertial Fusion, Edited by Stott *et al.*
Kluwer Academic/Plenum Publishers, New York, 2002

neutral pressures, as obtained by standard ion gauges located away from the plasma at the outer wall, range from 0.01 to 1mTorr. Since the various relevant edge scale-lengths typically are found to be smaller than 1cm, it is important that we measure with sufficient resolution the ionization rate and the neutral density in that region. By integrating the ionization rate over the volume at the edge, we obtain directly the neutral particle influx in the scrape-off layer and into the main discharge, and we can obtain directly the local effective diffusion of the main ions.

2. EXPERIMENTAL SETUP

2.1 Lyman alpha arrays

The diagnostic consists simply of a photodiode array, filter and aperture, which image the plasma edge. The outer edge detector looks tangentially (toroidally), 12.5 cm below the vacuum vessel midplane, with a radial span of 4 cm at the tangency point, centered near a typical magnetic separatrix location. The exact location of the detector and its corresponding view have been measured inside the vacuum vessel, and the chord locations are believed to be known within 2 mm. The inner view detector also looks tangentially, but 10 cm above the midplane, with a radial span of 6 cm. In this case, the chord locations have been verified by using a laser mounted on top of the tube, and again, believed to be known within 2 mm. We use a photodiode array which consists of a compact set of 20 channels, each measuring 0.75 mm x 4 mm. The spectral bandwidth is restricted to a very narrow band using a filter which transmission peaks near 1216Å, with a Full-Width-Half-Maximum (FWHM) of 70Å, and a maximum transmission of 6-8%. That spectral region has been examined with a VUV spectrometer and it has been found that the measured signal is largely dominated by the Lyman-alpha line, except when a relatively large quantity of nitrogen (in this case through Nv emission) is puffed during divertor performance studies.

A slit measuring 1 x 3 mm is used as an aperture, approximately 20 cm away from the detector, which, together with the individual channel size, define the radial resolution to be 2 and 3mm for the outer and inner systems respectively. The detector, filter and aperture are held in a stainless steel tube, 4 mm thick, which supports the diagnostic against disruption and prevents X-ray detection. In addition, the outer edge view detector has a mirror angled at 30°, located between the filter and the aperture, since the penetration flange is nearly perpendicular to the required view. This mirror has a 80% reflection coefficient at that wavelength and viewing angle. At the deuterium Lyman alpha wavelength, the photodiode has a sensitivity of 0.11 Amps/Watt. The electronics are composed of simple current to voltage amplifiers, and both systems have gains of 5×10^7 Volts/Amps. Typical signal levels of ~0.1 V and ~1.0 V have been observed for the outer and inner view respectively, with corresponding time responses of 10 msec and 1 msec.

2.2 Pressure gauges

In order to measure the neutral pressure near the plasma boundary, we installed inside the vacuum vessel a series of cold cathode (Penning) gauges. These gauges use a axial magnetic field inside a cylinder and a collecting cathode at one end, biased negatively at nominally 2500 V. Normally the magnetic field is generated by a permanent magnet attached to the cylinder, but in this application, we replaced it by the field generated by the tokamak coils during the discharge. Since the local magnetic field can be calculated using standard mapping techniques, we aligned the gauge along the local magnetic field, in the toroidal, radial and poloidal directions. They are installed at 4 locations, one very near the divertor, and 3 in the main chamber, 2 at the same toroidal location (approximately 25 cm above and below the midplane), and one displaced toroidally at the same poloidal location below the midplane.

3. EXPERIMENTAL RESULTS

The outer view detector has been in service routinely for more than one experimental campaign (beginning of 1999), and typical profiles are shown in Fig.1. Shown in the top plot is the local emissivity, obtained from the brightness profile using a standard Abel inversion technique. We notice immediately that the Lyman alpha emission is relatively narrow (1-2 cm), peaked near the separatrix. The magnitude of the emissivity has been observed to vary with density, and confinement regimes,[1] and no emission has been observed in a pure helium plasma.

With the local Lyman alpha emissivity, we can then obtain the local ionization rate and neutral density, using well-known excitation rates[2]. Electron density and temperature profiles are obtained via the high-resolution edge Thomson scattering system[3], which consists of 16 channels, with a nominal 1.3mm radial resolution, and are complemented with Langmuir probe measurements farther out in the scrape-off-layer. The ionization rate profile is very similar, as one would expect, to the Lyman alpha emissivity profile, while the neutral density exhibits the dramatic radial decay found inside or near the separatrix. The second array, which looks at the inner plasma edge was installed in winter 2000. In this case we must account for the contribution from the outer edge, which we measure, and which is expected to be small, due to the very small path length through that layer (nearly perpendicular compared to the tangent contribution at the inner edge). In addition, we may have to account for self-absorption; the plasma may not be completely optically thin[4] in this tangential view, especially at higher density. Consequently, the standard Abel inversion may not be appropriate for all cases of the inner edge view. First results indicated a much larger signal than the one obtained at the outer edge, by a factor of the order of 5-10.

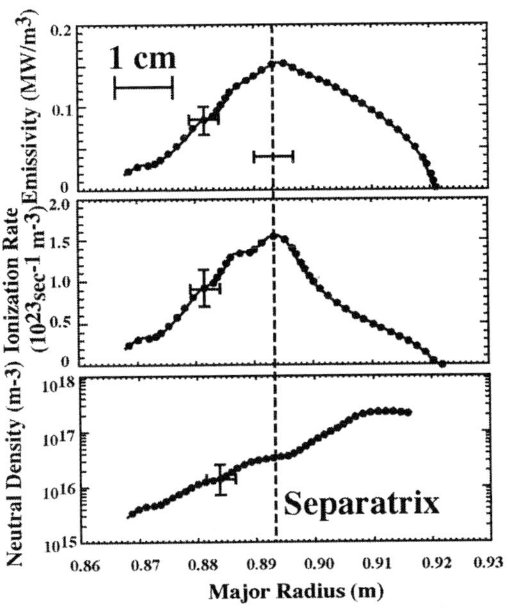

Figure 1. Lyman α emissivity, ionization and neutral density profiles measured at the outer edge of the plasma.

The first experimental operation of the pressure gauges indicate a strong signal (current), and the discharges inside the gauges were found to ignite easily when breakdown was initiated for the main discharge and the cathodes are operated at typically 2000V. The gauges were cross-calibrated with the other "external" gauges following a short plasma but with fields still remaining. Only small changes were found between the toroidally displaced gauges, but larger variations were found between the poloidally displaced pair, except with a double null discharge. The divertor gauge was quickly saturated and needed to be operated a lower voltage (typically 600V). A smaller gauge would reduce the current for a given voltage, and simplify the calibration.

4. REFERENCES

1. R.L. Boivin, J.A. Goetz, A.E. Hubbard, J.W. Hughes, I.H. Hutchinson, J.H. Irby, B. LaBombard, E.S. Marmar, D. Mossessian, C.S. Picher, and J.L. Terry, Effects of neutral particles on edge dynamics in Alcator C-Mod plasmas, Phys. Plasmas, 7, 1919 (2000)
2. L. Johnson, E. Hinnov, J. Quant. Spectrosc. Radiat. Transf. Ionization, recombination and population of excited levels in hydrogen plasmas, **13**, 333, (1973)
3. J. W. Hughes, D. Mossessian, A.E. Hubbard, E. S. Marmar, D. Johnson, and D. Simon, High resolution edge Thomson scattering measurements on the Alcator C-Mod tokamak, Rev. Scient. Instr., **72**, 1107, (2001)
4. Terry, B. Lipschultz, A. Yu Pigarov, S.I. Krasheninnikov, B. LaBombard, D. Lumma, H. Okhawa, D. Pappas, and M. Umansky, Volume recombination and opacity in Alcator C-Mod tokamak divertor plasmas, Phys. Plasmas, **5**, 1759 (1998)

INDEX